MW00844150

Recent Advancement in Prodrugs

Recent Advancement in Prodrugs

Edited by
Kamal Shah
Durgesh Nandini Chauhan
Nagendra Singh Chauhan
Pradeep Mishra

CRC Press is an imprint of the
Taylor & Francis Group, an **informa** business

First edition published 2020
by CRC Press
6000 Broken Sound Parkway NW, Suite 300, Boca Raton, FL 33487-2742

and by CRC Press
2 Park Square, Milton Park, Abingdon, Oxon, OX14 4RN

© 2020 by Taylor & Francis Group, LLC

CRC Press is an imprint of Taylor & Francis Group, an Informa business

No claim to original U.S. Government works

Printed on acid-free paper

International Standard Book Number-13: 978-0-367-34836-6 (Hardback)

This book contains information obtained from authentic and highly regarded sources. Reasonable efforts have been made to publish reliable data and information, but the author and publisher cannot assume responsibility for the validity of all materials or the consequences of their use. The authors and publishers have attempted to trace the copyright holders of all material reproduced in this publication and apologize to copyright holders if permission to publish in this form has not been obtained. If any copyright material has not been acknowledged please write and let us know so we may rectify in any future reprint.

Except as permitted under U.S. Copyright Law, no part of this book may be reprinted, reproduced, transmitted, or utilized in any form by any electronic, mechanical, or other means, now known or hereafter invented, including photocopying, microfilming, and recording, or in any information storage or retrieval system, without written permission from the publishers.

For permission to photocopy or use material electronically from this work, please access www.copyright.com (http://www.copyright.com/) or contact the Copyright Clearance Center, Inc. (CCC), 222 Rosewood Drive, Danvers, MA 01923, 978-750-8400. CCC is a not-for-profit organization that provides licenses and registration for a variety of users. For organizations that have been granted a photocopy license by the CCC, a separate system of payment has been arranged.

Trademark Notice: Product or corporate names may be trademarks or registered trademarks, and are used only for identification and explanation without intent to infringe.

Visit the Taylor & Francis Web site at
http://www.taylorandfrancis.com

and the CRC Press Web site at
http://www.crcpress.com

Contents

Preface

Prodrugs are said to be bioreversible derivatives which on biotransformation change into active pharmacophore. A prodrug approach has been extensively studied among drug design scientists for a wide range of applications. It has been successfully applied to encompassing a variety of various goals achieved not only for correction of pharmacokinetic behavior but also pharmaceutical, organoleptic, physical and chemical properties of parent drug compound which enhance the stability and patient compliance, improving the efficacy of therapy. An ideal prodrug has favorable administration, distribution, metabolism, excretion, and toxicity (ADMET) properties, is chemically stable in its dosage form, releases the active drug molecule at an appropriate rate at the desired site in the body, and releases a promoiety that is non-toxic. Since a large number of potential drugs or drug candidates have undesirable properties that may present pharmaceutical, pharmacokinetic, pharmacodynamic, or economical barriers to their clinical application, prodrugs should be considered as one potential solution in the early stage of drug design and development. They have to pass through enzymatic or chemical biotransformation before eliciting their pharmacological action. The prodrug concept may be used in case of solubility enhancement, bioavailability enhancement, chemical stability improvement, presystemic metabolism, site-specific delivery, toxicity masking, improving patient acceptance, or eradicating undesirable adverse effects. This book is aimed at an audience of advanced-level students, experts, and scientists working in prodrugs designing and synthesis. The book is a single window reference source to the industrial chemistry community. This book has an emphasis on application and its implementation in research as well as in pharmaceutical industries.

This book consists of 14 chapters. Chapter 1 has brief descriptions about different types of prodrugs, with examples and their potential application. This chapter is proposed to encourage scientists looking for innovative prodrug approaches and rationally designed drug discovery platforms. Chapter 2 describes challenges encountered during development, toxicological and regulatory aspects and some FDA-approved codrugs. Overall, a codrug approach is a versatile, powerful tool that can be applied to a wide range of drugs to improve their therapeutic efficacy, bioavailability, and minimize side effects.

Chapter 3 has recent examples of prodrugs from literature on the basis of linkages present in prodrugs. Their applications with the aim of its synthesis were discussed in this chapter. Chapter 4 highlights the applications of a prodrug approach in offering an attractive substitute for improving lipophilicity and thus eliminating various drug-related drawbacks like poor absorption, extensive first pass metabolism, poor bioavailability, foul odor, and taste, as illustrated by various examples. Lipidic prodrug approach is the focus of Chapter 5, and it presents a strategy of covalently binding a lipid carrier (fatty acids, triglycerides, steroids, or phospholipids) to the drug moiety. Chapter 6 describes the progress of modifying drug solubility using a prodrug approach. The chemical carriers used in the process and the examples of strategies that were successful in this regard are presented in order to comprehend better usage of this approach in the field of drug design. Chapter 7 covers valuable information regarding the challenges faced during the drug design of the central nervous system. The chapter is focused on delivery of drug by keeping basis of prodrugs by using examples so that the drug can be designed.

Chapter 8 discusses the activation, role, and applicability of prodrugs in the treatment of cancer. Chapter 9 discusses prodrugs for cancer therapy. In this chapter, emphasis is placed on oxidative stress biomarkers which could serve against cancer and simultaneously provide the basis for targeted pro-oxidant chemotherapy when combined with a biomarker for oxidized protein hydrolase. Chapter 10 focuses on the new approaches in prodrug therapies developed by nitroreductases. These nitroreductases have been developed for prodrug strategies and nitroaromatic compounds are mostly used in this enzyme/prodrug combinations. Chapter 11 describes the development of prodrugs, including an approach using carboxylesterase with different substrate specificities of human carboxylesterase isozymes. Chapter 12 familiarizes the reader with the various mathematical models for ADME of prodrugs. In addition, current challenges and future perspectives are discussed. Chapter 13 discusses secondary metabolites obtained from natural sources which serve as prodrugs for the advent of novel drugs, assuring a committed concept which can open new vistas for difficult and expensive tasks of new drug investigation. Chapter 14 highlights the biotechnological tools to be used in prodrug therapy. The chapter highlights antibody-directed enzyme prodrug therapy, gene prodrug therapy (bacterial vector-based gene prodrug therapy and viral vector-based enzyme prodrug therapy), and membrane transporters–assisted prodrug therapy.

We would like to express our sincere gratitude to all of the authors who have taken time from their busy schedules to be part of this project and written wonderful chapters that added to both the depth and value of this book. We welcome suggestions and criticisms from our readers. Special thanks to our families for their support and encouragement. We express our gratitude to the publishing and production team of CRC Press, Renu Upadhyay, Shikha Garg and Jyotsna Jangra, especially for their kind, proficient, and encouraging guidance.

Kamal Shah

Durgesh Nandini Chauhan

Nagendra Singh Chauhan

Pradeep Mishra

Editors

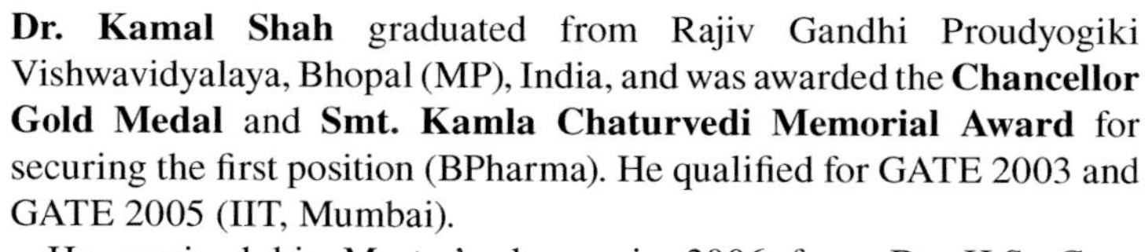

Dr. Kamal Shah graduated from Rajiv Gandhi Proudyogiki Vishwavidyalaya, Bhopal (MP), India, and was awarded the **Chancellor Gold Medal** and **Smt. Kamla Chaturvedi Memorial Award** for securing the first position (BPharma). He qualified for GATE 2003 and GATE 2005 (IIT, Mumbai).

He received his Master's degree in 2006 from Dr. H.S. Gour University Sagar and worked on method development for determination of pharmacological active components by UV and HPLC. He completed his doctorate from Dr. A.P.J. Abdul Kalam Technical University, Lucknow.

Currently he is working as an associate professor in the Institute of Pharmaceutical Research, GLA University Mathura (UP), India. He has published his research work in various journals of national and international repute, and has a total of 30 published research papers and more than 300 citations to his credit.

He is a member of various professional and academic bodies including the Association of Pharmaceutical Teachers of India (APTI), the Institutional Animal Ethical Committee (IAEC), and the Board of Studies (BOS).

He is a reviewer for various national and international journals, including *Medicinal Chemistry Research*, *Pakistan Journal of Pharmaceutical Sciences*, *Journal of Sexual Medicine*, *Natural Product Research*, *African Journal of Pharmacy and Pharmacology*, *International Journal of Pharmacy and Pharmacology*, *International Research Journal of Pure and Applied Chemistry*, *Pharmaceutical Biology* and *Asian Journal of Pharmaceutics*.

He has also written six chapters and edited one book. His area of interest is the synthesis and evaluation of prodrugs of NSAIDs.

Mrs. Durgesh Nandini Chauhan, MPharma, completed her BPharma degree in at the Rajiv Gandhi Proudyogiki Vishwavidyalaya, Bhopal, India, and her MPharma (pharmaceutical sciences) at Uttar Pradesh Technical University, currently Dr. A.P.J. Abdul Kalam Technical University, Lucknow, in 2006. Mrs. Durgesh Nandini Chauhan has 10 years of academic (teaching) experience from institutes in India. She taught subjects such as pharmaceutics, pharmacognosy, traditional concepts of medicinal plants, drug-delivery phytochemistry, cosmetic technology, pharmaceutical engineering, pharmaceutical packaging, quality assurance, dosage form designing, and anatomy and physiology.

She is a member of the Association of Pharmaceutical Teachers of India, SILAE: Società Italo-Latinoamericana di Etnomedicina (the Scientific Network on Ethnomedicine, Italy), and so forth. Her previous research work included "*Penetration Enhancement Studies on Organogel of Oxytetracycline HCL.*" She also attended an AICTE-sponsored staff development program on the "Effects of Teaching and Learning Skills in Pharmacy-Tool for Improvement of Young Pharmacy Teachers" and a workshop on analytical instruments. She has written more than 10 publications in national and international journals, 13 book chapters, and authored four books. She is also active as a reviewer for several international scientific journals and active participant in national and international conferences such as Bhartiya Vigyan Sammelan and the International Convention of Society of Pharmacognosy. Recently, she joined the Ishita Research Organization, Raipur, India as a freelance writer, guiding pharmacy, ayurvedic, and science students in their research projects.

Dr. Nagendra Singh Chauhan obtained his MPharma and PhD from the Department of Pharmaceutical Sciences, Dr. H.S. Gour University, Sagar in 2006 and 2011. He has around 14 years of research experience. He is presently working as Senior Scientific Officer Grade-II and Government Analyst at Drugs Testing Laboratory Avam Anusandhan Kendra, Raipur, Chhattisgarh, India. He has professional expertise in natural product isolation as well as hands-on experience in various techniques like extraction, isolation, and purification of natural products from plant extracts using various chromatographic techniques for separation of compounds includes column, LC-MS, chromatography, prep. TLC, Elucidation of structure of natural products with the help of various techniques such as UV, Mass, IR and NMR. He has published more than 50 articles in national and international journals as well as 25 book chapters. He has more than 2,000 citations. He is a member of various professional and academic bodies including the Society of Pharmacognosy, International Natural Product Sciences Taskforce (INPST) Society, SILAE: Società Italo-Latinoamericana di Etnomedicina (The Scientific Network on Ethnomedicine, Italy), Institutional Human Ethical Committee, and the Association of Pharmaceutical Teachers of India (APTI).

Prof. Pradeep Mishra has 42 years of experience in teaching undergraduate and postgraduate classes in pharmacy, 31 years of which have been with the Department of Pharmaceutical Sciences, Dr. Hari Singh Gour University, Sagar (MP), India. He received his BPharma, MPharma, and PhD from the Department of Pharmaceutical Sciences, Dr. Hari Singh Gour University, Sagar (MP). During this period he guided 41 MPharm students and 18 PhD students. He has published around 110 research papers in various Indian and international journals. During his stay at Sagar, he shouldered the additional responsibilities of Dean, Faculty of Technology. He has been invited to chair the Medicinal Chemistry/Pharmaceutical Analysis section of the Indian Pharmaceutical Congress (IPC) held at New Delhi, Pune, Indore, and Varanasi. He acts as a referee for a number of journals, including *Journal of Indian Chemical Society*, *Indian Journal of Pharmaceutical Sciences*, *Indian Journal of Natural Products* and *Journal of Scientific and Industrial Research*. In May 2006, he took up the responsibility of nurturing students at the Institute of Pharmaceutical Research, GLA University, Mathura (UP) as Director/Principal. Currently he is working the Director of the Institute. He was awarded the M.L. Khurana, IJPS Best Paper Award in 2006. He is a Fellow, Institution of Chemists, Calcutta, Indian Chemical Society, Calcutta, Life Member, Indian Society for Technical Education, New Delhi, Life Member, Association of Pharmaceutical Teachers of India (MP/LM018), Member, Indian Pharmaceutical Association, Bombay and Member, Pharmaceutical Society, Dr. Harisingh Gour Vishwavidyalaya, Sagar. He always works to improve the quality of teaching and learning. Although he has degrees in pharmaceutical chemistry, he has developed an interest in pharmaceutical analysis.

Contributors

Attia Afzal
Department of Pharmacy
The Islamia University of Bahawalpur
Bahawalpur, Pakistan

Mehmet Ay
Natural Products and Drug Research Laboratory
Faculty of Sciences and Arts
Department of Chemistry
Çanakkale Onsekiz Mart University
Çanakkale, Turkey

Shimon Ben-Shabat
Faculty of Health Sciences
Department of Clinical Pharmacology
School of Pharmacy
Ben-Gurion University of the Negev
Beer-Sheva, Israel

Vipin Bhati
Department of Regulatory Toxicology
National Institute of Pharmaceutical Education and Research (NIPER)-Raebareli
Lucknow, India

Neha V. Bhilare
Department of Pharmaceutical Chemistry
Arvind Gavali College of Pharmacy
Satara, India

Durgesh Nandini Chauhan
Columbia Institute of Pharmacy
Raipur, India

Nagendra Singh Chauhan
Drugs Testing Laboratory Avam Anusandhan Kendra
Raipur, India

Pooja Chawla
Department of Pharmaceutical Chemistry
ISF College of Pharmacy (ISFCP)
Moga, India

Viney Chawla
University Institute of Pharmaceutical Sciences and Research
Baba Farid University of Health Sciences
Faridkot, India

Chung Man Chin
School of Pharmaceutical Science
State University of São Paulo, UNESP
Araraquara, Brazil

Arik Dahan
Faculty of Health Sciences
Department of Clinical Pharmacology
School of Pharmacy
Ben-Gurion University of the Negev
Beer-Sheva, Israel

Suneela Dhaneshwar
Amity Institute of Pharmacy
Amity University Uttar Pradesh
Lucknow, India

Jean Leandro dos Santos
School of Pharmaceutical Science
State University of São Paulo, UNESP
Araraquara, Brazil

Guilherme Felipe dos Santos Fernandes
School of Pharmaceutical Science
State University of São Paulo, UNESP
Araraquara, Brazil

Vishal Gour
Department of Pharmaceutical Sciences
Dr. Harisingh Gour Central University
Sagar, India

Tuğba Güngör
Natural Products and Drug Research Laboratory
Faculty of Sciences and Arts
Department of Chemistry
Çanakkale Onsekiz Mart University
Çanakkale, Turkey

Jeetendra Kumar Gupta
Institute of Pharmaceutical Research,
GLA University
Mathura, India

Prem N. Gupta
PK-PD, Toxicology & Formulation Division
CSIR – Indian Institute of Integrative Medicine (IIIM)
Jammu, India

Tanweer Haider
Department of Pharmaceutical Sciences
Dr. Harisingh Gour Central University
Sagar, India

Masakiyo Hosokawa
Laboratory of Drug Metabolism and Biopharmaceutics
Graduate School of Pharmaceutical Sciences
Chiba Institute of Science
Choshi, Japan

Poorvashree P. Joshi
Drug Discovery Department
CSIR-URDIP
Pune, India

Gaurav Krishna
Institute of Pharmaceutical Research
GLA University
Mathura, India

K. Krishnan
Department of Internal Medicine
East Tennessee State University
Johnson City, Tennessee

Anoop Kumar
Department of Pharmacology and Toxicology
National Institute of Pharmaceutical Education and Research (NIPER)-Raebareli
Lucknow, India

Asadullah Madni
Department of Pharmacy
The Islamia University of Bahawalpur
Bahawalpur, Pakistan

Milica Markovic
Faculty of Health Sciences
Department of Clinical Pharmacology
School of Pharmacy
Ben-Gurion University of the Negev
Beer-Sheva, Israel

Sunita Minz
Department of Pharmacy
Indira Gandhi National Tribal University
Amarkantak, India

Pradeep Mishra
Institute of Pharmaceutical Research
GLA University
Mathura, India

Igor Muccilo Prokopczyk
School of Pharmaceutical Science
State University of São Paulo, UNESP
Araraquara, Brazil

Navyashree V.
Department of Pharmacology and Toxicology
National Institute of Pharmaceutical Education and Research (NIPER)-Raebareli
Lucknow, India

Sobia Noreen
Department of Pharmacy
The Islamia University of Bahawalpur
Bahawalpur, Pakistan

Ferah Cömert Önder
Natural Products and Drug Research LaboratoryFaculty of Sciences and Arts
Department of Chemistry
Çanakkale Onsekiz Mart University
Çanakkale, Turkey

Victoria E. Palau
Pharmaceutical Sciences
East Tennessee State University
Johnson City, Tennessee

Vikas Pandey
Department of Pharmaceutical Sciences
Dr. Harisingh Gour Central University
Sagar, India

Arun Singh Parihar
Drugs Testing Laboratory Avam Anusandhan Kendra (State Goverment lab of AYUSH)
Government Ayurvedic College
Raipur, India

Shubham Pawar
Poona College of Pharmacy
Bharati Vidyapeeth University
Pune, India

Madhulika Pradhan
Rungta College of Pharmaceutical Sciences and Research
Bhilai, India

Katalin Prokai-Tatrai
Department of Pharmacology and Neuroscience
Graduate School of Biomedical Sciences
University of North Texas Health Science Center
Fort Worth, Texas

Abdur Rahim
Department of Pharmacy
The Islamia University of Bahawalpur
Bahawalpur, Pakistan

Muhammad Sarfraz
Department of Pharmacy
The Islamia University of Bahawalpur
Bahawalpur, Pakistan

Afifa Shafique
Department of Pharmacy
The Islamia University of Bahawalpur
Bahawalpur, Pakistan

Kamal Shah
Institute of Pharmaceutical Research
GLA University
Mathura, India

Alok Sharma
Department of Pharmacognosy
ISF College of Pharmacy (ISFCP)
Moga, India

Deependra Singh
Department of Pharmaceutical Biotechnology
University Institute of Pharmacy
Pt. Ravishankar Shukla University
Raipur, India

Manju Rawat Singh
Department of Pharmaceutical Biotechnology
University Institute of Pharmacy
Pt. Ravishankar Shukla University
Raipur, India

Shamsher Singh
Department of Pharmacology
ISF College of Pharmacy (ISFCP)
Moga, India

Vandana Soni
Department of Pharmaceutical Sciences
Dr. Harisingh Gour Central University
Sagar, India

William L. Stone
Department of Pediatrics
East Tennessee State University
Johnson City, Tennessee

Masato Takahashi
Laboratory of Drug Metabolism and Biopharmaceutics
Graduate School of Pharmaceutical Sciences
Chiba Institute of Science
Choshi, Japan

Sudhir Kumar Thukral
Department of Pharmacognosy
ISF College of Pharmacy (ISFCP)
Moga, India

1

Recent Advancements in New Drug Design and Development of Prodrugs

Kamal Shah, Gaurav Krishna, Jeetendra Kumar Gupta, Durgesh Nandini Chauhan, Nagendra Singh Chauhan, and Pradeep Mishra

CONTENTS

1.1 Introduction

There are two major considerations in any drug design project. First of all, drug interacts with molecular target in the body and so it is important to choose the correct target for pharmacological activity. Secondly, a drug has to travel through the body in order to reach its target. Almost all drugs possess some undesirable side effects. The clinical efficiency can be enhanced by curtailing the unwanted side effects while accommodating the desired properties. A therapeutically promising drug may have some shortcomings so it cannot be used prominently (Remington and Gennaro 2000). A few of these shortcomings are poor aqueous solubility (corticosteroids), unpleasant taste (chloramphenicol), duration of action (estradiol, cytorabine, naloxone, testosterone), nonspecificity (carbidopa), poor bioavailability (carbencillin, ampicillin), gastric instability (erythromycin), etc. (Bhosle et al. 2006). These associated problems may be solved by different approaches. These are biological approaches that can be solved by the alteration in the route of administration or physical approach involves modification in the design of the dosage form such as controlled delivery of drugs and other is chemical approach, which is widely accepted and used mostly. The chemical approach may involve the design and development of novel drugs synthetically. These are synthesized which are essentially analogs of existing drugs and it can be termed as chemically derivatised prodrug. The term prodrug signifies that the molecule is not active or a chemical derivative of the drug. It becomes active after consumption. It generates or converts into active drug. The term prodrug was first coined by Albert (1958). He described prodrugs as compounds that undergo biotransformation before eliciting its activity. The chemical modification or transformation in prodrugs is done due to specific reasons. This process also referred as drug lamentation (Harper 1959) (Figure 1.1). It may be done to

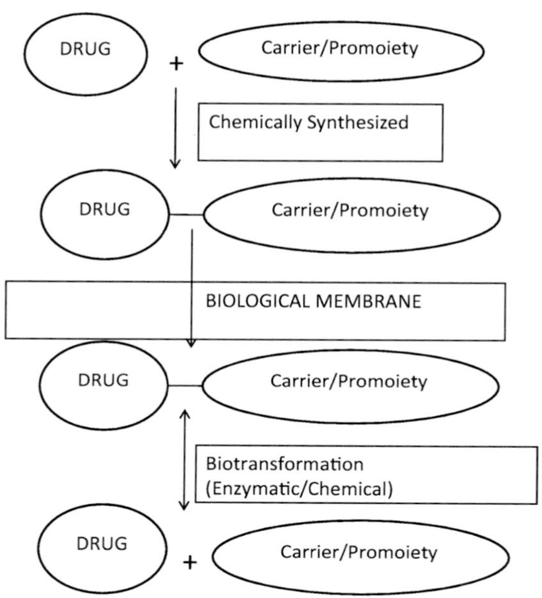

FIGURE 1.1 Prodrug concept.

increase patient acceptability in terms of taste, solubility, bioavailability, stability and to decrease toxicity. These compounds are also known as bioreversible derivatives. The chemical transformation or modification results in formation of a covalent bond between drug and carrier. According to the carrier attached with prodrugs, Wermuth classified the prodrugs into two broad classes, i.e., carrier-linked prodrugs and bioprecursors. Carrier-linked prodrugs should have carrier that should release the active drug on *in vivo* transformation. Bioprecursors do not have any carrier attached; therefore, they are formed by molecular modification of active drugs. The conversion of prodrugs to active drug involves number of reactions (Sloan and Wasdo 2003). The type of reaction involved in this varies according to the presence of bonding or functional group arrangement (Stella et al. 1985). The design of prodrug requires comprehensive study of parent drug. The functional group present in drug may decide the linkage present in resultant prodrug. The preclinical studies of the designed prodrug should always be kept in mind, as the designed prodrug might alter the tissue absorption, bio distribution, its metabolism and kinetics (ADME). This alteration ultimately had effect over efficacy, potency and toxicity of resultant prodrug. The promoiety or carrier which is taken with drug should be such that it may have synergetic effect as in codrugs or mutual prodrugs. The phenomenon of codrug relates with co-administrating two drugs. The objectivity of such codrugs is to enhance pharmacological activity by delivering the drugs at desired site at the same time or masking the side effects of parent drug. The choice of promoiety may vary according to the problem identified in parent drug. A complete study of ADME (absorption, distribution, metabolism and excretion) is required during designing of prodrugs. The by-products formed from the prodrug should also be taken into consideration. The formed prodrug must rapidly biotransform into its active form whether chemically or enzymatically. The design of desired prodrug involves the use of functional groups present in parent drug. The functional groups that form covalent linkages may be alcohols, phenols, acids, amine, amide, secondary amines, imide, phosphates, etc. The prodrugs were designed with the aim that they may alter solubility, bioavailability, reduce toxicity, enhance acceptability, increase site specific delivery, increase duration of action, increase patient compliance, etc. This chapter is going to cover the recent examples of prodrugs with their possible chemistry and application. The prodrugs can be discussed according to functional group present in the parent structure.

1.2 Prodrugs of Drugs Bearing Carboxylic Acids

The carboxylic acid is the common functional group present in the drugs. The pKa value of such drugs generally lies between 3.5 and 4.5. These drugs when taken get ionized (deprotonated). The logD values of these drugs decline down the range of logD values of strongly absorbed drugs (logD > 0). This problem has been identified over the many years and one of the common solutions of this is esterification. The pervasiveness of esterase, peptidases and other enzymes in the organisms as well as the easily availability of alcohols and phenols containing codrugs result in development of number of prodrugs (Bundgaard 1985, Beaumont et al. 2003). The added advantages with carboxylic acids containing prodrugs are to obtain desirable hydrophilicity or *in vivo* activity.

The general scheme for the hydrolysis of ester containing prodrugs are as follows (Figure 1.2).

There are enormous examples of prodrugs having ester linkage in the literature. Some recent examples are as follows (Figure 1.3).

The prodrug of propofol synthesized by Zhang and group having ester linkage (Zhang et al. 2019). Propofol is a water insoluble drug. It is commonly given intra venous and have rapid onset of action. During this study, the synthesized prodrug of propofol, ***HX0921*** (**1**) (sodium 2-(2-(2,6-diisopropylphenoxy)-2-oxoethoxy)acetate), showed good water-solubility. The pharmacological activities of this compound proof that it has short duration of action and onset also lowered. It was found that in prodrug form it hydrolyzed rapidly and showed its better activity. Zhou synthesized arbotegravir (CAB) nano-formulated prodrug (**Myristoyl CAB**) (**2**) to improve the properties of parent drug lipophilicity and its availability in circulation to have better antiretroviral profile (Zhou et al. 2018). The mutual prodrugs of aceclofenac with naturally occurring antioxidants, i.e., aceclofenac with quercetine (**ACF-Q**) (**3**), aceclofenac with vanillin (**ACF-V**) (**4**) and aceclofenac with L-tryptophan (**ACF-T**) (**5**). It was found these prodrugs were more lipophilic than aceclofenac with improved stability in acidic pH. The pharmacological activities, when compared with parent drug, were found to be better analgesic and anti-inflammatory agents with reduced ulcerogenicity (Rasheed et al. 2016). The amino acid esters of Nitazoxanide were verified for their animal activities. Its pharmacology, kinetics, and toxicity studies were carried. Among synthesized prodrugs, **RM5061 8a** (**6**) in rats showed sevenfold more blood concentration compared to previously reported drug (Stachulski et al. 2017). The synthesized prodrug had an excellent safety profile with least toxicity. Its bioavailability increased from 3% to 20%.

Redasani and Bari utilized natural phyto phenols like thymol, menthol and eugenol for acquiring the additive activity and lowers the gastric problems common with ibuprofen. The ester derivatives (**Ibuprofen-thymol ester** (**7**), **Ibuprofen-menthol ester** (**8**) and **Ibuprofen-eugenol ester** (**9**)) of ibuprofen were synthesized and evaluated (Redasani and Bari 2012). The prodrugs formed were found to

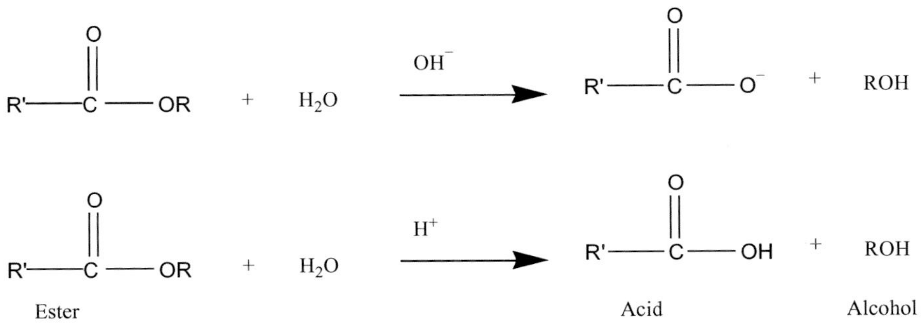

$$R'\text{—}C(=O)\text{—}OR + H_2O \xrightarrow{OH^-} R'\text{—}C(=O)\text{—}O^- + ROH$$

$$\underset{\text{Ester}}{R'\text{—}C(=O)\text{—}OR} + H_2O \xrightarrow{H^+} \underset{\text{Acid}}{R'\text{—}C(=O)\text{—}OH} + \underset{\text{Alcohol}}{ROH}$$

FIGURE 1.2 General scheme for the hydrolysis of ester.

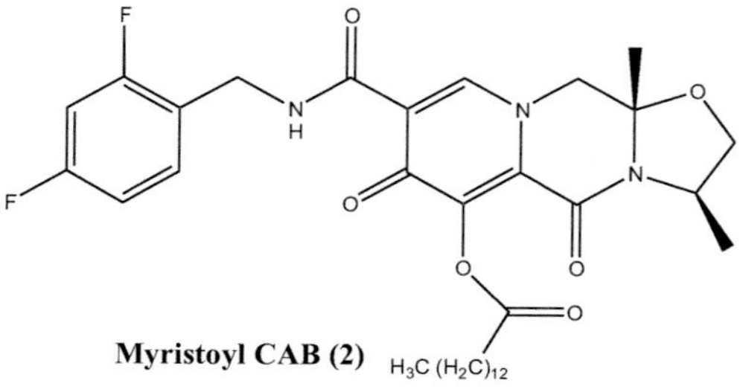

Myristoyl CAB (2)

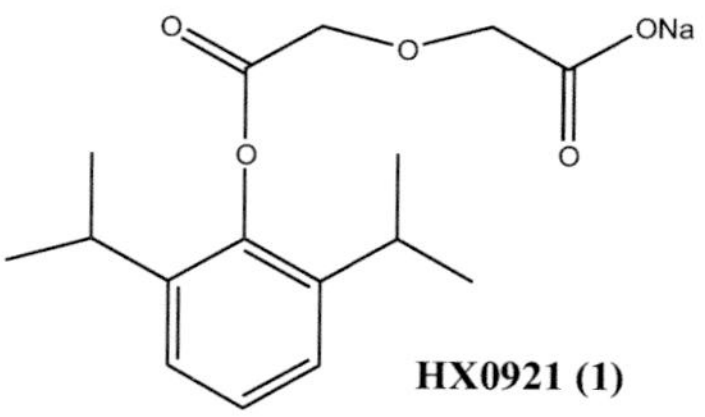

HX0921 (1)

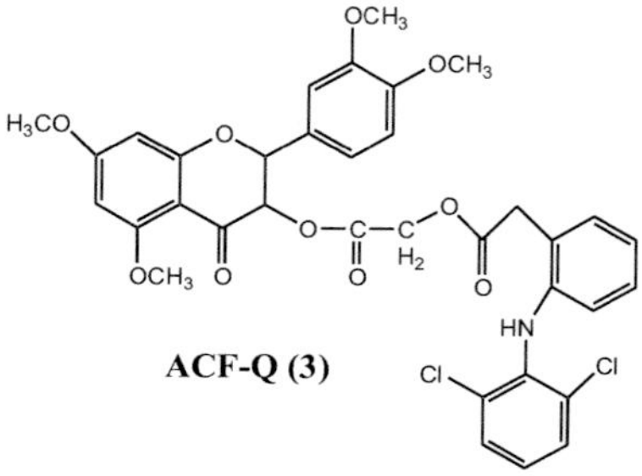

ACF-Q (3)

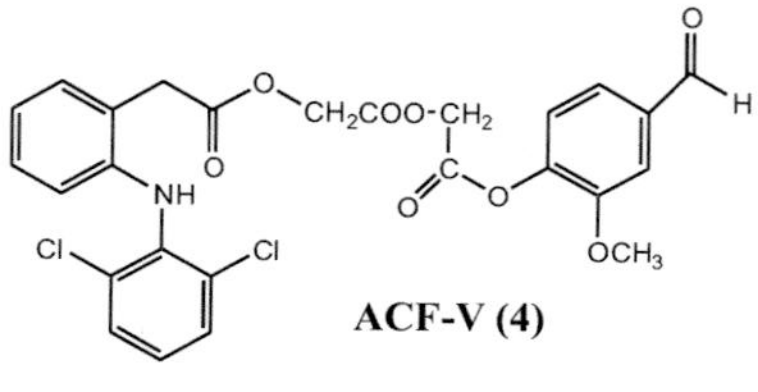

ACF-V (4)

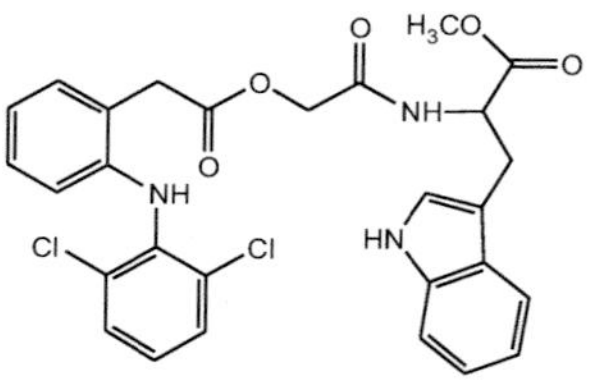

ACF-T (5)

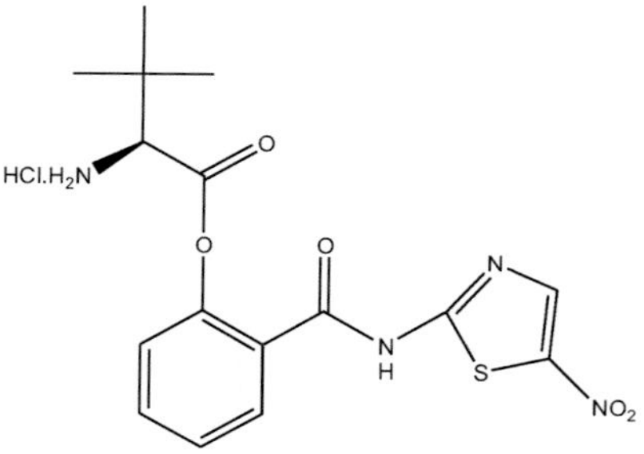

RM5061 8a (6)

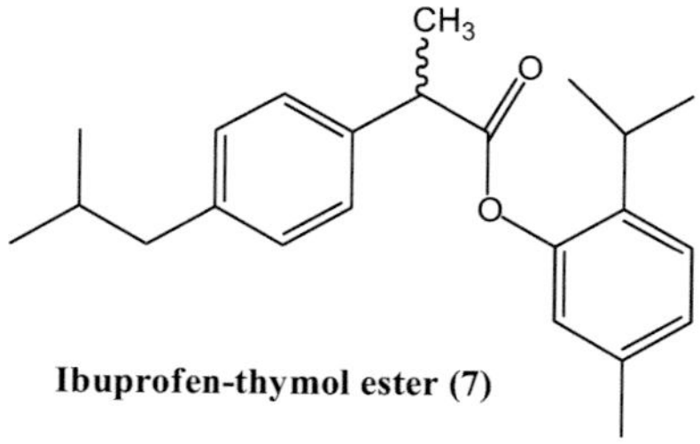

Ibuprofen-thymol ester (7)

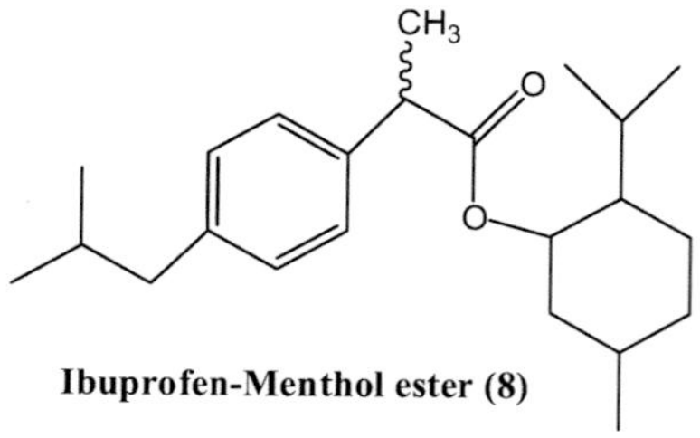

Ibuprofen-Menthol ester (8)

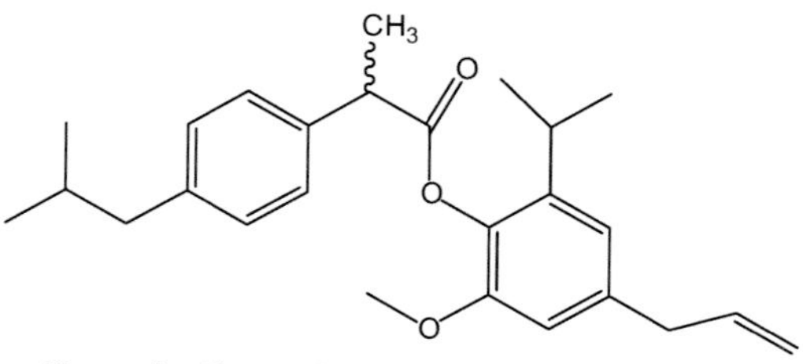

Ibuprofen-Eugenol ester (9)

FIGURE 1.3 Examples of prodrugs having carboxylic coupled linkage.

be lipophilic and stable at acidic pH. They elicited improved anti-inflammatory activity. These prodrugs showed additive effect, the reason of its activity might be conjugation of ibuprofen to natural analgesics. The result of ulcer index showed that the synthesized prodrug had lower gastric ulceration than the parent drug. This is a clear example of masked carboxylic acid.

1.3 Prodrugs of Alcohols and Phenols

The drug-bearing alcohols and phenols groups are polar in nature. They often undergo phase II metabolism. The modification of these functional groups leads to form prodrugs it may lead to change the properties of parent drug. The chemical reactions like alkylation, acylation or reduction would decrease the polarity, increases the lipophilicity so it increases membrane permeability however phosphorylation result in solubility enhancement. Many drugs that are sufficiently absorbed through gastro intestinal tract (GIT) have lower systemic concentration that is due to the first-pass metabolism. This prodrug approach circumvents the high first pass-metabolism and consequently increases the systemic concentration. The best possible way is esterification of the drugs having hydroxyl or phenolic group(s). Esterification may lead to form prodrugs of significant lipophilicity and have better *in vivo* lability (Longcope et al. 1985). Some of the recent examples are as follows (Figure 1.4).

The prodrugs of haloperidol were synthesized and evaluated for their hydrolysis in simulated environment corresponding to liver and small intestine. **Haloperidol pentanoate (10) and Haloperidol hexanoate (11**) showed high metabolic activation rates in the formulated esterprodrugs (Takahashi et al. 2019). Marinelli and his team synthesized 23 compounds and screened the antimicrobial activity. Among the synthesized compounds, compounds ***WSCP18 and 19*** **(12 and 13**) showed good antimicrobial activity (Marinelli et al. 2019). These were found to be more stable and lipophilic. The novel benzoic acid–based xanthine derivatives synthesized among that compound 3e (**BA-XN) (14**) sustained the antidiabetic effect for 48 h and result in improved glucose tolerance and found normal fasting blood glucose level. This led to a potential, efficacious, long-acting drug for type 2 diabetes mellitus (Li et al. 2019). The amino acid and peptide prodrugs of chalcones and 1,3-diphenylpropanones synthesized. These are positive allosteric modulators of α7 nicotinic receptors exhibiting analgesic activity. Various prodrugs were synthesized out of the synthesized prodrugs, **compound 21(Val-Pro-Val) (15**) showed significant activity (Balsera et al. 2018).

The prodrug of paclitaxel and vorinostat (**PAC-VOR**) (**16**) were significantly combining so far as cytotoxicity is concerned, by glycine and succinic acid. Out of these the prodrug with glycine shown better activity in cytotoxicity. *In vitro* evaluation was done by SRB assay in HCT-116 cells, MCF-7 and drug-resistant MCF7/ADR cells. The data obtained signified glycine had good activity than the parent drug principally in the MCF-7/ADR cells (Liu et al. 2019). All the data gave an idea that the synthesized prodrug has prominent activity in PTX resistance cancer.

The designed **hyaluronic acid-curcumin (HA-CUR) (17**) polymeric prodrug aimed to epithelial cells. This was designed to circumvent oxidative stress effects. The work was designed with the aim that natural polysaccharide HA worked as bio carrier for renal delivery of CUR. The rational here for designing the prodrug was to prolonged drug release and to enhance the solubility in circulation, increase the uptake via CD44 receptor in inflamed renal epithelial cells. This would promote the release of CUR at the inflamed site. This may result in eradication of kidney injury. The HA-CUR prodrug has 27 times more water solubility than that of curcumin (Hu et al. 2018).

The prodrug of acetaminophen with proline (**Proline-acetaminophen ester**) (**18**) was designed to mask the bitter taste of acetaminophen, which is due to the phenolic group. This phenolic group masked and *in vivo* studied were carried out. The data from different studies clearly indicated that formed prodrug was found to be stable and pharmacologically effective (Wu et al. 2010).

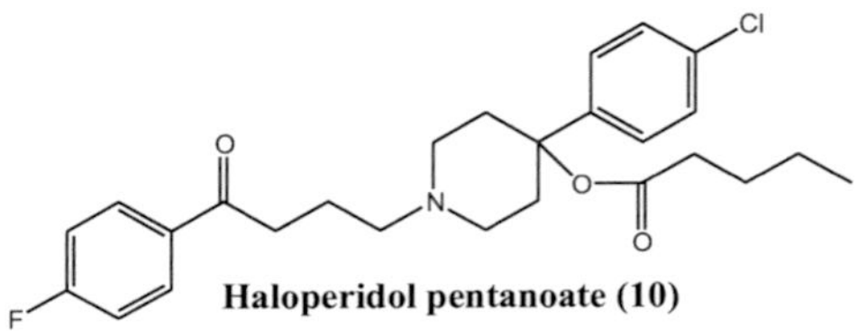

Haloperidol pentanoate (10)

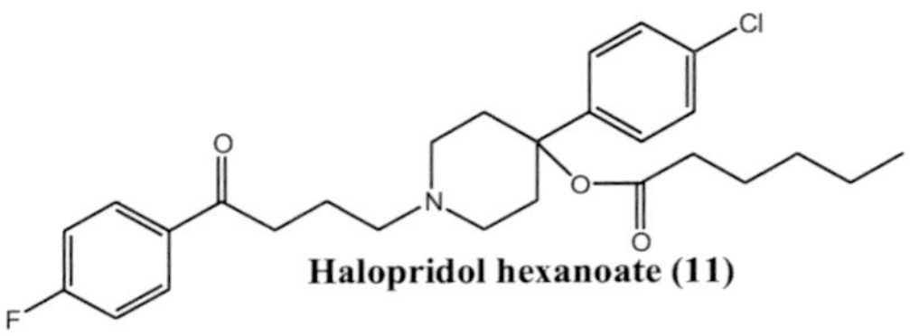

Halopridol hexanoate (11)

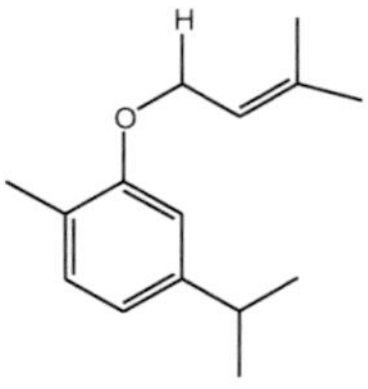

WSCP18 (12)

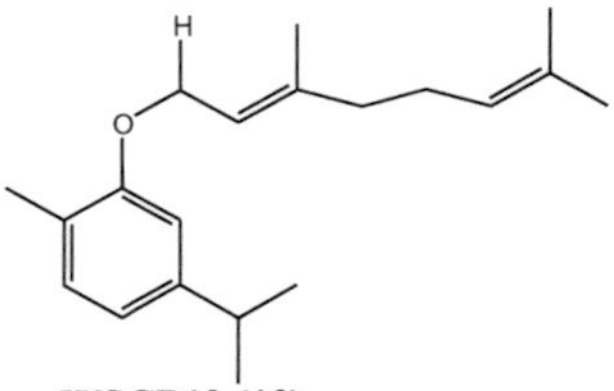

WSCP19 (13)

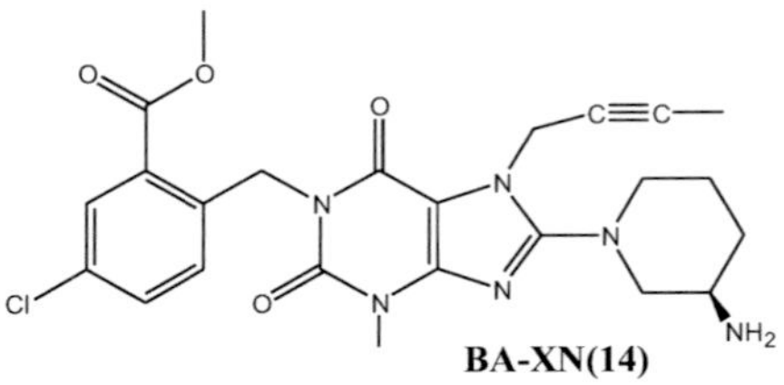

BA-XN(14)

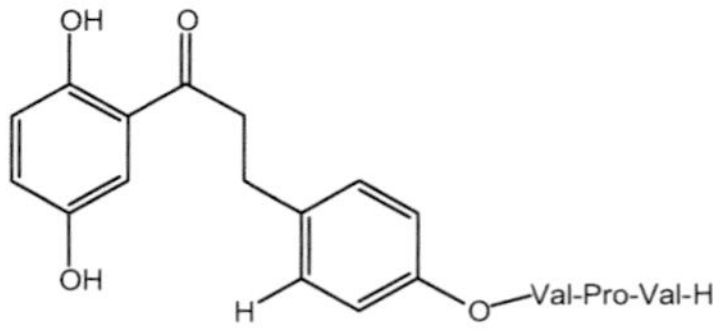

Compound 21 (Val-Pro-Val) (15)

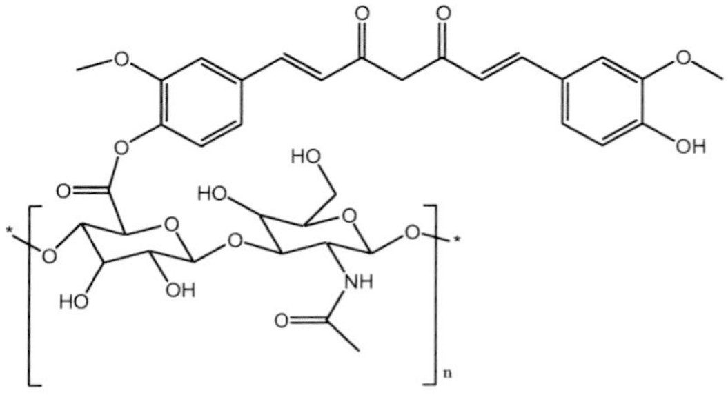

Hyaluronic acid-Curcumin prodrug (17)

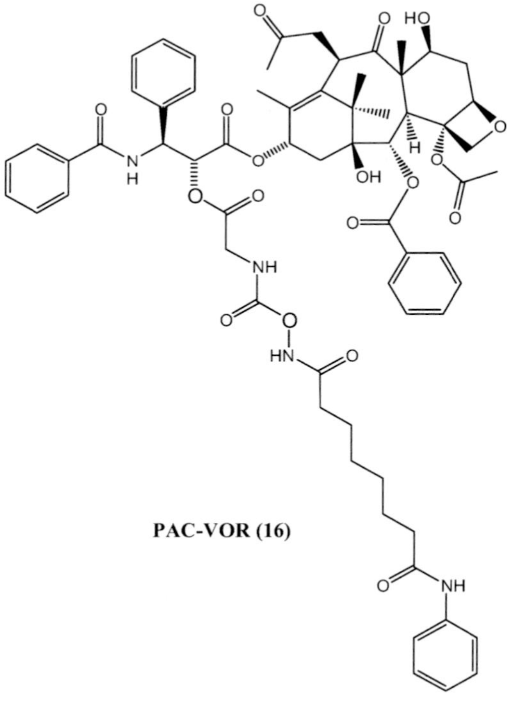

PAC-VOR (16)

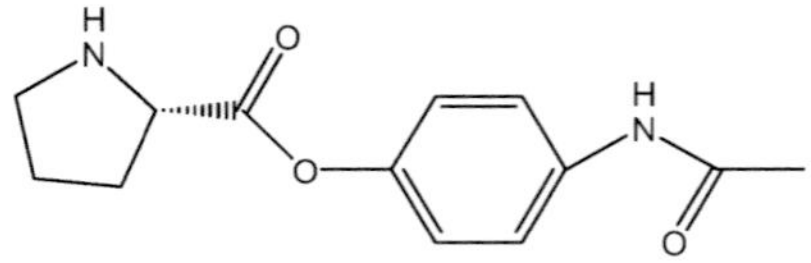

Proline-acetaminophen ester (18)

FIGURE 1.4 Examples of prodrugs having alcohol or phenol coupled linkage.

1.4 Prodrugs of Amines

A number of drugs have amino groups in them. The amino group may be present in the form of primary, secondary or tertiary amine or heterocyclic amine. The amino groups containing drugs got ionized at physiological conditions. This limits its use, as these drugs are unable to cross the lipoidal membrane. The second drawback is their instability. As the drugs-bearing amino groups undergo first-pass metabolism (Pitman 1981, Hu 2016). It may via *N*-acetylation and oxidation by monoamin oxidase. The prodrugs of amino group containing drugs may have *N*-acetyl (amide and simple carbamates), *N*-acyloxyalkyl, *N*-Mannich bases or imines (Testa and Mayer 2003) (Figures 1.5 and 1.6). These improve the lipophilicity, lower the pKa, or increase the probability of reaching the target. Generally, the amide prodrugs are not preferred due to slow hydrolysis of amide. But in case of activated amide such as *N*-benzoyl or *N*-pivaloyl derivatives this does not create problem (Testa and Mayer 1998). The amidines can be synthesized in the form of carbamates. Imines and enamines may have peptide derivatives.

Doxorubicin is one of common drug used for pancreatic cancer. The doxorubicin prodrug synthesized or chemically modified. The resultant compound has temporary masked group in form of arylboronic acids and their corresponding esters. These got activated to its desired form by reactive oxygen species (ROS). The synthesized prodrug **DOX-BRO (19)** had potent, effective and significant activity against cancer (Skarbek et al. 2019). The prodrug of phenol-bearing drug pterostilbene synthesized. The bioavailability of this drug affected due to the phenolic group that makes it to undergo phase II metabolism. The hydroxyl moiety got preserved as a carbamate ester linked with N terminal of amino acid. The synthesized prodrugs were further characterized and their pharmacological studies were done. The prodrug of pterostilbene with isoleucine (**pterostilbene-Ilu**) (**20**) showed good absorption, decreased metabolism, and greater concentrations of pterostilbene. It was also left for several hours and most of the organs tested for its toxicity (Azzolini et al. 2017). The prodrug of dopamine using carbamate linker synthesized, this formed prodrug released dopamine in brain. The synthesized prodrug DOPA-CBT was found to be show enhanced sustain release by 850 times than L-DOPA. Dopa-CBT had dopamine coupled via a secondary carbamate linker to L-tyrosine, a natural amino acid substrate of LAT1 and precursor of dopamine, at the para position. The carbamates groups were more easily hydrolyzed than amide bond. Due to this amide linkage containing **DOPA-AMD (21)** replaced by **DOPA-CBT (22)** (Thiele et al. 2018).

The Parthenolide has anticancer activity. But this drug is found to be water insoluble. The problem of solubility was solved by preparing its prodrug with other anticancer drug molecules (cytarabine and melphalan). It was found that these synthesized prodrugs (**parthabine (23) and parthalan (24)**) were found to be superior but the cytotoxicity was found to be higher than parent drug taken. The prodrug was prepared by using one of the name reaction aza-Michael addition. It involved the reaction of nitrogen present in anticancer drugs (cytarabine and melphalan) with the alpha-methylene-gamma-lactone group of parthenolide (Taleghani et al. 2017).

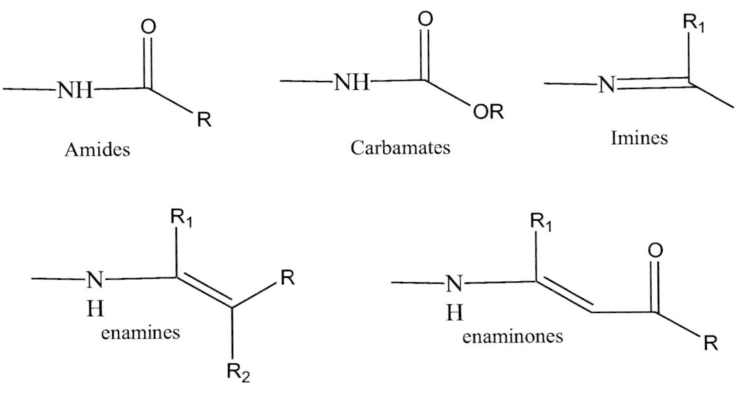

FIGURE 1.5 Amidines derivatives.

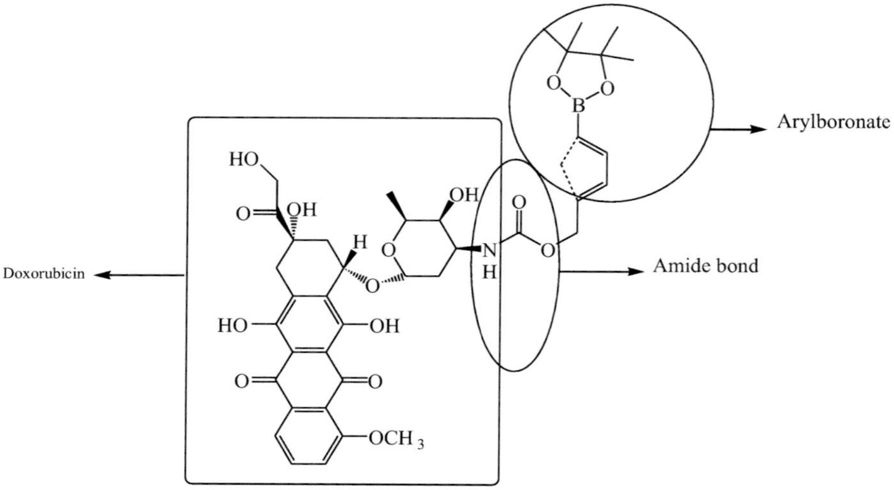

Doxorubicin

Arylboronate

Amide bond

DOX-BRO (19)

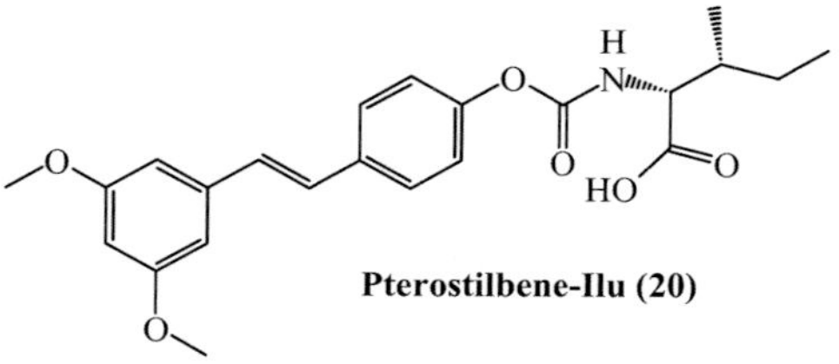

Pterostilbene-Ilu (20)

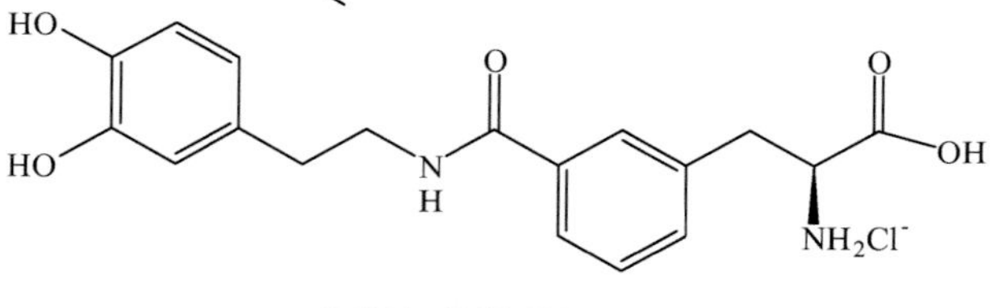

DOPA-AMD (21)

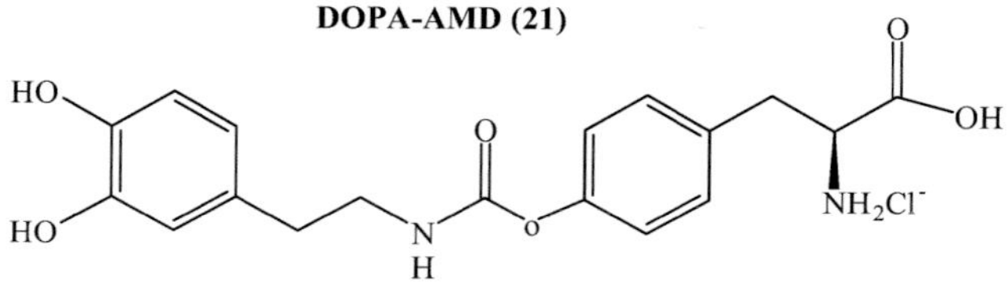

DOPA-CBT (22)

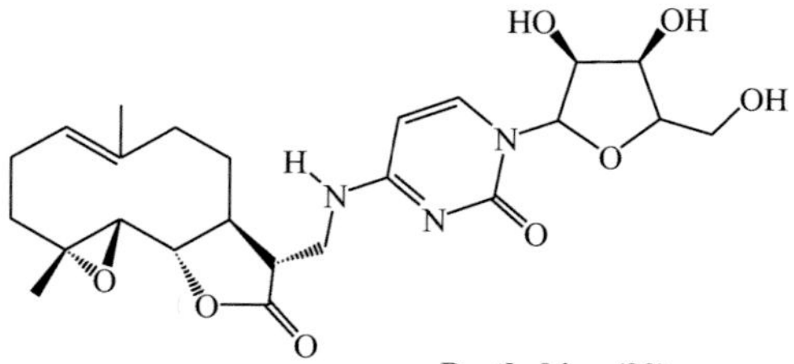

Parthabine (23)

FIGURE 1.6 Examples of prodrugs having amine coupled linkage. *(Continued)*

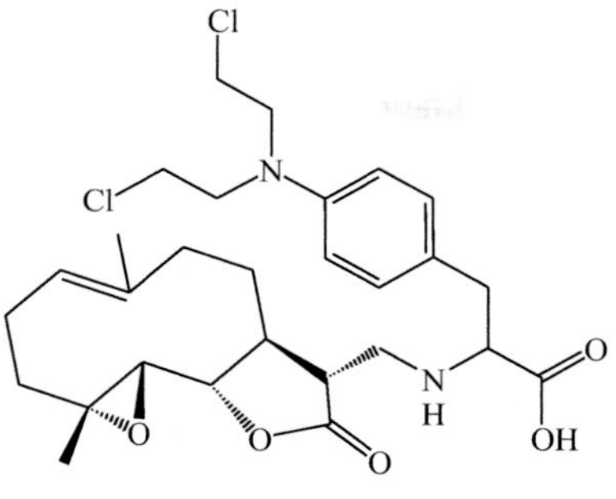

Parthalan (24)

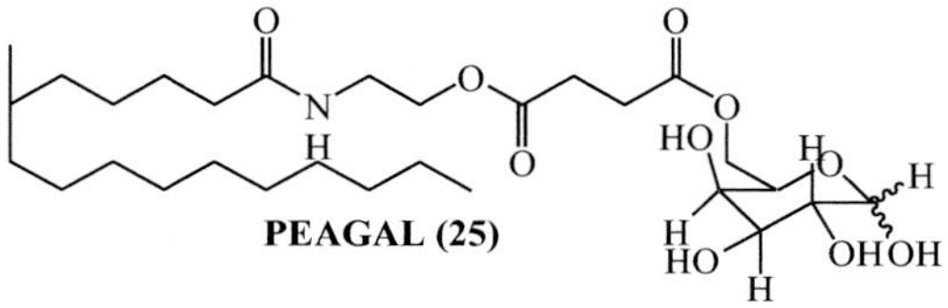

PEAGAL (25)

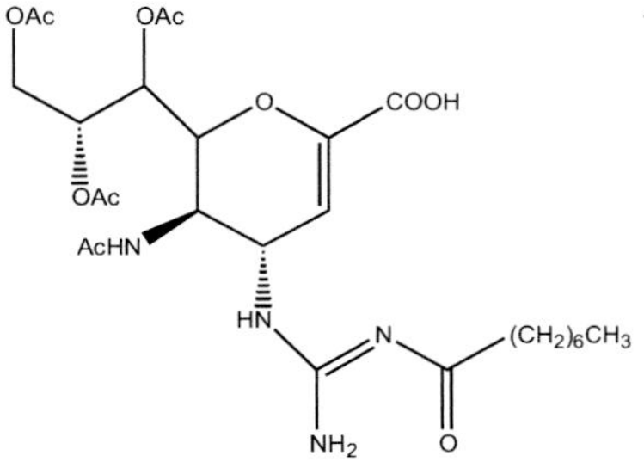

ZA-OG (26)

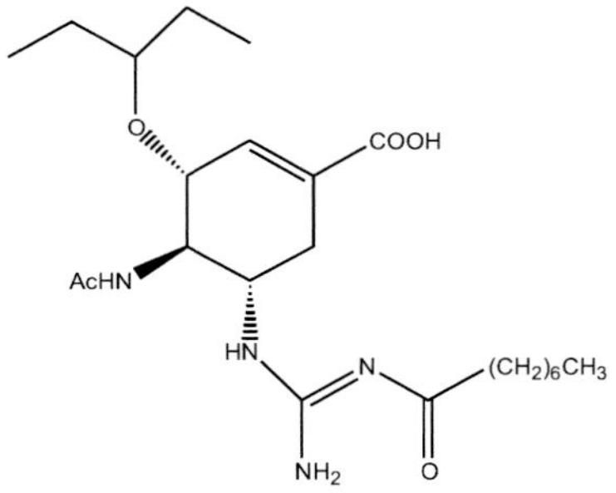

GOC-OG (27)

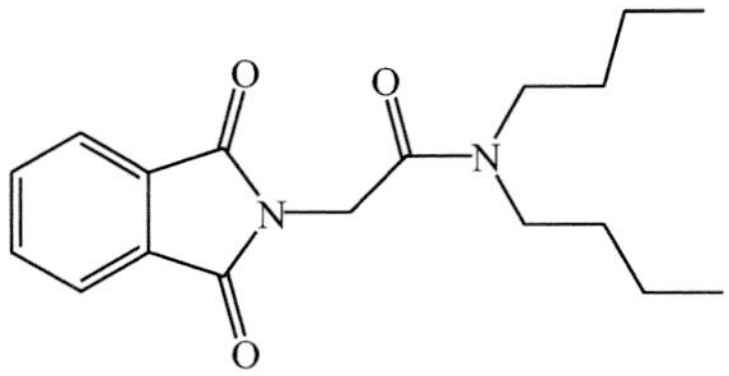

Amide phthaloylglycine (28)

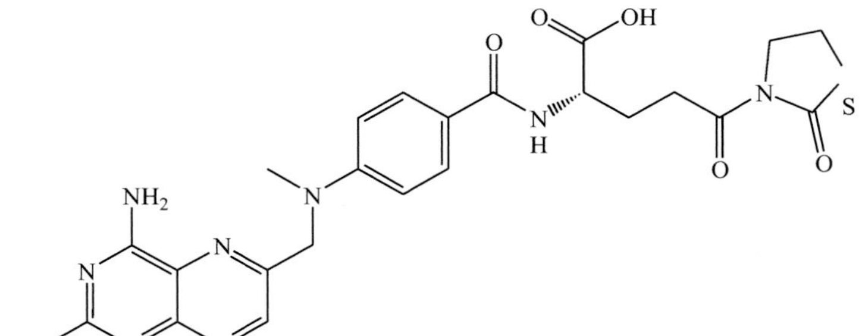

MTX-gamma-thiazolidinone (29)

FIGURE 1.6 (Continued) Examples of prodrugs having amine coupled linkage. *(Continued)*

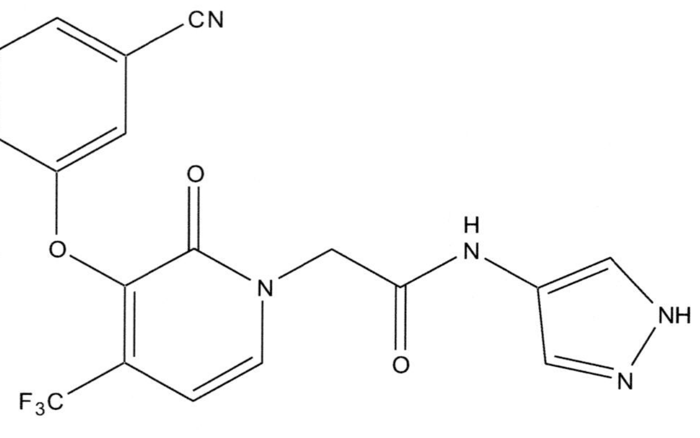

Aceta-doravirin, 8i (30)

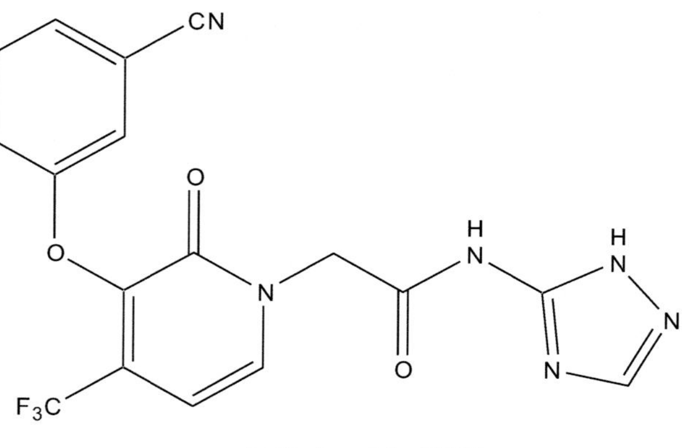

Aceta-doravirin, 8i (31)

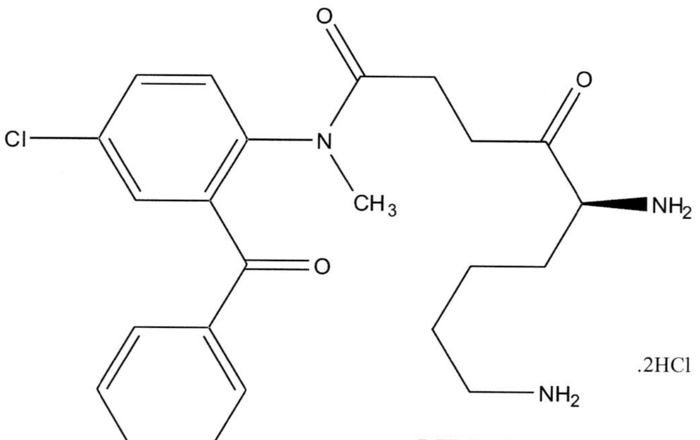

.2HCl

DZP-LYS (32)

FIGURE 1.6 (Continued) Examples of prodrugs having amine coupled linkage. *(Continued)*

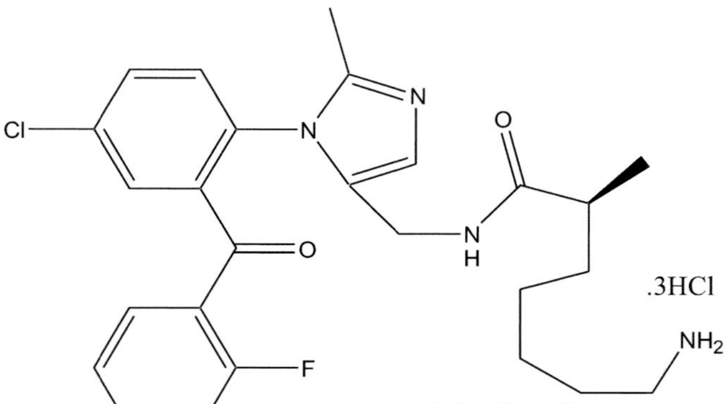

MDZ-LYS (33)

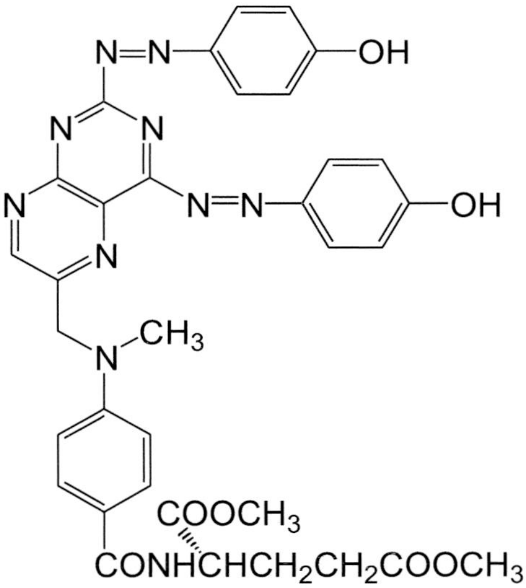

METHO10 (34)

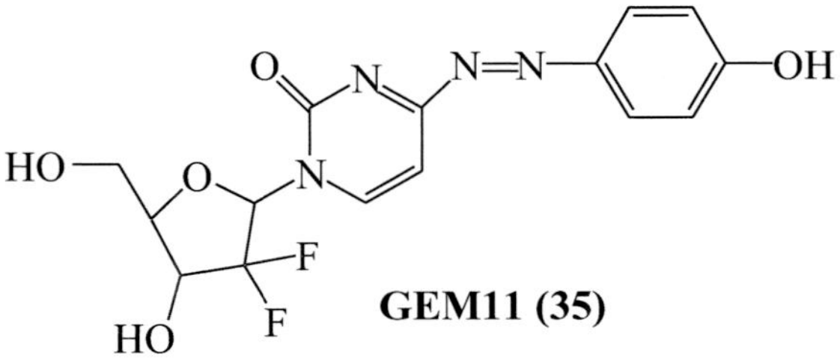

GEM11 (35)

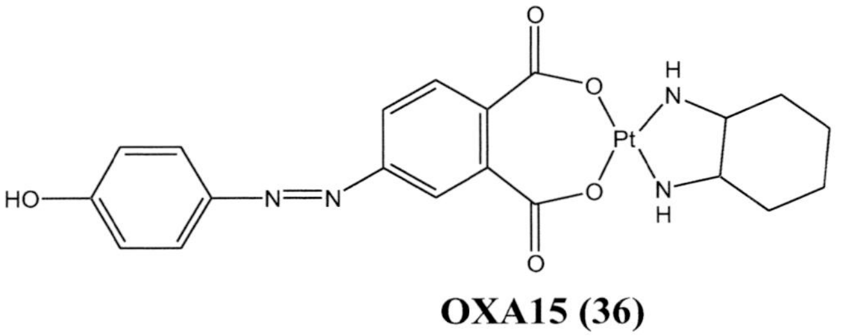

OXA15 (36)

FIGURE 1.6 (Continued) Examples of prodrugs having amine coupled linkage.

The prodrug for *N*-Palmitoylethanolamide (PEA) for the treatment of nerve pain and nerve degeneration diseases synthesized. As this parent drug PEA got inactivated by enzyme lipid amidases that bring its hydrolysis. This drawback got overcome by designing its prodrug, i.e., galactosyl prodrug of PEA (**PEAGAL**) (**25**). Its biological activities were observed and checked in neuroblastoma and in C6 glioma cells. The quantitative analysis of the same was also done by an LC–MS–MS technique. The results encouraged the scientists to go for studies using on the animal models of neuropathic pain and of neurological disorders and/or neurodegenerative diseases (Luongo et al. 2014). The prodrugs for zanamivir and oseltamivir as acylguanidine derivatives were designed. Zanamivir (ZA) and guanidino-oseltamivir carboxylic acid (GOC) were very strong inhibitors for neuraminidase (NA). These drugs work against influenza neuraminidase (NA). They have guanidinium moiety that has significant role in NA coupling. These drugs cannot be given orally due to their polar cationic nature. The acylguanidine derivatives of these drugs were synthesized with the aim that they can be given orally. The pharmacological experiments clearly indicated that ZA octanoylguanidine (**ZA-OG**) (**26**) derivative were better than ZA. The GOC octanoylguanidine (**GOC-OG**) (**27**) derivative had the same efficiency as GOC. This approach resulted in the development of oral drug for influenza (Hsu et al. 2018).

Begum and group designed *N*-phthaloylglycine amide derivatives. These were tested against butyrylcholinesterase inhibitor. The compound **Amide phthaloylglycine** (**28**) was found to be more active than standard galantamine. The same results were also evaluated *in silico* and parallel data were generated. Interaction studies clearly signified the result. The data suggested that the most active molecule **amide phthaloylglycine** can be tested for *in vivo* properties and this might be beneficial for the drug design and to discover new drugs in pharmaceutical industries (Begum et al. 2018). The two thiazolidinone-based Methotrexate (MTX) prodrugs were synthesized to mask the toxicity and to improve the patient acceptability. MTX belong to the category of anticancer but at lower concentration (5–25 mg/week) it works affectively for rheumatoid arthritis. At low dose also it has certain adverse effects like gastrointestinal toxicities, lethargy, fatigue, renal insufficiency and anemia. Most of the patients who take this medicine discontinue it within 3 years of therapy (Bannwarth et al. 1996). In this prodrug (**MTX–gamma-thiazolidonone**) (**29**) deliver the MTX locally and accumulate in inflammatory tissue and improved the safety profile and the efficacy of drug (Andersen et al. 2018).

Biological evaluation of novel acetamide substituted doravirin was done by MT-4 cell-based assays using the MTT method, the results of assay clearly exhibited the moderate to good activity of the prodrug against the wild-type HIV-1 strain and reported as active HIV-1 non-nucleoside reverse transcriptase inhibitors. The prodrug (**Aceta-doravirin,8i**) (**30**) and (**Aceta-doravirin,8k**) (**31**) displayed robust activity against wild-type HIV-1 with EC_{50} value 59.5 nm and 54.8 nm, respectively (Wang et al. 2019).

The prodrug for benzodiazepine (diazepam and midazolam) (**DZP-LYS** (**32**) and **MDZ-LYS**) (**33**) synthesized for intranasal rescue therapies as these drugs are found to be less water soluble causing practical problem with these drugs are administration. It is given parentally which limit its use. The nasal route has several challenges. First, the total volume that can be administered via the nose is quite small (~200 μL). Second, BZD has low aqueous solubilities, so aqueous vehicles cannot be used in nasal formulations. Here the prodrug developed has lysine residue with drug and converting enzyme in an aqueous vehicle. This method, besides giving an aqueous media to deliver benzodiazepines, may also result in more absorption. Such developed prodrugs would be safer, allow intranasal administration of proper doses in small volumes, and result in earlier termination of seizures (Siegel et al. 2015).

The azo linked prodrugs of drugs used against cancer like methotrexate, gemcitabine and potent analog of oxaliplatin were synthesized and characterized. Such drugs got cleaved in the colon with improved physicochemical properties, reduced toxicity and undesirable side effects. The prepared prodrugs showed stability at bot acidic and basic pH buffers and elicited their stability in upper GIT media. The analysis was done in rat fecal/cecum content and in intestinal mucosa, it occurred in presence of enzyme azoreductase which is responsible for the degradation and reduction of azo bond and drug release. The *in vitro* study was performed with rat fecal material. The prodrugs **METHO10** (**34**), **GEM11** (**35**), and **OXA15** (**36**) of methotrexate, gemcitabine and potent analog of oxaliplatin, respectively, had better release profile and found to be effective against colorectal cancer cell lines (COLO 205, COLO 320 DM, and HT-29). So, the designed prodrugs had azolinkage. This make to get effectively target the colon. So that the colon had maximal uptake of drug with minimal side effects (Sharma et al. 2013).

1.5 Prodrugs of Phosphonates, Phosphinates, and Phosphates

The designing of these prodrugs aimed to improve physical or chemical properties, bioavailability, lipophilicity, or targeting. These groups have negative charges at body pH making them polar (De Clercq et al. 1987). Due to their high polarity they have limited effect as these drugs show low volume of distribution and low bioavailability (Eisenberg et al. 2001).

The phosphate containing prodrugs are not in common use while phosphonate and phosphinate are available in market due to their chemical and enzymatic stability (Figure 1.7). The derivatization of these functional groups containing drugs may increase the lipophilicity. These drugs in systemic circulation on crossing biological membrane get bio transformed into its original form. So, if target is present inside the membrane the drug released exert its action. It can be shown diagrammatically as in Figure 1.8.

Some of the recent examples are as follows (Figure 1.9).

Eight prodrugs of (S)-3-(adenine-9-yl)-2-(phosphonomethoxy)propanoic acid (CPMEA) were synthesized, none of them showed any anti-HIV or anti-RSV action. However, two of the prodrugs, **bis-amidate (37)** and **phenyloxy monoamidite (38)**, thus represent the first examples of acyclic nucleoside phosphonates with natural nucleobase (adenine), exhibiting potent anti-HCV properties (Kaiser et al. 2016). The prodrug of a natural antibiotic, i.e., Fosmidomycin synthesized, has promising IspC (DXR, 1-deoxy-D-xylulose-5-phosphate reductoisomerase) inhibitory action. This enzyme helps in biosynthesis of non-mevalonate isoprenoids that is essential in *Plasmodium falciparum* and *Mycobacterium tuberculosis.* This natural antibiotic is unable to penetrate the bacterial cell so this trouble was overcome by converting

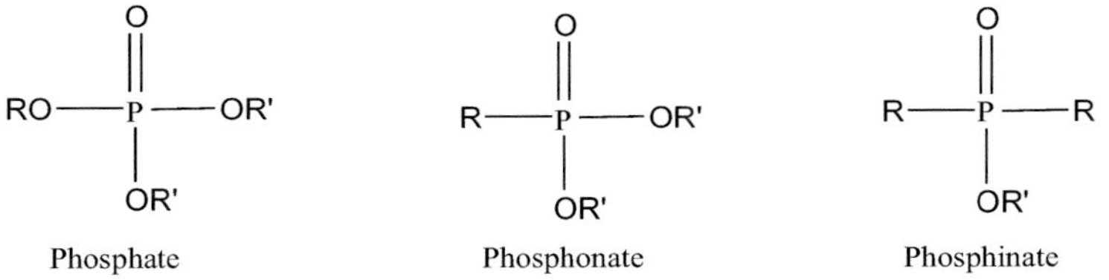

RO—P(=O)(OR')—OR' Phosphate   R—P(=O)(OR')—OR' Phosphonate   R—P(=O)(OR')—R Phosphinate

FIGURE 1.7 Basic structures of phosphates.

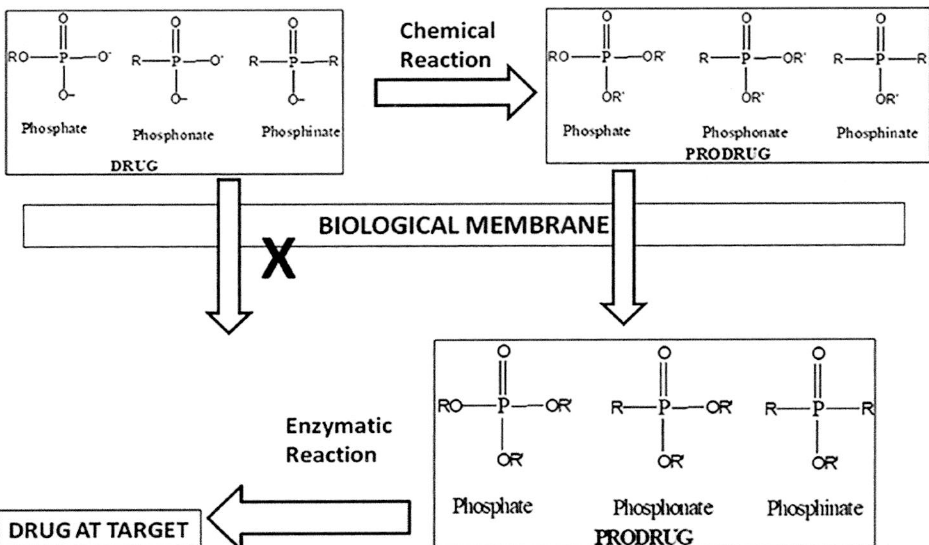

FIGURE 1.8 Mechanism of action for phosphates bearing prodrugs.

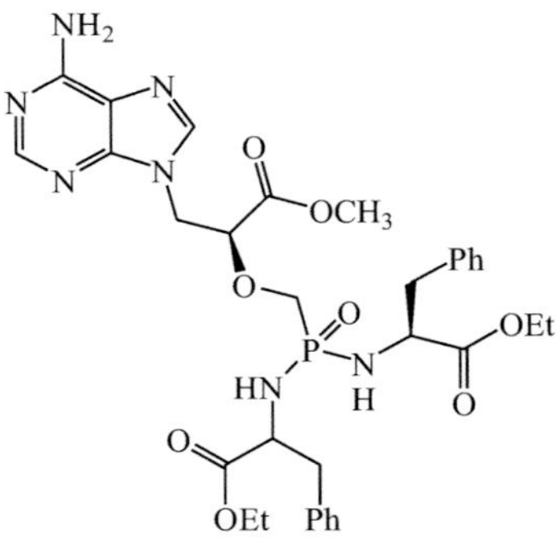

Bis-amidate (37)

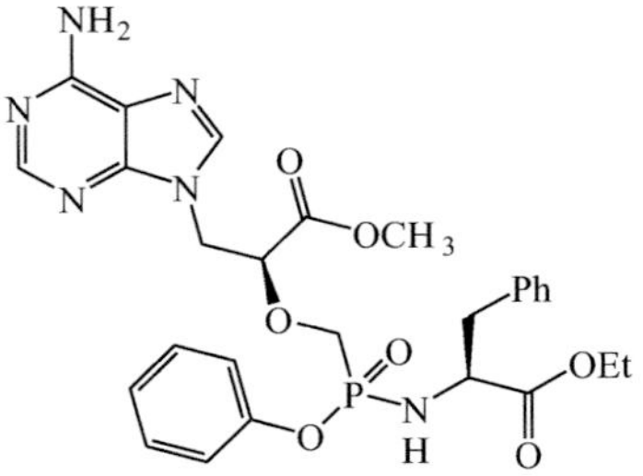

Phenyloxy monoamidate (38)

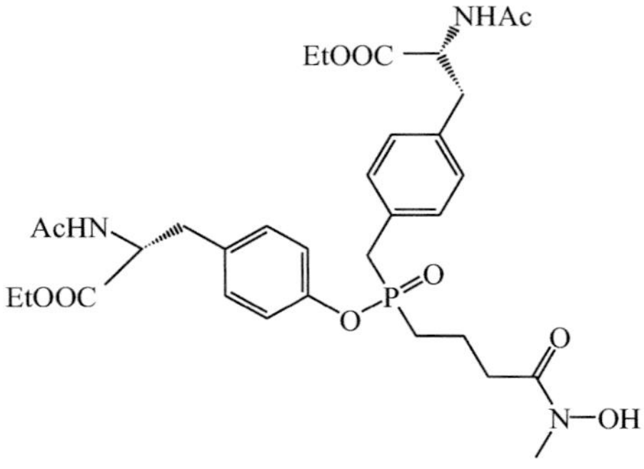

Phosphonodiamidate prodrug (39)

FIGURE 1.9 Examples of prodrugs bearing phosphate coupled linkage.

phosphonate moiety into tyrosine-derived esters. It was found with **phosphonodiamidate prodrug (39)** showed antitubercular activity. Among all the compounds **phosphonodiamidate prodrug** was found to be active (Courtens et al. 2019).

1.6 Peptides Bearing Prodrugs

A lot of peptides are known which have therapeutic activity. The peptides have limitation of having high polarity, poor membrane penetration and charged functional group which prevent it to develop an orally active agent. This limitation can be overcome by the prodrug design. The exposed functional groups in peptides are amino, carboxyl, hydroxyl, and guanidino. These groups were derivatized to get the effective prodrug. Like alkylation of amino group or formation of ester or amide. These derivatizations will lead to increase the lipophilicity and make it stable so that it must not degrade when come in circulation. Some of the recent examples in this category are shown in Figure 1.10.

The peptide prodrug of glucosamine with more gut permeability through gut peptide transporter-1was synthesized. Glucosamine is a naturally occurring amino sugar with mild anti-inflammatory properties. It has low oral bioavailability. The synthesized ester derivative, i.e., Glycine-Valine-Glucosamine **(Gly-Val-GluN) (40)** had increased gut permeability through peptide-1 transporter and reduced dose is required to achieve the therapeutic level. It also showed favorable stability in the gut and a fast cleavage to glucosamine after exposure to the liver homogenate (Gilzad Kohan et al. 2015). The bile acid-ribavirin conjugated as prodrug to deliver the drug specifically to the liver cell in hepatitis and avoid the off-target effect. Ribavirin is the drug of choice to treat hepatitis but it causes serious hemolytic anemia because

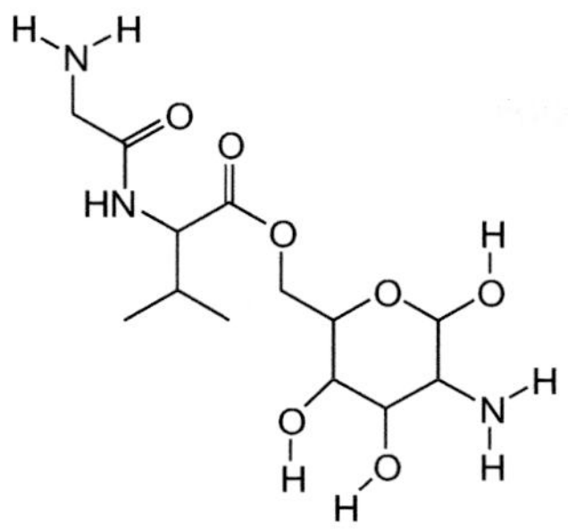

Gly-Val-GluN (40)

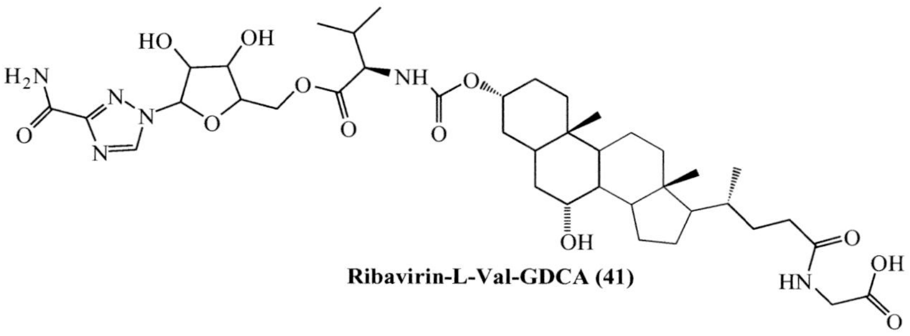

Ribavirin-L-Val-GDCA (41)

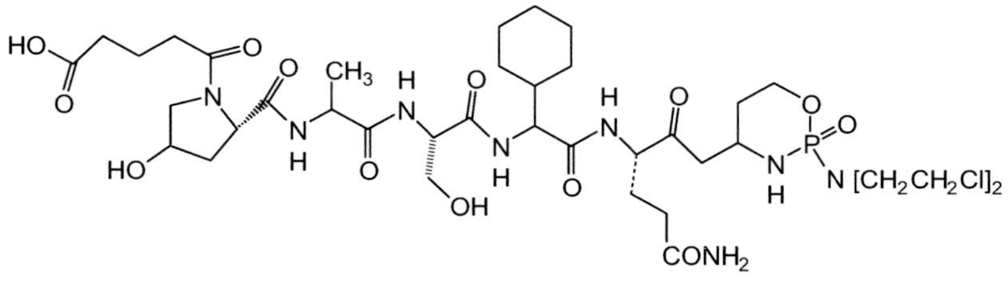

Glutaryl-Hyp-Ala-Ser-Chg-Gln-NHCPA (42)

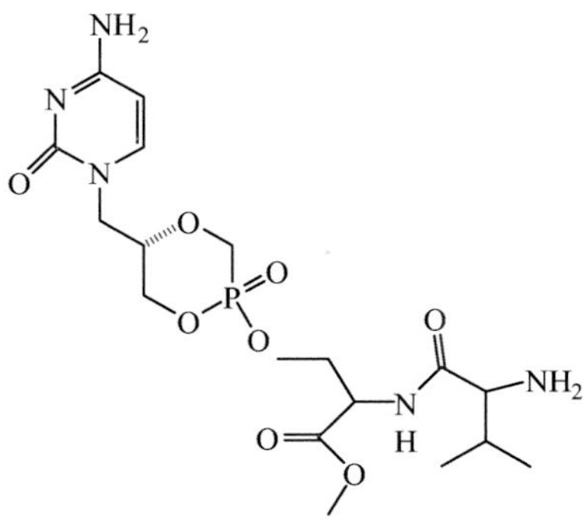

cHPMPC 7 (43)

FIGURE 1.10 Examples of prodrugs bearing peptide linkage. *(Continued)*

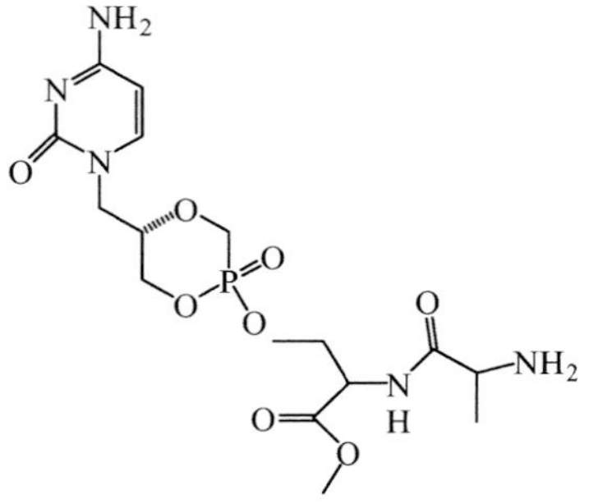

FIGURE 1.10 (Continued) Examples of prodrugs bearing peptide linkage.

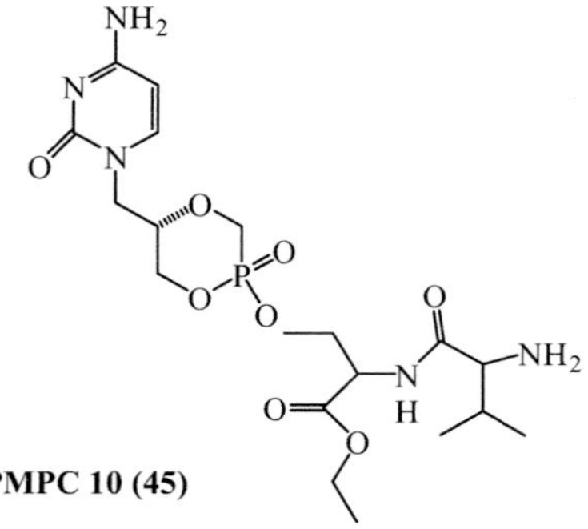

ribavirin accumulates in RBCs. The sodium taurocholate co-transporting polypeptide (NTCP) is a bile acid transporter predominantly present in liver. The synthesized prodrug **Ribavirin-L-Val-GDCA (41)** (glycochenodeoxycholic acid) is the substrate for NTCP so that the release of ribavin specifically in the liver cell takes place and 1.8-fold less revelation of ribavirin in RBCs, plasma and kidney of mice. *In vitro* studies gave the idea that ribavirin concentration in RBCs was lowered by 16.7-fold from prodrug compared with ribavirin alone (Dong et al. 2015). The peptide conjugates of 4-aminocyclophosphamide was synthesized. It acts as prodrug of phosphor amide mustard for selective activation by prostate specific antigen (PSA). In this study three PSA-specific peptides were synthesized which were cleaved by PSA effectively and fully after the expected glutamine residue to release 4-NH_2-CPA, the activated prodrug form of phosphor amide mustard. Among the three peptides, the **pentapeptide conjugate (42)** is most effective with half-life of 55 min. followed by hexapeptide conjugate and tetrapeptide conjugate having half-life of 6.5 hour and 12 hours, respectively. The **pentapeptide conjugate** was proved to be better prodrug against treatment of cancer (Jiang and Hu 2013). The prodrugs for cidofivir (HPMPC) synthesized to overcome its problem of lesser oral bioavailability and poor transportation into cells. It is a broad-spectrum antiviral drug. It is used to treat AIDS-related Cytomegalovirus (CMV) retinitis. To supersede its problem novel dipeptide prodrugs of HPMPC was synthesized via a cHPMPC phosphonate ester linked to the serine side chain hydroxy group. It was found that out of synthesized prodrugs **cHPMPC7 (43)**, **cHPMPC9 (44)** and **cHPMPC10 (45)** showed good oral bioavailability and transported prodrug by endogenous enzymes to its original form of drug (McKenna et al. 2005).

1.7 Future Prospects

It is said that the search for new drugs is still a difficult task. Sometimes it may involve serendipity, but nowadays some rational has developed. The rationale of developing a new drug may involve the study of specific receptor or enzyme that involve in pathological condition. The known receptors may lead to develop an effective drug of choice. These developed drugs should be studied *in vitro* enzymatically or through receptor binding then only *in vivo* testing should be promoted (DeVito 1990). The concept of prodrug has resulted in giving a lot of drugs in market which has clinical applications. This promising approach not only mask the problems (permeation, solubility of drug, bioavailability, stability, taste masking, target delivery, drug resistance, oral absorption and brain delivery) associated with the parent drug. It also has additional or synergetic effects by using molecular combination of existing drugs. This concept has an advantage to get the potent compound with lesser side effects. This concept may give the promising drugs from bench to bedside in near future.

REFERENCES

Albert, Adrien. 1958. Chemical Aspects of Selective Toxicity. *Nature* 182 (4633): 421–422.

Andersen, Nikolaj S., Jorge Peiró Cadahía, Viola Previtali, Jon Bondebjerg, Christian A. Hansen, Anders E. Hansen, Thomas L. Andresen, and Mads H. Clausen. 2018. Methotrexate Prodrugs Sensitive to Reactive Oxygen Species for the Improved Treatment of Rheumatoid Arthritis. *European Journal of Medicinal Chemistry* 156 (August): 738–46. doi:10.1016/j.ejmech.2018.07.045.

Azzolini, Michele, Andrea Mattarei, Martina La Spina, Michele Fanin, Giacomo Chiodarelli, Matteo Romio, Mario Zoratti, Cristina Paradisi, and Lucia Biasutto. 2017. New Natural Amino Acid-Bearing Prodrugs Boost Pterostilbene's Oral Pharmacokinetic and Distribution Profile. *European Journal of Pharmaceutics and Biopharmaceutics* 115 (June): 149–58. doi:10.1016/J.EJPB.2017.02.017.

Balsera, Beatriz, José Mulet, Salvador Sala, Francisco Sala, Roberto de la Torre-Martínez, Sara González-Rodríguez, Adrián Plata, et al. 2018. Amino Acid and Peptide Prodrugs of Diphenylpropanones Positive Allosteric Modulators of A7 Nicotinic Receptors with Analgesic Activity. *European Journal of Medicinal Chemistry* 143 (January): 157–65. doi:10.1016/j.ejmech.2017.10.083.

Bannwarth, Bernard, Fabienne Péhourcq, Thierry Schaeverbeke, and Joël Dehais. 1996. Clinical Pharmacokinetics of Low-Dose Pulse Methotrexate in Rheumatoid Arthritis. *Clinical Pharmacokinetics* 30 (3): 194–210. doi:10.2165/00003088-199630030-00002.

Beaumont, Kevin, Robert Webster, Iain Gardner, and Kevin Dack. 2003. Design of Ester Prodrugs to Enhance Oral Absorption of Poorly Permeable Compounds: Challenges to the Discovery Scientist. *Current Drug Metabolism* 4 (6): 461–85.

Begum, Samreen, Shaikh Sirajuddin Nizami, Uzma Mahmood, Summyia Masood, Sahar Iftikhar, and Summayya Saied. 2018. In-Vitro Evaluation and in-Silico Studies Applied on Newly Synthesized Amide Derivatives of *N*-Phthaloylglycine as Butyrylcholinesterase (BChE) Inhibitors. *Computational Biology and Chemistry* 74 (June): 212–17. doi:10.1016/j.compbiolchem.2018.04.003.

Bhosle, Deepak sadashiv, S. Bharambe, Neha Gairola, and Suneela S. Dhaneshwar. 2006. Mutual Prodrug Concept: Fundamentals and Applications. *Indian Journal of Pharmaceutical Sciences* 68 (3): 286. doi:10.4103/0250-474X.26654.

Bundgaard, Hans. 1985. *Design of Prodrugs*. Amsterdam, the Netherlands: Elsevier.

Clercq, Erik De, Takashi Sakuma, Masanori Baba, Rudi Pauwels, Jan Balzarini, Ivan Rosenberg, and Antonin Holý. 1987. Antiviral Activity of Phosphonylmethoxyalkyl Derivatives of Purine and Pyrimidines. *Antiviral Research* 8 (5–6): 261–72. doi:10.1016/s0166-3542(87)80004-9.

Courtens, Charlotte, Martijn Risseeuw, Guy Caljon, Louis Maes, Anandi Martin, and Serge Van Calenbergh. 2019. Amino Acid Based Prodrugs of a Fosmidomycin Surrogate as Antimalarial and Antitubercular Agents. *Bioorganic & Medicinal Chemistry* 27 (5): 729–47. doi:10.1016/j.bmc.2019.01.016.

DeVito, Stephen C. 1990. Computer-Aided Drug Design: Methods and Applications. *Journal of Pharmaceutical Sciences* 79 (12): 1125. doi:10.1002/jps.2600791218.

Dong, Zhongqi, Qing Li, Dong Guo, Yan Shu, and James E. Polli. 2015. Synthesis and Evaluation of Bile Acid–Ribavirin Conjugates as Prodrugs to Target the Liver. *Journal of Pharmaceutical Sciences* 104 (9): 2864–76. doi:10.1002/jps.24375.

Eisenberg, Eugene J., Gong-Xin He, and William A. Lee. 2001. Metabolism of GS-7340, A Novel Phenyl Monophosphoramidate Intracellular Prodrug of Pmpa, in Blood. *Nucleosides, Nucleotides and Nucleic Acids* 20 (4–7): 1091–98. doi:10.1081/NCN-100002496.

Gilzad Kohan, Hamed, Kamaljit Kaur, and Fakhreddin Jamali. 2015. Synthesis and Characterization of a New Peptide Prodrug of Glucosamine with Enhanced Gut Permeability. Edited by Mária A. Deli. *PLoS One* 10 (5): e0126786. doi:10.1371/journal.pone.0126786.

Harper, Norman J. 1959. Drug Latentiation. *Journal of Medicinal and Pharmaceutical Chemistry* 1 (5): 467–500. doi:10.1021/jm50006a005.

Hsu, Peng-Hao, Din-Chi Chiu, Kuan-Lin Wu, Pei-Shan Lee, Jia-Tsrong Jan, Yih-Shyun E. Cheng, Keng-Chang Tsai, Ting-Jen Cheng, and Jim-Min Fang. 2018. Acylguanidine Derivatives of Zanamivir and Oseltamivir: Potential Orally Available Prodrugs against Influenza Viruses. *European Journal of Medicinal Chemistry* 154 (June): 314–23. doi:10.1016/j.ejmech.2018.05.030.

Hu, Jing-Bo, Shu-Juan Li, Xu-Qi Kang, Jing Qi, Jia-Hui Wu, Xiao-Juan Wang, Xiao-Ling Xu, et al. 2018. CD44-Targeted Hyaluronic Acid-Curcumin Prodrug Protects Renal Tubular Epithelial Cell Survival from Oxidative Stress Damage. *Carbohydrate Polymers* 193 (August): 268–80. doi:10.1016/j.carbpol.2018.04.011.

Hu, Longqin. 2016. Prodrug Approaches to Drug Delivery. In *Drug Delivery*, 227–71. Hoboken, NJ: John Wiley & Sons. doi:10.1002/9781118833322.ch12.

Jiang, Yongying, and Longqin Hu. 2013. Peptide Conjugates of 4-Aminocyclophosphamide as Prodrugs of Phosphoramide Mustard for Selective Activation by Prostate-Specific Antigen (PSA). *Bioorganic & Medicinal Chemistry* 21 (23): 7507–14. doi:10.1016/j.bmc.2013.09.039.

Kaiser, Martin Maxmilian, Lenka Poštová-Slavětínská, Martin Dračínský, Yu-Jen Lee, Yang Tian, and Zlatko Janeba. 2016. Synthesis and Biological Properties of Prodrugs of (S)-3-(Adenin-9-Yl)-2-(Phosphonomethoxy) Propanoic Acid. *European Journal of Medicinal Chemistry* 108 (January): 374–80. doi:10.1016/j.ejmech.2015.12.009.

Li, Qing, Liuwei Meng, Siru Zhou, Xiaoyan Deng, Na Wang, Yi Ji, Yichun Peng, Junhao Xing, and Gongmei Yao. 2019. Rapid Generation of Novel Benzoic Acid–Based Xanthine Derivatives as Highly Potent, Selective and Long Acting DPP-4 Inhibitors: Scaffold-Hopping and Prodrug Study. *European Journal of Medicinal Chemistry* 180 (October): 509–23. doi:10.1016/J.EJMECH.2019.07.045.

Liu, Shuangxi, Kaili Zhang, Qiwen Zhu, Qianqian Shen, Qiumeng Zhang, Jiahui Yu, Yi Chen, and Wei Lu. 2019. Synthesis and Biological Evaluation of Paclitaxel and Vorinostat Co-Prodrugs for Overcoming Drug Resistance in Cancer Therapy in Vitro. *Bioorganic & Medicinal Chemistry* 27 (7): 1405–13. doi:10.1016/j.bmc.2019.02.046.

Longcope, Christopher, Sherwood Gorbach, Barry R. Goldin, Margo N. Woods, Johanna Dwyer, and James H. Warram. 1985. The Metabolism of Estradiol; Oral Compared to Intravenous Administration. *Journal of Steroid Biochemistry* 23 (6): 1065–70. doi:10.1016/0022-4731(85)90068-8.

Luongo, Elvira, Roberto Russo, Carmen Avagliano, Anna Santoro, Daniela Melisi, Nicola Salvatore Orefice, Giuseppina Mattace Raso, et al. 2014. Galactosyl Prodrug of Palmitoylethanolamide: Synthesis, Stability, Cell Permeation and Cytoprotective Activity. *European Journal of Pharmaceutical Sciences* 62 (October): 33–39. doi:10.1016/j.ejps.2014.05.009.

Marinelli, Lisa, Erika Fornasari, Piera Eusepi, Michele Ciulla, Salvatore Genovese, Francesco Epifano, Serena Fiorito et al. 2019. Carvacrol Prodrugs as Novel Antimicrobial Agents. *European Journal of Medicinal Chemistry* 178 (September): 515–29. doi:10.1016/J.EJMECH.2019.05.093.

McKenna, Charles E., Boris A. Kashemirov, Ulrika Eriksson, Gordon L. Amidon, Phillip E. Kish, Stefanie Mitchell, Jae-Seung Kim, and John M. Hilfinger. 2005. Cidofovir Peptide Conjugates as Prodrugs. *Journal of Organometallic Chemistry* 690 (10): 2673–78. doi:10.1016/J.JORGANCHEM.2005.03.004.

Pitman, Ian H. 1981. Pro-Drugs of Amides, Imides, and Amines. *Medicinal Research Reviews* 1 (2): 189–214.

Rasheed, Arun, G. Lathika, Yalavarthi Prasanna Raju, Kenza Mansoor, Abdul Karim Azeem, and Nija Balan. 2016. Synthesis and Pharmacological Evaluation of Mutual Prodrugs of Aceclofenac with Quercetin, Vanillin and l-Tryptophan as Gastrosparing NSAIDS. *Medicinal Chemistry Research* 25 (1): 70–82. doi:10.1007/s00044-015-1469-7.

Redasani, Vivekkumar K., and Sanjay B. Bari. 2012. Synthesis and Evaluation of Mutual Prodrugs of Ibuprofen with Menthol, Thymol and Eugenol. *European Journal of Medicinal Chemistry* 56 (October): 134–38. doi:10.1016/j.ejmech.2012.08.030.

Remington, Joseph P. (Joseph Price), and Alfonso R. Gennaro. 2000. *Remington: The Science and Practice of Pharmacy*, 20th ed. Baltimore, MD: Lippincott Williams & Wilkins.

Sharma, Rajiv, Ravindra K. Rawal, Tripti Gaba, Nishu Singla, Manav Malhotra, Sahil Matharoo, and T.R. Bhardwaj. 2013. Design, Synthesis and Ex Vivo Evaluation of Colon-Specific Azo Based Prodrugs of Anticancer Agents. *Bioorganic & Medicinal Chemistry Letters* 23 (19): 5332–38. doi:10.1016/j.bmcl.2013.07.059.

Siegel, Ronald A., Mamta Kapoor, Narsihmulu Cheryala, Gunda I. Georg, and James C. Cloyd. 2015. Water-Soluble Benzodiazepine Prodrug/Enzyme Combinations for Intranasal Rescue Therapies. *Epilepsy & Behavior* 49 (August): 347–50. doi:10.1016/j.yebeh.2015.05.004.

Skarbek, Charles, Silvia Serra, Hichem Maslah, Estelle Rascol, and Raphaël Labruère. 2019. Arylboronate Prodrugs of Doxorubicin as Promising Chemotherapy for Pancreatic Cancer. *Bioorganic Chemistry* 91 (October): 103158. doi:10.1016/J.BIOORG.2019.103158.

Sloan, Kenneth B., and Scott Wasdo. 2003. Designing for Topical Delivery: Prodrugs Can Make the Difference. *Medicinal Research Reviews* 23 (6): 763–93. doi:10.1002/med.10048.

Stachulski, Andrew V., Karl Swift, Mark Cooper, Stephan Reynolds, Daniel Norton, Steven D Slonecker, and Jean-Francois Rossignol. 2017. Synthesis and Pre-Clinical Studies of New Amino-Acid Ester Thiazolide Prodrugs. *European Journal of Medicinal Chemistry* 126: 154–59. doi:10.1016/J.EJMECH.2016.09.080.

Stella, Valentino J., William Charman, and Vijay H. Naringrekar. 1985. Prodrugs. *Drugs* 29 (5): 455–73. doi:10.2165/00003495-198529050-00002.

Takahashi, Masato, Tomoki Uehara, Minori Nonaka, Yuka Minagawa, Riona Yamazaki, Masami Haba, and Masakiyo Hosokawa. 2019. Synthesis and Evaluation of Haloperidol Ester Prodrugs Metabolically Activated by Human Carboxylesterase. *European Journal of Pharmaceutical Sciences* 132 (April): 125–31. doi:10.1016/j.ejps.2019.03.009.

Taleghani, Akram, Mohammad Ali Nasseri, and Mehrdad Iranshahi. 2017. Synthesis of Dual-Action Parthenolide Prodrugs as Potent Anticancer Agents. *Bioorganic Chemistry* 71 (April): 128–34. doi:10.1016/J.BIOORG.2017.01.020.

Testa, Bernard, and Joachim M. Mayer. 1998. Design of Intramolecularly Activated Prodrugs. *Drug Metabolism Reviews* 30 (4): 787–807. doi:10.3109/03602539808996330.

Testa Bernard, and Joachim M. Mayer. 2003. Hydrolysis in Drug and Prodrug metabolism, Chemistry, biochemistry and enzymology. Weinheim: Wiley-VCH. pp. 690–95.

Thiele, Nikki A., Jussi Kärkkäinen, Kenneth B. Sloan, Jarkko Rautio, and Kristiina M. Huttunen. 2018. Secondary Carbamate Linker Can Facilitate the Sustained Release of Dopamine from Brain-Targeted Prodrug. *Bioorganic & Medicinal Chemistry Letters* 28 (17): 2856–60. doi:10.1016/j.bmcl.2018.07.030.

Wang, Zhao, Zhao Yu, Dongwei Kang, Jian Zhang, Ye Tian, Dirk Daelemans, Erik De Clercq, Christophe Pannecouque, Peng Zhan, and Xinyong Liu. 2019. Design, Synthesis and Biological Evaluation of Novel Acetamide-Substituted Doravirine and Its Prodrugs as Potent HIV-1 NNRTIs. *Bioorganic & Medicinal Chemistry* 27 (3): 447–56. doi:10.1016/j.bmc.2018.12.039.

Wu, Zhiqian, Ashish Patel, Rutesh Dave, and Xudong Yuan. 2010. Development of Acetaminophen Proline Prodrug. *Bioorganic & Medicinal Chemistry Letters* 20 (13): 3851–54. doi:10.1016/j.bmcl.2010.05.050.

Zhang, Weiyi, Jun Yang, Jing Fan, Bin Wang, Yi Kang, Jin Liu, Wensheng Zhang, and Tao Zhu. 2019. An Improved Water-Soluble Prodrug of Propofol with High Molecular Utilization and Rapid Onset of Action. *European Journal of Pharmaceutical Sciences* 127 (January): 9–13. doi:10.1016/j.ejps.2018.09.024.

Zhou, Tian, Hang Su, Prasanta Dash, Zhiyi Lin, Bhagya Laxmi Dyavar Shetty, Ted Kocher, Adam Szlachetka, et al. 2018. Creation of a Nanoformulated Cabotegravir Prodrug with Improved Antiretroviral Profiles. *Biomaterials* 151 (January): 53–65. doi:10.1016/j.biomaterials.2017.10.023.

2

Codrugs: Optimum Use through Prodrugs

Suneela Dhaneshwar, Poorvashree P. Joshi, and Shubham Pawar

CONTENTS

2.1 Introduction

Prodrug technology is one of the most justified tools to address the unmet needs of key patient/prescriber regarding marketability of drugs. Prodrug is a better version of already known pharmaceutical drugs. Prodrug development leverages enhancement to improve onset of action, duration, and consistency of drug delivery for a more predictable therapeutic effect, and reduction of abuse potential (Huttunen and Jarkko 2011). Prodrugs make good drugs better with reduction of development time horizon than traditional drug development (Zawilska et al., 2013). The potential of prodrug technology is evidenced by its vast success in the past few years. Between years 2000 to 2008, approximately 20% of approved drugs are prodrugs of small molecular weight. Previously, 10% of prodrugs were marketed but now the share of prodrugs has been increased to 12% in the market between 2008 and 2017 (Najjar and Rafik, 2019).

This elegant concept was first coined by Adrien Albert as "prodrugs" which was later corrected by the more accurate and descriptive term, predrug (Albert, 1958). Clinically relevant prodrugs are abundant. The first purposefully synthesized site-specific prodrug was methenamine (or hexamine) in 1899 by Schering from the antibacterial agent formaldehyde and ammonium ions (Testa, 2004). Another early one is sulfasalazine, approved in the United States in 1950 as a first line therapeutic for rheumatoid arthritis or inflammatory bowel disease. Sulfasalazine is formed by linking sulfapyridine with 5-amino salicylic acid (Peppercorn, 1984).

A drug, whose usefulness is limited by adverse physicochemical properties, such that it is not capable of overcoming a particular barrier, is chemically modified via the attachment of a promoiety to generate a new chemical entity, the prodrug, whose properties are such that it is capable of traversing the limitation. The physicochemical and pharmacokinetics barriers generally involve poor aqueous solubility, low lipophilicity, chemical instability, incomplete and rapid absorption, low or variable bioavailability, poor site specificity. Ideally, the promoiety/drug bond will be designed to undergo *in vivo* activation by detaching the promoiety with enzyme/non-enzyme-assisted cleavage. So, a prodrug is inactive, bioreversible form of an active parent drug molecule that elicits desired pharmacological action after *in vivo* enzymatic or chemical transformation to parent drug (Figure 2.1). A new version of the existing drugs is created by modifying their physicochemical, biopharmaceutical and pharmacokinetic properties to improve absorption, distribution and enzymatic metabolism to give better performance, improving its usage against noncurable diseases and maintaining drug in developmental stages (Testa, 2007).

The functional group modification of drugs to overcome their varied pharmacokinetic/pharmacodynamic drawbacks is termed as drug latentiation. Harper (1959, 1962) described drug latentiation as a temporary modification of clinically significant drug by applying chemical approach (Harper, 1958). Kupchan et al. (1965) utilized the prodrug or latentiated form of drug to combat difficulties and included

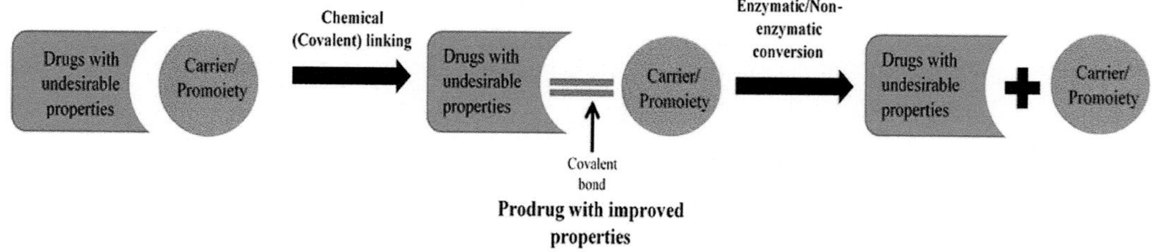

FIGURE 2.1 Schematic representation of the "prodrug" concept.

nonenzymatic regeneration of the parent compound as a consequence of nonenzyme-mediated hydrolytic, dissociative, and other reactions.

Though terms such as latentiated drugs, bioreversible derivatives and prodrugs are replaceable terms, "prodrug" remains the most commonly accepted term. In the review authored by Sinkula and Yalkowsky, the time lapse in regenerating active drug *in vivo* explains latentiation. So, latentiated drug derivatives contribute to the term prodrug that covers inactive substances which become active after administration and elicit the action by combining with receptors (Sinkula and Yalkowsky, 1975).

2.2 Need/Objective Behind Development of Prodrugs

Prodrugs are designed with an objective of improving biopharmaceutical, pharmacokinetic and physicochemical properties of active pharmaceutical ingredients. The main goal of prodrug design is to correct unfavorable physicochemical parameters, and increase *in vivo* stability and targeted drug delivery. So, the prodrug becomes more convenient, safer, effective form of administration of approved drug.

2.2.1 Pharmacodynamic Objectives

Pharmacodynamic objective of designing of prodrugs is achieved by masking reactive species of drug, by improving therapeutic index, in situ activation of drugs and by decreasing systemic toxicity. The best example of in situ activation of drugs is clopidogrel (Figure 2.2). Clopidogrel is a bioprecursor prodrug having antithrombotic properties. Oral delivery of clopidogrel leads to rapid absorption through intestine followed by complete liver metabolism, involving oxidation of clopidogrel to 2-oxoclopidogrel (cyclic thioester) by CYP450 enzyme. In the second step, there is rapid breakdown of 2-oxoclopidogrel to thiolated metabolite that exerts irreversible antagonism of platelet ADP receptors by formation of disulphide (S-S) bond.

2.2.2 Pharmaceutical Objectives

Modification of physicochemical properties involves making drug more hydrophilic, i.e., increase in aqueous solubility (e.g., cyclosporine) or lowering of lipophilicity, improving resistance to chemical degradation (e.g., dopamine), improving bitter taste or foul odor (e.g., chloramphenicol), decreasing pain or irritation at injection site or changing physical form of a drug (e.g., trichloroethanol). Improving physicochemical properties by pharmaceutical technology is an uncertain and time-consuming approach. Hence scientists prefer to take advantage of prodrug strategy. In case of dapsone, various efforts are taken to improve water solubility of drug by forming amide linkage between dapsone and various amino acids (Pochopin et al., 1994). Out of that, glycyl, alanyl, and lysyl prodrugs of dapsone showed good water solubility as represented in Figure 2.3. Palmitate ester of chloramphenicol is an example of improvement of unpleasant taste. Prodrug technology has yielded taste-free prodrug of erythromycin A as 2′-ethyl

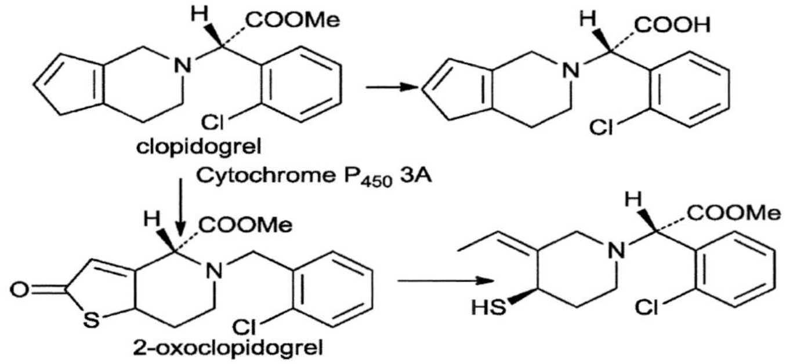

FIGURE 2.2 Bioactivation of clopidogrel.

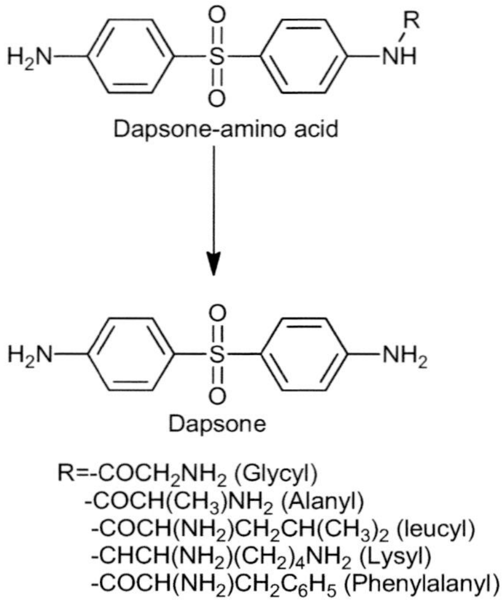

FIGURE 2.3 Bioactivation mechanism of dapsone-amino acid prodrug.

succinate ester. This prodrug undergoes slow, significant amount of hydrolysis in containers hence double prodrug of 2′-ethyl succinate prodrug is synthesized.

2.2.3 Pharmacokinetic Objectives

Improvement in pharmacokinetic properties involves increasing per-oral bioassimilation (oseltamivir, epinephrine and ampicillin) decreasing biotransformation in liver (propranolol), improving absorption by routes other than oral, providing organ or tissue-targeted delivery of active agents. Oseltamivir is ethyl ester prodrug of Ro-64-0802 which had shown significant inhibitory activity in cell lines but poor oral bioavailability (Lew et al., 2000). To overcome this problem Ro-64-0802 was derivatized into ester prodrug, i.e., oseltamivir which undergoes rapid cleavage by carboxylesterases to provide extended and high Ro-64-0802 release as depicted in Figure 2.4. Prodrug strategy is also useful tool to provide slow and prolonged release of steroid hormones. This concept is applied to provide depot effect of bambuterol. Bambuterol is β_2-adrenoreceptor agonist prodrug of terbutaline with ester linkage. Prodrug approach reduced dosing frequency of terbutaline from thrice a daily to once daily. Indeed, bambuterol is hydrolyzed by blood-cholinesterase and undergoes monooxygenase-catalyzed chemical oxidation in different tissues (Persson and Pahlm, 1995); it is represented in Figure 2.5. Another pharmacokinetic objective achieved by prodrugs is site-specific delivery. This approach is attracting lot of attention of scientists working in cancer therapy. For example, capecitabine (Tsukamoto et al., 2001) an orally administered prodrug of 5-fluorouracil (5-FU), undergoes three activation steps and results in higher accumulation of 5-FU in tumor cells (Figure 2.6).

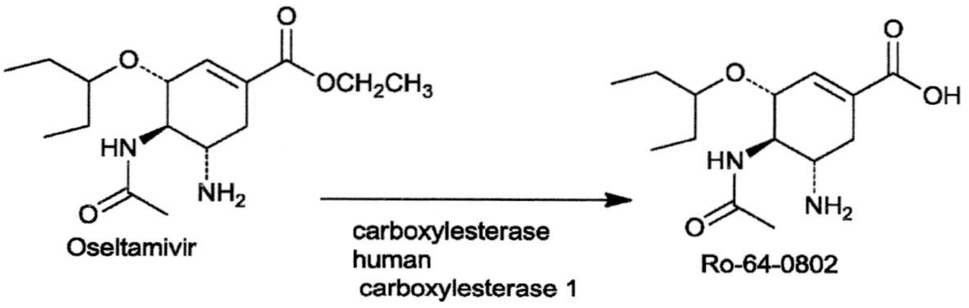

FIGURE 2.4 Bioactivation of oseltamivir.

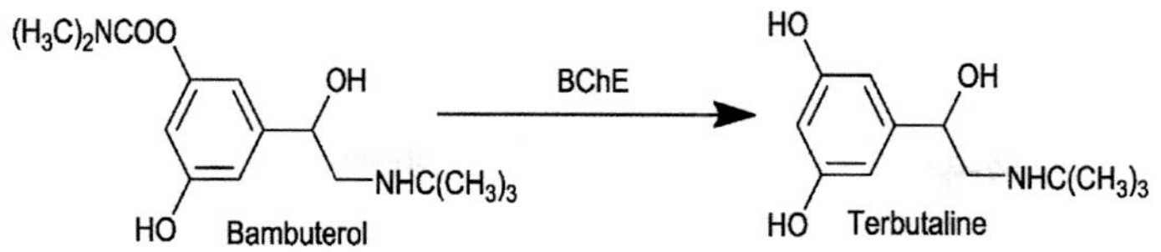

FIGURE 2.5 Bioactivation of bambuterol.

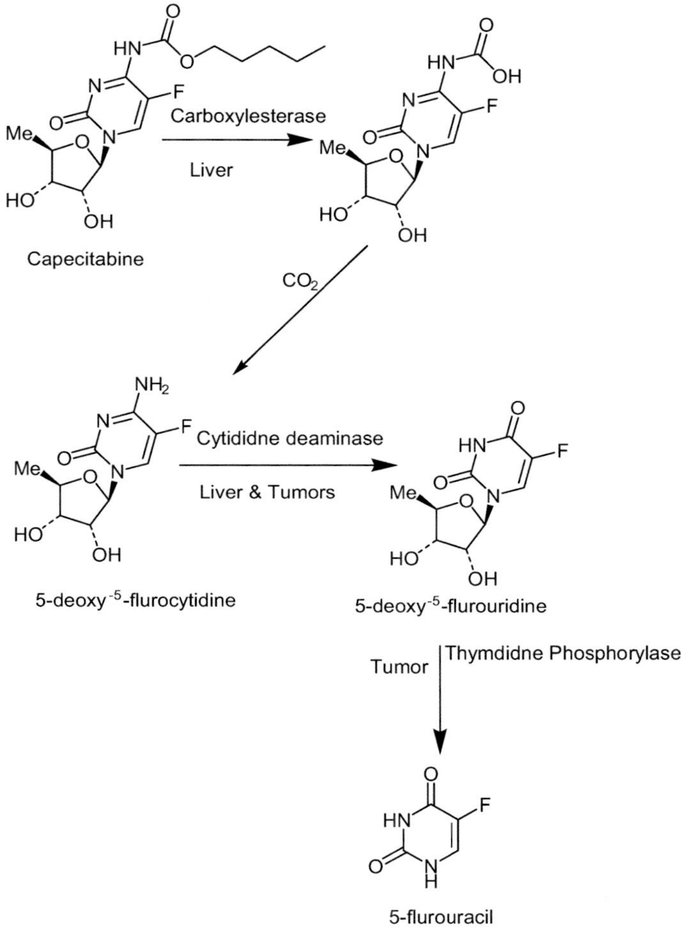

FIGURE 2.6 Bioactivation of capecitabine.

2.3 Ideal Characteristics of a Prodrug

Prodrugs are designed with the purpose to overcome unattractive physicochemical properties that limit a drug's utility. It has been recommended that some ideal properties must be fulfilled while designing prodrugs for optimized drug delivery (Rao, 2003).

Ideal characteristics of prodrugs are represented in Figure 2.7:

1. Bioreversible
2. Ready transport to site of action
3. Nontoxic

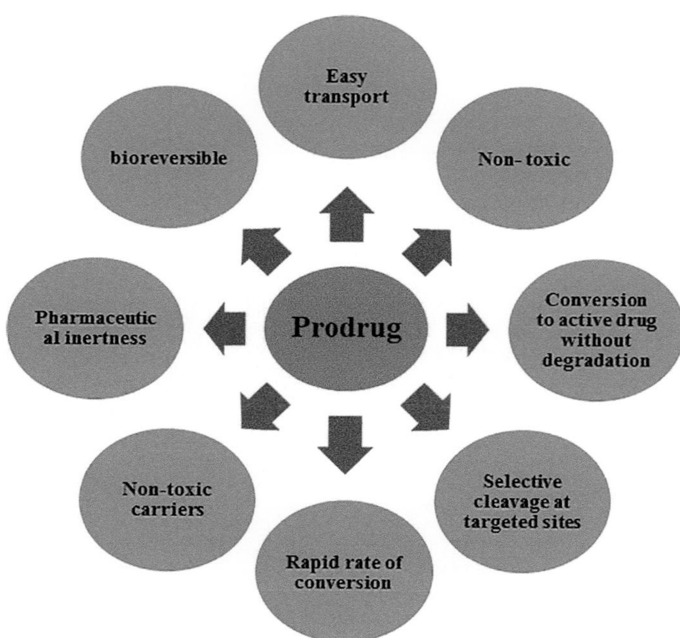

FIGURE 2.7 Ideal characteristics of prodrugs.

4. Selective cleavage at site of action by enzymatic mechanism
5. Inactivity or less active before bioactivation
6. After conversion to active form tissue must retain the drug without degradation
7. Rapid rate of conversion to maintain therapeutic concentration
8. Nontoxicity of released carrier/promoiety

2.4 Classification of Prodrugs

Schematic representation of classification of prodrugs is presented in Figure 2.8.

2.4.1 Current Classification of Prodrugs

There are various ways to classify prodrugs:

1. **Based on therapeutic categories:** mainly classified depending on therapeutic application of prodrug; for example, prodrugs used in HIV, cancer, bacterial infection, cardiovascular diseases, etc (Beale and Block, 2011).
2. **Based on mechanism of activation:** Mainly classified into prodrugs that undergo enzymatic activation and/or nonenzymatic (chemical) activation. Hydrolysis, reduction, oxidation and pH-dependent activation are different mechanisms employed in bioconversion of prodrugs (Rautio et al., 2010).
3. **Based on carriers, covalently connected to drugs:** Includes ester prodrugs, glycosidic prodrugs, antibody-, gene-, virus-directed enzyme prodrugs, bipartate prodrugs, tripartate prodrugs (De Albuquerque et al., 2005).

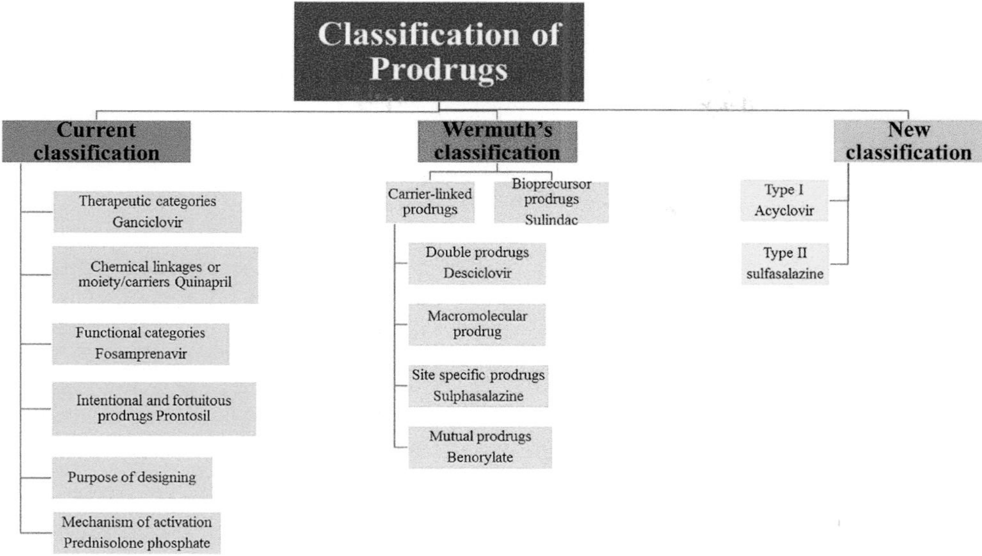

FIGURE 2.8 Classification of prodrugs.

4. **Based on functional categories:** To overcome inherent shortcomings of active drug. For example, prodrugs for improvement in site specificity, bypassing high first-pass metabolism, improving absorption, reducing adverse effects (Redasani and Bari 2015).
5. **Intentional and fortuitous prodrugs:** Intentional prodrugs are formed by purposeful chemical modification of known active agent to improve major pharmacokinetic and/or pharmaceutical drawbacks associated with active agent. Most prodrugs used clinically are intentionally designed prodrugs. Fortuitous prodrugs are few and developed serendipitously (e.g., aspirin, prontosil) (Karaman, 2014).
6. **Based on purpose of designing:** Post-hoc design involves modifications to overcome some undesirable properties of active drug and Ad-hoc design is meant for modifications to overcome serious issues of active but withdrawn drug (Testa, 2009).

2.4.2 Wermuth's Classification

Wermuth (2011) classified prodrugs under two broad classes:

1. Carrier-linked prodrugs
2. Bioprecursors

- **Carrier-linked prodrugs:** Carrier-linked prodrug comprises of covalent linkage between carrier group or promoiety and the active drug to modify its physicochemical properties followed by enzymatic or nonenzymatic activation to give its active form. Thus, this class of prodrugs are modified drugs with covalent tethering between specialized nontoxic protective carriers in a temporary manner to expel unwanted properties in the parent molecule. Characteristics that should be taken into consideration regarding choice of carrier attached to active drug include presence of complimentary functional groups for linkage with parent molecule, nonimmunogenicity, low cost, easy synthesis, *in vivo* stability at desired site and nonactive cleavage products after biodegradation. Major functional groups involved in designing of carrier-linked prodrugs are amides, esters, phosphates, glycosides, oximes, imines, carbamates, carbonates, N-mannich base, and azo derivatives. For example, dipivaloyladrenaline is a diester prodrug of adrenalin which hydrolyzes to adrenaline in presence of esterase enzymes (Shirke et al., 2015). Carrier-linked prodrugs are further categorized as:

- **Pro-prodrugs, double prodrugs, or cascade-latentiated prodrugs:** where prodrug is further derivatized by attachment to one more carrier that on two cycles of enzymatic and or chemical activation releases previous prodrug and then drug respectively (Bundgaard, 1989).
- **Macromolecular prodrug:** where carriers are macromolecules like cyclodextrins, peptides, proteins, polysaccharides, dextrans (Soyez et al., 1996).
- **Site-specific prodrugs:** where carrier acts as a transporter or homing device for site-specific delivery (Jiang and Dolphin, 2008).
- **Mutual prodrugs:** where pharmacologically active carrier is used in place of inactive moiety. Mutual prodrugs may be designed for additive or synergistic action by selecting carrier with same or complementary pharmacological action. Another aim behind designing mutual prodrugs could be targeting the active drug to a specific site or to reduce side effects associated with active drug. In mutual prodrug design one biologically active molecule behaves as a promoiety for other and *vice versa* (Abet et al., 2017).

- **Bioprecursor prodrugs:** Bioprecursor prodrug is designed by molecular modification of active agent without attachment of any promoiety so that it acts as a substrate for various metabolizing enzymes of one or more Phase-I metabolic pathways to active moiety such as hydration, oxidation and reduction. For example, sulindac, an anti-inflammatory drug, undergoes biotransformation to sulphone (inactive metabolite) and sulfide (active metabolite) by irreversible oxidation and reversible reduction respectively and thus masks exposure of active species to gastric and intestinal environment providing therapeutic advantage (Dhaneshwar et al., 2014a).

2.4.3 New Classification of Prodrugs: Based on Cellular Site of Activation

Prodrugs are classified in view of regulatory perspective by taking into account their risk-benefit ratio. This classification system provides insights into kinetics of prodrug and bioactivation mechanism. So, prodrugs are classified depending on site of activation into Type I and II classes (Table 2.1) (Wu, 2009).

Type I: Intracellular prodrugs whose activation occurs inside the cells (e.g., lipid-lowering statins and anti-viral nucleoside analogs).

Type II: Extracellular prodrugs whose location of activation is outside the cellular space mainly in the systemic circulation and in digestive fluids (e.g., gene- or virus-directed enzyme prodrugs [ADEP/GDEP/VDEP], valganciclovir, fosamprenavir, antibody-directed drugs for chemotherapy or immunotherapy).

TABLE 2.1

Classification of Prodrugs Depending on Activation Site

Prodrug Types	Site of Conversion	Subtypes	Tissue Location of Conversion	Examples
Type I	Intracellular	A	Tissues/cells	Cyclophosphamide, acyclovir, 5-flurouracil, mitomycine C, L-dopa, diethylstilbestrol, diphosphate, 6-mercaptopurine, zidovudine
		B	Hepatic cells, mucosal cell of GI and lung	Carisoprodol, heroin molsidomine, phenacetin, Cabamazepine, captopril, paliperidone, primidone, psilocybin suldinac, tetrahydrofurfuryl disulphide
Type II	Extracellular	A	GI fluids	Lisdexamfetamine, sulfasalazine loperamide oxide oxyphenisatin
		B	Blood and other extracellular fluid compartments	Acetylsalicylate, dipivefrin, fosphenytoin, acampicillin, bambuterol, chloramphenicol succinate, dihydropyridine, pralixoxime
		C	Tissues/cells	ADEPs, GDEPs, VDEPs

2.5 Common Prodruggable Handles and Subsequent Functional Group Conversion

A prodrug approach can be applied to drugs having functional groups that can be modified, keeping in view the basic objective of development of prodrugs. The literature is flooded with examples of esters prodrugs among all other chemical modification types like amides, glycosides, azo derivatives or mannich bases. One more approach is modification of functional groups into alcohol or carboxylic acid then convert them into ester prodrug form. Another strategy is addition of spacer group that separates or attaches drug and promoiety. A spacer group extends the spatial distance between reactive section of the promoiety and sterically hindered groups to avoid effect of bulkiness. Some common prodrug strategies are described in Figure 2.9 (Dhokchawle et al., 2014, Guarino, 2010).

2.5.1 Aliphatic and Aromatic Alcohols

This is the favorite and most easily accessible handle for development of prodrugs. Drugs containing hydroxyl functional groups are directly functionalized as ester with promoiety which can be cleaved by esterase enzymes. Also, hydroxyl is good leaving group so aldehyde-based spacer moieties can also be used to modify the drug. Some common modifications of hydroxyl group involve phosphate monoesters, acyl esters and amino acid esters.

1. **Phosphate monoesters:** Phosphate monoester prodrugs are synthesized by phosphorylating a hydroxyl group. Newly entered phosphate group improves solubility due to ionization afforded by phosphate groups. A bioconversion of phosphate ester prodrugs occurs by alkaline phosphatases which are found in all parts of body like liver, kidney, apical membrane of enterocytes. This type of prodrugs is extremely useful to fulfill requirements of oral delivery of drugs instead of parenteral administration. Due to occurrence of alkaline phosphatases at intestinal lining cells, phosphate prodrugs create supersaturated solution by concentrating poorly soluble parent drug at oral absorption site. Best example is fosamprenavir (Lexiva®) which is a

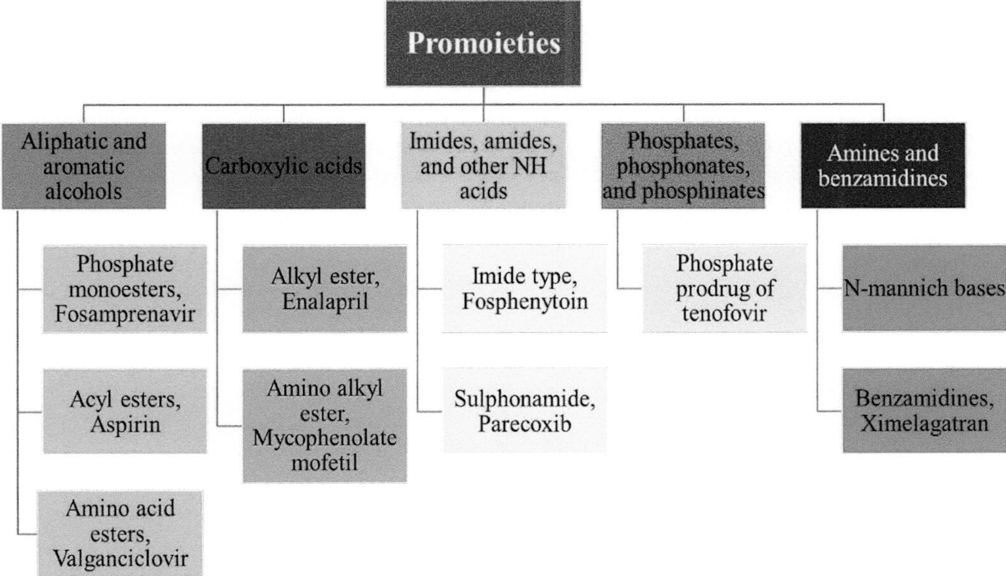

FIGURE 2.9 Prodrug strategies for most common functional groups.

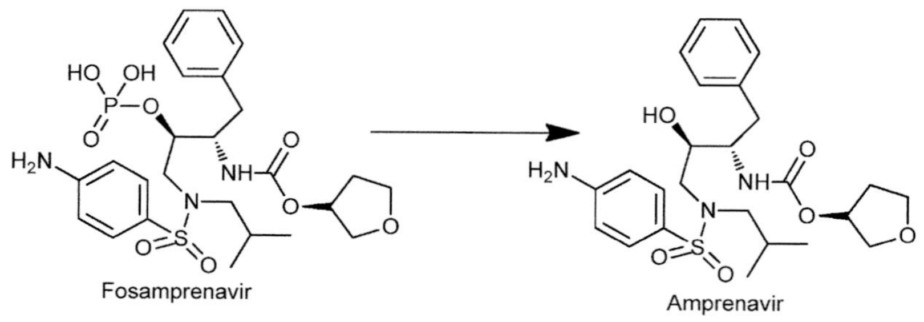

FIGURE 2.10 Bioactivation mechanism of fosamprenavir.

phosphate prodrug of HIV protease inhibitor amprenavir. Bioactivation mechanism is depicted in Figure 2.10. Amprenavir has poor aqueous solubility and a very high dose (1200 mg twice daily) which is formulated as a capsule taken twice a daily. Its phosphate prodrug has improved solubility and delivery attributes which is available as a tablet of calcium salt with same dose (Furfine et al., 2004).

2. **Acyl esters:** Phosphate esters improve aqueous solubility of drug but do not improve lipophilicity or membrane permeability of drug. Aliphatic and aromatic esters due to their ability to undergo easy solid-state breakdown of crystal structure serve as a good choice for prodrug modification. In this approach, alcohol is attached to acetyl group which is converted into acetic acid after bioconversion. Example of this approach is aspirin which is acetyl prodrug of salicylic acid. Another example is O-acetyl propranolol. O-acetyl propranolol avoids first-pass metabolism and enhances the drug delivery of propranolol after oral administration. Here prodrug needs to bypass the location containing esterases to provide stability (Anderson et al., 1988). As compared to short-chain alkyl esters, esters of fatty acids provide superior lipophilicity improvement and also modulate the *in vivo* transport of the prodrug to different compartments that affect elimination half-life of the compound. Example is haloperidol deaconate (Beresford and Ward, 1987), as illustrated in Figure 2.11.
3. **Amino acid esters:** Amino acid esters are built by introducing ionizable amine into promoiety to improve solubility. Here endogenous amino acids are preferred due to safety concerns. But they have two drawbacks that result in formulation challenges due to extra chiral center being contributed by amino acid expect for glycine and less hydrolytic stability. But these amino acid esters have some potential benefits as easy permeation through enterocytes. Example is valganciclovir (Valcyte®), a valine-based prodrug of ganciclovir. Valganciclovir permeates through enterocytes and helps in bypassing the biotransformation in liver or elimination transporters in the enterocytes (Sugawara et al., 2000). Bioactivation of valganciclovir is represented in Figure 2.12.

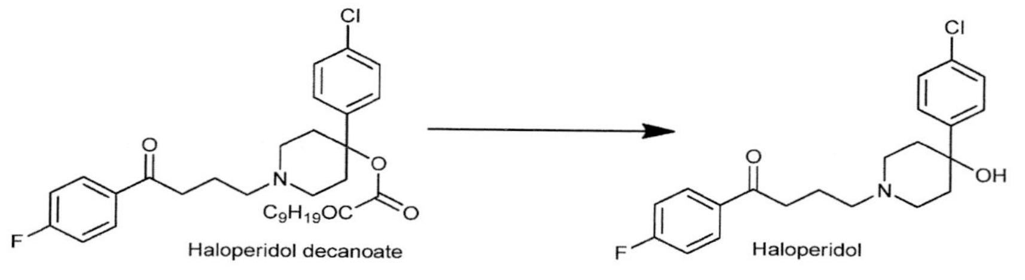

FIGURE 2.11 Bioactivation mechanism of haloperidol decanoate.

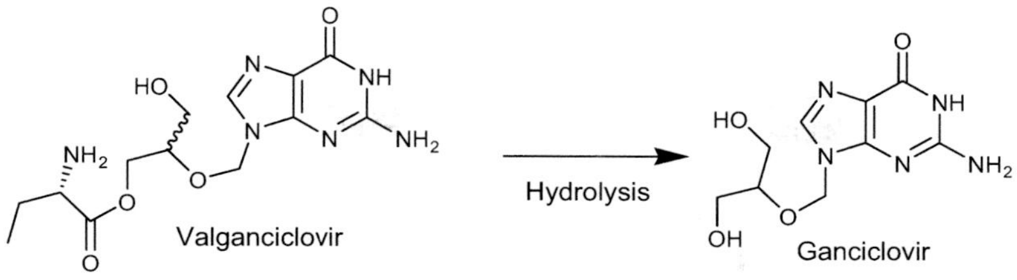

FIGURE 2.12 Bioactivation mechanism of valganciclovir.

2.5.2 Carboxylic Acids

Carboxylic acids can be readily derivatized into esters, which are promptly hydrolyzed to parent drugs by abundantly present esterases, making them most favorable bioreversible derivatives for prodrug development. The main aim behind development of carboxylate ester is to mask the ionization of carboxylate to increase membrane permeability. As carboxylic acids have pKa in the range of 4–5, they are in ionized form in intestinal tract which increases its aqueous solubility and limits passive diffusion through enterocytes. This is a major limitation for oral delivery strategy. Some approaches for designing prodrugs of carboxylic acid involve alkyl ester and amino alkyl esters:

1. **Alkyl ester:** Methyl ester is the easiest strategy used for making prodrugs of carboxylic acid. In this case methanol is reconversion product obtained but due to its safety concern it is avoided. So, ethyl ester of carboxylic acid is more preferred choice as the generated ethanol is considered safe at generated levels. For example, oseltamivir (Tamiflu®) involves masking of carboxylate group of the active moiety by ethyl ester (He et al., 1999). This led to enhancement of oral bioavailability to 80%. Another example is enalapril (Vasotec®) where two carboxylic functionalities are available for protection as ethyl esters but one carboxylic group is needed intact for recognition by PEPT1 (active intestinal influx transporter) (Todd and Heel, 1986). Therefore, it is desirable that only one carboxylic group out of two be esterified as ethyl ester (Figure 2.13).
2. **Amino alkyl ester:** The main issue in masking carboxylate group is increasing too much lipophilicity and decreasing aqueous solubility that limits oral delivery of drug. The amino alkyl ester potentially prevents the situation by maintaining ionization potential to maintain balance of aqueous solubility. Example is mycophenolate mofetil in which carboxylate group is masked by morpholinoethyl ester (Allison and Elsie, 2000). The ester prodrug has pKa of 5.6 so it reaches target, i.e., small intestine for absorption as neutral drug molecule, as depicted in Figure 2.14.

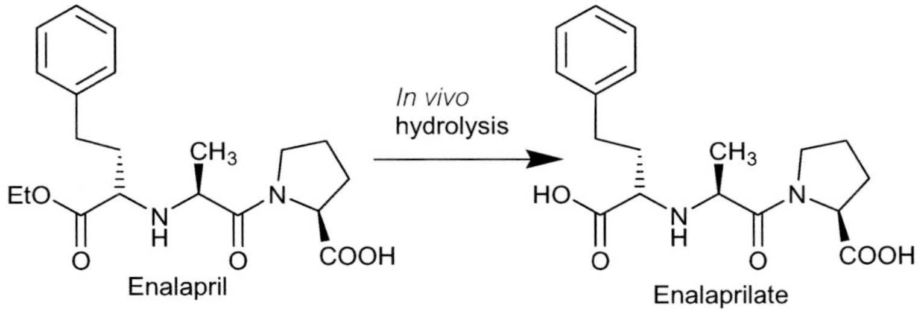

FIGURE 2.13 Bioactivation mechanism of enalapril.

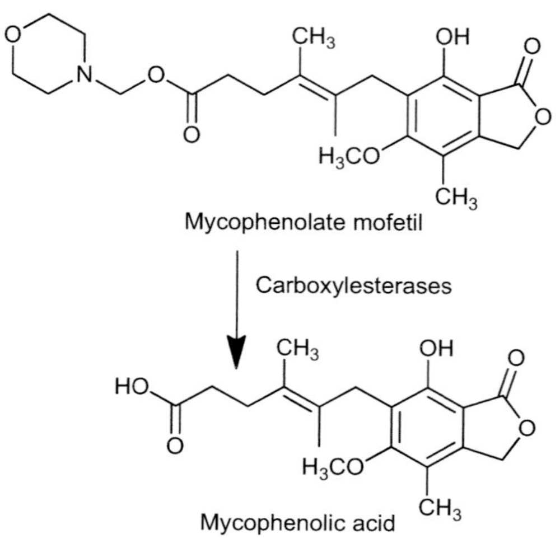

FIGURE 2.14 Bioactivation mechanism of mycophenolate mofetil.

2.5.3 Imides, NH Acids, and Amides

1. **Imide type:** NH acids are most acidic and good leaving groups. So, acyloxyalkyl ester spacers can be used which form hydroxyalkyl intermediate after initial hydrolysis. In this approach NH acids are converted into hydroxyl esters which undergo bioactivation by enzymes. Best example of imide type NH acid containing prodrug is fosphenytoin (Cerebyx®). It is phosphoryloxymethyl phenytoin prodrug with increased aqueous solubility. Bioactivation of prodrug is catalyzed by enzyme alkaline phosphatase into hydroxymethyl intermediate which further gets converted into phenytoin and formaldehyde (Stella, 1996). Bioactivation is depicted in Figure 2.15.
2. **Sulfonamide:** Parecoxib (Dynastat®) is a highly water-soluble sulfonamide NH prodrug of COX-2 inhibitor valdexcoxib (Talley et al., 2000). N-propanoylation of parent drug decreases the pKa and becomes a good choice for parenteral administration. Here solubility improvement occurs due to N-acylation of promoiety which makes –NH group of sulfonamides more readily ionizable (Figure 2.16).

2.5.4 Phosphates, Phosphonates, and Phosphinates

Drugs containing phosphate, phosphonate and phosphinate functional groups have ionization potential higher than carboxylate group due to anionic species. This ionized form does not support oral delivery and passive membrane permeability leading to limited efficacy at site of action. Therefore, restricting this ionization potential through esterification is a common goal in designing prodrugs of these functional groups. Related to their poor passive membrane permeability phosphates, phosphonates, and phosphinates can be good candidates for drug targeting to an intracellular space because if they can be released into the intracellular space, there is a reasonable chance that they would remain trapped in that cell because of a lower likelihood of diffusing back to the extracellular space; however, this aspect can also raise concerns if this same trapping phenomenon would result in an unintended toxicity for the cell. Simple alkyl and aryl esters, acyloxyalkyl and alkoxycarbonyloxyalkyl esters, aryl phospho(n/r)amidates and phospho(n/r)diamides are some approaches used while designing prodrugs of phosphates, phosphonates, and phosphinates drugs. For example, phosphate prodrug of tenofovir (Barditch-Crovo et al., 2001) enhances oral absorption of tenofovir (Figure 2.17).

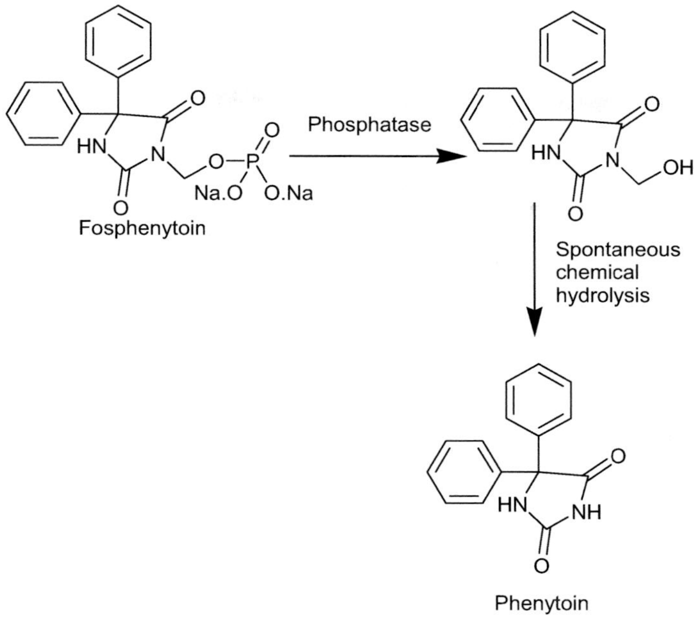

FIGURE 2.15 Bioactivation mechanism fosphenytoin.

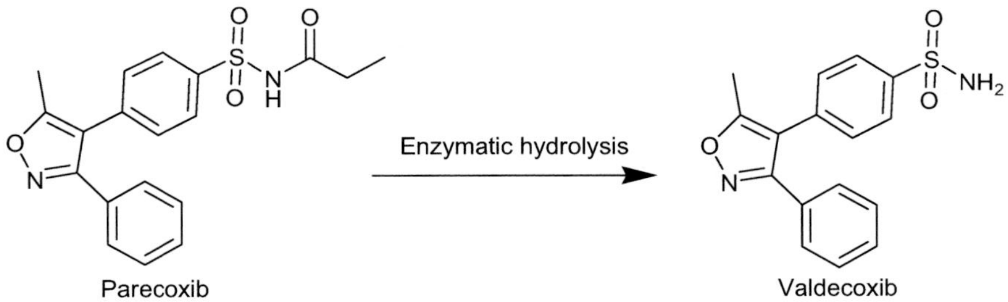

FIGURE 2.16 Bioactivation mechanism of parecoxib.

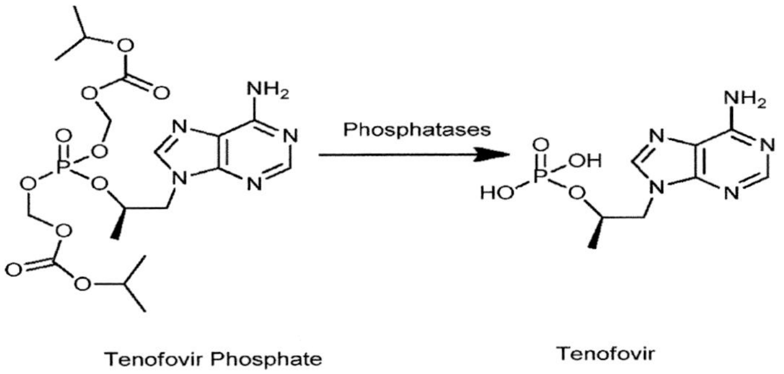

FIGURE 2.17 Bioactivation mechanism of tenofovir phosphate.

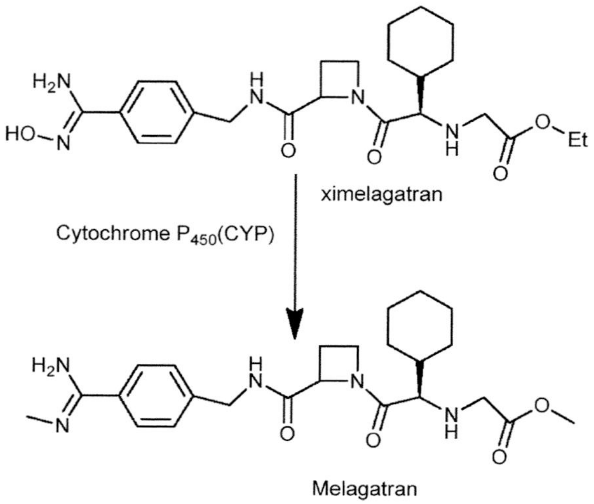

FIGURE 2.18 Bioactivation of ximelagatran.

2.5.5 Amines and Benzamidines

pKa of amines plays a major role in their oral administration. Higher pKa value contributes to protonation in the intestine that allows enhanced solubility and reduction in passive membrane permeability. Derivatization of amines is an easy task. It produces amides which are highly stable than esters but the challenging task is to design prodrugs with desired stability and bioconversion. Some approaches for designing prodrugs of amines are N-acyloxyalkoxycarbonyl, N-acyloxyalkyl, N-mannich bases, and N-phosphoryloxyalkyl prodrugs of N-hydroxy, tertiary amines and benzamidines.

1. **N-mannich bases:** Hydrophilicity of NH acidic drugs is generally enhanced by their conversion into N-mannich bases. The implication of reduced pKa is better oral absorption from intestine and membrane permeability. Rolitetracycline is a mannich base prodrug of tetracycline with enhanced aqueous solubility.
2. **Benzamidines** are most basic groups present in drugs. High basicity hampers oral absorption of drugs. Basicity of melagatran has been improved by its conversion into ximelagatran prodrug (Sorbera et al., 2001) having significant lower pKa which improves the per-oral absorption (Figure 2.18).

2.6 Mutual Prodrugs

Mutual prodrug is a type of carrier-linked prodrug in which the drug is covalently linked with another biologically active carrier or drug so that those two components act as "transporter" for each other. Like prodrug, mutual prodrug also undergoes enzymatic or nonenzymatic cleavage to two active drug moieties. So, mutual prodrug is mainly composed of two biologically active drugs that may be complementary or additive in action tethered by a bioreversible linkage. The promoiety provides advantages of site-specific delivery of drug to organ or cells and gets rid of adverse effects associated with parent drugs. Selection of candidates for codrug designing is from same or different therapeutic categories. Similarly, the constituents of codrug can act on the same or different biological targets with similar or different mechanism of action. Mutual prodrugs are also referred as codrugs or latentiated chimeras. Estramustine sodium phosphate (Emcyt, Pharmacia, La Roche) was developed in 1970 as a mutual prodrug as antineoplastic agent. It showed properties of mutual prodrug as it was formed by urethane linkage between two therapeutically active drugs, i.e., 17-alpha-estradiol and

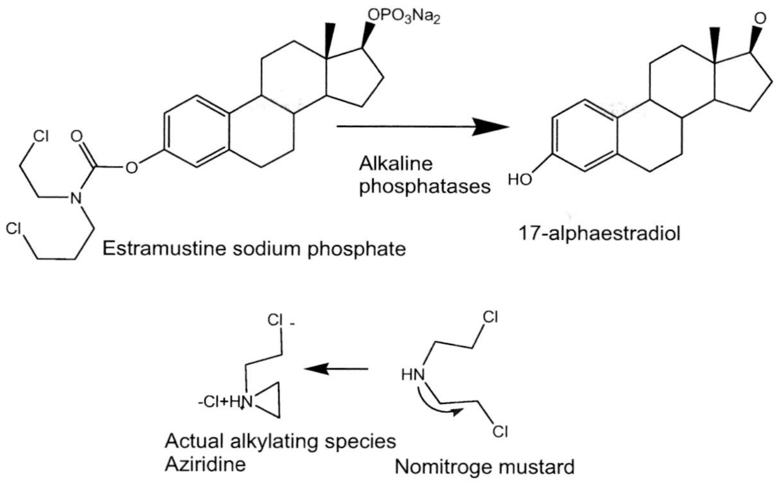

FIGURE 2.19 Bioactivation of estramustine sodium phosphate.

nitrogen mustard. Estramustine sodium phosphate is taken up by estrogen receptor positive cells where urethane linkage breaks down to release two parent drugs which show synergistic action against prostate cell growth. 17-Alpha-estradiol acts as a homing device and has tendency to accumulate in the prostate glands due to which it transports the nitrogen mustard which has been linked with it also to the same site. Figure 2.19 depicts bioconversion of estramustine sodium phosphate (Capomacchia et al., 2013, Wang et al., 1998).

Benorylate is a gastrosparing analgesic codrug of aspirin and paracetamol tethered via ester linkage (Ohlan et al., 2011). Mutual prodrugs of tolmetin with paracetamol and aspirin with salicylamide have been designed with the objective of eliminating the gastrointestinal (GI) irritation of these drugs (Manon and Sharma, 2009). Novel mutual prodrugs of ibuprofen (NSAID) coupled with sulfa drugs such as sulfanilide, sulfacetamide, sulfamethoxazole and sulfisoxazole are used to overcome drawback of GI irritation and ulceration produced due to free carboxylic group of ibuprofens by converting it into amide linkage (Wright, 1976, Das et al., 2010, Aljuffali et al., 2016, Bhosle et al., 2006).

2.6.1 Designing and Basic Structure of Codrugs

Mutual prodrugs are also classified further as bipartate mutual prodrugs and tripartate mutual prodrugs. Bipartate mutual prodrug consists of pharmacologically active carrier drug which is directly attached to parent drug whereas in tripartate mutual prodrugs the pharmacologically active carrier drug is not linked directly to the original drug but via a bridge. Tripartate prodrug provides benefit over bipartate prodrug by decreasing steric hindrance experienced by active moiety through linker, also helpful during enzymatic cleavage. In tripartate system, first carrier drug is enzymatically cleaved from linker followed by spontaneous cleavage of linker (Ohlan et al., 2011).

2.6.2 Rational Behind Codrug Development

Intention behind development of a codrug is to improve therapeutic efficacy of drug by manipulating physicochemical properties of drug. Codrug is successful approach to improve physicochemical properties as per needed bioactivity. Basic requirement behind development of codrug is to deliver therapeutically active drugs to their respective target site simultaneously and at appropriate time provided that enzymatic or nonenzymatic cleavage should be feasible. Codrugs may also show some intrinsic biological activity if they remain intact inside body. Codrugs made out of two drugs with similar activity show synergistic or additive activity. Objective behind development of codrugs could also be improvement of clinical effectiveness of drug with minimization of undesired effects of each other (Huttunen et al., 2011, Ottenhaus and Rosemeyer, 2015, Fujii et al., 2009, Aljuffali et al., 2016). Some advantages of codrugs over parent drug are given below:

1. In codrugs, mostly unwanted promoities are absent as both drugs exert required pharmacological action.
2. Codrugs show increased stability by masking autodegradation of liable functional groups in drug structure.
3. Codrugs show enhanced lipophilicity than parent drugs hence they are easily permeable through bio membranes (GI wall, cornea and skin). After entry into biomembrane, codrugs are converted to hydrophilic parent drugs by *in vivo* activation. This helps in overcoming shortcomings like poor bioavailability, first-pass effects and poor absorption.
4. Codrugs help in improving organoleptic properties of drugs, including odor and taste, by masking responsible functional groups.
5. Codrugs show superior bioactivity due to synergistic biological action of two drugs linked covalently.
6. Codrugs show limited toxicity as compared to parent drugs as due to synergistic action dose reduction is possible for drugs.
7. Targeted delivery of drugs is possible using codrug approach where site-specific (tissue and organ) activation is mediated by targeting a specific enzyme or antigen present on the surface.

2.6.3 Selection Criteria for Codrugs

Some important characteristics of drugs are necessary to be taken into account while selecting drugs for designing of codrugs that are summarized in Figure 2.20 (Das et al., 2010).

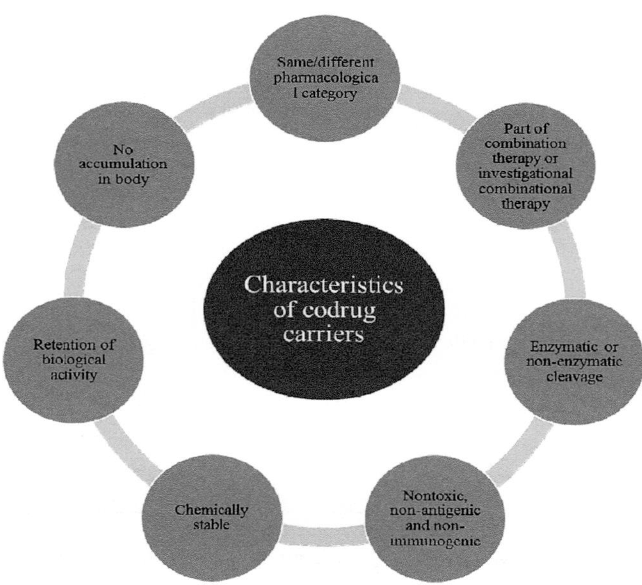

FIGURE 2.20 Characteristics for carriers of codrugs.

2.6.4 Methods for Evaluation of Mutual Prodrugs

1. **Solubility measurements:** Solubility of mutual prodrugs is evaluated by dissolving mutual prodrug in different solvents in vials. Vials are centrifuged for 5 min. Solution is filtered. Filtrate is tested for UV absorption at wavelength of maximum absorbance for solubility concentration determination and compared with parent drugs. These values give idea about change in solubility in polar as well as nonpolar solvents and helps in predicting their absorption on a preliminary basis. Aqueous solubility is another important parameter that helps in predicting the *in vivo* absorption and distribution of the prodrug as also in preformulation studies. A saturated solution of the prodrug is prepared in distilled water and stirred for 24 h, filtered and unknown concentration is measured by UV or HPLC.
2. **Partition coefficient determination:** Partition coefficient of a mutual prodrug (log P) is determined by shake flask method in n-octanol: water at 37°C ± 0.5°C. Concentrations are determined by UV-Visible double beam spectrophotometer to afford rapid estimation and better reliability. The sample is suspended in n-octanol: distilled water (10:10) and shaken for 24 h on wrist shaker to reach distribution equilibrium. After separating two layers in separating funnel, aqueous layer is estimated by UV spectroscopy and ratio of concentration of drug in aqueous and oil phase is calculated. Partition coefficient gives idea about affinity of mutual prodrug toward two heterogeneous phases which helps in predicting their oral bioavailability and peroral absorption (Zhang and Wang, 2010).
3. ***In vitro* kinetic study:** The basic aim behind designing of prodrugs is to modify pharmacokinetics (absorption, distribution, metabolism, excretion and unwanted toxicity) of the parent drug. It is a biologically inactive derivative of drug molecules that undergoes an *in vivo* activation [enzymatic and/or chemical (pH-dependent)] to release biologically active form. So, it is necessary to assess their suitability with respect to stability and release at the desired site of action. So, after synthesis and characterization of mutual prodrug, its stability and release profile is studied in aqueous buffers of different pH, simulated fluids and homogenates of various parts of gastrointestinal tract (Rai and Dhaneshwar, 2015).
4. ***In vivo* kinetic study:** To ensure establishment of bioreversibility of prodrugs, it is essential that prodrugs as well as released parent drugs have enhanced pharmacokinetic properties. Hence accurate estimation of pharmacokinetic properties (*in vivo*) as well as factors affecting it acts as contributing factor for prodrug design. It is hoped that a better understanding of the fate of the prodrugs and a balanced *in vitro/in vivo* pharmacokinetic behavior can contribute significantly to the development of safe and more efficacious codrugs. After designing a prodrug system, designed prototype is tested for *in vitro* as well as *in vivo* properties to get information regarding drug release and pharmacokinetics of the delivery system. *In vivo* pharmacokinetics and behavior of any drug candidate for its delivery system can be estimated by chromatographic techniques for prodrug and released drugs in blood/plasma, urine and feces of rodents.
5. **Pharmacological evaluation:** It is very essential to screen mutual prodrugs for evaluation of their pharmacological action in different animal models to establish whether the activity is retained or enhanced due to synergism or additive effect as the carrier is a drug with its own pharmacological activity which will be released in the vicinity of the parent drug and hence is bound to affect the activity produced by the parent drug. Modulation of biological activity will be demonstrated by the codrug depending on extent of change in physicochemical and pharmacokinetic properties of parent drug. Therefore, it is essential to establish whether biological activity is retained or improved, is synergistic or additive, side effects are minimized or increased, the codrug is safe or toxic and also the probable mechanism by which synergism or additive effect is produced. The dose calculation of the codrug poses a major challenge and two active drugs are covalently linked, both of which will be released after the codrug is activated. It is generally calculated on an equimolar basis to the parent drug. But it needs to be handled cautiously as judiciously. Despite equimolar calculation

(equimolar to dose of parent drug) assuring effective concentration of the parent drug, there is no guarantee that effective concentration of the carrier drug will be delivered as the doses of parent drug and carrier drug will be different. It is also advisable to compare the pharmacological effect of physical mixture of the parent and carrier drug with the codrug which will throw light on effectiveness of codrug over their physical mixture and justify the rationale behind mutual prodrug design. Acute toxicity studies are also required to be carried out since two active drugs will be released which may or may not synergize.

2.6.5 Limitations of Mutual Prodrug Design

Though mutual prodrugs are advantageous to overcome undesired properties of drugs they show some difficulties in pharmacological, pharmacokinetic, toxicological and clinical assessment (Bhosle et al., 2006, Strømgaard et al., 2017).

1. **Pharmacological assessment:** Codrugs cannot be tested for *in vitro* binding studies and enzyme inhibition measurements as formation of active species by bioactivation is a prerequisite.
2. **Toxicological assessment:** Codrugs are constructed by conjugating two pharmacologically active molecules which more than often are FDA-approved drugs with known toxic profile. Still it is necessary to study toxicity of codrugs due to possibilities of codrug being toxic as intact molecule till it gets activated, formation of toxic metabolites, generation of toxic derivatives, release of modifiers that alter pharmacokinetic properties by inducing an enzyme or by altering elimination of drug.
3. **Clinical assessment:** Sometimes values from pharmacological study of codrugs give questionable results as an active dose of a mutual prodrug may appear to be quite different in clinical investigations than seen in preclinical investigations.
4. **Pharmacokinetic assessment:** Codrugs may not act as ideal substrates for activating enzymes due to lack of electron withdrawing or donating groups to facilitate the hydrolysis or due to steric hindrance. So, misinterpretations in results may appear in pharmacokinetic study.

2.6.6 Challenges in Codrug Development

1. Access rate and extent of codrug activation
2. Toxicity differences of codrug and promoiety as compared to parent drug
3. Optimization of effective dose of a codrug
4. Unpredictability of toxicity profile of the codrug

2.6.7 Comparison between Codrugs/Hybrid Drugs/Prodrugs

In case of simple carrier-linked prodrugs, two components are linked by a covalent, bioreversible linkage like esteric or glycosidic, through a bipartate, or tripartate system, out of which one is a pharmacologically inactive carrier and the other one has therapeutic activity. In case of hybrid drugs, two pharmacophores are linked directly or by spacer molecule to give a new hybrid molecule which as a whole intact molecule is expected to bind with a receptor or a biomolecule to elicit its pharmacological response. However, codrugs consist of two active moieties coupled by covalent, bioreversible linkage. Prodrugs are inactive in *ex vivo* condition and become active after *in vivo* activation by enzymatic or nonenzymatic system. Hybrid drugs are active during *ex vivo* condition and do not undergo enzymatic activation, whereas codrugs can have activity *ex vivo* which undergoes *in vivo* enzymatic activation into their active metabolites having inherent biological activity. From designing point of view, prodrugs have huge flexibility with respect to selection of a carrier over hybrid drugs and codrugs as in case of hybrid drugs, selection depends on nonlabile linkers while in codrug design selection is based on functional groups available on linkers or promoiety (Das et al., 2010).

TABLE 2.2
Details of Codrugs and Prodrugs in Clinical Trials

Drug	Clinical Phase Status	Parent Drugs	Indication and MOA	Results of Clinical Trial
CYP-1020, Perphenazine 4-aminobutyrate mesylate (Gras, 2012)	Phase IIb/III	GABA and perphenazine	Schizophrenia, dopamine D2 receptor antagonist and GABAA receptor agonist activity	Increase in dopamine efflux, increase of prolactin levels, delivery of GABA to brain, improved psychotic symptoms and cognition, minimum cataleptic manifestations
AN-9, Pivanex[58]	Phase II	Butyric acid, *all-trans*-retinoic acid (ATRA)	Non-small-cell lung carcinoma, histone deacetylase inhibitor	Enhanced differentiation activity in leukemic cells, low toxicity, protection to hair follicles

2.6.8 Codrugs in Clinical Trials

Codrugs which are currently in clinical trials are given in Table 2.2.

2.6.9 Toxicology and Regulatory Concerns Related to Codrugs

It is necessary to look into toxicity of codrugs during their designing and development. Toxicity profiles of intact codrug and its active components need to be established. By performing comparative toxicological profiling, codrug scientist can get idea about the source of toxicity as it can be contributed by promoiety itself or byproduct released after bioconversion or metabolism. In case of some codrugs, enzymes responsible for bioconversion are fast acting, so leave a small fraction of intact codrug and intermediates of active drug that remain in systemic circulation. That fraction may be responsible for toxicity so it is necessary to understand the toxicities associated with codrug when given alone. For example, formaldehyde is a byproduct of many bioconversions and it affects regular physiology. In case of adefovir, dipivoxil and pivampicillin, promoiety pivalic acid is released. Pivalic acid depletes carnitine homeostasis. So, supplementation of pivalic acid is recommended. In the United States, terfenadine is removed from market and replaced by active drug fexofenadine as higher concentration of prodrug inhibits its conversion into active form. Hence in evaluating overall risk of codrugs, toxicity of codrug as well as promoiety, dose and duration of treatment should be evaluated (Wu, 2009, Strømgaard et al., 2017).

2.6.10 Regulatory and IPR Perspectives for Codrug Development

To become patentable, codrugs must be novel, should have inventive step, and should possess nonobviousness with respect to prior art. Novelty and inventive step are easy to be proved. But it's difficult to establish nonobviousness of codrugs due to very small structural differences between parent drug and codrug. From some recent cases regarding invalidity of codrug's claim on nonobviousness basis, Patent Trial and Appeal Board (PTAB) illustrated advantages of codrug patents. Nonobviousness decision is mainly dependent on unpredictable results of codrugs over parent drugs. Reasonable expectation of success by person having ordinary knowledge about relevant art is key finding about obviousness determination. In the context of patents concerning codrugs, patent owners and challengers should pay careful attention to processes involved in converting the codrugs to active compounds, particularly effect of disease condition on conversion process. Patentability depends on the extent to which such codrug conversion processes yield predictable results. According to patent office guidelines regarding secondary inventions such as codrugs, enantiomers, polymorphs; codrugs generally fail to fulfill standards of inventive step until submission of proof of overcoming shortcomings (pharmacokinetic and pharmaceutical) of active drugs (Strømgaard et al., 2017, Holman et al., 2018).

The following are some examples of patent applications for codrugs:

1. Monophosphates codrugs of anti-inflammatory signal transduction modulators (AISTMs) and beta-agonists for the treatment of pulmonary inflammation and bronchoconstriction: Inventive step involves synthesis, formulation as either liquids or dry powders and delivery of mutual prodrugs of AISTM-β monophosphates agonist (Baker et al., 2010).
2. Mutual prodrugs and methods to treat cancer: Inventive step involves compounds, compositions, manufacturing methods thereof, as well as uses thereof in treatment and/or prevention by mutual prodrugs of one or more histone deacetylase inhibitors (HDACIs) to one or more retinoids (Njar et al., 2010).
3. Amilodipine-atorvastatin codrugs: Inventive step involves codrugs of amlodipine-atorvastatin, its pharmaceutically acceptable salts thereof, compositions thereof and its use, manufacture of medicaments against angina pectoris, atherosclerosis, hyperlipidemia, combined hypertension and other cardiac risk (Crook et al., 2004).
4. Monophosphates as mutual prodrugs of muscarinic receptor antagonists and β-agonists against chronic obstructive pulmonary disease (COPD) and chronic bronchitis: Preparation of novel, mutual prodrugs of muscarinic receptor antagonists (MRA) and β-agonists for delivery to the lung by aerosolization (Baker et al., 2010).
5. Hepatoprotectant Acetaminophen Mutual Prodrugs: Invention relates to a compound comprising acetaminophen moiety and a hepatoprotectant moiety, wherein the hepatoprotectant moiety is capable of inactivating N-acetyl-p-benzoquinone imine (NAPQI) (Sloan et al., 1991).

2.6.11 Applications of Codrugs

2.6.11.1 Codrugs for CNS Delivery

Parkinson disease (PD) involves deficiency of dopamine and L-dopa (3, 4-dihydroxyphenyl-L-alanine) which is an analog of dopamine and can penetrate blood-brain barrier which dopamine cannot. The anti-Parkinson drug entacapone [(E)-2-cyano-N, N-diethyl-3-(3, 4-dihydroxy-5-nitrophenyl) propenamide] is a new, potent inhibitor of catechol-O-methyltransferase (COMT), approved as adjunct to L-dopa. Co-administration of entacapone, L-dopa and an amino acid decarboxylase (AADC) inhibitor increases therapeutic concentration of L-dopa for longer period of time although research has showed low bioavailability of L-dopa (5%–10%) after co-administration with entacapone. A novel codrug of L-dopa and entacapone (Figure 2.21) has been synthesized by linking both the drugs through the carbamate spacer.

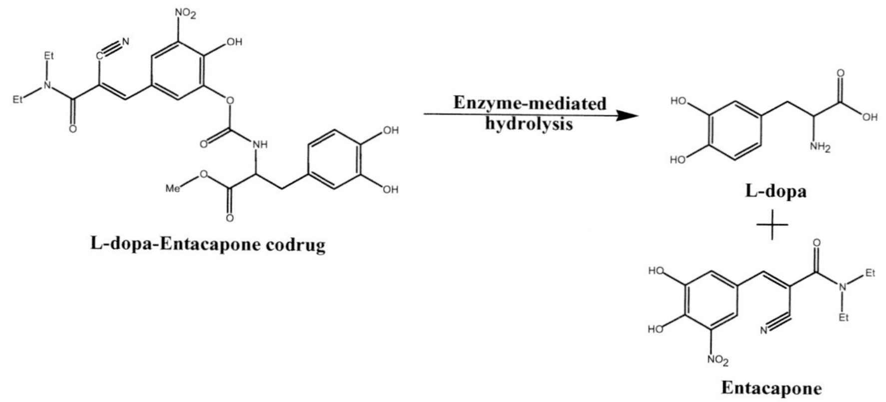

L-dopa-Entacapone codrug

Enzyme-mediated hydrolysis

L-dopa

+

Entacapone

FIGURE 2.21 Bioactivation mechanism of L-DOPA-entacapone codrug.

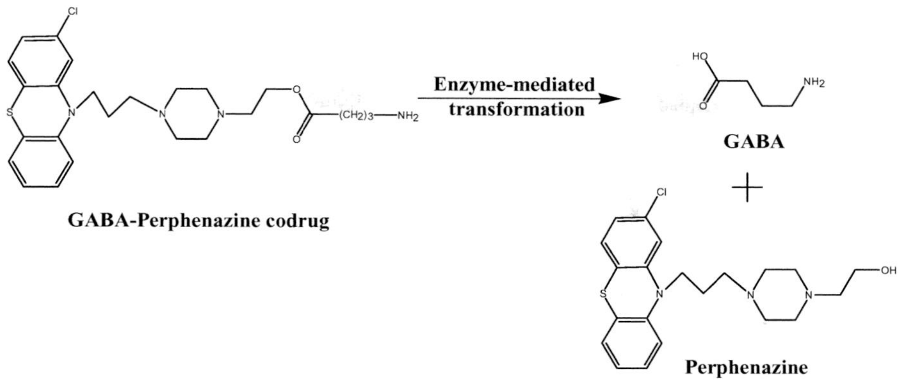

FIGURE 2.22 Bioactivation of perphenazine-GABA codrug.

This codrug provided adequate stability [$t_{(1/2)}$ 12.1 h (pH 1.2); 1.4 h (pH 5.0); 1.1 h (pH 7.4)] against chemical hydrolysis but rapidly hydrolyzes to L-dopa and entacapone in liver homogenate [$t_{(1/2)}$ 7 min; pH 7.4) at 37°C (Leppänen et al., 2002, Kaakkola et al., 1994).

Neuroleptics are commonly prescribed treatment options against schizophrenia and related psychotic disorders. Phenothiazines and haloperidol cause extrapyramidal side effects (EPS). γ-Aminobutyric acid (GABA), the major inhibitory neurotransmitter in the brain, possesses anticonvulsant, anxiolytic, mood-stabilizing, hypnotic, and muscle relaxant properties. In addition, imbalance of GABA system in the brain contributes to the pathology of the disease. First-generation, typical antipsychotic drugs such as perphenazine, fluphenazine, and haloperidol are responsible for antagonism of dopamine receptors, especially D2. Nudelman et al. studied the effect of ester codrug of GABA and perphenazine on neuroleptic efficacy, oral bioavailability, and EPS over parent molecules and showed that the codrug had enhanced BBB crossing capability with efficient hydrolysis to GABA and perphenazine in brain (Shukla et al., 1985, Gunes et al., 2007, Marsden, 2006, Nudelman et al., 2008) (Figure 2.22).

2.6.11.2 Codrugs for Cancer

Vitamin A owes its activity to its bioactive metabolite *all-trans*-retinoic acid (ATRA) which belongs to retinoid family. Endogenous retinoids play important role in cell growth, differentiation and apoptosis as well as development of dermal tissues. They have shown a great potential in cancer chemotherapy by competitive binding to retinoic acid receptor (RAR) followed by activation of target receptor by receptor-ligand complex, causing differentiation and inhibition of proliferation in promyelocytic leukemia cells and neuroblastoma. Butyric acid is a fatty acid, prevalent in digestive organs as a product of bacterial fermentation. Butyric acid has been reported to promote multiple biological effects in cell line studies like cytodifferentiation, relaxation of the chromatin structure, inhibition of histone deacetylase (HDAC), and also arresting cell proliferation. However, butyric acid has demonstrated low potency during *in vivo* studies due to fast metabolism and short half-life. A series of long acting prodrugs of butyric acid and ATRA called RN1 codrugs have been shown to affect a distinctive cellular target synergistically with more potent induction of cancer cell differentiation and inhibition of proliferation (40-fold decrease in ED50) than the parent acids (Mangelsdorf et al., 1995, Minucci and Pelicci, 1999). The order of differentiation activity was found to depend on functional groups attached to retinoyl moiety and was in the order RN1 > RN2 > RN3 > ATRA _ RN4 (Figure 2.23).

A 5-Florouracil (5-FU) and cytarabin (Ara-C) mutual prodrug has been designed and synthesized, linking the two anticancer agents through hydrolyzable amide bond. 5-FU has broad-spectrum antitumor activity against cancers of various organs while (ara-C) is mainly indicated in the treatment of acute myeloid leukemia, acute lymphocytic leukemia, chronic myelogenous leukemia, and non-Hodgkin's lymphoma. Mutual prodrug releases both drug components under physiological conditions within an hour by enzymatic hydrolysis (Menger and Rourk, 1997).

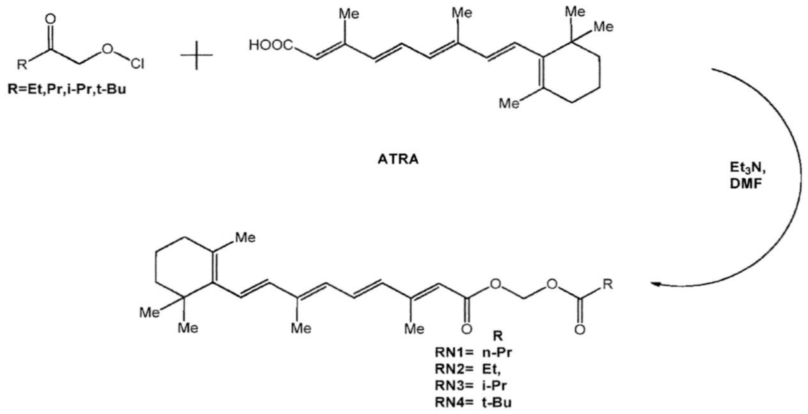

FIGURE 2.23 Bioactivation of RN1/2/3/4.

NSAIDs though popular for their analgesic and anti-inflammatory effect are now in focus due to their potential to inhibit tumor growth by restoration of apoptosis and cell-cycle arrest. Phosphoramidate strategy has been applied to NSAIDs to synthesize novel NSAID codrugs for chemopreventive therapy with less toxic side effects (Figure 2.24). Proposed activation is by enzymes whose concentration becomes elevated in tumor tissues compared with normal tissues (Wittine et al., 2009).

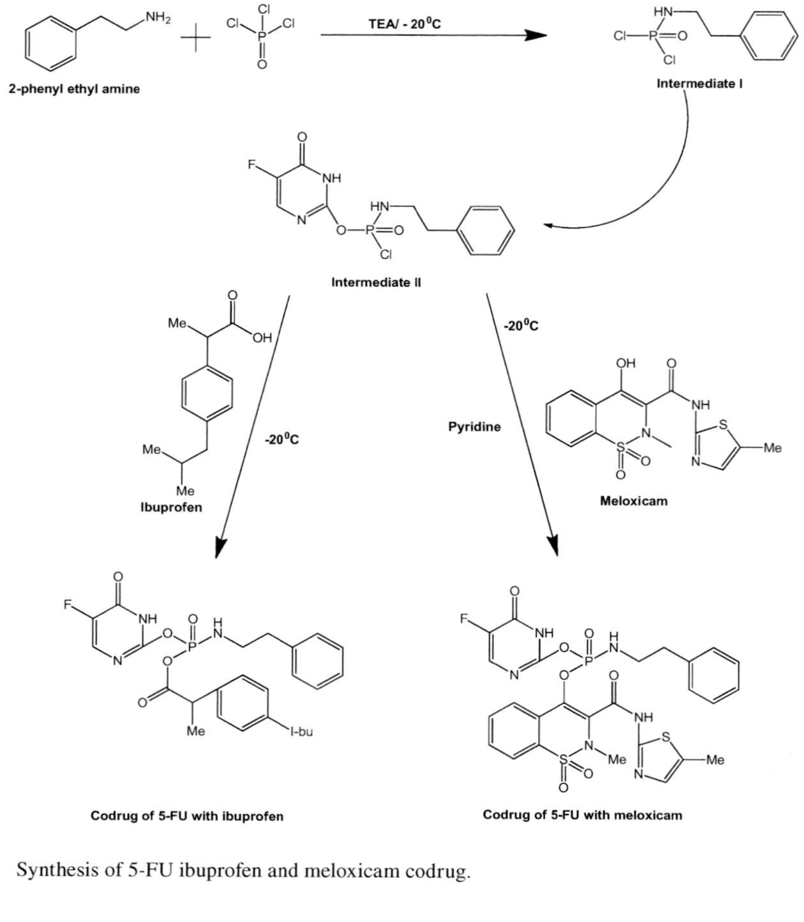

FIGURE 2.24 Synthesis of 5-FU ibuprofen and meloxicam codrug.

2.6.11.3 Codrugs for Diseases of Blood

Brown et al. (1996) designed a codrug by linking diamide and diester groups of thromboxane receptor antagonist (TXA2) and hexanoic acid side chain of thromboxane synthase inhibitor. *In vitro* testing of compounds (Figure 2.25) showed enzyme inhibitory properties on human platelets (Brown et al., 1996).

Dabigatran is an orally administered anticoagulant showing antithrombotic activity by reversible and selective binding to human thrombin. Ferulic acid methyl ester shows antagonizing effect on aggregation of platelets. Drawbacks with therapy of both the drugs is need for frequent dosing due to short duration of antiplatelet action. So, for combined benefits dabigatran and ferulic acid were chemically combined (Figure 2.26). Pharmacological testing of codrug showed potent venous thrombosis inhibition than individual drugs. In addition, ADP-induced platelet aggregation test showed stronger contribution of ferulic acid (Yang et al., 2013).

2.6.11.4 Codrugs for Antihypertensive Therapy

Hypertension contributes to higher economic as well as health burden in society with highest risk. It is prevalent in 4.5% of population worldwide with occurrence in third to fifth decades of life. Risk of hypertension increases with age. Other contributing factors for hypertension are obesity, physical inactivity and an unhealthy diet. Various anti-hypertensives are available for treatment option. Combination therapy is recommended by WHO as first-line therapy to lower blood pressure more quickly. Monotherapies are often failing to produce intended results and pose enhanced stroke and death risk. Low oral bioavailability, short duration of action, first-pass metabolism and variable lipohilicities are the main problems associated with antihypertensive drugs. Codrug strategy has been applied to overcome these limitations (Dhaneshwar et al., 2011a).

Obesity is a major risk factor for hypertension. So, the combination therapy regimen includes anti-hyeprlipidemics in combination with antihypertensive drugs. Crook et al. patented a codrug of atorvastatin-amlodipine (Figure 2.27) for the management of angina pectoris, atherosclerosis, and

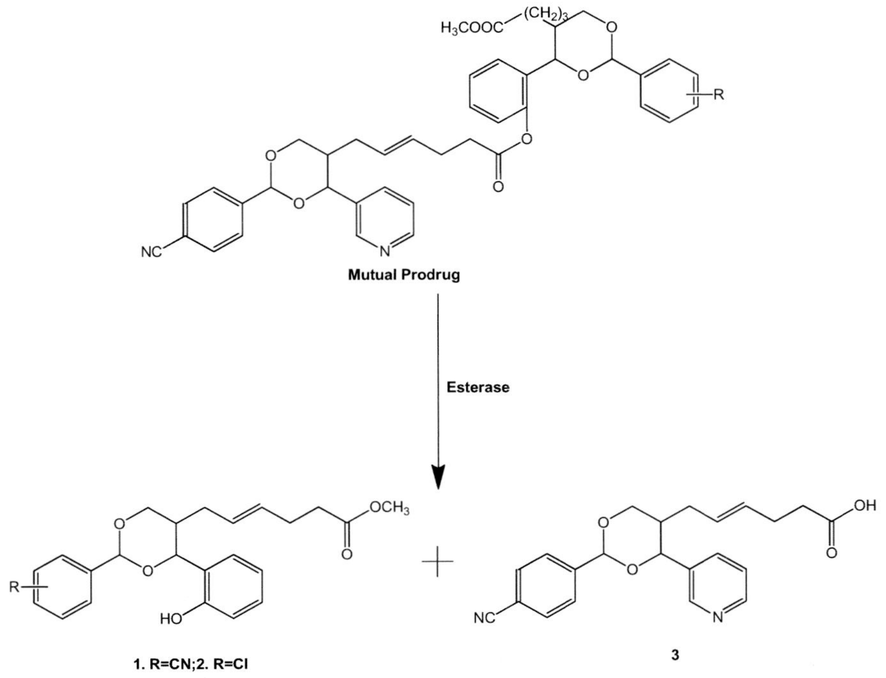

H_3COOC $(CH_2)_3$ O O R O O O O NC N

Mutual Prodrug

Esterase

OCH_3 O O O R HO + O O OH O NC N

1. R=CN;2. R=Cl **3**

FIGURE 2.25 Bioactivation mechanism of prodrug acting on blood disease.

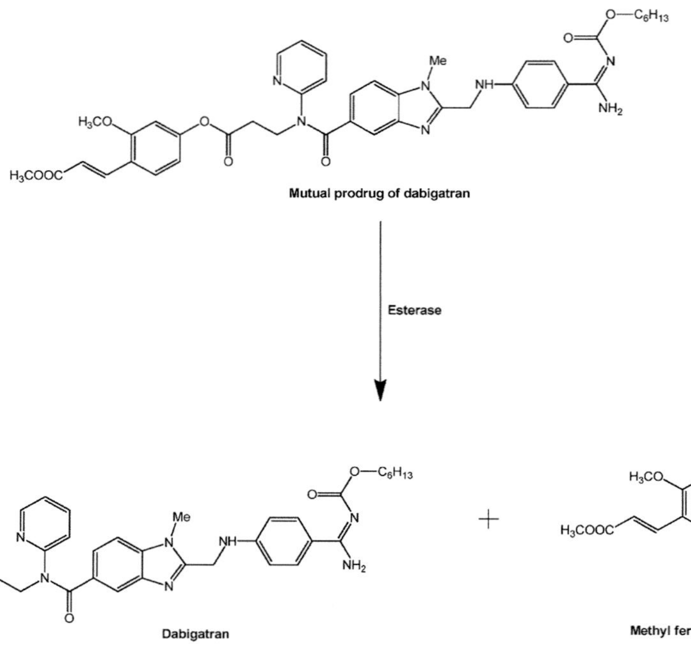

FIGURE 2.26 Bioactivation of dabigatran-ferulic acid.

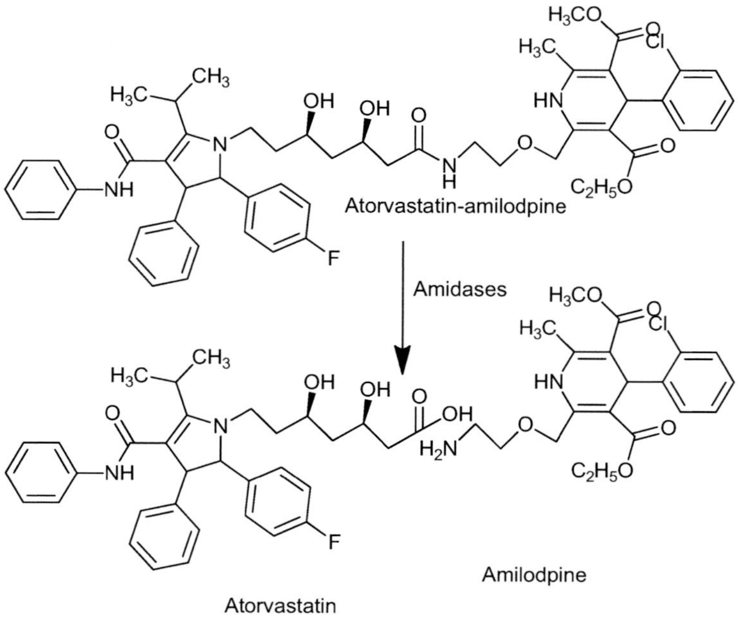

FIGURE 2.27 Atorvastatin-amilodipine codrug.

combined hypertension with hyperlipidemia and cardiac risk. Codrug is formed by amide linkage between both the drugs which release atorvastatin and amlodipine by hydrolytic cleavage *in vivo* (Crook et al., 2004).

Potential mutual prodrugs of β-blocker, propanolol and calcium channel blocker analogs of nifedipine conjugated by ester linkage have been designed aiming at additive action and decreasing hydrophilicity

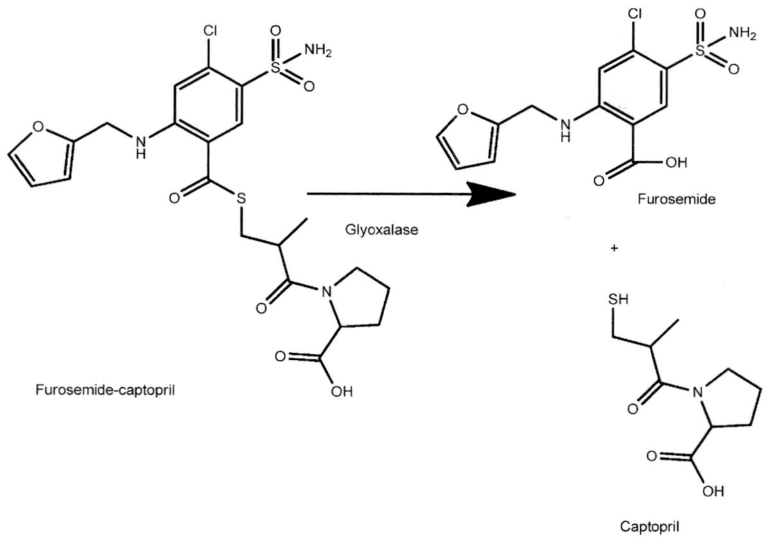

FIGURE 2.28 Bioactivation mechanism of furosemide-captopril codrug.

of the drugs by Nudelman et al. It provided benefit of reduction of frequency of episodes and duration of painful ischemia (Nudelman and Rachel, 1993).

Byrne and Rynne (1998) have a US patent of codrug captopril-S-aspirinate to their credit against hypertension, arrhythmias and angina pectoris with anticlotting effect. It is an ester prodrug consisting of aspirin and captopril having dual action. Along with antihypertensive effect of captopril, aspirin would provide antiplatelet effect by stopping agglomeration of platelets and in turn thrombosis formation to prevent heart attack and its complications. The patent has also covered pindolol-O-aspirinate, atenolol-O-aspirinate and metoprolol-O-aspirinate prodrugs.

Dhaneshwar et al. have designed a mutual prodrug of captopril and furosemide to achieve synergistic antihypertensive effect, overcome short duration of action of furosemide, maintain potassium levels in body, reduce thiol-induced side effects of captopril and gastric irritant effect of furosemide (Budhalakoti et al., 2014). Synthesized prodrug showed enhanced efficacy and delivery properties than furosemide and captopril independently or their physical mixture (Figure 2.28).

2.6.11.5 Codrugs for Rheumatoid Arthritis

Rheumatoid arthritis (RA) is a systemic auto-immune disease invading connective tissues of joints, muscles, tendons and fibrous tissue with chronic disability. It involves inflammatory and destructive events such as pannus formation, synovial hyperplasia, joint malformation, and cartilage and bone erosions (Du et al., 2008). Treatment for RA aims at controlling inflammation and pain, slowing down or preventing the progression of joint destruction. Current therapeutic interventions available for RA include NSAIDs, analgesics, corticosteroids, disease modifying antirheumatic drugs (DMARDs) such as leflunomide, hydroxychloroquine (HCQ), sulfasalazine (SLZ), methotrexate (MTX) and biological DMARDs such as abatacept, anakinara, adalimumab, etanercept, rituximab and infliximab (Donahue et al., 2008). Codrug concept has been used widely to overcome pharmaceutical, pharmacokinetic and pharmacodynamic drawbacks of drugs prescribed for the management of RA. Some examples are cited below.

Mahdi et al. synthesized and kinetically studied ester codrugs of NSAIDs with gabapentin by using glycol spacer with aiming at reducing the gastric irritation and obtaining synergistic analgesic effects. Preliminary kinetic studies of codrugs showed a significant enzymatic hydrolysis rate in plasma (Figure 2.29).

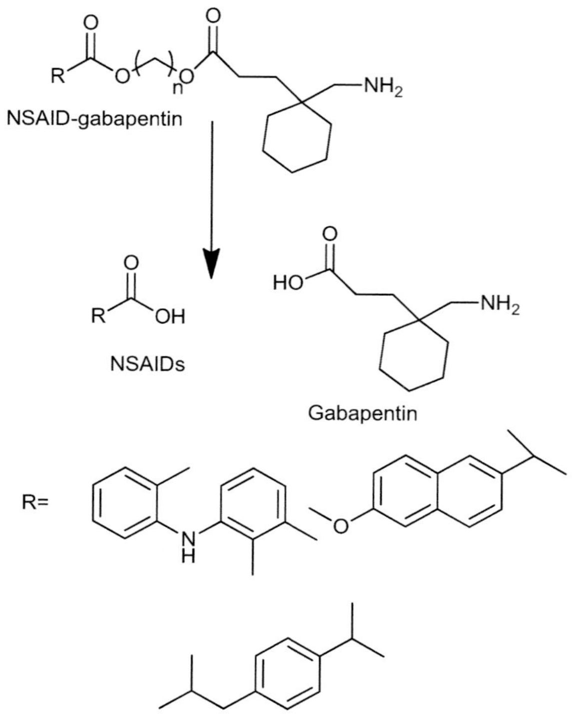

NSAID-gabapentin

NSAIDs

Gabapentin

R=

FIGURE 2.29 Bioactivation mechanism of gabapentin-NSAID codrug.

HCQ is used as second-line, slow-acting DMARD for RA. Drawbacks of HCQ therapy are slow redistribution into blood that results in slow onset of action. So, the therapy with HCQ demands for high doses but increase in dose is associated with dose related side effects due to its accumulation in nontargeted sites. Frequent use of NSAIDs in arthritis treatment is associated with gastrointestinal side effects. Dhaneshwar and Joshi reported mutual prodrugs of HCQ-NSAIDs containing aryl acetic acid (Joshi and Dhaneshwar, 2017). These two classes of antiarthritic drugs resolved problems of accumulation of HCQ in nontargeted sites, slow onset and NSAID-associated local gastric intolerance through their dual action by different mechanisms. The striking outcome of this study was broad protection against inflammation associated with RA than HCQ/NSAID alone (Figure 2.30).

Dhaneshwar et al. designed and synthesized novel codrugs of NSAID, i.e., biphenylacetic acid with two nonessential amino acids (D-phenylalanine and glycine) through covalent ester linkage having analgesic, wound healing and anti-inflammatory properties (Dhaneshwar et al., 2014b). Codrugs showed promising action in terms of longer duration of analgesia, enhanced/prolonged anti-inflammatory activity, reduced ulcerogenic propensity and superior antiarthritic effect (Figure 2.31).

Dhaneshwar et al. have reported a codrug of diacerein (slow-acting DMARD for osteoarthritis) with beta-boswellic acid (anti-inflammatory). The codrug exhibited quick onset, improved antiarthritic effect with minimized ulcerogenic potential when screened in collagenase type-II-induced osteoarthritis in Wistar rats (Dhaneshwar et al., 2013).

2.6.11.6 Colon-Targeting Codrugs

Colon-targeted delivery is gaining importance for the effective management of disorders such as ulcerative colitis (UC), Crohn's disease (CD), amebiasis, colorectal cancer, irritable bowel syndrome, and constipation. SLZ is the first colon-targeting codrug used in clinical practice, which is comprised of azo-linked sulfapyridine and 5-aminosalicylic acid (5-ASA); the reductive activation of which is mediated by azoreductases.

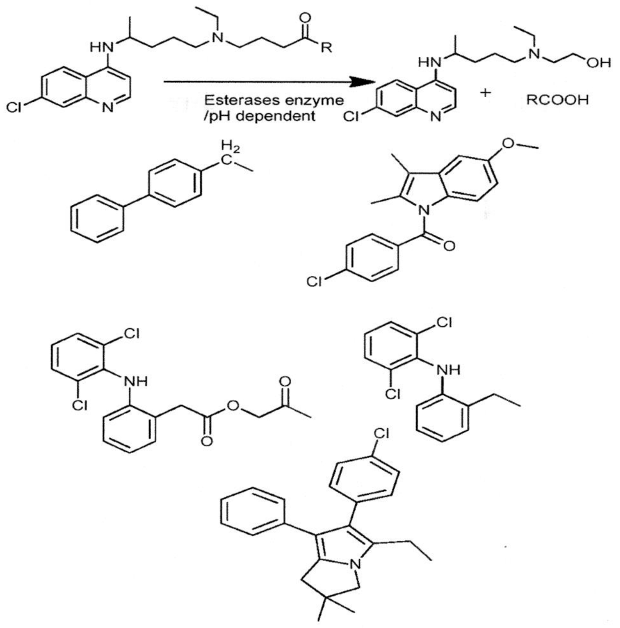

FIGURE 2.30 Bioactivation mechanism of hydroxychloroquine-NSAIDs codrug.

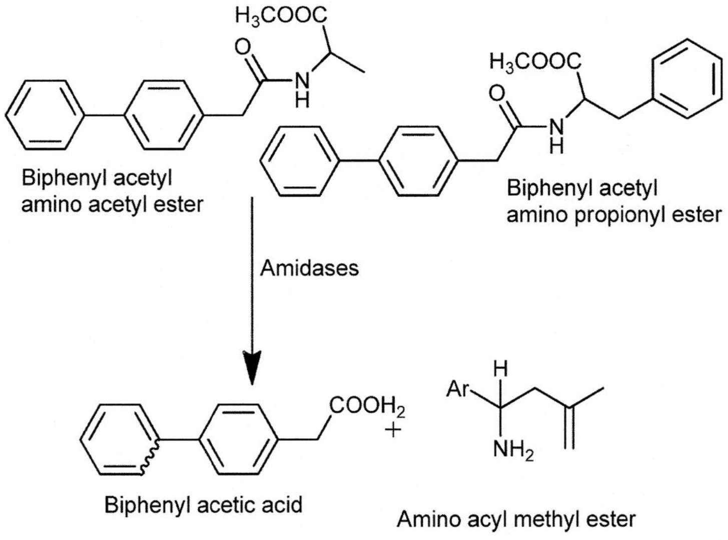

FIGURE 2.31 Bioactivation of biphenylacetyl amino acetyl/propyl ester.

In spite of development of most efficient biological products, most of the IBD patients rely on aminosalicylates as treatment option. Best treatment for UC patients is of 5-ASA but is associated with many undesired side effects. Dhaneshwar et al. have analyzed azo and amide codrug strategies for 5-ASA and its positional isomer 4-aminosalicylic acid (4-ASA) linked to amino acids like D-phenylalanine, L-tryptophan, L-glutamine, L-histidine and aminosugars like D-glucosamine, which are activated by azoreductases and N-acyl amidases enzymes, respectively, contributed by colonic bacteria (Dhaneshwar et al., 2009a, Nagpal et al., 2006, Dhaneshwar et al., 2011b, Dhaneshwar et al., 2009b, 2009c, Dhaneshwar, 2014). The chosen amino acids and amino sugars have demonstrated wound healing and anti-inflammatory properties.

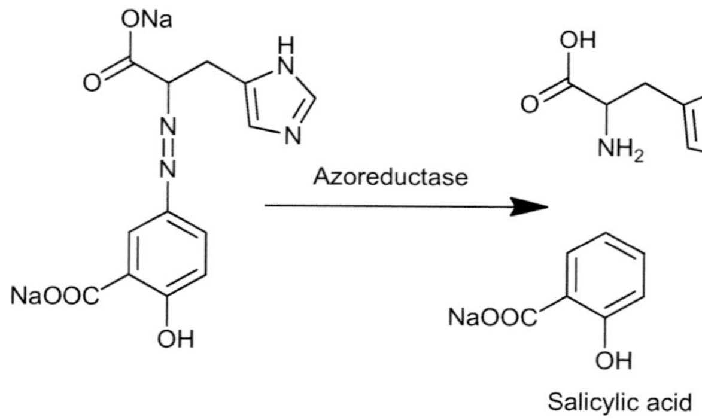

FIGURE 2.32 Bioactivation mechanism of 5-amino salicylic acid codrug.

The synthesized codrugs showed improved safety profile on stomach, liver and pancreas than oral SLZ, 4-ASA and 5-ASA, with equivalent efficacy to SLZ (Figure 2.32).

Making use of drug repurposing approach Dhaneshwar et al. have reported colon-specific codrugs of antihistaminic fexofenadine, immunosuppressant mycophenolic acid using amino acids and amino sugars as active promoities, respectively (Dhaneshwar and Gautam, 2012, Singh and Dhaneshwar, 2017, Chopade and Dhaneshwar, 2018).

Generally, absorption of NSAIDs and antibiotics/antibacterials occurs in upper part of GIT, i.e., stomach and jejunum or distal ileum, respectively. So, the treatment regimen for IBD with these drugs is not so effective due to their poor bioavailability in colon. Hussain et al. (2014) synthesized codrugs of indomethacin (NSAID) with norfloxacin and trimethoprim. Amide linkage between codrugs proved to release the drugs in colon without any appreciable release in upper GIT and enhancing the therapeutic utilization of drugs (Figure 2.33).

Research has shown that beta-boswellic acid (BA) improved disease condition in IBD patients. Gupta et al. (2001) confirmed efficacy of boswellic acid in UC as well as CD patients. Use of BA in IBD patients is restricted due to poor bioavailability and water solubility. Sarkate and Dhaneshwar (2017) reported that BA is good candidate for development of colon-targeted prodrugs which can improve bioavailability by enhancing its hydrophilicity. Amide codrugs of BA with various amino

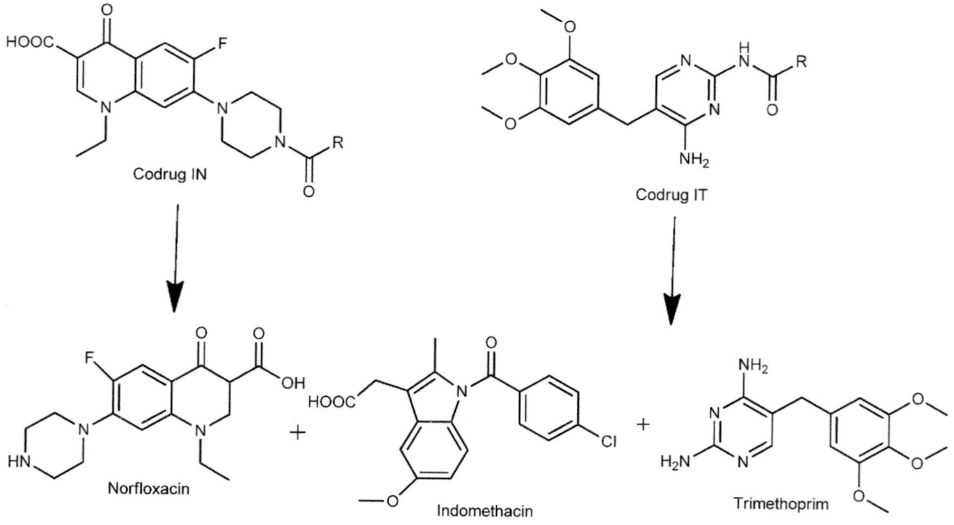

FIGURE 2.33 Synthesis of indomethacin codrug.

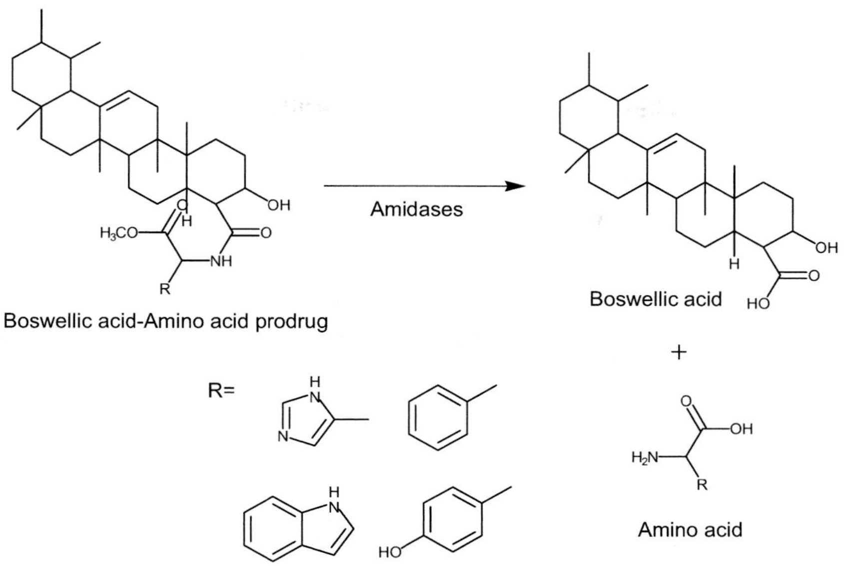

FIGURE 2.34 Bioactivation mechanism of boswellic acid-amino acid codrug.

acids demonstrated significant mitigating effect in experimental colitis model in rats without any adverse effects on nontargeted sites (Figure 2.34).

Use of celecoxib as COX-2 inhibitor is under dispute due its cardiac adverse reactions. However, Lee and co researchers (2015) have designed colon-specific codrugs of celecoxib and aspart-4-yl glycine and glycine and assayed its anti-inflammatory potential on NFκB. Codrugs inhibited NFκB in colon carcinoma cells, demonstrating potential of colon targetability (Figure 2.35).

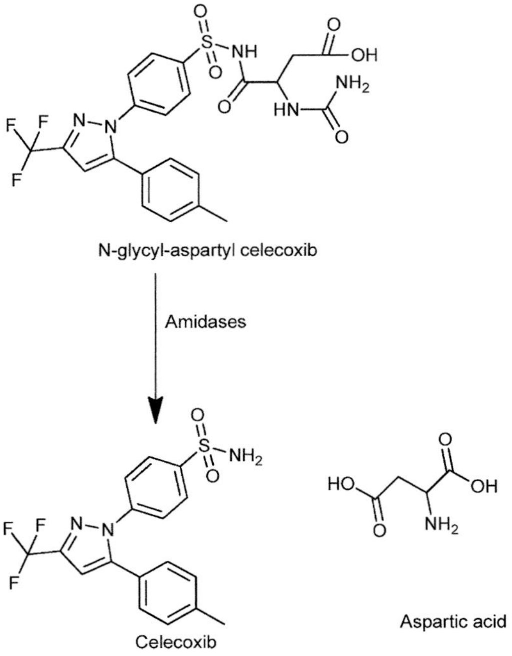

FIGURE 2.35 Bioactivation mechanism of N-glycyl-aspartyl celecoxib.

2.6.11.7 Codrugs for Tuberculosis

Treatment to tuberculosis appears to be the greatest challenge worldwide. Tuberculosis is the complex disease caused by *Mycobacterium tuberculae*. Treatment involves use of pyrazinamide (PZA), isoniazid (INH), and rifampin (RIF) as the first line and combination of INH and RIF as the second line of treatment. But infection of TB caused by extensively drug-resistant (XDR-TB) and totally drug-resistant (TDR-TB) strains is untreatable and hence calls for extremely shorter therapeutic regimen with least toxicity. Several antitubercular compounds include bioprecursor prodrugs which are activated by *M. tuberculosis* enzymes to attain bactericidal effect and some undergo host-mediated bioactivation. Hepatotoxicity is the hallmark of all antimycobacterial drugs. Advancement in treatment strategies using codrug approach has made outstanding progress in improving ADMET of TB drugs (Mori et al., 2017). Some innovative efforts undertaken to design codrugs of anti-TB drugs are given below.

Husain et al. (2015) have synthesized a mutual prodrug by clubbing isoniazid with nalidixic acid with an objective of getting a compound which may act with effectiveness on both the gram-positive and gram-negative bacteria including *M. tuberculosis*. The codrug showed remarkable antitubercular activity against *M. tuberculosis* H37Rv with MIC 10 μg/mL (Figure 2.36).

In 1996, Chung et al. (2008) synthesized primaquine (PQ)-nitrofurazone (NF) codrug with dipeptide linkers. Cleavage of this prodrug occurs specifically by enzymes present in *Trypanosoma cruzi* (i.e., cysteinyl-protease). This approach was hypothesized to effectively make use of the direct and indirect oxidative stress in parasite caused by PQ and NF respectively which is responsible for their antitubercular effects (Figure 2.37).

The conventional treatment with INH is associated with 4%–11% incidence of hepatotoxic events. INH is metabolized by acetylation at N2 center of hydrazinic chain and released free radicals are responsible for hepatotoxicity. No full proof treatment is available for reducing hepatotoxicity except discontinuation of therapy. But due to such discontinuation, chances of morbidity increase. Also INH and/or its biotransformation products increase risk for systemic lupus erythematous, steatosis, peripheral neuropathy, hepatic necrosis and neurological disorders. In past, efforts have been focused on designing INH-conjugates with improved hydrolytic stability, antitubercular activity and bioavailability with lowered hepatotoxicity. By keeping this in mind Bhilare et al. (2018) designed a novel series of codrugs by conjugating INH with antioxidant aminothiols like N-(2-mercaptopropionyl) glycine (MPG), N-acetyl cysteine (NAC) and L-methionine (Met) with lower hepatotoxicity. Objective behind blocking the hydrazinic functional group of INH was to prevent oxidation by CYP2E1 and acetylation by NAT. Significant refinement in the concentration of enzymes involved in the antioxidant defense system was observed. Liver function markers and biochemical parameters were substantially reduced by these hepatoprotective codrugs (Figure 2.38).

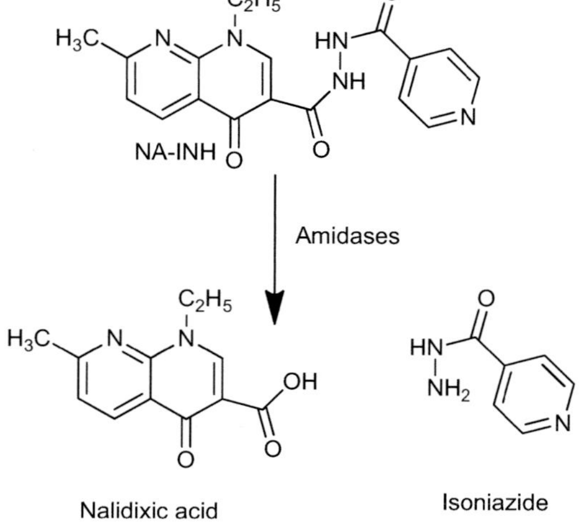

NA-INH

Amidases

Nalidixic acid

Isoniazide

FIGURE 2.36 Bioactivation mechanism of nalidixic acid-isoniazid codrug.

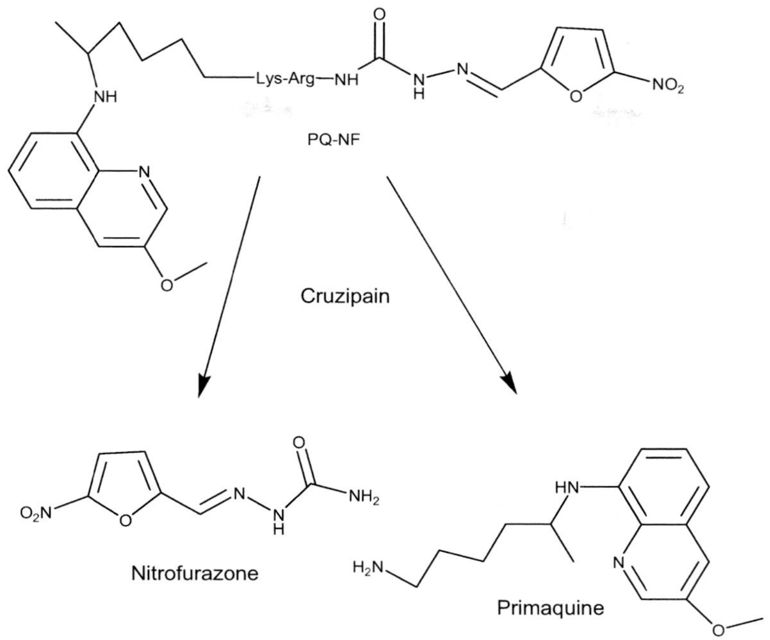

PQ-NF

Cruzipain

Nitrofurazone

Primaquine

FIGURE 2.37 Bioactivation mechanism of primaquine-nitrofurazone codrug.

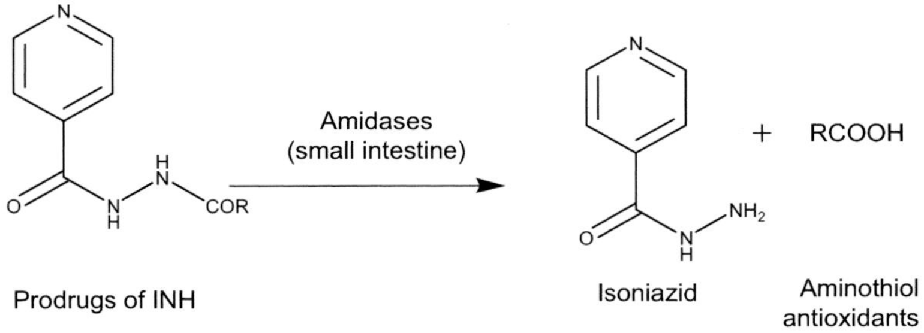

Amidases (small intestine)

Prodrugs of INH

Isoniazid + RCOOH

Aminothiol antioxidants

FIGURE 2.38 Bioactivation mechanism of isoniazid codrug.

2.6.11.8 Codrugs for Dermatological Delivery

Easy accessibility and large size of administration site provided large attention to topical route of administration over other routes. Dermal and transdermal route of administration are preferred in case of limited oral route of delivery due to a hepatic first-pass effect and gastrointestinal irritation. Despite attractiveness, dermal delivery restricts passage of some drugs. As permeation occurs through the subcutaneous layer, drug molecule must possess balanced oil in water partition coefficient. Then only pharmaceuticals can pass with ease through the epidermal and dermal layers. For effective transdermal delivery, a drug has to compartmentalize itself between formulation and skin, followed by proper entry into systemic circulation. In order to achieve this, a proper balance should exist between drug's tendency to bind with viable epidermis/dermis layers and subcutaneous layer. Physicochemical properties of drug also take part in topical curatives like molecular size, ionization, and hydrogen bonding. Higher number of hydrogen bonds hampers the transport of drugs in sizable amount due to its interaction with composition of subcutaneous layer. Out of ionized and unionized composite, unionized drug molecules travel through the skin more easily. Molecular size of 50 dalton of drug shows successful transdermal transport. On account of tremendous benefits of transdermal delivery, enormous efforts were undertaken to overcome stringent physicochemical barriers to optimize efficient topical delivery. Some physical measures such as iontophoresis, chemical enhancers, sonophorosis and electroporation have been adopted to improve passive permeation of drug into body but associated with adverse reaction at the application site. Codrug strategy is used for decades to overcome physicochemical, pharmacokinetic and pharmacodynamic problems. The success of this approach led scientist to develop prodrugs that offer tremendous advantages to improve topical as well as transdermal delivery of pharmaceuticals.

NSAIDs are the most commonly prescribed as analgesics, anti-inflammatory, antipyretic drugs. Naproxen has been used in treatment of skin conditions involving inflammation such as psoriasis and dermatitis. Naproxen shows poor bioavailability when administered dermally over oral route of administration. Only 1%–2% of drug reaches systemic circulation after dermal administration. This is due to ionizable groups present in naproxen that results in extremely low partition in oil leading to low percutaneous absorption. Codrugs corrected this issue by masking ionization group. Dithranol is an anti-proliferative hydroxyanthrone for antipsoriatic therapy. Dithranol is unstable drug which undergoes oxidative degradation to danthron and dithranol dimer which are associated with undesired effects. Dithranol is generally prescribed in combination with other drug therapies like cyclosporine. Undesired effects and instability of dithranol hampers its potential use in disease condition. Lau et al. designed codrugs of naproxen and dithranol (Nap-DTH) by utilizing hydroxyl and carboxyl functional moieties of two drugs. They hypothesized that masking of functional groups will improve physicochemical nature of parent compounds and stability issues and side effects of dithranol. Experiments showed increased log P (5.5) indicating accumulation in lipid layers to provide depot release. Also, codrug (Nap-DTH) exhibited advantage of less skin staining as compared to dithranol, 2.6% permeation and higher rate of *in vivo* hydrolysis. Appearance of codrug in skin was significantly higher than dithranol without presence of any dithranol metabolite. The quantity of naproxen released from codrug was equivalent to when administered alone (Lau et al., 2010). This could propose that the codrug was successful in reducing extent of autodegradation and entry into blood (Figure 2.39).

Exposure of epidermal keratinocytes to oxidative damage and UV rays results in formation of intracellular hydrogen peroxide (H_2O_2), reactive oxygen species (ROS) and the lipid peroxidation of the plasma membrane which in turn damages the skin. This could be initiator of skin cancer and photoaging.

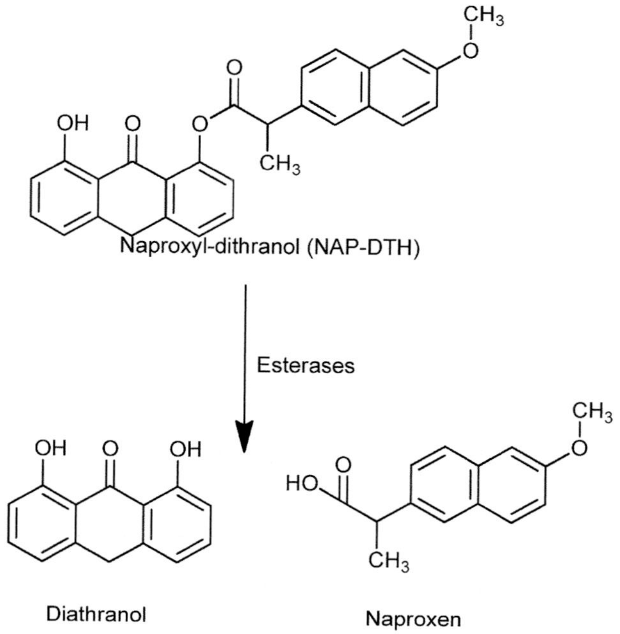

Naproxyl-dithranol (NAP-DTH)

Esterases

Diathranol

Naproxen

FIGURE 2.39 Bioactivation of naproxyl-diathranol codrug.

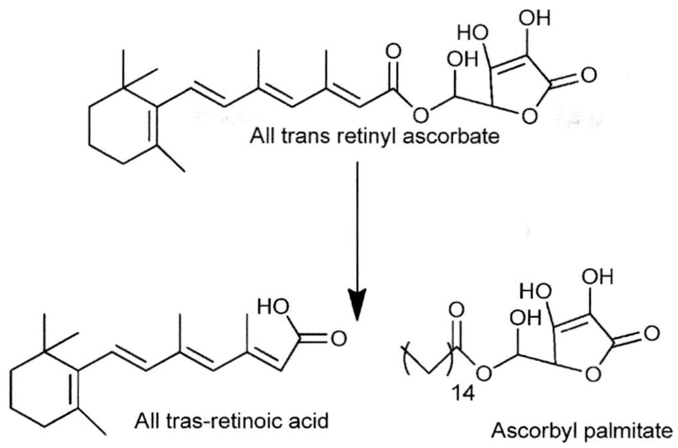

FIGURE 2.40 Bioactivation mechanism of *all-trans*-retinyl ascorbate.

Antioxidants provide protection to cutaneous system by eliminating ROS species. Retinoid in its potent form, i.e., retinoic acid has been demonstrated to regulate G1 phase of cell division and possesses prominent antioxidant properties. L-Ascorbic acid (ASA) is a strong and powerful water-soluble antioxidant that provides protection against oxidative degradation of skin and also induction to collagen synthesis. L-Ascorbic acid works well in combination with other antioxidants by synergistic effects. But due to low log P values it is not suitable candidate for dermal delivery. In addition, retinoic acid and ASA showed photo- and thermo- instability. Retinoic acid and ASA both possess carboxylic and hydroxyl group which can be modified into ester codrugs. Jones and group first synthesized codrug of retinoic acid with L-Ascorbic acid that undergoes enzymatic/chemical bioactivation with no toxic byproduct. In this study they hypothesized that the codrug would provide better uptake in epidermis, i.e., in lipoidal domains of the stratum corneum due to enhanced lipophilicity (log P = 2.2). Furthermore, the ester bond formed between retinoic acid-ASA could give higher stability, deep epidermis penetration and synergistic antioxidant properties after in situ enzymatic/chemical hydrolysis. Thus, ester functionality would provide better dermal protection against oxidative stress caused by exposure of UV rays to skin and superior stability over parent compounds. Ester codrugs possess optimal log P required for improved skin delivery as well as greater degree of skin interaction (Abdulmajed et al., 2004). This provided benefit of limiting the dose entering the system (Figure 2.40).

Global statistics shows that dependence on nicotine, caffeine and alcohol is on the rise. Current therapy for alcohol abuse and tobacco dependence involves simultaneous treatment so while treating the alcoholic person other factors also need to be considered. Dependence is mediated through activation and reinforcement of central nicotinic acetylcholine receptors. Hamad et al. designed codrug of naltrexone and bupropion, both used in therapies against nicotine dependence. Naltrexone is opiate antagonist with low bioavailability and GI side effects as vomiting, nausea, constipation, and pain in abdomen. β-Naltrexol is active metabolite of naltrexone. Bupropion is an antidepressant that acts by antagonism of nicotinic receptor where hydroxybupropion; a major metabolite contributes to antidepressant activity. Hamad and group (2006) synthesized and tested tripartate codrug of naltrexone and β-naltrexol with bupropion or hydroxybupropion as linker. This tripartate codrug was designed with objective of dual treatment for both alcohol abuse and tobacco dependence. Results showed synthesized mutual prodrugs were efficiently bio-activated to the parent drugs at pH 7.35–7.45 *in vitro* as well as *in vivo*, proving to be potential candidates for treating patients of alcohol withdrawal and tobacco dependency (Figure 2.41).

2.6.11.9 Codrugs for Ocular Delivery

Most of the ailments of eye are treated by administration of eye drops. But only 5% of instilled dose reaches inner eye where actual drug site is located. So, a major challenge in ocular delivery includes

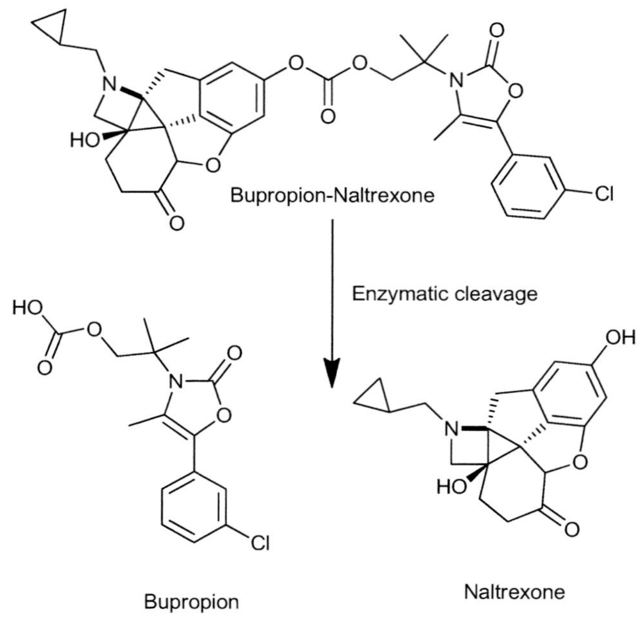

Bupropion-Naltrexone

Enzymatic cleavage

Bupropion

Naltrexone

FIGURE 2.41 Bioactivation mechanism of bupropion-naltrexone codrug.

tightness of corneal epithelium, precorneal drug elimination and limited systemic absorption from conjuctival layer. Approaches used for improving bioavailability of ocular drugs are aimed at extending drug's presence in conjuctival sac and enhancing diffusion of drug corneal membrane. Ocular absorption of drugs is also increased by increasing lipophilicity of drugs with the help of prodrug approach. Key requirements for ocular prodrugs involve stability and aqueous solubility in epithelium and stroma as a preformulation requisite, sufficient lipophilicity for corneal penetration, least trouble to eye, and therapeutically adequate rate of release of parent drug in eye. Preferred way of achieving it is to generate codrug which will release hydrophilic drug in the epithelium. Biotransformation of ocular prodrugs occurs by utilizing ocular enzymes. Developed ocular prodrugs are tested in rabbit's eye which has various esterases enzymes, like cholinesterase (BuChE, AChE), carbonic anhydrase, peptidases and phosphatases. Most of occular therapeutics containing hydroxyl (-OH) or carboxyl (-COO) functional groups are modified as ester to lipophilic promioeties with the aim of bioactivation by AChE and BuChE. BuChE is the dominant esterase present in ocular tissue. Esterase activity is highest in the order of iris-ciliary body, cornea, aqueous humor and not prevalent in tears. This absence of enzyme in tears ensures intact delivery of ocular drug in cornea and it acts as a major site for bioactivation of ester prodrugs of ocular drugs.

Penetrating ocular trauma associated with visual loss, proliferative vitreoretinopathy (PVR) and disability are the growing eye-related conditions. Despite of development of various pharmacological agents against incidence of tractional retinal detachment due to PVR, very less are effective against ocular trauma. Anti-proliferative drug 5-FU is recommended for treatment of PVR. Beneficial effect of 5-FU and corticosteroid is already proved but associated with some risks. Repeated injections of FU are observed to cause retinal problems as endophtalmitis and detachment of retina, as well as patient incompatibility. High doses of drug instillation are destructive to ocular organs like cornea and retina. NSAIDs provide better alternative with additional advantages while controlling inflammation. Main aim of PVR treatment is providing prolonged rather than transient effects. So, Cardillo et al. (2004) investigated codrugs of NSAIDs/5-FU for the prevention of an experimental trauma model of PVR. Codrug effectively inhibited progression of PVR than individual therapy with least toxicity due to simultaneous down-regulation of wound healing process involving inflammation and proliferative maturation of components (Figure 2.42).

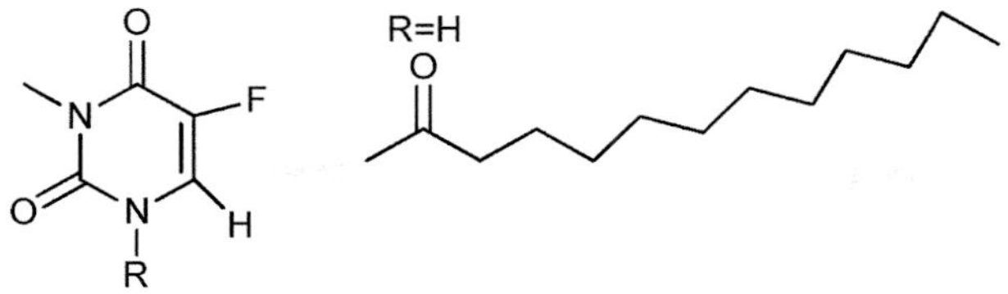

FIGURE 2.42 5-Fluorouracil codrug.

3α, 17α, 21-Trihydroxy-5β-pregnan-20-one (trihydroxy steroid, THS) is an angiostatic steroidal drug that inhibits angiogenesis similar to cortisol but devoid of corticosteroid side effects. THS exhibits anti-angiogenic activity by suppressing protease activity. THS shows less bioavailability due to its poor solubility (12 μg/mL). So, problems of THS related to bioavailability could be overcome by designing its codrugs. 5FU is a known cytostatic, antineoplastic agent approved for management of PVR. Unfortunately, 5-FU possesses high aqueous solubility and rapid clearance rate from vitreous so multiple frequencies of doses are required to achieve effective therapeutic concentration of 5FU. Howard-Sparks and group (2007) have designed a mutual prodrug of anti-angiostatic steroid (THS) linked to 5FU. This codrug was formed by incorporating carbonic ester linkage between two molecules of 5-FU and two molecules THS. Researchers reported that codrug THS-BIS-5FU fulfilled goal of improved drug delivery overcoming solubility problems of each other with enhanced bioavailability. The codrug proved to be better angiostatic than THS, 5FU or their combination (Figure 2.43).

Diuretics like ethacrynic acid (ECA) inhibit active transport of sodium/potassium/chloride. It is administered in glaucoma for maintaining intraocular pressure. It is achieved by injecting ECA into intracameral region of human eyes. ECA lowers intraocular pressure (IOP) by chemically modifying

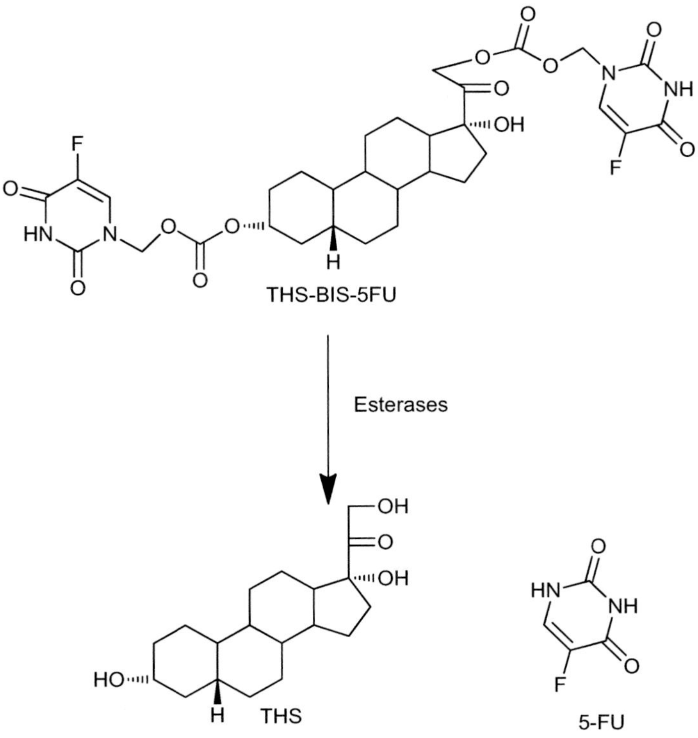

FIGURE 2.43 Bioactivation mechanism of THS-BIS-5FU.

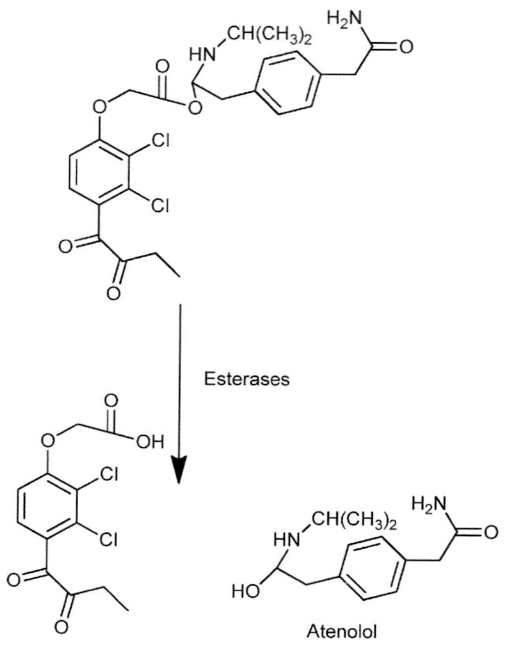

FIGURE 2.44 Bioactivation mechanism of ethacrynic acid-atenolol codrug.

cellular sulfhydryl groups in trabecular meshwork (TM) cells that alters outflow of aqueous humor from the TM. Animal studies reported that long term use of higher doses of topical or intracameral ECA cause serious adverse effect such as moderately diffused superficial corneal erosion, prominent eyelid edema, and conjunctival hyperemia. This serious limitation of ECA is due to its physicochemical properties. At physiological pH, ECA exists as carboxylate anion having unfavorable lipophilicity contributing to limited or inability to penetrate into cornea. That lowers therapeutic concentration at the site. So, there is a need to develop the prodrug of ECA that will overcome this barrier and penetrate into cornea with higher therapeutic index and greater ocular safety. β-blockers such as timolol (TML) and atenolol (ATL) provide effective protection against cardiovascular disease like ischemic heart disease, hypertensive disorders and certain arrhythmias. ATL and TML can also lower IOP. Increased outflow of aqueous humor lowers IOP which is possible by ECA treatment whereas β-blockers lower IOP by decreasing aqueous humor formation. Cynkowska et al. (2005) synthesized ester codrugs of ECA with atenolol and timolol to achieve benefits of improved delivery to the eye for apparent synergistic effect. They reported longer half-life with ECA-ATL because of hindrance offered by bulky group around the ester linkage in the ECA-ATL that provided superior binding in target site and shielded ester linkage from base hydrolysis. So, this novel tethering of ECA to β-adrenergic blockers provided less toxic and more efficient delivery of ECA for longer duration (Figure 2.44).

2.6.11.10 *Examples of Marketed Codrugs*

Since last several decades, many prodrugs were discovered, launched and are available in market for therapeutic use. Due to changing approach for tackling the deadly diseases, prodrugs are slowly becoming a part of drug discovery process rather than development, capturing the markets fast. In the period of 2008–2017, 249 new molecular entities were approved and 31 out of them were prodrugs. Except 2012 and 2016, each year at least two prodrugs have been approved by FDA. Noticeable approval to prodrugs were given by FDA in the years 2008 and 2015. In 2015, out of total FDA-approved small molecules, 20% were prodrugs while among total FDA-approved drugs, 15% were prodrugs. In 2008 the percentage for

TABLE 2.3
FDA-Approved Codrugs

Codrug Name	Brand Name	Promoieties	Application	Rational
Benorylate (Croft et al., 1972)	Beilijin	Aspirin and Paracetamol	Rheumatoid arthritis, osteoarthrosis and other musculoskeletal al conditions	To decrease gastric irritation and for synergistic action
Estramustin (Heimbach et al., 2003)	Emcyt and Estracyt	Phosphorylated steroid, 17 α-estradiol and normustard [HN $(CH_2CH_2Cl)_2$]	Prostate cancer	Site-specific release of drug and synergistic action
Sultamicillin (Lode et al., 1989)	Unasyn	Ampicillin and Sulbactam	Skin and skin structure infections	To increase absorption and reduce side effects like diarrhea and dysentery
Fenethylline (Katselou et al., 2016)	Captagon, Biocapton, and Fitton	Amphetamine and Theophylline	Treatment of hyperkinetic children	Synergistic cation
Sulfasalazine (Sandborn, 2002)	Azulfidine	Sulfapyridine and 5-aminosalicylic acid	Ulcerative colitis	Site-specific release at colon

approved prodrugs increased from 20% to 33% among small molecules approved by FDA (Stella, 2010). Some of these prodrugs are compiled in Table 2.3.

2.7 Some Recent Codrug Examples

2.7.1 Hypoxia-Activated Anticancer Codrugs

Codrugs that undergoes bioactivation under hypoxic condition to release active drugs are hypoxia-activated prodrugs. Nitroaromatics is one of the most researched chemical structures due to its activation under hypoxic conditions. Nitroaromatics undergo reductive bioactivation under tumor hypoxic condition to release hydroxylamine or amine derivatives. Released nucleophiles trigger release of cytotoxic drugs. In 1994, first Denny and group reported tumor hypoxia-activated release of aminoaniline mustard cytotoxin from 2-nitrophenyl acetamides by spontaneous cyclization. They synthesized mutual ester or amide prodrug scaffold which showed hypoxia-mediated activation to active cytotoxin, i.e., floxuridine cytotoxin and indolin-2-one like semaxanib as waste fragment. Mechanism behind activation was (1) presence of invariable hydrogen bond between amine function of pyrrole with-C=O functional group in amides and esters of 2-nitrophenylacetic acid and (2) bioreductive activation and spontaneous cyclization in tumor cell to indoline-2-one and aniline. Side effects of drugs were avoided by these types of mutual prodrugs, also provided higher tumor concentrations of actives. Same concept was applied further to develop rational design of codrugs by Sansom et al. who synthesized hypoxia-activated amide and ester-linked codrug prototypes (Figure 2.45) by providing chemical proof of concept (Figure 2.46). Here aryl nitro group reduction led to release of two anticancer

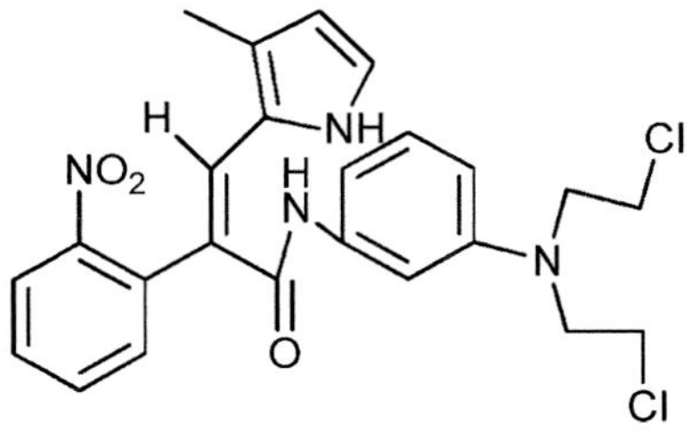

FIGURE 2.45 Ester-linked hypoxia-activated mutual prodrug prototypes.

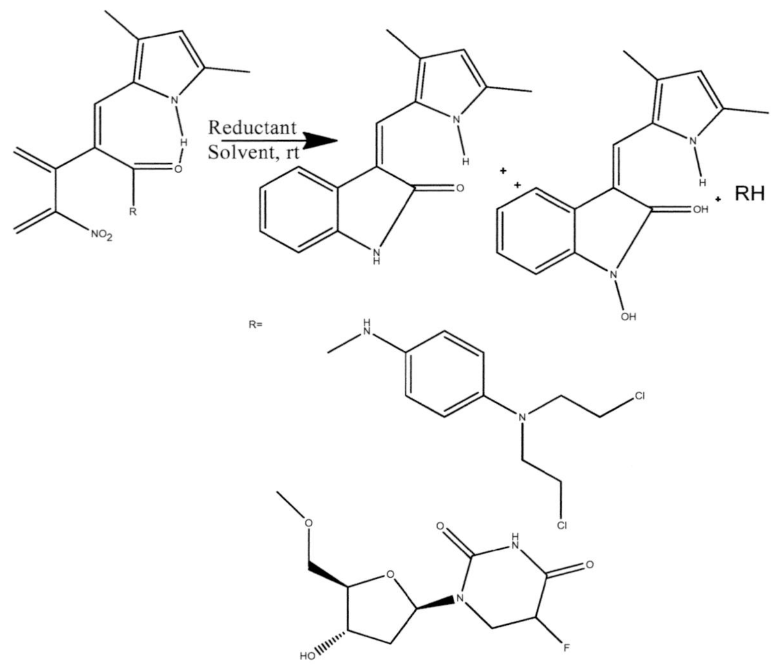

FIGURE 2.46 Bioactivation mechanism of hypoxia-activated codrug.

drugs and *in vivo* production of N-hydroxy semaxanib. They also reported that the attachment of electron withdrawing group to nitroaryl ring increases the reduction potential toward the target (Sansom et al., 2019).

2.7.2 Human Carboxylesterases (hCES1 and hCES2)-Activated Codrugs

Epalrestat is the only aldose reductase (ALR2) inhibitor which reduces accumulation of sorbitol intracellularly and used in India, Japan, and China against neuropathy, nephropathy and retinopathy. In such diabetic complications, polyol pathway flux is popular hypothesis which involves sorbitol dehydrogenase and aldose reductase as rate limiting enzymes. In diabetic complications, blocking all pathways instead of complete inhibition of one is recommended strategy. Drugs currently in clinical trials are ALR2 inhibitors for diabetic complications. Out of that epalrestat is the only effective drug available but not approved by FDA yet. Treatment with epalrestat is associated with some drawbacks like targets only one pathway in diabetic complication so chances of progression of disease via other pathways, need for frequent dosing, side effects is always there. So, in one patent it is claimed that prodrug designing of epalrestat could overcome these problems. In light of this, Choudhary and Silakari put the efforts to design ester or amide codrugs of epalrestat with antioxidants by using amino acids with antioxidant property (Figure 2.47). Antioxidant amino

Epalrestatt-linker-Antioxidant

↓

NH2-Linker-COOH + HO-Antioxidants

Linker=Glycine, Alanine, Phenylalanine

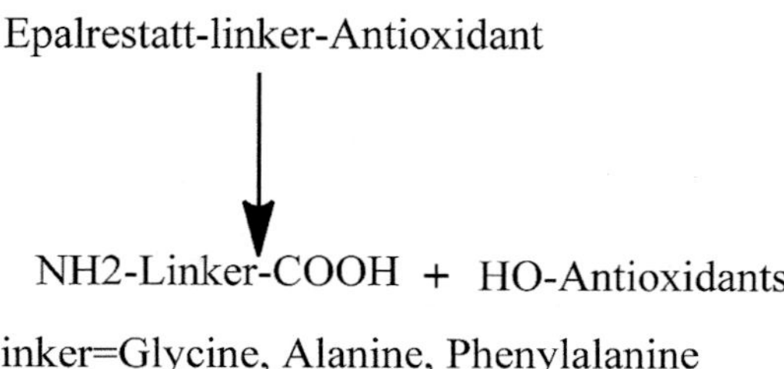

FIGURE 2.47 Bioactivation mechanism of epalrestat-antioxidant codrug.

acids have been reported to have helped in attenuating free radicals with promising gastrosparing effect. Proposed bioactivation mechanism of synthesized prodrugs was by carboxylesterase (CES) enzymes. Here researcher studied CES-mediated activation of ester and amide codrugs by *in silico* techniques like molecular dynamics (MD), molecular docking, quantum mechanics (QM)/molecular mechanics (MM). Results suggested that generated mutual prodrugs undergo better activation by hCES1 than that of hCES2 with larger acyl group than its alcoholic counterpart (Choudhary and Silakari, 2019).

2.7.3 Antitubercular Codrug Nanoparticles

Ethionamide (ETH) is most demanded second-line antitubercular treatment against multi-drug-resistant TB (MDR-TB). ETH interrupts bacterial wall synthesis by inhibiting mycobacterial enzyme enoyl–acyl carrier protein reductase InhA. ETH treatment is hampered by low therapeutic index and dose-dependent side effects. Co-administration of small molecule inhibitors of EthR, i.e., transcriptional repressor is general practice in today's treatment regimen. EthR controls flavin-dependent monooxygenase enzyme (EthA), which is responsible for conversion of ETH to active antitubercular form nicotinamide adenine dinucleotide (NAD). BDM43266 is potent ligand of EthR. However, co-administration of ETH and BDM43266 is affected by low hydrophilicity of both compounds due to crystallization of ETH. So, Pastor et al. (2019) thought of designing codrugs that provide solution to pharmacokinetic problems. So, after synthesis, ETH/BDM43266 codrug was embedded in assembly that increases pulmonary delivery. They accepted the challenge to synthesize ETH coupled with its booster as a novel conjugate in the form of nanoparticles. Therapeutic efficacy of the codrug nanoparticles against TB has been documented as first proof of concept in the literature. Results showed that nanoparticle codrugs were effectively administered by intrapulmonary route (Figure 2.48).

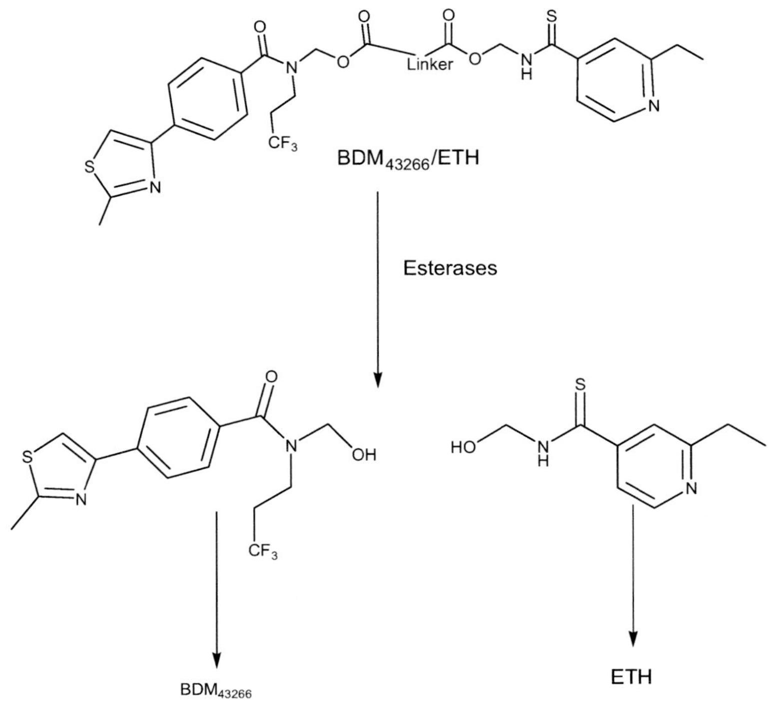

FIGURE 2.48 Bioactivation mechanism of ETH-BDM_{43266} codrug.

2.8 Conclusion

Codrug or mutual prodrug approach provides a multifaceted roadmap to tackle defects of clinically feasible or accepted drugs. Though earlier the cardinal significance of prodrugs in general was questioned by the pharma industries, it is now being considered as a promising and inherent part of drug discovery process and not only as the last option of developmental stage. It should clearly be taken into account in early stages of lead optimization as it has been successful in offering the benefits of combination therapy by demonstrating synergistic or additional complimentary effects to that of the parent drug, minimizing side effects and toxic effects of drugs and improving utility and bioavailability of both, i.e., the parent drug and the biologically active carrier drug. Coming times will see more and more codrugs in the market as combination therapy which would be more effective than physical combination of two drugs.

REFERENCES

Abdulmajed, Kasem, Charles M Heard, C McGuigan, and WJ Pugh. 2004. Topical delivery of retinyl ascorbate co-drug. *Skin Pharmacology and Physiology* 17 (6):274–282.

Abet, Valentina, Fabiana Filace, Javier Recio, Julio Alvarez-Builla, and Carolina Burgos. 2017. Prodrug approach: An overview of recent cases. *European Journal of Medicinal Chemistry* 127:810–827.

Albert, Adrien. 1958. Chemical aspects of selective toxicity. *Nature* 182 (4633):421.

Aljuffali, Ibrahim A, Chwan-Fwu Lin, Chun-Han Chen, and Jia-You Fang. 2016. The codrug approach for facilitating drug delivery and bioactivity. *Expert Opinion on Drug Delivery* 13 (9):1311–1325.

Allison, Anthony C, and Elsie M Eugui. 2000. Mycophenolate mofetil and its mechanisms of action. *Immunopharmacology* 47 (2–3):85–118.

Anderson, Bradley D, WW Chu, and Raymond E Galinsky. 1988. Reduction of first-pass metabolism of propranolol after oral administration of ester prodrugs. *International Journal of Pharmaceutics* 43 (3):261–265.

Baker, William, Marcin Stasiak, Sundaramoorthi Swaminathan, and Musong Kim. 2010. Monophosphates as mutual prodrugs of anti-inflammatory signal transduction modulators (AISTM's) and beta-agonists for the treatment of pulmonary inflammation and bronchoconstriction. U.S. Patent Application No. 12/519,305.

Barditch-Crovo, Patricia, Steven G Deeks, Ann Collier, et al. 2001. Phase I/II trial of the pharmacokinetics, safety, and antiretroviral activity of tenofovir disoproxil fumarate in human immunodeficiency virus-infected adults. *Antimicrobial Agents and Chemotherapy* 45 (10):2733–2739.

Beale, John M, and John H Block. 2011. *Wilson and Gisvold's Textbook of Organic Medicinal and Pharmaceutical Chemistry*. Wolters Kluwer Company.

Beresford, R, and Alan Ward. 1987. Haloperidol decanoate. *Drugs* 33 (1):31–49.

Bhilare, Neha Vithal, Suneela Dhaneshwar, and Kakasaheb Ramoo Mahadik. 2018. Amelioration of hepatotoxicity by biocleavable aminothiol chimeras of isoniazid: Design, synthesis, kinetics and pharmacological evaluation. *World Journal of Hepatology* 10 (7):496.

Bhosle, Deepak, Saket Bharambe, Neha Gairola, and Suneela Dhaneshwar. 2006. Mutual prodrug concept: Fundamentals and applications. *Indian Journal of Pharmaceutical Sciences* 68 (3):284–294.

Brown, GR, DS Clarke, AW Faull, AJ Foubister, and MJ Smithers. 1996. Design of dual-acting thromboxane antagonist-synthase inhibitors by a mutual prodrug approach. *Bioorganic & Medicinal Chemistry Letters* 6 (3):273–278.

Budhalakoti, Latika, Suneela Dhaneshwar, Shakuntala Chopade, Hemant Kamble, and SL Bodhankar. 2014. Exploring thioester chemistry in mutual prodrug design for combination antihypertensive therapy. *World Journal of Pharmaceutical Research* 4 (1):740–767.

Bundgaard, Hans. 1989. The double prodrug concept and its applications. *Advanced Drug Delivery Reviews* 3 (1):39–65.

Byrne, William, and Andrew Rynne. 1998. Pharmaceutical product comprising a salicylate of an esterifiable ACE-inhibitor. U.S. Patent 5,852,047.

Capomacchia, Anthony C, Solomon T Garner Jr, and J Warren Beach. 2013. Glucosamine and glucosamine/anti-inflammatory mutual prodrugs, compositions, and methods. US Patent 8,361,990.

Cardillo, JA, ME Farah, J Mitre, et al. 2004. An intravitreal biodegradable sustained release naproxen and 5-fluorouracil system for the treatment of experimental post-traumatic proliferative vitreoretinopathy. *British Journal of Ophthalmology* 88 (9):1201–1205.

Chopade, Shakuntala Santosh, and Suneela Dhaneshwar. 2018. Determination of the mitigating effect of colon-specific bioreversible codrugs of mycophenolic acid and aminosugars in an experimental colitis model in Wistar rats. *World Journal of Gastroenterology* 24 (10):1093.

Choudhary, Shalki, and Om Silakari. 2019. hCES1 and hCES2 mediated activation of epalrestat-antioxidant mutual prodrugs: Unwinding the hydrolytic mechanism using in silico approaches. *Journal of Molecular Graphics and Modelling* 91:148–163.

Chung, Man, Elizabeth Ferreira, Jean Santos, et al. 2008. Prodrugs for the treatment of neglected diseases. *Molecules* 13 (3):616–677.

Croft, DN, JHP Cuddigan, and Carole Sweetland. 1972. Gastric bleeding and benorylate, a new aspirin. *British Medical Journal* 3 (5826):545–547.

Crook, Robert James, and Alan John Pettman. 2004. Mutual prodrug of amlodipine and atorvastatin. U.S. Patent 6,737,430.

Cynkowska, Grazyna, Tadeusz Cynkowski, Abeer A Al-Ghananeem, Hong Guo, Paul Ashton, and Peter A Crooks. 2005. Novel antiglaucoma prodrugs and codrugs of ethacrynic acid. *Bioorganic & Medicinal Chemistry Letters* 15 (15):3524–3527.

Das, Nirupam, Meenakshi Dhanawat, Biswajit Dash, Ramesh C Nagarwal, and Sushant K. Shrivastava. 2010. Codrug: An efficient approach for drug optimization. *European Journal of Pharmaceutical Sciences* 41 (5):571–588.

De Albuquerque, Silva, Tavoro Antonio, Man Chin Chung, et al. 2005. Advances in prodrug design. *Mini Reviews in Medicinal Chemistry* 5 (10):893–914.

Dhaneshwar, Suneela, Jain Astha and Tewari Kunal. 2014a. Design and applications of bioprecursors: A retro metabolic approach. *Current Drug Metabolism* 15 (3):291–335.

Dhaneshwar, Suneela, and Harjeet Gautam. 2012. Exploring novel colon-targeting antihistaminic prodrug for colitis. *Journal of Physiology & Pharmacology* 63 (4):327–337.

Dhaneshwar, Suneela, Metreyi Sharma, and Gaurav Vadnerkar. 2011a. Co-drugs of aminosalicylates and nutraceutical amino sugar for ulcerative colitis. *Journal of Drug Delivery Science and Technology* 21 (6):527–533.

Dhaneshwar, Suneela, Metreyi Sharma, Vivek Patel, Upasana Desai, and Jiten Bhojak. 2011b. Prodrug strategies for antihypertensives. *Current Topics in Medicinal Chemistry* 11 (18):2299–2317.

Dhaneshwar, Suneela, Mukta Chail, Mahavir Patil, Salma Naqvi, and Gaurav Vadnerkar. 2009a. Colon-specific mutual amide prodrugs of 4-aminosalicylic acid for their mitigating effect on experimental colitis in rats. *European Journal of Medicinal Chemistry* 44 (1):131–142.

Dhaneshwar, Suneela, Mukta Kandpal, and Gaurav Vadnerkar. 2009b. L-glutamine conjugate of meselamine: A novel approach for targeted delivery to colon. *Journal of Drug Delivery Science and Technology* 19 (1):67–72.

Dhaneshwar, Suneela, Neha Gairola, Mini Kandpal, et al. 2009c. Synthesis, kinetic studies and pharmacological evaluation of mutual azo prodrugs of 5-aminosalicylic acid for colon-specific drug delivery in inflammatory bowel disease. *European Journal of Medicinal Chemistry* 44 (10):3922–3929.

Dhaneshwar, Suneela. 2014. Colon-specific prodrugs of 4-aminosalicylic acid for inflammatory bowel disease. *World Journal of Gastroenterology: WJG* 20 (13):3564.

Dhaneshwar, Suneela, Manisha Kusurkar, Subhash Bodhankar, and Gopal Bihani. 2014b. Carrier-linked mutual prodrugs of biphenylacetic acid as a promising alternative to bioprecursor fenbufen: design, kinetics, and pharmacological studies. *Inflammopharmacology* 22 (4):235–250.

Dhaneshwar, Suneela, Patil Dipmala, Harsulkar Abhay, and Bhondave Prashant. 2013. Disease-modifying effect of anthraquinone prodrug with boswellic acid on collagenase-induced osteoarthritis in Wistar rats. *Inflammation & Allergy-Drug Targets (Formerly Current Drug Targets-Inflammation & Allergy)* 12 (4):288–295.

Dhokchawle, BV. Gawad, JB. Kamble, MD. Tauro, SJ and Bhandari, AB. 2014. Promoieties used in prodrug design: A review. *Indian Journal of Pharmaceutical Education and Research* 48 (2):35–40.

Donahue, Katrina E, Gerald Gartlehner, Daniel E Jonas, et al. 2008. Systematic review: comparative effectiveness and harms of disease-modifying medications for rheumatoid arthritis. *Annals of Internal Medicine* 148 (2):124–134.

Du, Fang, Liang-jing Lü, Qiong Fu, et al. 2008. T-614, a novel immunomodulator, attenuates joint inflammation and articular damage in collagen-induced arthritis. *Arthritis Research & Therapy* 10 (6):R136.

Fujii, Hideaki, Akio Watanabe, Toru Nemoto, et al. 2009. Synthesis of novel twin drug consisting of 8-oxaendoethanotetrahydromorphides with a 1, 4-dioxane spacer and its pharmacological activities: μ, κ, and putative ε opioid receptor antagonists. *Bioorganic & Medicinal Chemistry Letters* 19 (2):438–441.

Furfine, Eric S, Christopher T Baker, Michael R Hale, et al. 2004. Preclinical pharmacology and pharmacokinetics of GW433908, a water-soluble prodrug of the human immunodeficiency virus protease inhibitor amprenavir. *Antimicrobial Agents and Chemotherapy* 48 (3):791–798.

Gras, Jordi. 2012. Perphenazine 4-aminobutyrate mesylate Dopamine D-2 Receptor Antagonist GABA (A) Receptor Agonist Treatment of Schizophrenia. *Drugs of the Future* 37 (9):645–650.

Guarino, Victor R. 2010. The molecular design of prodrugs by functional group. *Prodrugs and Targeted Delivery: Towards Better ADME Properties* 47:31–60.

Gunes, Arzu, Maria Gabriella Scordo, Peeter Jaanson, and Marja-Liisa Dahl. 2007. Serotonin and dopamine receptor gene polymorphisms and the risk of extrapyramidal side effects in perphenazine-treated schizophrenic patients. *Psychopharmacology* 190 (4):479–484.

Hamad, Mohamed O, Paul K Kiptoo, Audra L Stinchcomb, and Peter A Crooks. 2006. Synthesis and hydrolytic behavior of two novel tripartate codrugs of naltrexone and 6β-naltrexol with hydroxybupropion as potential alcohol abuse and smoking cessation agents. *Bioorganic & Medicinal Chemistry* 14 (20):7051–7061.

Harper, Norman J. 1958. Drug latentiation. *Journal of Medicinal Chemistry* 1 (5):467–500.

Harper, Norman J. 1959. Drug latentiation. *Journal of Medicinal Chemistry* 1:467–500.

Harper, Norman J. 1962. Drug latentiation. In: *Progress in drug research*, New York: Wiley 4:221–294.

He, George, Joseph Massarella, and Penelope Ward. 1999. Clinical pharmacokinetics of the prodrug oseltamivir and its active metabolite Ro 64-0802. *Clinical Pharmacokinetics* 37 (6):471–484.

Heimbach, Tycho, Doo-Man Oh, Lilian Y Li, Naír Rodríguez-Hornedo, George Garcia, and David Fleisher. 2003. Enzyme-mediated precipitation of parent drugs from their phosphate prodrugs. *International Journal of Pharmaceutics* 261 (1–2):81–92.

Holman, Christopher M, Timo Minssen, and Eric M Solovy. 2018. Patentability standards for follow-on pharmaceutical innovation. *Biotechnology Law Report* 37 (3):131–161.

Howard-Sparks, Michelle, Abeer M Al-Ghananeem, Peter A Crooks, and Andrew P Pearson. 2007. A novel chemical delivery system comprising an ocular sustained release formulation of a 3α, 17α, 21-trihydroxy-5β-pregnan-20-one-BIS-5-flourouracil codrug. *Drug Development & Industrial Pharmacy* 33 (6):677–682.

Husain, Asif, Ahmed, Aftab, and Khan Shah, 2015. Studies on an amide based mutual prodrug: Synthesis and evaluation. *Journal of Biomedical and Pharmaceutical Research*, 4 (2):43–46.

Hussain, Arshi, Anil Kumar Shrivastava, and Pradeep Parashar. 2014. Synthesis, characterization and release studies of mutual prodrugs of norfloxacin and trimethoprim with indomethacin for colon-specific drug delivery. *International Journal Pharmaceutical Sciences* 3 (5):7–11.

Huttunen, Kristiina M, and Jarkko Rautio. 2011. Prodrugs-an efficient way to breach delivery and targeting barriers. *Current Topics in Medicinal Chemistry* 11 (18):2265–2287.

Huttunen, Kristiina M, Hannu Raunio, and Jarkko Rautio. 2011. Prodrugs—from serendipity to rational design. *Pharmacological Reviews* 63 (3):750–771.

Jiang, Michael Y, and David Dolphin. 2008. Site-specific prodrug release using visible light. *Journal of the American Chemical Society* 130 (13):4236–4237.

Joshi, Poorvashree, and Suneela Dhaneshwar. 2017. Novel drug delivery of dual acting prodrugs of hydroxychloroquine with aryl acetic acid NSAIDs: Design, kinetics and pharmacological study. *Drug Delivery and Translational Research* 7 (5):709–730.

Kaakkola, Seppo, Heikki Teräväinen, Sirpa Ahtila, Heli Rita, and Ariel Gordin. 1994. Effect of entacapone, a COMT inhibitor, on clinical disability and levodopa metabolism in parkinsonian patients. *Neurology* 44 (1):77–77.

Karaman, Rafik. 2014. Using predrugs to optimize drug candidates. *Expert Opinion on Drug Discovery* 9 (12):1405–1419.

Katselou, Maria, Ioannis Papoutsis, Panagiota Nikolaou, Samir Qammaz, Chara Spiliopoulou, and Sotiris Athanaselis. 2016. Fenethylline (captagon) abuse–local problems from an old drug become universal. *Basic & Clinical Pharmacology & Toxicology* 119 (2):133–140.

Kupchan, S Morris, Alan F Casy, and Joseph V Swintosky. 1965. Drug latentiation: Synthesis and preliminary evaluation of testosterone derivatives. *Journal of Pharmaceutical Sciences* 54 (4):514–524.

Lau, Wing Man, Alex W White, and Charles M Heard. 2010. Topical delivery of a naproxen-dithranol co-drug: In vitro skin penetration, permeation, and staining. *Pharmaceutical Research* 27 (12):2734–2742.

Lee, Sunyoung, Yonghyun Lee, Wooseong Kim, et al. 2015. Evaluation of glycine-bearing celecoxib derivatives as a colon-specific mutual prodrug acting on nuclear factor-κB, an anti-inflammatory target. *Drug Design, Development and Therapy* 9:4227.

Leppänen, Jukka, Juhani Huuskonen, Tapio Nevalainen, Jukka Gynther, Hannu Taipale, and Tomi Järvinen. 2002. Design and synthesis of a novel L-dopa–entacapone codrug. *Journal of Medicinal Chemistry* 45 (6):1379–1382.

Lew, Willard, Xiaowu Chen, and Choung U Kim. 2000. Discovery and development of GS 4104 (oseltamivir) an orally active influenza neuraminidase inhibitor. *Current Medicinal Chemistry* 7 (6):663–672.

Lode, H, B Hampel, G Bruckner, and P Koeppe. 1989. The pharmacokinetics of sultamicillin. *APMIS. Supplementum* 5:17–22.

Mangelsdorf, David J, Carl Thummel, Miguel Beato, et al. 1995. The nuclear receptor superfamily: the second decade. *Cell* 83 (6):835.

Manon, Benu, and Pritam D Sharma. 2009. Design, synthesis and evaluation of diclofenac-antioxidant mutual prodrugs as safer NSAIDs. *CSIR-NISCAIR* 1279–1287.

Marsden, Charles A. 2006. Dopamine: The rewarding years. *British Journal of Pharmacology* 147 (S1):S136–S144.

Menger, Fredric M, and Michael J Rourk. 1997. Synthesis and reactivity of 5-fluorouracil/cytarabine mutual prodrugs. *The Journal of Organic Chemistry* 62 (26):9083–9088.

Minucci, Saverio, and Pier Giuseppe Pelicci. 1999. Retinoid receptors in health and disease: Co-regulators and the chromatin connection. Paper read at *Seminars in Cell & Developmental Biology*. 10 (2):215–225.

Mori, Giorgia, Laurent Roberto Chiarelli, Giovanna Riccardi, and Maria Rosalia Pasca. 2017. New prodrugs against tuberculosis. *Drug Discovery Today* 22 (3):519–525.

Nagpal, Deepika, R Singh, Neha Gairola, SL Bodhankar, and Suneela Dhaneshwar. 2006. Mutual azo prodrug of 5-aminosalicylic acid for colon targeted drug delivery: Synthesis, kinetic studies and pharmacological evaluation. *Indian Journal of Pharmaceutical Sciences* 68 (2):171–178.

Najjar, Anas, and Rafik Karaman. 2019. The prodrug approach in the era of drug design. Taylor and Francis, 16 (1):1–5.

Njar, Vincent CO, Lalji K Gediya, and Aakanksha Khandelwal. 2010. Mutual prodrugs and methods to treat cancer. U.S. Patent Application No. 12/663,194.

Nudelman, Abraham, and Rachel Kelner. 1993. Approaches to mutual prodrugs: Calcium-β-Blockers. *Archiv der Pharmazie* 326 (11):907–909.

Nudelman, Abraham, Irit Gil-Ad, Nava Shpaisman, et al. 2008. A mutual prodrug ester of GABA and perphenazine exhibits antischizophrenic efficacy with diminished extrapyramidal effects. *Journal of Medicinal Chemistry* 51 (9):2858–2862.

Ohlan, Sucheta, Sanju Nanda, Dharam Pal Pathak, and Moksh Jagia. 2011. Mutual prodrugs-A swot analysis. *International Journal of Pharmaceutical Sciences and Research* 2 (4):719.

Ottenhaus, Vanessa, and Helmut Rosemeyer. 2015. Mitsunobu alkylation of cancerostatic 5-fluorouridine with (2E)-10-hydroxydec-2-enoic acid, a fatty acid from royal jelly with multiple biological activities. *Chemistry & Biodiversity* 12 (9):1307–1312.

Pastor, Alexandra, Arnaud Machelart, Xue Li, et al. 2019. A novel codrug made of the combination of ethionamide and its potentiating booster: Synthesis, self-assembly into nanoparticles and antimycobacterial evaluation. *Organic & Biomolecular Chemistry* 17 (20):5129–5137.

Peppercorn, Mark A. 1984. Sulfasalazine: Pharmacology, clinical use, toxicity, and related new drug development. *Annals of Internal Medicine* 101 (3):377–386.

Persson, Gunnar, and Olle Pahlm. 1995. Oral bambuterol versus terbutaline in patients with asthma. *Current Therapeutic Research* 56 (5):457–465.

Pochopin, Nancy L, William N Charman, and Valentino J Stella. 1994. Pharmacokinetics of dapsone and amino acid prodrugs of dapsone. *Drug Metabolism and Disposition* 22 (5):770–775.

Rai, Himanshu, and Suneela Dhaneshwar. 2015. Amide-linked ethanolamine conjugate of gemfibrozil as a profound HDL enhancer: Design, synthesis, pharmacological screening and docking study. *Current Drug Discovery Technologies* 12 (3):155–169.

Rao, H Surya Prakash. 2003. Capping drugs: development of prodrugs. *Resonance* 8 (2):19–27.

Rautio, Jarkko, Raimund Mannhold, Hugo Kubinyl, and Gerd Folkers. 2010. *Prodrugs and Targeted Delivery*. Hoboken, NJ: Wiley VCH.

Redasani, Vivekkumar K, and Sanjay B Bari. 2015. *Prodrug Design: Perspectives, Approaches and Applications in Medicinal Chemistry*. Amsterdam, the Netherlands: Academic Press.

Sandborn, William J. 2002. Rational selection of oral 5-aminosalicylate formulations and prodrugs for the treatment of ulcerative colitis. *American Journal of Gastroenterology* 97 (12):2939–2941.

Sansom, Geraud N, Nicholas S Kirk, Christopher P Guise, et al. 2019. Prototyping kinase inhibitor-cytotoxin anticancer mutual prodrugs activated by tumour hypoxia: A chemical proof of concept study. *Bioorganic & Medicinal Chemistry Letters* 29 (10):1215–1219.

Sarkate, Ajinkya, and Suneela Dhaneshwar. 2017. Investigation of mitigating effect of colon-specific prodrugs of boswellic acid on 2, 4, 6-trinitrobenzene sulfonic acid-induced colitis in Wistar rats: Design, kinetics and biological evaluation. *World Journal of Gastroenterology* 23 (7):1147.

Shirke, Supriya, Sheet al Shewale, and Manik Satpute. 2015. Prodrug design: An overview. *International Journal of Pharmaceutical, Chemical & Biological Sciences* 5 (1):232–241.

Shukla, VK, SK Garg, VS Mathur, and SK Kulkarni. 1985. GABAergic, dopaminergic and cholinergic interactions in perphenazine-induced catatonia in rats. *Archives Internationales de Pharmacodynamie et de Therapie* 278 (2):236–248.

Singh, Priyanka, and Suneela Dhaneshwar. 2017. Investigating drug repositioning approach to design novel prodrugs for colon-specific release of fexofenadine for ulcerative colitis. *Current Drug Delivery* 14 (4):543–554.

Sinkula, AA, and Samuel H Yalkowsky. 1975. Rationale for design of biologically reversible drug derivatives: prodrugs. *Journal of Pharmaceutical Sciences* 64 (2):181–210.

Sloan, Kenneth B. 1991. Prodrugs of biologically active hydroxyaromatic compounds. U.S. Patent 5,001,115.

Sorbera, LA, M Bayes, J Castaner, and J Silvestre. 2001. Melagatran and ximelagatran. *Drug Future* 26:1155–1170.

Soyez, Heidi, Etienne Schacht, and Sylvie Vanderkerken. 1996. The crucial role of spacer groups in macromolecular prodrug design. *Advanced Drug Delivery Reviews* 21 (2):81–106.

Stella, Valentino J. 1996. A case for prodrugs: Fosphenytoin. *Advanced Drug Delivery Reviews* 19 (2):311–330.

Stella, Valentino J. 2010. Prodrugs: Some thoughts and current issues. *Journal of Pharmaceutical Sciences* 99 (12):4755–4765.

Strømgaard, Kristian, Povl Krogsgaard-Larsen, and Ulf Madsen. 2017. *Textbook of Drug Design and Discovery*. Boca Raton, FL: CRC Press.

Sugawara, Mitsuru, Wei Huang, You-Jun Fei, Frederick H Leibach, Vadivel Ganapathy, and Malliga E Ganapathy. 2000. Transport of valganciclovir, a ganciclovir prodrug, via peptide transporters PEPT1 and PEPT2. *Journal of Pharmaceutical Sciences* 89 (6):781–789.

Talley, John J, Stephen R Bertenshaw, David L Brown, et al. 2000. N-[[(5-methyl-3-phenylisoxazol-4-yl)-phenyl] sulfonyl] propanamide, sodium salt, parecoxib sodium: A potent and selective inhibitor of COX-2 for parenteral administration. *Journal of Medicinal Chemistry* 43 (9):1661–1663.

Testa, Bernard. 2004. Prodrug research: Futile or fertile? *Biochemical Pharmacology* 68 (11):2097–2106.

Testa, Bernard. 2007. Prodrug objectives and design. *Comprehensive Medicinal Chemistry II* 5:1009–1041.

Testa, Bernard. 2009. Prodrugs: Bridging pharmacodynamic/pharmacokinetic gaps. *Current Opinion in Chemical Biology* 13 (3):338–344.

Todd, Peter, and Heel, Rennie. 1986. Enalapril: A review of its pharmacodynamic and pharmacokinetic properties, and therapeutic use in hypertension and congestive heart failure. *Drugs* 31 (3):198.

Tsukamoto, Yuko, Yukio Kato, Masako Ura, et al. 2001. A physiologically based pharmacokinetic analysis of capecitabine, a triple prodrug of 5-FU, in humans: The mechanism for tumor-selective accumulation of 5-FU. *Pharmaceutical Research* 18 (8):1190–1202.

Wang, Long G, Xiao M Liu, Willi Kreis, and Daniel R Budman. 1998. Androgen antagonistic effect of estramustine phosphate (EMP) metabolites on wild-type and mutated androgen receptor. *Biochemical Pharmacology* 55 (9):1427–1433.

Wermuth, Camille Georges. 2011. *The Practice of Medicinal Chemistry*. Amsterdam, the Netherlands: Academic Press.

Wittine, Karlo, Krešimir Benci, Zrinka Rajić, et al. 2009. The novel phosphoramidate derivatives of NSAID 3-hydroxypropylamides: Synthesis, cytostatic and antiviral activity evaluations. *European Journal of Medicinal Chemistry* 44 (1):143–151.

Wright, V. 1976. A review of benorylate–A new antirheumatic drug. *Scandinavian Journal of Rheumatology* 4 (sup13):5–8.

Wu, Kuei-Meng. 2009. A new classification of prodrugs: Regulatory perspectives. *Pharmaceuticals* 2 (3):77–81.

Yang, Xiao-Zhi, Xiao-Juan Diao, Wen-Hui Yang, et al. 2013. Design, synthesis and antithrombotic evaluation of novel dabigatran prodrugs containing methyl ferulate. *Bioorganic & Medicinal Chemistry Letters* 23 (7):2089–2092.

Zawilska, Jolanta B, Jakub Wojcieszak, and Agnieszka B Olejniczak. 2013. Prodrugs: A challenge for the drug development. *Pharmacological Reports* 65 (1):1–14.

Zhang, Cong-liang, and Yan Wang. 2010. Determination and correlation of 1-octanol/water partition coefficients for six quinolones from 293.15 K to 323.15 K. *Chemical Research in Chinese Universities* 26:636–639.

3

Role of Prodrugs in Drug Design

Asadullah Madni, Sobia Noreen, Afifa Shafique, Abdur Rahim, Muhammad Sarfraz, and Attia Afzal

CONTENTS

3.1 Introduction

In modern research of medicine, the development of prodrugs attained remarkable value as an efficacious method for present-day therapy. Prodrug approach is a practical approach used to amend the fundamental as well as biopharmaceutical properties in order to enhance deliverable payload of therapeutic agent (Rautio et al., 2018). At present, approximately 10% of globally approved drugs are categorized as prodrugs and approximately 1/2 of these prodrugs are converted to an active form by hydroxylation, especially by ester hydrolysis. Furthermore, prodrugs are 1/3rd of worldwide accepted marketed small-molecule drugs (Zawilska et al., 2013, Huttunen et al., 2011). In fact, in the preceding ten years, numerous books on prodrug have been published and thousands of articles in scientific databases are exploring and scrutinizing new potential candidates (Rautio et al., 2010, Redasani and Bari, 2015). Additionally, now lots of prodrugs are under clinical trials. Regardless of this extraordinary record, pharmaceutical technologists have merely initiated to recognize the full prospective of prodrug strategy as an integral part in the modern era of drug innovation and development and numerous versatile prodrug novelties are anticipating their discovery (Vig et al., 2013).

In medical practice, Acetanilide; an antipyretic agent (under the brand name Antifebrin®) is considered the first well-known prodrug, converted to acetaminophen *in vivo* endowed the property of both analgesic and antipyretic by aromatic hydroxylation (Hajnal et al., 2016, Zawilska et al., 2013, Bertolini et al., 2006). Another example of a historical prodrug is Sulfasalazine approved in 1950, still, use as first-line treatment in autoimmune diseases, e.g., Crohn's disease and ulcerative colitis. Sulfasalazine metabolized in the colon by bacteria into its active metabolites 5-aminosalicylic acid and sulfapyridine (Zawilska et al., 2013, Peppercorn, 1984).

A prodrug approach has presented a versatile strategy to increase utility of clinical practicality and developability of voluminous pharmacologically potent molecules by optimizing their pharmacokinetic (ADMET), physicochemical or biopharmaceutical properties. Most common functional groups that can be amenable to the prodrug designing are hydroxyl (–OH), carboxylic (–COOH), carbonyl (–RCOR), amine ($–NH_3$) and phosphate ($–PO_4$) groups. Prodrugs usually formed through alteration of these functional groups including amides, carbonates, oximes, carbamates, esters, and phosphate. Other exceptional functional groups, e.g., thiols, have also been considered in prodrug strategy as theoretically beneficial structures (Simplício et al., 2007, Majumdar and Sloan, 2006, Peyrottes et al., 2004).

Prodrug approach offers opportunities to overcome innumerable barriers for drug delivery and drug formulation. For example, poor water solubility, unpleasant taste or odor, inadequate oral absorption, insufficient chemical instability, presystematic metabolism, concomitant toxicity profile, inappropriate blood-brain barrier penetration, multidrug resistance and local irritation or pain (Rautio et al., 2008, Stella, 2007, Huttunen et al., 2011, Testa and Mayer, 2003, Hajnal et al., 2016). Other terms used for prodrugs are latentiated drugs, biolabile drug carrier conjugates or bioreversible/reversible derivatives but the word prodrug is considered as standard (Notari, 1981, Karaman, 2013c). Prodrug strategy shares an essential part in drug design and discovery paradigm. In fact, prodrugs increase efficacy and reduce associated toxicity of parent drug molecule, that's why must be deliberately contemplate at initial phases of preclinical development and would not be regarded as latter possibility (Vig et al., 2013). The main purpose of designing prodrug is to enhance bioavailability of drug having harsh physical properties, undesirable pharmacokinetic profile and unfavorable physicochemical properties in order to improve metabolic and chemical stability to accomplish intended delivery of a drug. Administration of prodrugs is considered more sophisticated, safe, effective and convenient than conventional dosage forms (Stella, 2011, Stella et al., 2007a). A conventional drug candidate with undesirable properties can also be modified into prodrug with improved characteristics may signify life cycle management opportunities (Rautio et al., 2008). The high importance of the prodrug approach is due to its rational designing for selectivity of the active drug to their intended cell or tissue targets (Ettmayer et al., 2004). Even though the prodrug strategy is considered very challenging, it is more realistic and faster than probing an exclusively new therapeutically active compound with appropriate ADMET properties. The prodrug approach is more attractive and successful consequently fetching more attention of scientists (Rautio et al., 2008).

3.2 Prodrug Concept

The need for prodrug is to incite potential for modification or eradication of shortcomings in parent molecule which are poor water solubility, nonselectivity to target, instability, associated toxicity, disgusting taste and irritation after local administration (Stella, 2011, Testa, 2009, Rautio et al., 2008, Zawilska et al., 2013). Since late nineteenth century, the prodrug approach was used to modify deficient properties of drugs but at the end of the 1950s, Adrien Albert was the first person who formally recognized the term prodrug or pro agent as: "Pharmacologically inactive, bioreversible derived active drug which endures physicochemical alteration to enhance its effectiveness and diminution of its related toxicity" (Albert, 1958, Hajnal et al., 2016, Karaman, 2013a). Bundgaard defined the prodrug as: "A prodrug is formulated to overcome the hindrance in the way to optimal use of the active principle by attaching pro moiety to the active pharmaceutical ingredient." To some extent, this statement is considered narrow because it did not include bioprecursors, modular prodrugs (comprise of trigger, linker and active agent) and conjugated metabolites (Krogsgaard-Larsen, 1991). In 1959, the concept of prodrug was finalized by Harper which presented the new term of "drug latentiation" which mean the drugs specially designed to necessitate bioactivation (Hajnal et al., 2016). This definition is still accepted and fitted in the IUPAC definition of prodrug which states, "A prodrug is a compound that undergoes biotransformation before exhibiting its pharmacological effects" (Wermuth et al., 1998). Prodrugs are significantly less active or inactive cunning precursors of parent drugs invoked to improve deficiencies in the pharmacokinetic profile of a therapeutic drug that has limited formulation choices and consequently deplorable biopharmaceutical performance. Figure 3.1 shows the basic prodrug concept.

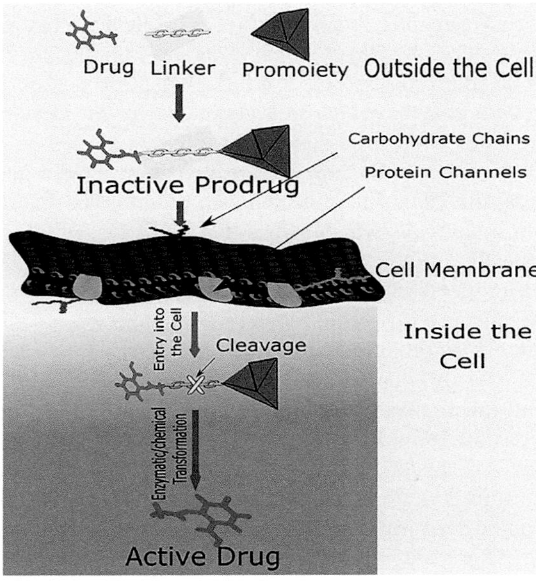

FIGURE 3.1 Basic concept of prodrug.

Prodrug offers a versatile approach to mask the pharmacological activity of the drug and restored *in vivo* before eliciting a therapeutic effect by bioconversion (activation) mediated by enzymes or chemical reactions (Najjar and Karaman, 2019, Walther et al., 2017). A prodrug is not a formulation approach; rather, it's a chemical approach aimed to overcome barrier to the drug's clinical usefulness by improving physicochemical, biopharmaceutical properties. That's why it is considered as an alternative to an analog approach (redesign of drug molecule) (Stella, 2007, Huttunen et al., 2011). We can say that prodrugs are the distinctive nontoxic group comprising drugs that are used to modify or eradicate the objectionable properties from the parental molecule. There is frequent confusion related to terms prodrugs, soft drugs, and codrugs but can be differentiated. In comparison to the prodrug, soft drugs (also known as antedrugs) are pharmacological active isosteric–isoelectronic analogs of lead molecules which after attaining its therapeutic role converted to less pharmacological active form or even entirely deactivated in a controlled or predictable way (Bodor and Buchwald, 2000, Testa, 2004). Soft drug approach is used for versatile applications to decrease the systemic exposure and adverse effects. Cleviprex (Clevidipine butyrate; ultra-short acting calcium channel blocker) is a good example of a soft drug, after intravenous administration in hypertensive patient readily hydrolyzed into pharmacologically inactive form with a half-life of about 1 minute helps attaining the predicted reduction in blood pressure (Maloberti et al., 2018, Ghawanmeh et al., 2018). In contrast, codrugs or mutual prodrugs are pharmacologically active forms of lead compounds as prodrugs. Regardless of being similar to the prodrug, codrugs have two therapeutic compounds bonded directly or indirectly through a covalent chemical linkage and both are promoiety to each other. Salazopyrin (Sulfasalazine; derivative of mesalazine) first clinically approved codrug intended for the treatment of rheumatoid arthritis (RA), which is a combination of 5-amino salicylic acid (5-ASA) and sulfapyridine coupled via azo linkage which is broken down by azoreductase enzyme (Das et al., 2010, Akhani et al., 2017).

3.3 Historical Perspective of Prodrugs

Prodrugs approach has been arguably effective for discovery of numeral clinically used pharmacologically active agents. The very first prodrug was invented in 1899 by Schering which is most probably methenamina (or hexamine). Meanwhile, Aspirin; acetyl-salicylic acid less irritating prodrug approach of sodium salicylate was introduced by Bayer which is an anti-inflammatory drug. But it is still a part of argument that aspirin is a true prodrug or not although aspirin worked as a prodrug by irreversibly inhibiting cyclooxygenase enzyme which is accountable for the development of important biological mediators, thromboxanes and prostaglandins by acetylation of one hydroxyl group of serine at active site of enzyme. Whereas, sodium salicylate is a weak reversible inhibitor of cyclooxygenase (Vane, 2014). That's why aspirin cannot be reflected as a prodrug but readily hydrolyzed to salicylic acid in blood, liver and intestinal walls which shows that aspirin is, in fact, a prodrug (Gilmer et al., 2013). After a long time in 1935, Bayer presented their next prodrug named prontosil; an antibiotic. At that time prontosil prodrug was unintentionally developed but late in the same year by the action of reductive enzyme an active agent; *para*-amino phenylsulfonamide was released made prontosil a true prodrug. Discovery of prontosil as prodrug leads to the discovery of the second generation of sulfonamide because sulfonamide is easily coupled with other promoiety (Bentley, 2009). In 1952, the prodrug activity of isoniazid; an anti-tuberculosis drug was introduced by Roche. The mycobacterial catalase-peroxidase enzyme is responsible for bioactivation of isoniazid in the human body after administration which inhibits the biosynthesis of mycolic acid for the synthesis of the mycobacterial cell wall (Timmins and Deretic, 2006). Heroin (diacetylmorphine) was unintentionally discovered prodrug, and was marketed as nonaddictive morphine derivative used to suppress cough as well as to cure addiction caused by both cocaine and morphine. After a long time, the appealing truth of heroin that after oral administration it is rapidly metabolized into the active form (morphine) was appreciated by Bayer (Karaman, 2013b, Huttunen et al., 2011).

Prodrug approach has been attained explosive success in drug discovery and development since 1960. In 1963, out of 43 newly approved drugs, five were registered as prodrugs (Stella, 2007). The real breakthrough in prodrugs discovery and development process was started at the beginning of the twenty-first century when property-centered drug design came to be the crucial part of the discovery progression (Van De Waterbeemd et al., 2001). It is revealed by primary and secondary literature sources such as paper journals, publications, scientific literatures, reviews, conference proceedings, patents and marketed medicines. In addition, heaps of prodrugs are undertaking clinical trials (Hajnal et al., 2016).

In 2001, 25 new therapeutic biological or chemical entities were reported as approved drugs and three were clearly reported as prodrugs while two other compounds may behave as prodrugs. Thirty-three NCEs and NBEs were approved, four were prodrugs while one almost certainly behaved as a prodrug. Therefore, during the last two years (2001 and 2002), approximately 14% of total chemical and biological entities were prodrugs and 22% possible prodrugs were approved worldwide (Bernardelli et al., 2002, Boyer-Joubert et al., 2003).

In 2005, 24 of the new chemical entities were approved by FDA, 19 were labeled as small molecular weight drugs in which the number of prodrugs was 2 (10.5%). Instead of these two other antiviral small molecular weight drugs was probably acting as a prodrug. Technically these were also prodrugs. Thus, in 2005, 26.3% of overall approved drugs were considered prodrugs (Stella et al., 2007b). During the year 2008, eight approved drugs were labeled as prodrugs account 33% of all FDA approved drugs. Successful implementation of prodrug strategy was seen in 2009, as approximately 15% of 100 best marketed drugs were prodrugs (Rautio, 2011, Hughes, 2009, 2010). Majority of prodrugs were approved and marketed in 2010 to enhance the bioavailability (Mullard, 2011). Between the five-year period of 2010 to 2014, 127 new small molecular weight drugs were approved worldwide and 13 were clearly declared as prodrugs. These 13 prodrugs are Fingolimod, Dabigatran etexilate, Ceftarolinefosamil, Azilsartanmedoximil, Gabapentin enacarpil, Abiraterone acetate, Fidaxomicin, Tafluprost, Dimethyl fumarate, Eslicarbazepine Acetate, Sofosbuvir, Droxidopa and Tedizolid phosphate (Mullard, 2012, 2013, 2014, 2015, Rautio et al., 2008).

A substantial increase in the prodrugs over time can be perceived, signifying an enlarged interest in prodrug strategy. The prevalence of prodrugs in 2015 was 15% among all the FDA approved drugs while 20% among the small molecular weight drugs. Two novel prodrugs strategies approved in 2015 were Isavuconazonium sulfate and Sacubitril prepared to improve the solubility and permeability through increased lipophilicity respectively. While other prodrugs approved by FDA in 2015 were Uridine triacetate, Aripiprazole lauroxil, Tenofovir alafenamide, Ixazomib, Selexipag and Telotristatetiprate (Rautio et al., 2017, Mullard, 2016). No prodrug was approved in 2016 (Mullard, 2017). In 2017, five new chemical entities were listed as prodrugs (Mullard, 2018a, 2018b). Three drugs (tafenoquine and fosnetupitant Baloxavirmarboxil) have been registered as prodrugs in 2018. Numerous prodrugs are going through the process of phase II clinical trials. There were a considerable number of prodrugs approved and marketed during last 8 years shown in Table 3.1.

However, the gap lays in the challenges during the discovery of prodrugs has to keep into consideration.

3.4 Classification of Prodrugs

3.4.1 Conventional Prodrugs

3.4.1.1 Carrier-Linked Prodrugs

In the carrier-linked prodrugs, active drug and carrier is temporarily linked by a bioreversible covalent linkage and release the drug into the body through biotransformation. For the attachment of carrier moiety, as a minimum there must be 1 functional group in the drug molecule. Most commonly hydroxyl or amino groups are employed as functional groups; however, carbonyl or carboxylic acids functional

TABLE 3.1

Prodrugs Approved by the FDA from 2010 to 2018

Approval Date	Prodrug Name (Trade Names)	Active Metabolite	Indications	Improved Property	References
21 Sep 2010	Fingolimod (Gilenya®)	Fingolimod phosphate	Multiple sclerosis	*In vivo* phosphorylation	Gonzalez-Cabrera et al. (2012)
19 Oct 2010	Dabigatran Etexilate (Pradaxa®)	Dabigatran	Thromboembolism acute coronary syndrome	Increase in permeation and solubility	Eisert et al. (2010), Sorbera et al. (2005)
29 Oct 2010	Ceftaroline Fosamil (Teflaro®)	Ceftaroline	Acute bacterial skin, skin structure infections & community acquired pneumonia	Improve aqueous solubility	Mullard (2011), Ishikawa et al. (2003)
25 Feb 2011	Azilsartan Medoxomil (Edarbi®)	Azilsartan	Hypertension	Changes to metabolic profile	Perry (2012)
6 Apr 2011	Gabapentin Enacarbil (Horizant®)	Gabapentin	Restless leg syndrome, post-herpetic neuralgia and calcium channel modulator	Increased permeation	Mullard (2012), Agarwal et al. (2010)
28 Apr 2011	Abiraterone Acetate (Zytiga®)	Abiraterone sulfate and N-oxide abiraterone sulfate	Hormone refractory prostate cancer	Bioavailability (probably increase in permeation)	Acharya et al. (2013)
27 May 2011	Fidaxomicin (Dificid®)	OP-1118	*Clostridium difficile* infection (rCDI)	Naturally occurring macrocycle to form its microbiologically active metabolite OP-1118	Yanagihara et al. (2018)
10 Feb 2012	Tafluprost (Zioptan®)	Tafluprost acid	Elevated intraocular pressure	Increased permeation	Mullard (2013), Fukano and Kawazu (2009)
27 Mar 2013	Dimethyl Fumarate (Tecfidera®)	Monomethy fumarate	Multiple sclerosis	Increased permeation	Rosenkranz et al. (2015)
08 Nov 2013	Eslicarbazepine Acetate (Aptiom®)	Eslicarbazepine	Focal epilepsy	Avoid the formation of epoxide	Soares-da-Silva et al. (2015)
6 Dec 2013	Sofosbuvir (Sovaldi®)	GS-461203	Hepatitis C infection	Phosphoramidate prodrug	Mullard (2014)
20 June 2014	Tedizolid Phosphate (Sivextro®)	Tedizolid	Acute bacterial skin & skin infections	SAR design, improved aqueous solubility	Mullard (2015)
6 Mar 2015	Isavuconazonium Sulfate (Cresemba®)	Isavuconazole	Invasive aspergillosis & invasive mucormycosis	Improved aqueous solubility	Kovanda et al. (2016)
7 July 2015	Sacubitril (Entresto®)	LBQ657	Heart failure	Increased permeability	Hanna et al. (2018)

(*Continued*)

TABLE 3.1 (*Continued*)

Prodrugs Approved by the FDA from 2010 to 2018

Approval Date	Prodrug Name (Trade Names)	Active Metabolite	Indications	Improved Property	References
5 Nov 2015	Tenofovir alafenamide (Genvoya Odefseym Descovy®)	Tenofovir	HIV	Phosphoramidate prodrug & Increased lipophilicity	Walker et al. (2016)
20 Nov 2015	Ixazomib citrate (Ninlaro®)	Active boronic form	Multiple myeloma	Improved affinity	Shirley (2016)
22 Dec 2015	Selexipag (Uptravi®)	ACT-333679 (MRE-269)	Pulmonary arterial hypertension	Improved bioavailability	Mullard (2016)
2016	No drug approved				Najjar and Karaman (2019)
9 Feb 2017	Deflazacort (Emflaza®)	21-desacetyldeflazacort	Duchenne muscular dystrophy	Improved efficacy profile	
28 Feb 2017	Telotristat etiprate (Xermelo®)	Telotristat	Carcinoid syndrome Diarrhea	Increased permeability	Pavel et al. (2015)
28 Feb 2017	Telotristat ethyl (Xermelo®)	Lp-778902	Carcinoid syndrome diarrhea	Improved bioavailability	Kulke et al. (2017)
3 Apr 2017	Deutetrabenazine (Austedo®)	α-dihydrotetrabenazine and β-dihydrotetrabenazine	Tourette syndrome	Deuterated to retard hepatic metabolism	Jankovic et al. (2016)
11 Apr 2017	Valbenazine Tosylate (Ingrezza®)	α-dihydrotetrabenazine	Tardive dyskinesia	Improved pharmacokinetics	Uhlyar and Rey (2018)
29 Aug 2017	Benznidazole (Benznidazole®)	Various electrophilic metabolites	Chagas disease	Improve aqueous solubility	Soy et al. (2015)
15 Sep 2017	Secnidazole (Solosec®)	(Active metabolite)	Bacterial vaginosis	Improve solubility	Nyirjesy and Schwebke (2018)
2 Nov 2017	Latanoprostene Bunod (Vyzulta®)	Latanoprost acid Butanediol mononitrate	Glaucoma	Improve corneal penetration	Hoy (2018)
17 Apr 2018	Fosnetupitant (Akynzeo®)	Netupitant	Chemotherapy induced nausea and vomiting	Phosphorylation for IV injection	Hossain (2018)
20 July 2018	Tafenoquine (Krintafel®)	5,6 ortho-quinone tafenoquine	Malaria		Frampton (2018)
1 Oct 2018	Baloxavir Marboxil (Xofluza®)	S-033447	Influenza A and B	Inhibits viral shedding through the inhibition of viral CAP endonuclease	Mullar (2019), O'Hanlon and Shaw (2019)

groups are also in practice. Among these functionalities hydrolysis conditions and attachment, chemistry can vary markedly. To deliver the parent drug into the body the linkage is cleaved either nonenzymatically or enzymatically (e.g., labile amide or an ester). Carriers are mostly hydrophobic in nature and usually a small molecule, e.g., a fatty chain, a polymer or PEG, or a macromolecule, like an albumin or an antibody. The promoiety (active drug) must be labile for in vivo efficient activation while pharmacologically inactive, nontoxic, and nonimmunogenic groups are removed (Jana et al., 2010, Abet et al., 2017).

Depending upon the nature of the carrier used, the carrier-linked prodrug may further be classified into the followings. **(Scheme 3.1 shows the classification of prodrugs.)**

Bipartite**:** In bipartite prodrugs a carrier, i.e., a group, is directly attached to the drug. This prodrug carrier complex has significantly altered the lipophilicity of the prodrugs. As a result of chemical or enzymatically cleavage active drug is released into the body, e.g., Tolmetin-glycine prodrug.

Tripartite**:** If a spacer or group is sandwiched between the drugs and promoiety then the prodrug is of tripartite type. Occasionally, an intrinsic action of the drug-promoiety linkage leads to the instability of bipartite type prodrugs. So the problem is overcome by the formation of tripartite prodrug.

Mutual prodrug**:** In mutual prodrugs the carrier is interactively work with the drug to which it is linked. The carrier employed here is another drug used in synergistic manner with the associated drug. In mutual prodrugs, both bipartite and tripartite strategy can be used. These drugs are so designed that each one of them functions as a promoiety for the other one and vice versa. The second drug (carrier) aids the parent drug to target at a specific site. In mutual prodrug

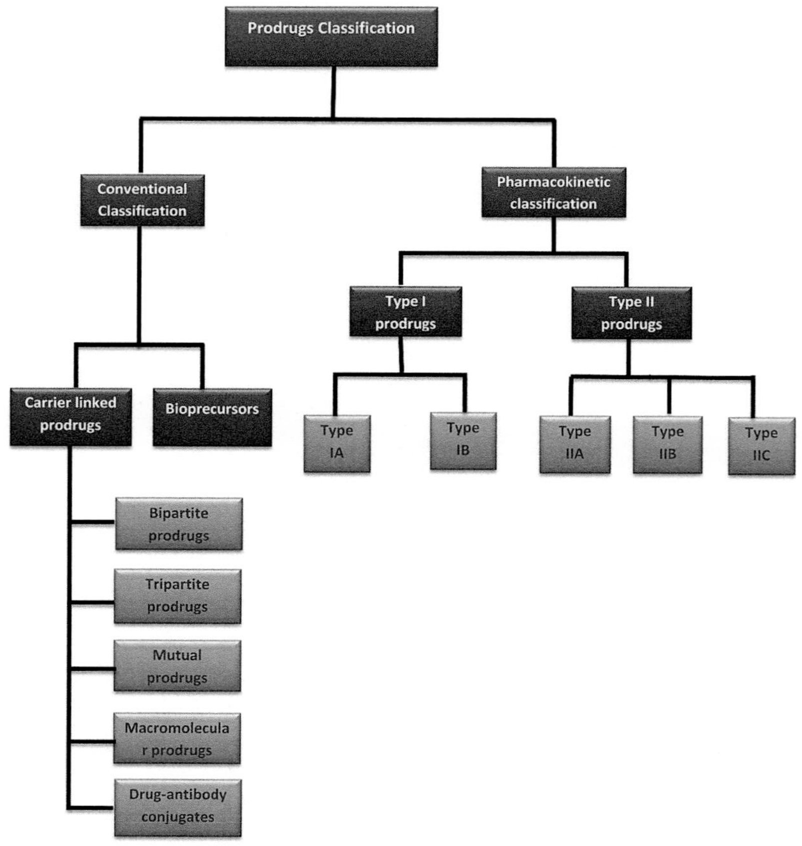

SCHEME 3.1 Schematic illustration of classification of prodrugs.

type the next used drug can be selected from same or different therapeutic class. Numbers of side effects are decreased by mutual prodrug type. In mutual prodrug, the carrier drug either have a similar MOA to the parent drug and act on the same biological target or have different mechanisms of action by acting on different biological targets.

***Macromolecular prodrugs*:** When a macromolecule is used as a carrier then the type is known macromolecular prodrugs. PEG (polyethyleneglycol) is used as a carrier in this type of prodrug.

***Drug–antibody conjugates*:** For tumor cells, mostly different types of treatment protocols are used. In some cases, antibodies are used to target tumor cells. When antibodies are used a carrier then this is drug–antibody conjugates type prodrug (Dubey and Valecha, 2014).

3.4.1.2 Bioprecursor Prodrugs

Bioprecursors don't hold a promoiety and trigger active drug itself. Activation of bioprecursor prodrug is done via metabolic modification of functional group, e.g., oxidation, reduction or hydrolysis. Chemically or metabolic transformation of active drug leads to the formation of new compounds. Such as, to produce the drug having carboxylic functional group, bioprecursor with alcoholic group is employed. It is then metabolized to aldehydes through oxidation and after that into the carboxylic acid drug. As a result of phase I (oxidation, reduction or phosphorylation) and phase II conjugation reactions pharmacologically active compounds are produced. Bioprecursor prodrugs are highly used for antibacterial and chemotherapeutic agents that are activated by reductases. Examples of these prodrugs are quinones, nitroarenes amidoximes, Pt(IV) complexes and N-oxides (Silverman, 2004, Wermuth, 2008, Choi-Sledeski and Wermuth, 2015, Kokil et al., 2010).

3.4.2 Based on Pharmacokinetic Properties

Another system of classification of prodrugs is based on PK properties of drugs. Kinetic analysis and conversion of prodrug into active drug in the body is the basis principle of this classification system. Furthermore, information about the safety and efficacy profile of prodrug and active drug is discussed under this heading. Cellular sites of conversion further classify the types into subtypes, i.e., type I and type II (Wu, 2009).

3.4.2.1 Type I prodrugs

Type I prodrugs are metabolized intracellularly. A prodrug that is transformed simultaneously in both target cells and metabolic tissues is nominated as a Type I A/I B prodrug (e.g., antiviral nucleoside analogs, HMG Co-A reductase inhibitors, statins).

***Type I A prodrugs*:** Type I A prodrugs are metabolized at the cellular site actions. These prodrugs are converted into active drug at therapeutic action site. Examples include many antibacterial and cancerous drugs (e.g., acyclovir, cyclophosphamide, 5-fluorouracil, zidovudine and Levo-DOPA).

***Type I B prodrugs*:** Type I B agents are metabolized into the liver by using hepatic enzymes. Prodrug is converted intracellularly into active drugs via liver cells. CCB (captopril), anticonvulsant medication(carbamazepine), Barbiturates (primidone) are few examples of this type (Wu, 2009).

3.4.2.2 Type II prodrugs

Type II prodrugs are activated extra-cellularly depending on some common enzymes. Phosphatases and esterases are the enzymes used to metabolize type II prodrugs. Type II prodrugs are converted successively, such as primarily in the gastrointestinal fluids and secondly into the target cells (e.g., antibody-, gene-, or virus-directed enzyme prodrugs [ADEP/GDEP/VDEP] for chemotherapy or immunotherapy) (Wu and Farrelly, 2007).

***Type II A*:** In this type of prodrugs, extracellular conversion takes place in the setting of the gastrointestinal fluid (e.g., loperamide oxide, sulfsalazine).

***Type II B*:** At extracellular fluid compartments metabolic conversion of prodrugs takes place. Sometimes conversion process occurs inside the circulatory system (e.g., aspirin, bambuterol, fosphenytoin).

***Type II C*:** In type II C prodrugs, active drug is formed near or inside therapeutic target/cells (ADEPT, GDEPT) (Wu and Farrelly, 2007, Jana et al., 2010).

3.5 Approaches of Prodrugs in Drug Designing

Prodrug is the end product of active drug moieties, they are so designed that release the parent drug (active form) in the body by metabolic conversion. In a prodrug approach a covalent bond is placed between carrier and drug in this fashion that upon oral administration, the particle retains its integrity in the harsh environment of the body and breakdown into the active drug molecule as it reaches the desired site. For the delivery of drug at specific site prodrug activation is achieved through a number of strategies, for example, by modifying pH at target site, by enzymatic activation at target site for the prodrug–drug conversion (Jain et al., 2006, Arik et al., 2014).

Following are some ideal characteristics for a prodrug

- Must be inactive against any pharmacological target
- At different environments (pH range) it must maintain its chemical stability
- Aqueous solubility of prodrug must be higher
- Prodrug should have excellent transcellular absorption
- Rapid conversion of prodrug into active drug
- In the course of absorption phase, it must be resistance to hydrolysis
- One of the most important purpose of a prodrug approach is to increase the oral absorption by increasing lipophilicity and overall membrane permeability
- There should be quantitative breakdown of prodrug to produce high amount of the active moieties after post absorption (Jana et al., 2010).

For improving oral drug delivery some common prodrug approaches are discussed with detail mechanisms in the subsequent sections, including enhanced aqueous solubility, improved lipophilicity and permeability, prodrug for prolonged duration of action and target-based prodrug design.

3.5.1 Prodrugs with Improved Aqueous Solubility

One of the critical parameters for drug development is solubility and can be exploited thru a prodrug strategy. Drugs having poor dissolution rate and intestinal solubility are more prone to this strategy (solubility enhancement). Hepatic first pass metabolism has no concern to this strategy. BCS Class II drugs with low solubility and high permeability are idea candidates for improving solubility. Data according to combinatorial screening programs reveals that nearly about 40% of the drugs candidates have poor aqueous solubility (<5 μg/mL) (Lipinski, 2002). Clinically, the majority of prodrugs are used with the aim of enhancing drug absorption by increasing lipophilicity or by improving water solubility. Frequently, during pharmaceutical formulation development scientists faced serious problems related to organoleptic properties and poor drug solubility. However, pharmaceutical sciences resolve such issues by using different technologies to improve solubility such as following.

3.5.1.1 Techniques for Solubility Enhancement

Solubility enhancement can be achieved through a number of techniques, i.e., physical modifications, chemical modifications, and miscellaneous techniques (Savjani et al., 2012).

3.5.1.1.1 Physical Modifications

- Reduction in particle size reduction by using micronization and nanosuspension technology
- Alteration in crystal habits by using polymorphs, amorphous form, and co-crystallization
- Using a carrier for drug dispersion for example eutectic mixtures, solid dispersions, solid solutions, and cryogenic techniques

3.5.1.1.2 Chemical Modifications

- By using buffer
- Formation of drug derivatives
- pH modification
- Formation of salts
- Complex formation

3.5.1.1.3 Miscellaneous Methods

- Prodrug approach
- SCF (Supercritical fluid) process
- Functional polymer technology
- Cryogenic techniques
- Precipitation porous
- Microparticles technology
- Nanoparticle approach
- By using specific adjuvants like
    - Surface active agent
    - Co-solvents methods
    - Solubilizing agents
    - Hydrotrophy
    - Novel excipient

Prodrug approach for solubility enhancement is briefly described here as it is our topic of concern. Oral absorption and water solubility of drugs can be increased by integrating ionic groups into the prodrug design as shown in Figure 3.2.

Though, produced molecules probably have poor permeation through the intestinal epithelial cells of cellular membrane (Mueller, 2009). Another technique is phosphate ester prodrug strategy having alcohol group in its structure. In order to improve solubility of poor water-soluble parent drugs after oral administration phosphate ester bonds have been added. The intention for the attainment of this strategy (phosphate esters) is that phosphate ester prodrugs (R–OPO-3) are converted to parent active drug (R–OH) by the brush-border enzyme, alkaline phosphatase. Latterly, passive absorption of R–OH occurs leads to increased oral absorption (Heimbach et al., 2003a).

3.5.2 Prodrugs with Improved Lipophilicity and Permeability

The permeability of the prodrugs is directly linked with HLB value. In general, the drugs with high HLB value or hydrophilicity are associated with low permeability owing to the highly lipophilic behavior of biological membrane (Tojo et al., 1987). To overcome the low permeability of the polar drugs, the polar or ionic structural parts of the parent drugs are masked or made lipophilic. In instance of prodrugs the parent drugs are made lipophilic by masking the parent drug. The prodrugs are mostly linked with alkyl esters as a functional group in the parent drugs

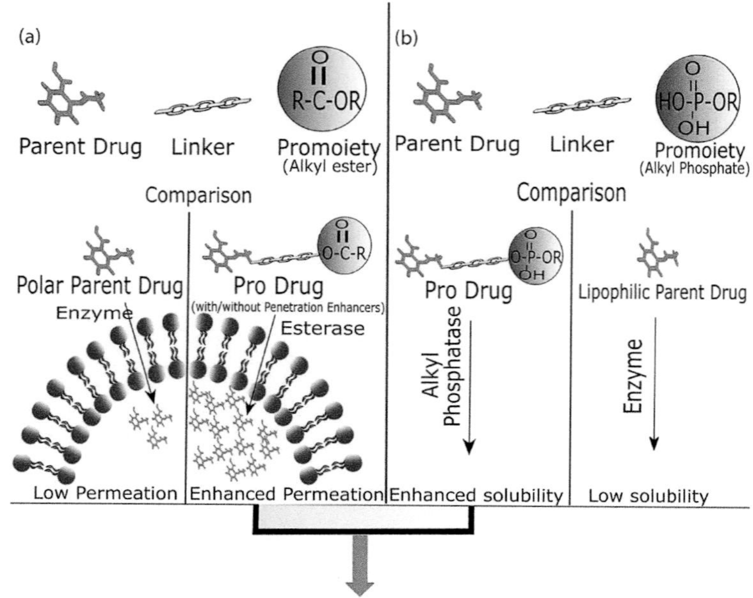

FIGURE 3.2 (a) Prodrug approach for permeation enhancement and (b) Prodrug approach for aqueous solubility enhancement.

of BCS class III drugs (Stinchcomb et al., 1996). The activation of the prodrugs by biological fluids are done by esterases and collectively the 2–20-folds enhancement in oral bioavailability of the parent drugs has been reported (Hodge et al., 1989, Ehrnebo et al., 1979). The morphine contains two hydroxyl groups at position 3 and 6 and it has been reported that esterification at position 3 through prodrug results in successful enhancement in the permeation of prodrug across the skin (Mignat et al., 1996). Likewise, the permeation of the naltrexone was also enhanced through prodrug technique by esterification at position 3 and in turn enhancement in bioavailability was achieved (Stinchcomb et al., 2002). The permeation of parent drug nalbuphine was increased by the prodrug nalbuphine benzoate (Wang et al., 2006). The alkyl ester and amide forms of ketorolac showed higher permeation across the skin than parent drug ketorolac. More than 11-fold enhancement of permeation of one propyl ester of ketorolac across the skin was achieved through prodrug (Doh et al., 2003, Kim et al., 2005). The polyethylene glycol esters also have taken part in the improvement in the permeability of various NSAIDS (Bonina et al., 2001). The permeation of various antiviral drugs like Acyclovir and IDU (5-iodo-2′-deoxyuridine) across the skin has been successfully enhanced through prodrugs with/without incorporation of penetration enhancers (Gupta et al., 2005, Bando et al., 1996). Similarly, permeation enhancement of ACE inhibitors, vitamins, antibiotics, anti-influenza, corticosteroids and antiacne drugs through prodrugs have also been reported (Fang and Leu, 2006, Rautio et al., 2008).

3.5.3 Prodrug for Prolonged Duration of Action

Although limited data is available regarding prolonged duration of action through prodrugs, still the reported data reveals that prolonged duration of action could be achieved through oral, parenteral and topical drug delivery routes (Gomez-Orellana, 2005). In general, chemical modification causes the improvement in half-life and prolonged duration of action (Oliyai and Stella, 1993).

through oxidation and reduction reactions (Guengerich, 2001). Hepatic enzymes cytochrome P450 metabolized 75% of drugs and prodrug and release the active drug into the hepatocyte. Cytochrome P450 metabolism is detected as another major challenge for prodrug development leads to poor oral bioavailability. Furthermore, hepatocyte contains Phase I and II metabolizing enzymes which metabolize the prodrug in a nonproductive manner. Consequently, this metabolism the active drug does not transfer to the blood and persuade decrease in oral bioavailability (Beaumont et al., 2003). The latest development to progress local activation of CYP-activated prodrugs is the use of enzyme antibodies (antibody-directed enzyme prodrug therapy, ADEPT), or enzyme encoding genes (gene-directed enzyme prodrug therapy, GDEPT). These strategies are mostly employed for anticancerous drugs, to decrease their toxicity at other body organs. In these types of approaches, prodrugs are activated by specific exogenous enzymes, which are either conjugated to target cell by selective monoclonal antibodies or delivered to the target cell as a gene that encodes the prodrug activating enzyme (Roy and Waxman, 2006, Chen and Waxman, 2002, Huttunen et al., 2008).

3.6.6 Parenteral Formulations Challenges

As compared to oral formulations IV formulations have some exceptional challenges. These formulations require sterile environment for production process. Second major challenge is that a parenteral formulation essentially needs to be free from significant particulates. Another major challenge is chemical stability, shelf-life maintenance and precipitation limit of many prodrug solutions. To overcome these problems parenteral formulations are lyophilized for reconstitution (Shi et al., 2009, Baker and Naguib, 2005).

3.7 Applications of the Prodrug in Drug Designing

3.7.1 Pharmaceutical Applications

Prodrug strategy is magnificently used to overwhelm the innumerable pharmaceutical barriers, enlightening therapeutic worth of pharmacologically active drugs. Some pharmaceutical applications of the prodrug approach are as follows:

Prodrug for concealing taste and odor: Obnoxious taste and odor are one of the problems associated with patient noncompliance, emerge as a result of drug solubility and interaction with that of taste receptors, and can be resolved by formation of prodrug (Vummaneni and Nagpal, 2012). Alteration of Chloramphenicol to Chloramphenicol palmitate makes it tasteless and converted to active drug *in vivo* by pancreatic lipase, is one of the examples of prodrug used for masking taste. Other drugs with bitter taste are nonsteroidal anti-inflammatory drugs (NSAIDs) and antibiotics (penicillin, macrolide) (Ayenew et al., 2009, Shelke et al., 2019, Karaman, 2013d).

Prodrug for reducing local pain or irritation: Another objective of prodrug usage is to restrict drugs for causing local pain or irritation at injection site specifically after IM injection, instigated via precipitation and allocation of administered drug into adjoining cells. Intramuscular injection of clindamycin and phenytoin was painful but the formation of phosphate ester prodrugs helps to overcome this problem (Grandhi and Abd-Elsayed, 2019, Shah et al., 2017, Morozowich and Karnes, 2007).

Prodrug for better chemical stability: Pharmaceutical companies are struggling to formulate therapeutic agents for producing its pharmacological activity for persist duration. Modern methods are used for the designing of innovative prodrugs of active pharmacological agents with chemical instability, achieved by modifying the functional groups or by altering the physical properties of parent drugs. Hetacillin, a prodrug of ampicillin, obstructs auto aminolysis *in vivo* is a good approach for chemical stability of antibiotic ampicillin (Lohar et al., 2012, Hu, 2016).

3.7.2 Pharmacokinetic Applications

Utilizing the prodrug approach has the potential to improve pharmacokinetic barriers in the way of the effectiveness of drugs by optimizing ADME properties. Some pharmacokinetic applications of prodrug approach are listed:

Prodrugs for improving drug absorption: One of the significant applications of the prodrug is the advancement of absorption of drugs administered by the oral, percutaneous and ophthalmic route. The main reason for the low absorption of therapeutical agents is high polarity and meager lipophilicity consequently results in a reduction of bioavailability. Prodrugs are successfully implanted to enhance water solubility and oral absorption of poor water-soluble drugs by integrating ionic groups into prodrug designing (Fleisher et al., 1996, Jana et al., 2010). Fosamprenavir is the phosphate ester of amprenavir (HIV protease inhibitor), illustrate better aqueous solubility and oral bioavailability which is greater than parent drug (Brouwers et al., 2007). Enhanced therapeutic usefulness of epinephrine by ophthalmic route is achieved by using the prodrug approach by the acylation of phenolic hydroxyl groups (Hussain and Truelove, 1976, Barot et al., 2012). Mefenide hydrochloride and mefenide acetate salts formed by prodrug approach showed better results in inflammatory, burn therapy, allergy, and pruritic conditions than the parent drug molecule (Ghadiri et al., 2014).

Prodrugs for increasing aqueous solubility: A prodrug approach is aimed to boost water solubility of poor water-soluble drugs to a level by decreasing melting point and tallying ionizable or polar promoiety to parent drug molecule (Krishnaiah, 2010, Järvinen et al., 2010, Lohar et al., 2012). For example, estramustine phosphate is phosphate ester prodrug of estramustine used for the treatment of prostate carcinoma (Inoue et al., 2016). The water-soluble prodrug is metabolized into two active metabolites *in vivo*, namely estradiol and estrone. The bioconversion of intravenously administered estramustine phosphate prodrug produces a dual antimicrotubule and antigonadotropic activity (Heimbach et al., 2003b, Zawilska et al., 2013).

Prodrugs for prolong duration of drug action: Sustained plasma levels and consequently prolonged of duration of action might be achieved by applying the prodrug approach by formation of long chain aliphatic esters. Esters are prone to hydrolyzed slowly and mostly used for intramuscular injection formulations. Fluphenazine (neuroleptic agent) which is between 24 to 72 hours duration of action, while the release of fluphenazine decanoate (ester prodrug) gradually sustains for 7 days to 2 months with an average period of 1 month. Prodrugs of various steroids (e.g., testosterone nandrolone) are slowly released by intramuscularly injection and consequently prolonged duration of action is achieved (Verma et al., 2009, Gaekens et al., 2016, Lohar et al., 2012).

Prodrugs for targeted delivery: The prodrug strategy conquers attention as a module for advancement in site-specific or targeted drug delivery. Site-specific bioactivation/bioconversion of drugs or site-directed delivery of drugs are two phenomena used to achieve targeted delivery of therapeutic agents. In prodrug strategy, both site specific-drug bioactivation and site-directed drug delivery has been combined to accomplish effective targeted dug delivery. Tumor, kidney, liver, colon and brain targeting are the most common prodrug applications in field of targeted drug delivery. Levodopa is a substrate of neutral amino acid transporter, available in market with brand name Dopar® works as the targeted delivery into central dopaminergic neurons, rapidly converted into dopamine by the action of enzyme facilitating pharmacodynamic effects (Gomes et al., 2007). Prodrug sulphasalazine is an example of colonic targeting; delivered in colon as intact form and converted to its active metabolite 5-amino salicylic acid by azo reductase enzyme present in colonic bacterial microflora prevented precolonic absorption related side effects (Bhosle et al., 2006, Shinde et al., 2015).

3.8 Conclusion and Future Prospects

A prodrug approach is one of the favorable strategies used for pharmacologically active agents to improve their solubility, permeability, stability, bioavailability and for sustained and targeted drug delivery. Hence, not surprisingly, prodrugs are turning into a suitable candidate for the drug discovery archetype. Nonetheless, developing a prodrug can still be a more practical and productive approach rather than investigating for a completely new drug moiety with suitable ADME properties. It is anticipated that at initial phases of the drug discovery prodrug approaches and their strategies give rise to the progress of prodrugs with better potential applications. Despite the incredible evolution made in the domain of prodrug development, additional studies are evidently needed for prodrugs design and development. Furthermore, understanding of prodrug development processes and challenges during formulation development is another milestone experienced in this field. It is envisioned that prodrug technology holds the potential to produce harmless and successful provision of a variety of active therapeutic moieties in the upcoming arena.

REFERENCES

Abet, V., Filace, F., Recio, J., Alvarez-Builla, J. & Burgos, C. 2017. Prodrug approach: An overview of recent cases. *European Journal of Medicinal Chemistry* 127:810–27.

Acharya, M., Gonzalez, M., Mannens, G., et al. 2013. A phase I, open-label, single-dose, mass balance study of 14C-labeled abiraterone acetate in healthy male subjects. *Xenobiotica* 43:379–89.

Agarwal, P., Griffith, A., Costantino, H.R. & Vaish, N. 2010. Gabapentin enacarbil–clinical efficacy in restless legs syndrome. *Neuropsychiatric Disease and Treatment* 6:151.

Akhani, P., Thakkar, A., Shah, H., et al. 2017. Correlation approach of pro-drug and co-drug in biotransformation. *European Journal of Pharmaceutical and Medical Research* 4:488–500.

Albert, A. 1958. Chemical aspects of selective toxicity. *Nature* 182:421.

Arik, D., Zimmermann, E. & Shimon, B.S. 2014. Modern prodrug design for targeted oral drug delivery. *Molecules* 19:16489–505.

Ayenew, Z., Puri, V., Kumar, L. & Bansal, A.K. 2009. Trends in pharmaceutical taste masking technologies: A patent review. *Recent Patents on Drug Delivery & Formulation* 3:26–39.

Baker, D.C., Haskell, T.H. & Putt, S.R. 1978. Prodrugs of 9-.beta.-D-arabinofuranosyladenine. 1. Synthesis and evaluation of some 5′-(O-acyl) derivatives. *Journal of Medicinal Chemistry* 21:1218–21.

Baker, M.T. & Naguib, M. 2005. Propofol The Challenges of Formulation. *Anesthesiology: The Journal of the American Society of Anesthesiologists* 103:860–76.

Bando, H., Takagi, T., Yamashita, F., Takakura, Y. & Hashida, M. 1996. Theoretical design of prodrug-enhancer combination based on a skin diffusion model: Prediction of permeation of acyclovir prodrugs treated with l-geranylazacycloheptan-2-one. *Pharmaceutical Research* 13:427–32.

Barot, M., Bagui, M., Gokulgandhi, M.R. & Mitra, A.K. 2012. Prodrug strategies in ocular drug delivery. *Medicinal Chemistry* 8:753–68.

Beaumont, K., Webster, R., Gardner, I. & Dack, K. 2003. Design of ester prodrugs to enhance oral absorption of poorly permeable compounds: Challenges to the discovery scientist. *Current Drug Metabolism* 4:461–85.

Bentley, R. 2009. Different roads to discovery; Prontosil (hence sulfa drugs) and penicillin (hence β-lactams). *Journal of Industrial Microbiology & Biotechnology* 36:775–86.

Bernardelli, P., Gaudillière, B. & Vergne, F. 2002. Trends and perspectives, Chapter 28: To market, to market-2000. *Annual Reports in Medicinal Chemistry* 36:293–318.

Bertolini, A., Ferrari, A., Ottani, A., et al. 2006. Paracetamol: New vistas of an old drug. *CNS Drug Reviews* 12:250–75.

Bhosle, D., Bharambe, S., Gairola, N. & Dhaneshwar, S.S. 2006. Mutual prodrug concept: Fundamentals and applications. *Indian Journal of Pharmaceutical Sciences* 68(3):286–94.

Bobeck, D.R., Schinazi, R.F. & Coats, S.J. 2010. Advances in nucleoside monophosphate prodrugs as anti-HCV agents. *Antiviral Therapy* 15:935–50.

Bodor, N. & Buchwald, P. 2000. Soft drug design: General principles and recent applications. *Medicinal Research Reviews* 20:58–101.

Bonina, F.P., Puglia, C., Barbuzzi, T., et al. 2001. In vitro and in vivo evaluation of polyoxyethylene esters as dermal prodrugs of ketoprofen, naproxen and diclofenac. *European Journal of Pharmaceutical Sciences* 14:123–34.

Boyer-Joubert, C., Lorthiois, E. & Moreau, F. 2003. Trends and perspectives, Chapter 33: To market, to market-2002. *Annual Reports in Medicinal Chemistry* 38:347–74.

Brass, E.P. 2002. Pivalate-generating prodrugs and carnitine homeostasis in man. *Pharmacological Reviews* 54:589–98.

Brouwers, J., Tack, J. & Augustijns, P. 2007. In vitro behavior of a phosphate ester prodrug of amprenavir in human intestinal fluids and in the Caco-2 system: Illustration of intraluminal supersaturation. *International Journal of Pharmaceutics* 336:302–9.

Chen, L. & Waxman, D.J. 2002. Cytochrome P450 gene-directed enzyme prodrug therapy (GDEPT) for cancer. *Current Pharmaceutical Design* 8:1405–16.

Chin, C.F., Tian, Q., Setyawati, M.I., et al. 2012. Tuning the activity of platinum (IV) anticancer complexes through asymmetric acylation. *Journal of Medicinal Chemistry* 55:7571–82.

Chin, C.F., Yap, S.Q., Li, J., Pastorin, G. & Ang, W.H. 2014. Ratiometric delivery of cisplatin and doxorubicin using tumour-targeting carbon-nanotubes entrapping platinum (IV) prodrugs. *Chemical Science* 5:2265–70.

Choi-Sledeski, Y.M. & Wermuth, C.G. 2015. Designing prodrugs and bioprecursors. *The Practice of Medicinal Chemistry*. Elsevier.

da Silva, S.S., Igne, F.E. & Giarolla, J. 2016. Dendrimer prodrugs. *Molecules (Basel, Switzerland)* 21 (6):686.

Das, N., Dhanawat, M., Dash, B., Nagarwal, R. & Shrivastava, S. 2010. Codrug: An efficient approach for drug optimization. *European Journal of Pharmaceutical Sciences* 41:571–88.

Deng, T., Wang, J., Li, Y., et al. 2018. Quantum dots-based multifunctional nano-prodrug fabricated by ingenious self-assembly strategies for tumor theranostic. *ACS Applied Materials & Interfaces* 10:27657–68.

Dhareshwar, S.S. & Stella, V.J. 2008. Your prodrug releases formaldehyde: Should you be concerned? No! *Journal of Pharmaceutical Sciences* 97:4184–93.

Diasio, R.B. & Harris, B.E. 1989. Clinical pharmacology of 5-fluorouracil. *Clinical Pharmacokinetics* 16:215–37.

Doh, H.-J., Cho, W.-J., Yong, C.-S., et al. 2003. Synthesis and evaluation of ketorolac ester prodrugs for transdermal delivery. *Journal of Pharmaceutical Sciences* 92:1008–17.

Dong, Q., Wang, X., Hu, X., et al. 2018. Simultaneous application of photothermal therapy and an anti-inflammatory prodrug using pyrene–aspirin-loaded gold nanorod graphitic nanocapsules. *Angewandte Chemie International Edition* 57:177–81.

Dubey, S. & Valecha, V. 2014. Prodrugs: A review. *World Journal of Pharmaceutical Research* 3:277–97.

Ehrnebo, M., Nilsson, S.-O. & Boréus, L.O. 1979. Pharmacokinetics of ampicillin and its prodrugs bacampicillin and pivampicillin in man. *Journal of Pharmacokinetics and Biopharmaceutics* 7:429–51.

Eisert, W.G., Hauel, N., Stangier, J., et al. 2010. Dabigatran: An oral novel potent reversible nonpeptide inhibitor of thrombin. *Arteriosclerosis, Thrombosis, and Vascular Biology* 30:1885–9.

Ettmayer, P., Amidon, G.L., Clement, B. & Testa, B. 2004. Lessons learned from marketed and investigational prodrugs. *Journal of Medicinal Chemistry* 47:2393–404.

Fang, J.-Y. & Leu, Y.-L. 2006. Prodrug strategy for enhancing drug delivery via skin. *Current Drug Discovery Technologies* 3:211–24.

Fasinu, P., Pillay, V., Ndesendo, V.M., du Toit, L.C. & Choonara, Y.E. 2011. Diverse approaches for the enhancement of oral drug bioavailability. *Biopharmaceutics & Drug Disposition* 32:185–209.

Fleisher, D., Bong, R. & Stewart, B.H. 1996. Improved oral drug delivery: Solubility limitations overcome by the use of prodrugs. *Advanced Drug Delivery Reviews* 19:115–30.

Frampton, J.E. 2018. Tafenoquine: First global approval. *Drugs* 78:1517–23.

Fukano, Y. & Kawazu, K. 2009. Disposition and metabolism of a novel prostanoid antiglaucoma medication, tafluprost, following ocular administration to rats. *Drug Metabolism and Disposition* 37:1622–34.

Gaekens, T., Guillaume, M., Borghys, H., et al. 2016. Lipophilic nalmefene prodrugs to achieve a one-month sustained release. *Journal of Controlled Release* 232:196–202.

Ghadiri, M., Chrzanowski, W. & Rohanizadeh, R. 2014. Antibiotic eluting clay mineral (Laponite®) for wound healing application: An in vitro study. *Journal of Materials Science: Materials in Medicine* 25:2513–26.

Ghawanmeh, A.A., Chong, K.F., Sarkar, S.M., et al. 2018. Colchicine prodrugs and codrugs: Chemistry and bioactivities. *European Journal of Medicinal Chemistry* 144:229–42.

Gilmer, J.F., Clune-Moriarty, L. & Lally, M. 2013. Efficient aspirin prodrugs. Google Patents.

Gomes, P., Vale, N. & Moreira, R. 2007. Cyclization-activated prodrugs. *Molecules* 12:2484–506.

Gomez-Orellana, I. 2005. Strategies to improve oral drug bioavailability. *Expert Opinion on Drug Delivery* 2:419–33.

Gonzalez-Cabrera, P.J., M Cahalan, S., Ferguson, J. & Rosen, H. 2012. S1P receptor modulators in cell trafficking and therapeutics. *Current Immunology Reviews* 8:170–80.

Grandhi, R.K. & Abd-Elsayed, A. 2019. Post-operative pain management in spine surgery. *Textbook of Neuroanesthesia and Neurocritical Care*. Singapore: Springer.

Guengerich, F.P. 2001. Common and uncommon cytochrome P450 reactions related to metabolism and chemical toxicity. *Chemical Research in Toxicology* 14:611–50.

Gupta, R., Hill, E.L., McClernon, D., et al. 2005. Acyclovir sensitivity of sequential herpes simplex virus type 2 isolates from the genital mucosa of immunocompetent women. *The Journal of Infectious Diseases* 192:1102–7.

Hajnal, K., Gabriel, H., Aura, R., Erzsébet, V. & Blanka, S.S. 2016. Prodrug strategy in drug development. *Acta Medica Marisiensis* 62:356–62.

Hanna, I., Alexander, N., Crouthamel, M.H., et al. 2018. Transport properties of valsartan, sacubitril and its active metabolite (LBQ657) as determinants of disposition. *Xenobiotica* 48:300–13.

Heimbach, T., Oh, D.-M., Li, L.Y., et al. 2003a. Absorption rate limit considerations for oral phosphate prodrugs. *Pharmaceutical Research* 20:848–56.

Heimbach, T., Oh, D.-M., Li, L.Y., et al. 2003b. Enzyme-mediated precipitation of parent drugs from their phosphate prodrugs. *International Journal of Pharmaceutics* 261:81–92.

Hodge, R.V., Sutton, D., Boyd, M., Harnden, M. & Jarvest, R. 1989. Selection of an oral prodrug (BRL 42810; famciclovir) for the antiherpesvirus agent BRL 39123 [9-(4-hydroxy-3-hydroxymethylbut-l-yl) guanine; penciclovir]. *Antimicrobial Agents and Chemotherapy* 33:1765–73.

Hossain, M.A. 2018. New drug approvals. *Bangladesh Pharmaceutical Journal* 21:173–5.

Hoy, S.M. 2018. Latanoprostene bunod ophthalmic solution 0.024%: A review in open-angle glaucoma and ocular hypertension. *Drugs* 78:773–80.

Hu, L. 2016. Prodrug approaches to drug delivery. *Drug Delivery: Principles and Applications* 2:227–71.

Hughes, B. 2009. 2008 FDA drug approvals. *Nature Reviews Drug Discovery* 8:93–96.

Hughes, B. 2010. 2009 FDA drug approvals. *Nature Reviews Drug Discovery* 9:89–92.

Hussain, A. & Truelove, J. 1976. Prodrug approaches to enhancement of physicochemical properties of drugs IV: Novel epinephrine prodrug. *Journal of Pharmaceutical Sciences* 65:1510–12.

Huttunen, K.M., Mahonen, N., Raunio, H. & Rautio, J. 2008. Cytochrome P450-activated prodrugs: Targeted drug delivery. *Current Medicinal Chemistry* 15:2346–65.

Huttunen, K.M., Raunio, H. & Rautio, J. 2011. Prodrugs—from serendipity to rational design. *Pharmacological Reviews* 63:750–71.

Inoue, T., Ogura, K., Kawakita, M., et al. 2016. Effective and safe administration of low-dose estramustine phosphate for castration-resistant prostate cancer. *Clinical Genitourinary Cancer* 14:e9–e17.

Ishikawa, T., Matsunaga, N., Tawada, H., et al. 2003. TAK-599, a novel N-phosphono type prodrug of anti-MRSA cephalosporin T-91825: Synthesis, physicochemical and pharmacological properties. *Bioorganic & Medicinal Chemistry* 11:2427–37.

Jain, A., Gupta, Y. & Jain, S.K. 2006. Azo chemistry and its potential for colonic delivery. *Critical Reviews™ in Therapeutic Drug Carrier Systems* 23:349–400.

Jana, S., Mandlekar, S. & Marathe, P. 2010. Prodrug design to improve pharmacokinetic and drug delivery properties: Challenges to the discovery scientists. *Current Medicinal Chemistry* 17:3874–908.

Jankovic, J., Jimenez-Shahed, J., Budman, C., et al. 2016. Deutetrabenazine in tics associated with Tourette syndrome. *Tremor and Other Hyperkinetic Movements* 6:422.

Järvinen, T., Rautio, J., Masson, M. & Loftsson, T. 2010. Design and pharmaceutical applications of prodrugs. *Pharmaceutical Sciences Encyclopedia: Drug Discovery, Development, and Manufacturing* 1–64.

Jornada, D., dos Santos Fernandes, G., Chiba, D., et al. 2015. The prodrug approach: A successful tool for improving drug solubility. *Molecules* 21:42.

Karaman, R. 2013a. Prodrug design vs. drug design. *Journal of Drug Design* 2:e114.

Karaman, R. 2013b. The prodrug naming dilemma. *Drug Design* 2:e115.

Karaman, R. 2013c. Prodrugs design by computation methods-a new era. *Journal of Drug Designing* 1:e113.

Karaman, R. 2013d. Prodrugs for masking bitter taste of antibacterial drugs—a computational approach. *Journal of Molecular Modeling* 19:2399–412.

Kawakami, S., Yamamura, K., Mukai, T., et al. 2001. Sustained ocular delivery of tilisolol to rabbits after topical administration or intravitreal injection of lipophilic prodrug incorporated in liposomes. *Journal of Pharmacy and Pharmacology* 53:1157–61.

Kim, B.-Y., Doh, H.-J., Le, T.N., et al. 2005. Ketorolac amide prodrugs for transdermal delivery: Stability and in vitro rat skin permeation studies. *International Journal of Pharmaceutics* 293:193–202.

Kimura, H., Asano, R., Tsukamoto, N., Tsugawa, W. & Sode, K. 2018. Convenient and universal fabrication method for antibody–enzyme complexes as sensing elements using the spycatcher/spytag system. *Analytical Chemistry* 90:14500–6.

Kokil, G.R. & V Rewatkar, P. 2010. Bioprecursor prodrugs: Molecular modification of the active principle. *Mini Reviews in Medicinal Chemistry* 10:1316–30.

Kovanda, L.L., Maher, R. & Hope, W.W. 2016. Isavuconazonium sulfate: A new agent for the treatment of invasive aspergillosis and invasive mucormycosis. *Expert Review of Clinical Pharmacology* 9:887–97.

Kratz, F., Müller, I.A., Ryppa, C. & Warnecke, A. 2008. Prodrug strategies in anticancer chemotherapy. *ChemMedChem* 3:20–53.

Krishnaiah, Y.S. 2010. Pharmaceutical technologies for enhancing oral bioavailability of poorly soluble drugs. *Journal of Bioequivalence BioAvailability* 2:28–36.

Krogsgaard-Larsen, P. 1991. *A Textbook of Drug Design and Development*. Amsterdam, the Netherlands: Harwood Academic Publishers.

Kulke, M.H., Hörsch, D., Caplin, M.E., et al. 2017. Telotristat ethyl, a tryptophan hydroxylase inhibitor for the treatment of carcinoid syndrome. *Journal of Clinical Oncology* 35:14.

Li, F., Maag, H. & Alfredson, T. 2008. Prodrugs of nucleoside analogues for improved oral absorption and tissue targeting. *Journal of Pharmaceutical Sciences* 97:1109–34.

Lian, X., Huang, Y., Zhu, Y., et al. 2018. Enzyme-MOF nanoreactor activates nontoxic paracetamol for cancer therapy. *Angewandte Chemie International Edition* 57:5725–30.

Lipinski, C. 2002. Poor aqueous solubility—an industry wide problem in drug discovery. *American Pharmaceutical Review* 5:82–5.

Lohar, V., Singhal, S. & Arora, V. 2012. Research article prodrug: Approach to better drug delivery. *International Journal of Pharmaceutical Research* 4:15–21.

Luo, Q., Wang, P., Miao, Y., He, H. & Tang, X. 2012. A novel 5-fluorouracil prodrug using hydroxyethyl starch as a macromolecular carrier for sustained release. *Carbohydrate Polymers* 87:2642–7.

Majumdar, S. & Sloan, K.B. 2006. Synthesis, hydrolyses and dermal delivery of N-alkyl-N-alkyloxycarbonylaminomethyl (NANAOCAM) derivatives of phenol, imide and thiol containing drugs. *Bioorganic & Medicinal Chemistry Letters* 16:3590–4.

Maloberti, A., Cassano, G., Capsoni, N., et al. 2018. Therapeutic Approach to hypertension urgencies and emergencies in the emergency room. *High Blood Pressure & Cardiovascular Prevention* 25:177–89.

McLeod, A.D., Friend, D.R. & Tozer, T.N. 1993. Synthesis and chemical stability of glucocorticoid-dextran esters: Potential prodrugs for colon-specific delivery. *International Journal of Pharmaceutics* 92:105–14.

Mignat, C., Heber, D., Schlicht, H. & Ziegler, A. 1996. Synthesis, opioid receptor affinity, and enzymatic hydrolysis of sterically hindered morphine 3-esters. *Journal of Pharmaceutical Sciences* 85:690–4.

Morozowich, W. & Karnes, H. 2007. Case study: Clindamycin 2-phosphate, a prodrug of clindamycin. *Prodrugs*. New York: Springer.

Mueller, C.E. 2009. Prodrug approaches for enhancing the bioavailability of drugs with low solubility. *Chemistry & Biodiversity* 6:2071–83.

Mullar, A. 2019. 2018 FDA drug approvals. *Nature Reviews Drug Discovery* 18:85–89.

Mullard, A. 2011. 2010 FDA drug approvals. *Nature Reviews Drug Discovery* 10:82.

Mullard, A. 2012. 2011 FDA drug approvals. *Nature Reviews Drug Discovery* 11:91.

Mullard, A. 2013. 2012 FDA drug approvals. *Nature Reviews Drug Discovery* 12:87.

Mullard, A. 2014. 2013 FDA drug approvals. *Nature Reviews Drug Discovery* 13:85.

Mullard, A. 2015. 2014 FDA drug approvals. *Nature Reviews Drug Discovery* 14:77.

Mullard, A. 2016. 2015 FDA drug approvals. *Nature Reviews Drug Discovery* 15:73.
Mullard, A. 2017. 2016 FDA drug approvals. *Nature Reviews Drug Discovery* 16:73.
Mullard, A. 2018a. 2017 FDA drug approvals. *Nature Reviews Drug Discovery* 17:81–85.
Mullard, A. 2018b. 2017 FDA drug approvals. *Nature Reviews Drug Discovery* 17:150.
Najjar, A. & Karaman, R. 2019. The prodrug approach in the era of drug design. *Journal Expert Opinion on Drug Delivery* 16(1):1–5
Ngawhirunpat, T., Kawakami, N., Hatanaka, T., Kawakami, J. & Adachi, I. 2003. Age dependency of esterase activity in rat and human keratinocytes. *Biological and Pharmaceutical Bulletin* 26:1311–14.
Notari, R.E. 1981. Prodrug design. *Pharmacology & Therapeutics* 14:25–53.
Nyirjesy, P. & Schwebke, J.R. 2018. Secnidazole: Next-generation antimicrobial agent for bacterial vaginosis treatment. *Future Microbiology* 13:507–24.
O'Hanlon, R. & Shaw, M.L. 2019. Baloxavir marboxil: The new influenza drug on the market. *Current Opinion in Virology* 35:14–18.
Oliyai, R. & Stella, V.J. 1993. Prodrugs of peptides and proteins for improved formulation and delivery. *Annual Review of Pharmacology and Toxicology* 33:521–44.
Pavel, M., Hörsch, D., Caplin, M., et al. 2015. Telotristat etiprate for carcinoid syndrome: A single-arm, multicenter trial. *The Journal of Clinical Endocrinology & Metabolism* 100:1511–19.
Peppercorn, M.A. 1984. Sulfasalazine: Pharmacology, clinical use, toxicity, and related new drug development. *Annals of Internal Medicine* 101:377–86.
Perry, C.M. 2012. Azilsartan medoxomil. *Clinical Drug Investigation* 32:621–39.
Peyrottes, S., Egron, D., Lefebvre, I., et al. 2004. SATE pronucleotide approaches: An overview. *Mini reviews in Medicinal Chemistry* 4:395–408.
Prueksaritanont, T., Gorham, L.M., Hochman, J.H., Tran, L.O. & Vyas, K.P. 1996. Comparative studies of drug-metabolizing enzymes in dog, monkey, and human small intestines, and in Caco-2 cells. *Drug Metabolism and Disposition* 24:634–42.
Rains, C.P., Noble, S. & Faulds, D. 1995. Sulfasalazine. *Drugs* 50:137–56.
Rautio, J. 2011. Prodrug strategies in drug design. *Prodrugs and Targeted Delivery* 1–30.
Rautio, J., Kärkkäinen, J. & Sloan, K.B. 2017. Prodrugs–Recent approvals and a glimpse of the pipeline. *European Journal of Pharmaceutical Sciences* 109:146–61.
Rautio, J., Kumpulainen, H., Heimbach, T., et al. 2008. Prodrugs: Design and clinical applications. *Nature Reviews Drug Discovery* 7:255.
Rautio, J., Mannhold, R., Kubinyl, H. & Folkers, G. 2010. *Prodrugs and Targeted Delivery.* Hoboken, NJ: Wiley VCH.
Rautio, J., Meanwell, N.A., Di, L. & Hageman, M.J. 2018. The expanding role of prodrugs in contemporary drug design and development. *Nature Reviews Drug Discovery* 17:559.
Redasani, V.K. & Bari, S.B. 2015. *Prodrug Design: Perspectives, Approaches and Applications in Medicinal Chemistry.* Amsterdam, the Netherlands: Academic Press.
Rosenblum, D., Joshi, N., Tao, W., Karp, JM., & Peer, D. 2018. Progress and challenges towards targeted delivery of cancer therapeutics. *Nature communications* 9(1):1–2.
Rosenkranz, T., Novas, M. & Terborg, C. 2015. PML in a patient with lymphocytopenia treated with dimethyl fumarate. *New England Journal of Medicine* 372:1476–8.
Roy, P. & Waxman, D.J. 2006. Activation of oxazaphosphorines by cytochrome P450: Application to gene-directed enzyme prodrug therapy for cancer. *Toxicology in Vitro* 20:176–86.
Satsangi, A., Roy, S.S., Satsangi, R.K., Vadlamudi, R.K. & Ong, J.L. 2014. Design of a paclitaxel prodrug conjugate for active targeting of an enzyme upregulated in breast cancer cells. *Molecular Pharmaceutics* 11:1906–18.
Savjani, K.T., Gajjar, A.K. & Savjani, J.K. 2012. Drug solubility: Importance and enhancement techniques. *International Scholaarly Research Network* 1–10.
Schiavon, O., Pasut, G., Moro, S., et al. 2004. PEG–Ara-C conjugates for controlled release. *European Journal of Medicinal Chemistry* 39:123–33.
Shah, K., Gupta, J.K., Chauhan, N.S., et al. 2017. Prodrugs of NSAIDs: A review. *The Open Medicinal Chemistry Journal* 11:146.
Shan, L., Zhuo, X., Zhang, F., et al. 2018. A paclitaxel prodrug with bifunctional folate and albumin binding moieties for both passive and active targeted cancer therapy. *Theranostics* 8.

Sharma, S.K. & Bagshawe, K.D. 2017. Antibody directed enzyme prodrug therapy (ADEPT): Trials and tribulations. *Advanced Drug Delivery Reviews* 118:2–7.

Shelke, M.M., Bidkar, S. & Dama, G. 2019. A review: Masking of taste unique approaches for bitter drug.

Shi, Y., Porter, W., Merdan, T. & Li, L.C. 2009. Recent advances in intravenous delivery of poorly water-soluble compounds. *Expert Opinion on Drug Delivery* 6:1261–82.

Shinde, S.C., Mahale, N., Chaudhari, S. & Thorat, R. 2015. Recent advances in brain targeted drug delivery system: A review. *World Journal of Pharmaceutical Research* 4:542–59.

Shirley, M. 2016. Ixazomib: First global approval. *Drugs* 76:405–11.

Silverman, R. 2004. *The Organic Chemistry of Drug Design and Drug Action*, pp. 1–617. London, UK: Elsevier Academic Press.

Simplício, A.L., Clancy, J.M. & Gilmer, J.F. 2007. β-Aminoketones as prodrugs with pH-controlled activation. *International Journal of Pharmaceutics* 336:208–14.

Soares-da-Silva, P., Pires, N., Bonifácio, M.J., et al. 2015. Eslicarbazepine acetate for the treatment of focal epilepsy: An update on its proposed mechanisms of action. *Pharmacology Research & Perspectives* 3:e00124.

Sorbera, L., Bozzo, J. & Castaner, J. 2005. Dabigatran/dabigatran etexilate. *Drugs of the Future* 30:877.

Soy, D., Aldasoro, E., Guerrero, L., et al. 2015. Population pharmacokinetics of benznidazole in adult patients with Chagas disease. *Antimicrobial Agents and Chemotherapy* 59:3342–9.

Stella, V., Borchardt, R., Hageman, M., et al. 2007a. *Prodrugs: Challenges and Rewards*. New York: Springer Science & Business Media.

Stella, V.J. 2007. A case for prodrugs. *Prodrugs*. New York: Springer.

Stella, V.J. 2011. Prodrugs: Some thoughts and current issues. *Journal of Pharmaceutical Sciences* 99:4755–4765.

Stella, V.J. & Himmelstein, K.J. 1980. Prodrugs and site-specific drug delivery. *Journal of Medicinal Chemistry* 23:1275–82.

Stella, V.J. & Nti-Addae, K.W. 2007b. Prodrug strategies to overcome poor water solubility. *Advanced Drug Delivery Reviews* 59:677–94.

Stinchcomb, A.L., Paliwal, A., Dua, R., et al. 1996. Permeation of buprenorphine and its 3-alkyl-ester prodrugs through human skin. *Pharmaceutical Research* 13:1519–23.

Stinchcomb, A.L., Swaan, P.W., Ekabo, O., et al. 2002. Straight-chain naltrexone ester prodrugs: Diffusion and concurrent esterase biotransformation in human skin. *Journal of Pharmaceutical Sciences* 91:2571–8.

Szewczuk, M., Boguszewska, K., Żebrowska, M., et al. 2017. Virus-directed enzyme prodrug therapy and the assessment of the cytotoxic impact of some benzimidazole derivatives. *Tumor Biology* 39:1010428317713675.

Testa, B. 2004. Prodrug research: Futile or fertile? *Biochemical Pharmacology* 68:2097–106.

Testa, B. 2009. Prodrugs: Bridging pharmacodynamic/pharmacokinetic gaps. *Current Opinion in Chemical Biology* 13:338–44.

Testa, B. & Mayer, J.M. 2003. *Hydrolysis in Drug and Prodrug Metabolism*. Zürich, Switzerland: John Wiley & Sons.

Timmins, G.S. & Deretic, V. 2006. Mechanisms of action of isoniazid. *Molecular Microbiology* 62:1220–7.

Tojo, K., Chiang, C. & Chien, Y. 1987. Drug permeation across the skin: Effect of penetrant hydrophilicity. *Journal of Pharmaceutical Sciences* 76:123–6.

Uhlyar, S. & Rey, J.A. 2018. Valbenazine (Ingrezza): The first FDA-approved treatment for tardive dyskinesia. *Pharmacy and Therapeutics* 43:328.

Van De Waterbeemd, H., Smith, D.A., Beaumont, K. & Walker, D.K. 2001. Property-based design: Optimization of drug absorption and pharmacokinetics. *Journal of Medicinal Chemistry* 44:1313–33.

Vane, J.R. 2014. Inhibition of prostaglandin biosynthesis as the mechanism of action of aspirin-like drugs. *Advances in the Biosciences* 9:395–411.

Verma, A., Verma, B., Prajapati, S.K. & Tripathi, K. 2009. Prodrug as a chemical delivery system: A Review. *Asian Journal of Research in Chemistry* 2:100–3.

Vig, B.S., Huttunen, K.M., Laine, K. & Rautio, J. 2013. Amino acids as promoieties in prodrug design and development. *Advanced Drug Delivery Reviews* 65:1370–85.

Vummaneni, V. & Nagpal, D. 2012. Taste masking technologies: An overview and recent updates. *International Journal of Research in Pharmaceutical and Biomedical Sciences* 3:510–24.

Walker, R., Zeuli, J. & Temesgen, Z. 2016. Isavuconazonium sulfate for the treatment of fungal infection. *Drugs Today (Barc)*52:7–16.

Walther, R., Rautio, J. & Zelikin, A.N. 2017. Prodrugs in medicinal chemistry and enzyme prodrug therapies. *Advanced Drug Delivery Reviews* 118:65–77.

Wang, J.-J., Sung, K., Hu, O.Y.-P., Yeh, C.-H. & Fang, J.-Y. 2006. Submicron lipid emulsion as a drug delivery system for nalbuphine and its prodrugs. *Journal of Controlled Release* 115:140–9.

Wei, Y., Yan, Y., Pei, D. & Gong, B. 1998. A photoactivated prodrug. *Bioorganic & Medicinal Chemistry Letters* 8:2419–22.

Wermuth, C.G. 2008. Designing prodrugs and bioprecursors. *The Practice of Medicinal Chemistry*. Elsevier.

Wermuth, C.G., Ganellin, C., Lindberg, P. & Mitscher, L. 1998. Glossary of terms used in medicinal chemistry (IUPAC Recommendations 1998). *Pure and Applied Chemistry* 70:1129–43.

Wu, K.-M. 2009. A new classification of prodrugs: Regulatory perspectives. *Pharmaceuticals* 2:77–81.

Wu, K.-M. & Farrelly, J.G. 2007. Regulatory perspectives of Type II prodrug development and time-dependent toxicity management: Nonclinical Pharm/Tox analysis and the role of comparative toxicology. *Toxicology* 236:1–6.

Yanagihara, K., Akamatsu, N., Matsuda, J., et al. 2018. Susceptibility of Clostridium species isolated in Japan to fidaxomicin and its major metabolite OP-1118. *Journal of Infection and Chemotherapy* 24:492–5.

Yang, C., Gao, H. & Mitra, A.K. 2001. Chemical stability, enzymatic hydrolysis, and nasal uptake of amino acid ester prodrugs of acyclovir. *Journal of Pharmaceutical Sciences* 90:617–24.

Zawilska, J.B., Wojcieszak, J. & Olejniczak, A.B. 2013. Prodrugs: A challenge for the drug development. *Pharmacological Reports* 65:1–14.

Zhang, J., Kale, V. & Chen, M. 2015. Gene-directed enzyme prodrug therapy. *The AAPS Journal* 17:102–10.

Zhang, Y., Yang, C., Wang, W., et al. 2016. Co-delivery of doxorubicin and curcumin by pH-sensitive prodrug nanoparticle for combination therapy of cancer. *Scientific Reports* 6:21225.

4

Prodrugs for Enhancement of Lipophilicity

Suneela Dhaneshwar and Neha V. Bhilare

CONTENTS

4.1 Introduction

The capability of a drug to reverse a pathological condition on administration and elicit the desired pharmacological response is the ultimate goal of every drug therapy. This response is subject to bioavailability of the drug at the site of action, which is directly affected by the amount of drug in the blood. For a drug delivery system to be effective, it should be able to accomplish as well as sustain effective levels of drug in plasma. The route of a drug to its target site involves transport across a number of lipid membranes and hence membrane permeability or lipophilicity has a substantial impact on drug efficacy.

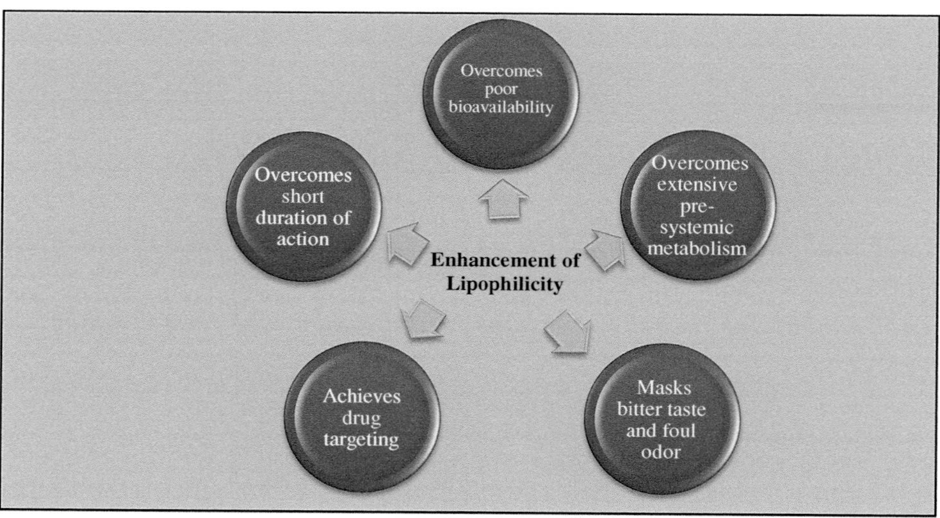

ILLUSTRATION 4.1 Applications of lipophilicity enhancement through prodrug approach.

Modification of hydrocarbon moieties of poorly permeable drugs results in increased lipophilicity. However, structure required for good activity is sometimes far from the structure required ideally for sufficient trans-membrane penetration. Under these circumstances the prodrug tool can prove be an enormously beneficial alternative. Enhancements of lipophilicity have been the most extensively explored and hence presently the most fruitful area of prodrug research. It has been accomplished by masking functional groups which are ionizable, nonionizable or polar in order to improve oral as well as topical absorption (e.g., especially in case of drugs that are usually administered by ocular and transdermal route) (Beaumont et al. 2003). Usually a hydrophilic carboxyl, thiol, phosphate, amine, or hydroxyl group present in the parent drug can be transformed to more lipophilic aryl or alkyl esters, and such prodrugs are then readily biocleaved by ubiquitous esterases (present throughout the body) to their active species (Liederer and Borchardt 2006). Several prodrug methodologies have been advanced for the polar phosphates, phosphonates and carboxylates; in most cases, conversion to prodrugs with ester group has been fruitful (Huttunen and Rautio 2011).

By employing prodrug strategy for lipophilicity enhancement, various drug-related issues like poor bioavailability, extensive pre-systemic metabolism, bitter taste, foul odor, short duration of action and drug-targeting have been successfully achieved (Illustration 4.1) (Rautio et al. 2008). These applications have been discussed section-wise in this chapter.

4.2 Prodrugs for Enhancement of Oral Bioavailability

In the field of discovery of a new chemical entity and its consequent development, oral bioavailability is one of the crucial considerations. It is widely accepted that poor oral bioavailability is one of the foremost causes of variability in therapeutics which in turn depends upon the differences observed in the amount of drug that is exposed. This is predominantly significant for drugs with lower therapeutic index or drugs with capability for development of resistance (cytotoxic drugs and antibiotics) (Bardelmeijer et al. 2008). Hellriegel et al. (1996) reported that "oral bioavailability of drugs from several therapeutic classes and the coefficient of inter-individual variability in their oral bioavailability are inversely proportional to each other" (Hellriegel et al. 1996). Unfortunately, poor oral bioavailability restricts the selection of oral route of administration of several drugs and in many cases considerable window is kept for losses in the dosage regimen design. As evident from the number of cases, more than 90% of drug after administration is lost to pre-systemic metabolism. Poor control of therapeutic effects and plasma concentrations and high extent of differences are outcome of low oral bioavailability. Especially, poor oral bioavailability is progressively a hurdle in the dosage regimen design as well as drug discovery process.

Oral bioavailability is a product of portion absorbed, portion bypassing excretion through liver and gut-wall; and the factors that impact bioavailability may be classified into biopharmaceutical factors, physiological, and physicochemical. It has been well recognized that oral absorption and drug metabolism is determined by physicochemical properties. The "rule-of-five" conceived by Lipinski and coworkers introduced an important breakthrough, displaying that compounds within certain physicochemical dimensions were bound to be clinically more significant than others (Abrahamsson and Lennernäs 2005).

Bioavailability of a drug can be influenced by wide range of factors. Generally, three principal factors govern the availability of the drug and/or metabolite to the target receptor or organ, these are as follows:

1. The speed and concentration of drug released from its formulation, and its subsequent absorption.
2. The first-pass metabolism during its passage through the liver following absorption.
3. The cumulative effect of drug distribution to different body fluids, plasma protein binding, metabolism and excretion (Brahmankar and Jaiswal 2006).

Application of prodrug strategy in enhancing lipophilicity and consequently overcoming bioavailability issues by minimizing first-pass metabolism and facilitating biological membrane permeation have been discussed in detail here. Prodrugs that increase lipophilicity by transiently masking the charge are employed for the enhancement of permeability of molecules. Active consideration of prodrug approach is crucial for researchers working in the field of drug discovery and development as it can yield candidates with anticipated pharmacokinetic and pharmaceutical properties. Advantages *versus* disadvantages of prodrug approach are shown in Illustration 4.2 (Baudy et al. 2009, Landowski et al. 2006, Donghi et al. 2009, Prichard et al. 2008, Ulrika et al. 2008).

4.2.1 Minimizing Pre-Systemic Metabolism through Prodrugs

For drugs that are administered orally, first-pass metabolism and hepatic uptake is a challenging barrier to efficient delivery. In addition to dropping the dose reaching predicted site of action, extensive first-pass metabolism often results in serious fluctuations in bioavailability, demanding cautious monitoring of drug levels in patient blood. Although pre-systemic metabolism can be circumvented by choosing alternate routes of administration (transdermal, intravenous, sublingual, buccal, rectal, etc.), oral route is usually the preferred one. Rational approach like prodrug design for sidestepping first-pass metabolism

Advantages of lipophilicity enhancement

- **Drugs are not dependent on active transport into the cells and enter by passive diffusion and can thereby bypass active transporters;**
- **Possibility to administer the drug orally rather than intravenously (IV)**
- **Enhancement of drug bioavailability by preventing enzymatic breakdown**
- **Passage through the blood-brain barrier (BBB)**
- **Reduced toxicity**

Disadvantages of lipophilicity enhancement

- **Poor water solubility and decreasing drug uptake**
- **Accumulation into liver tissue, causing increased hepatic toxicity**

ILLUSTRATION 4.2 Advantages and disadvantages of lipophilicity enhancement.

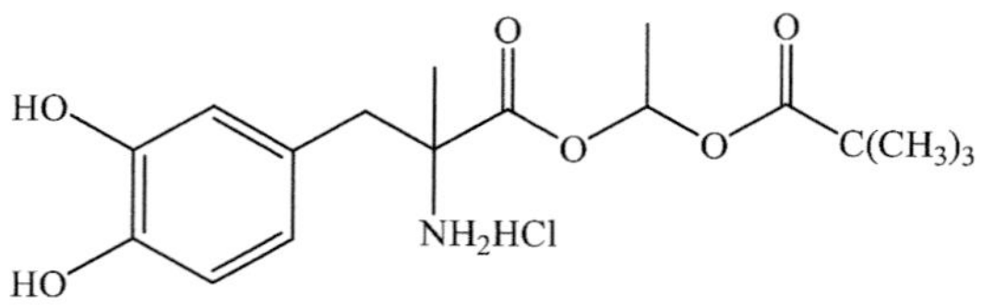

HO, HO, NH$_2$HCl, O, C(CH$_3$)$_3$

FIGURE 4.1 Structure of pivaloyloxyethyl ester prodrug of methyldopa.

can be applied to drug candidates who demonstrate promising pharmacological activity but undergo extensive first-pass metabolism when administered orally would be rather valuable.

There are instances of the use of the prodrug approach to minimize first-pass effect and thus enhance oral bioavailability. For example, Dobrinska et al. (1982) observed that the highly lipophilic pivaloyloxyethyl ester prodrug of methyldopa (Figure 4.1) displayed a 2.3 times higher systemic availability of methyldopa after oral administration compared to corresponding oral dose of plain methyldopa, this occurred due to decreased first-pass metabolism (Dobrinska et al. 1982).

Bodor et al. (1977) demonstrated that diacetyl ester, benzyl ester and various dipeptides of anti-Parkinson's agent levodopa result in significantly higher bioavailability of the drug (Bodor et al. 1977). This effect may be ascribed to the fact that transient derivatives of levodopa could effectively offer protection against metabolism that occurred during the first passage through the liver and in the gastrointestinal tract.

4.2.1.1 Gemcitabine

Another example is gemcitabine which is chemically 2,2-difluoro-2-deoxycytidine belonging to the class of pyrimidine antimetabolite, proven to be effective against wide range of solid tumors is well accepted in clinical trials. Major shortcoming of this agent is its extensive and rapid deamination by cytidine deaminase to its inert metabolite 2′,2′-difluorodeoxyuridine in the liver, kidney, blood and its consequent urinal excretion; the plasma half-life is very short (8–17 min) (Castelli et al. 2007). Forming amide prodrugs of gemcitabine for protecting its amine group is the method employed to improve its related cytotoxic activity and the metabolic stability (Figure 4.2).

Tethering of gemcitabine with acyl side chains of varying number of carbon atoms provided a bunch of molecular tools that possibly would be useful in investigating the impact of enhancing lipophilicity on interaction of drug with physiological membranes.

4.2.1.2 Epitiostanol

Epitiostanol (EP) is chemically epithiosteroid which is lipophilic in nature and possesses anti-estrogenic activity; making it beneficial in breast cancer therapy. It suffers from a serious drawback of extensive first-pass metabolism in liver and intestinal mucosa which limits its clinical effectiveness. It can only be

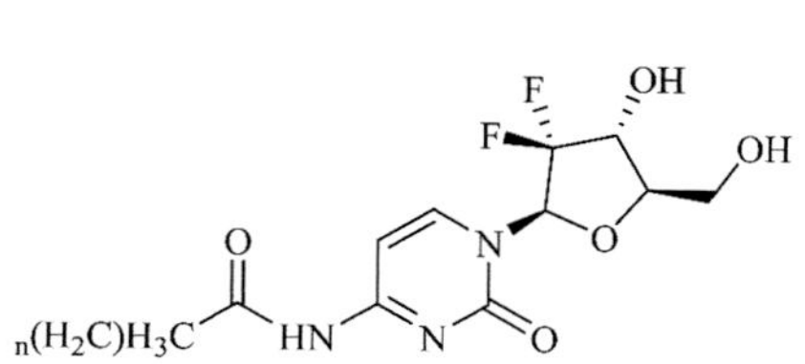

F, OH, F, OH, N, O, O, $_n$(H$_2$C)H$_3$C, HN, N, O

n=3 4-(N)-valeroyl gemcitabine

n=10 4-(N)-lauroyl gemcitabine

n=16 4-(N)-stearoyl gemcitabine

FIGURE 4.2 Lipophilic prodrugs of gemcitabine.

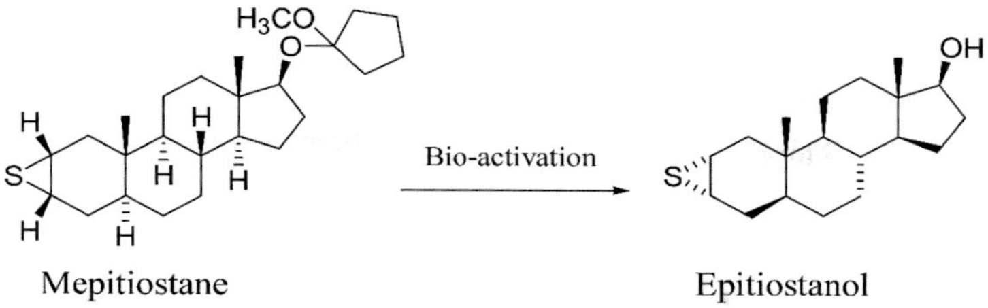

Mepitiostane → Bio-activation → Epitiostanol

FIGURE 4.3 Bioactivation of prodrug mepitiostane to active drug epitiostanol.

administered in the form of intramuscular injection as the first-pass effect prohibits its oral administration (Ichihashi et al. 1991). To overcome this shortcoming, a 17-substituted methoxycyclopentane prodrug of EP-mepitiostane (MP) was designed. This derivative is activated to epitiostanol by the enzymes esterases (Figure 4.3).

4.2.1.3 Ketobemidone

Ketobemidone (Figure 4.4) is a potent narcotic analgesic found to be equivalent to morphine. Owing to variable and extensive first-pass metabolism, it displays varied and incomplete bioavailability after subsequent rectal and per-oral administration in humans.

Ester prodrugs facilitated permeation of ketobemidone across porcine buccal mucosa and prominently enhanced lipophilicity and susceptibility to undergo transformation *via* mucosal esterases to parent drug (Hansen et al. 1992) (Table 4.1).

4.2.2 Prodrugs for Improvement of Oral Absorption of Polar Drugs

While inadequate aqueous solubility of drug molecules can be a serious issue, extreme polarity and hence limited solubility precludes absorption in lipidic environments. The transport of a drug to its target

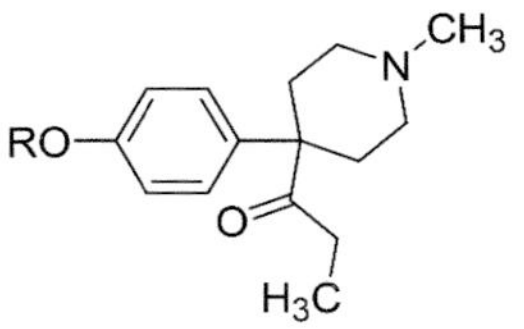

FIGURE 4.4 Ketobemidone.

TABLE 4.1

Prodrugs of Ketobomidone with Their log *P* Values

Sr.No.	Name	R	Log P
1	Ketobomidone	H	–0.40
2	*o*-Acetyl ketobomidone	$-COCH_3$	–0.43
3	*o*-Pivalyl ketobomidone	$-COC(CH_3)_2$	2.11
4	*o*-3,3-Dimethyl butyryl ketobomidone	$-COCH_2C(CH_3)_2$	2.55
5	*o*-Benzoyl ketobomidone	$-COC_6H_5$	2.41
6	*o*-Ethoxy carbonyl ketobomidone	$-COOC_2H_5$	1.11
7	*o*-Isopropyl carbonyl ketobomidone	$-COOCH(CH_3)_2$	1.54
8	*o*-Isobutoxy carbonyl ketobomidone	$-COOCH_2CH(CH_3)_2$	2.20
9	*o*-Butoxy carbonyl ketobomidone	$-COOC_4H_9$	2.24

site generally involves passage across a number of lipid membranes; thus, drug efficacy is significantly influenced by membrane permeability (Chan et al. 1996).

Orally administered drugs are mostly absorbed by passive diffusion which is neither specific nor facilitated. Ester prodrugs are most commonly used to improve the lipophilicity, and consequently the passive membrane permeability of hydrophilic drugs by masking charged groups such as phosphates and carboxylic acids (Beaumont et al. 2003). It has been found out that nearly 49% of the prodrugs available in market are bioactivated by enzymatic hydrolysis (Ettmayer et al. 2004). Ester prodrug synthesis is often facile and straightforward. *In vivo*, ubiquitous esterases (acetylcholinesterases, butyrylcholinesterases, carboxylesterases, arylesterases and paraoxonases) found in the tissues, blood, liver and other organs rapidly hydrolyze the ester bond (Liederer and Borchardt 2006).

4.2.2.1 (Deoxy) Nucleoside and Fluoropyrimidine Analogs

The majority of drugs that are presently utilized for cancer therapy have limitations, for example, resistance induction or lower biological half-life which is responsible for reducing their therapeutic effectiveness. Prodrug strategy has been widely explored to overcome these limitations. Chemically modifying gemcitabine [(deoxy) nucleoside analog] by conjugating it with moieties that are lipophilic in nature (Figure 4.5) enhances the plasma half-life, changes the bio-distribution, and improves cellular uptake of the drug (Peters et al. 2011). Enhancing lipophilicity of phosphorylated (deoxy) nucleoside by linking it with lipophilic promoiety improves the drug activity by sidetracking activation step of (deoxy) nucleoside analogs.

4.2.2.2 SN38 (7-ethyl-10-hydroxy Camptothecin)

SN38 (Figure 4.6), a member of camptothecin family, is a strong anticancer agent, but its poor transmucosal permeability and low solubility in excipients limit its use in the form of oral delivery. To overcome these barriers, a class of lipophilic prodrugs were developed by converting SN38 (C_{10} and/or C_{20}) into fatty acid esters with varying degrees of carbon atoms (Bala et al. 2015).

Among these synthesized prodrugs alkyl ester derivative of SN38 at position(s) demonstrated better lipid solubility, controlled reconversion rates, and cytotoxicity profiles.

4.2.2.3 Metformin

Metformin is presently used as a first-line antidiabetic drug for treatment of patients with type 2 diabetes. However, the undesirable gastrointestinal side effects and moderate absorption related with metformin therapy restricts its use. Metformin was transformed into two different prodrugs of biguanidine

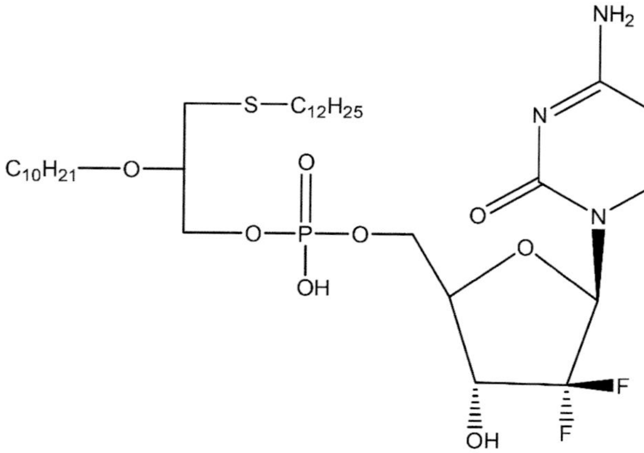

$C_{10}H_{21}$—O— ; S—$C_{12}H_{25}$; O—P(=O)(OH)—O ; NH_2 ; N ; O ; F ; OH

FIGURE 4.5 Gemcitabine prodrug with a phospholipid moiety.

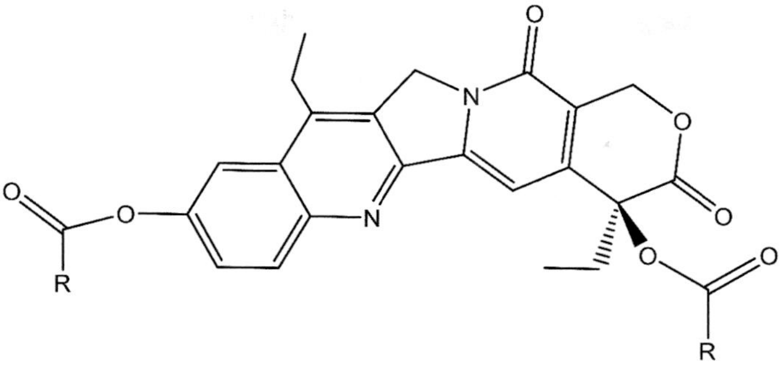

FIGURE 4.6 SN38 prodrugs (R is palmitic acid or propionic acid or undecanoic acid).

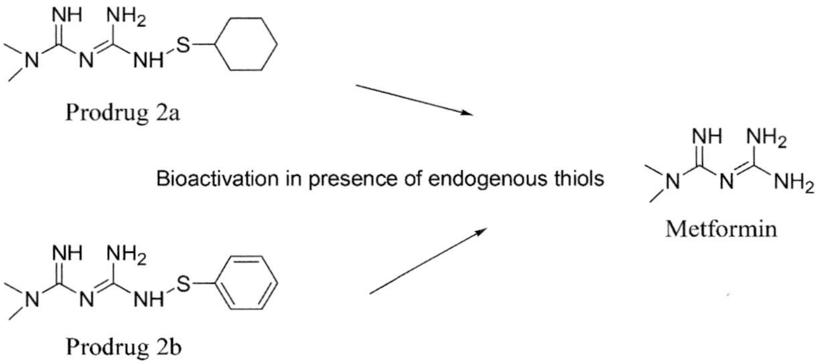

FIGURE 4.7 Prodrugs of metformin and their bioactivation.

TABLE 4.2
Prodrugs of Metformin and Their Distribution Coefficients

Sr.No.	Log D in Buffers	Metformin	Prodrug 2a	Prodrug 2b
1	Log D at pH 4	−3.4 ± 0.71	−1.10 ± 0.02	−0.70 ± 0.01
2	Log D at pH 7.4	3.37 ± 0.39	0.49 ± 0.01	−0.76 ± 0.02

functionality by Huttunen et al. (2009) and these synthesized prodrugs were investigated *in vitro* and *in vivo* to achieve enhanced lipophilicity and subsequently, improved oral absorption of this highly hydrophilic drug (Huttunen et al. 2009). The outcomes of this study demonstrated that the prodrug 2a (Figure 4.7) which was more lipophilic got bioactivated quantitatively to active drug metformin primarily after absorption in presence of endogenous thiols. The improved oral absorption in turn improved the bioavailability of metformin from 43% to 65% in rats (Table 4.2).

4.2.2.4 Adefovir Dipifoxil

Adefovir dipifoxil is a pivaloyloxymethyl (POM) ester prodrug of nucleosidic reverse transcriptase inhibitor, adefovir (Figure 4.8). It is basically used in retro-, herpes-, and hep-adnaviruses treatment. The bioavailability of adefovir dipifoxil which was found to be more lipophilic is nearly 45% as compared to adefovir which is as low as 12% in humans (Benzaria et al. 1996). Unfortunately, pivalate esters are responsible for lowering free carnitine levels in serum and therefore supplementation of L-carnitine becomes essential with adefovir dipifoxil (Noble and Goa 1999).

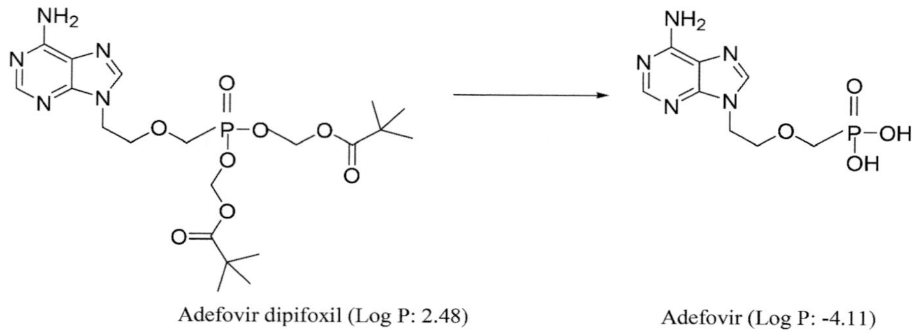

Adefovir dipifoxil (Log P: 2.48)

Adefovir (Log P: -4.11)

FIGURE 4.8 Pivalate esters of adefovir.

4.2.2.5 Olmesartan

Olmesartan is an angiotensin II receptor blocker and a potent antihypertensive agent (Bardsley-Elliot and Noble 1999, Aoki and Doucette 2001). It is prescribed for the treatment of high blood pressure; however, it lacks sufficient oral bioavailability due to poor lipophilicity.

Olmesartan medoxomil which is cyclic carbonate ester of olmesartan (Figure 4.9) has oral bioavailability 26% in humans. It is completely bioactivated during absorption (designed for paraoxonases) (McClellan and Perry 2001, Shi et al. 2006)

4.2.2.6 Candesartan

Candesartan is an antihypertensive agent. It lowers hypertension by antagonizing the angiotensin II effect which is primarily aldosterone secretion and vasoconstriction by obstructing the angiotensin II receptor (McClellan and Goa 1998). When administered orally, this drug is incompletely absorbed; therefore, candesartan cilexetil (ester prodrug) was prepared. It is biotransformed to candesartan *via* ester hydrolysis during absorption from the GI tract (Figure 4.10) (Ali and Hussein 2017).

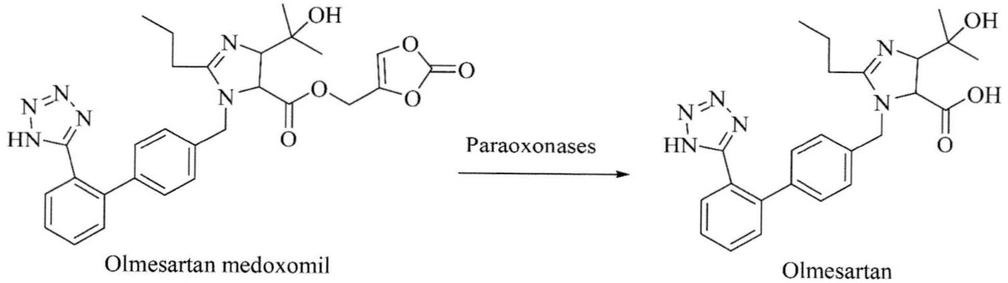

Olmesartan medoxomil

Olmesartan

FIGURE 4.9 Bioactivation of olmesartan medoxomil.

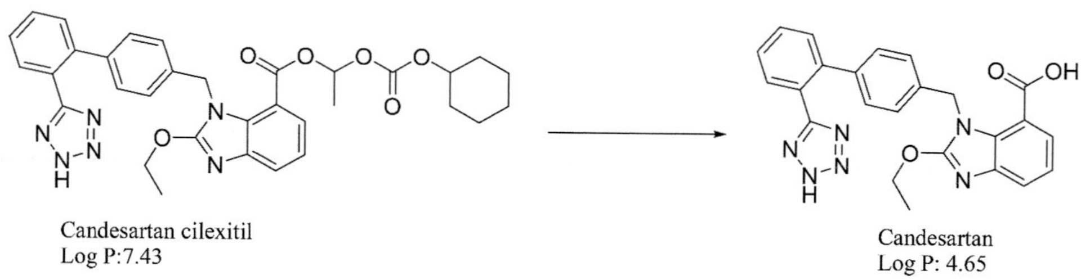

Candesartan cilexitil
Log P:7.43

Candesartan
Log P: 4.65

FIGURE 4.10 Bioactivation of candesartan cilexitil.

4.2.2.7 Melagatran

Melagatran was recognized in the year 1990 as an effective platelet aggregation and thrombin inhibitor. It is amphoteric in nature due to presence of one carboxylic acid group and two strongly basic groups, which results in oral bioavailability of only 5%; further this is greatly reduced when dosed with food. AstraZeneca plc, USA has launched ximelagatran (Exanta, AstraZeneca plc; Figure 4.11) that was designed as a double prodrug of melagatran with goal of enhancing permeability and sustaining sufficient pharmacological properties (Hu 2004). COOH functionality was derivatized into an ester while imidine was hydroxylated to lessen its basic nature. Ximelagatran is bioconverted to melgatran (Figure 4.11) *in vivo*.

4.2.2.8 Oseltamavir

The widely used remedy for influenza A and B is oseltamivir (Tamiflu), an orally active selective inhibitor of viral neuraminidase glycoprotein. This was designed and developed as an ethyl ester prodrug (McClellan and Perry 2001). Oseltamivir is rapidly biotransformed into its parent drug in presence of enzyme carboxylesterase-1 post-absorption. The bioavailability of the new lipophilic oseltamivir is approximately 80%, while the value of corresponding free carboxylate is just 5% (Shi et al. 2006) (Figure 4.12).

4.2.2.9 Dabigatran Etixilate

The active drug dabigatran indicated in thromboembolism acute coronary syndrome is highly polar. It has log P of 2.4 in n-octanol/buffer, pH 7.4 and is permanently charged molecule, therefore after oral administration it has zero bioavailability (Stangier et al. 2007). In the dabigatran etixilate (lipophilic bifunctional ester prodrug), COOH and amidinium groups are derivatized into simple and carbamate esters, respectively. These modifications lead to improved per-oral absorption and bioavailability (Blech et al. 2008), activation of this prodrug is mediated by carboxylesterases (Figure 4.13).

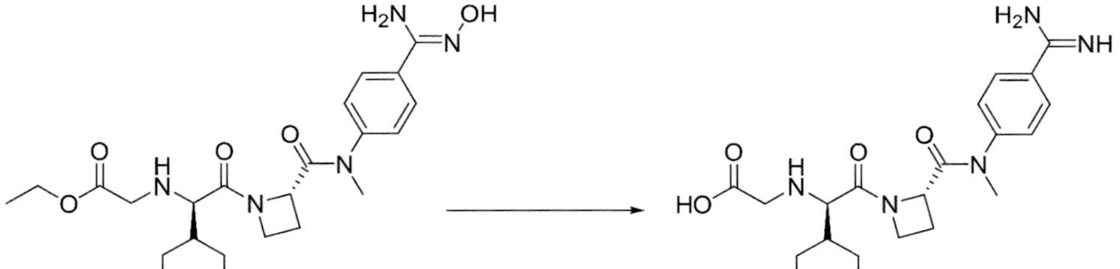

Ximelagatran (Log P: 0.9)

Melagatran (Log P: -1.3)

FIGURE 4.11 Bioactivation of prodrug ximelagatran.

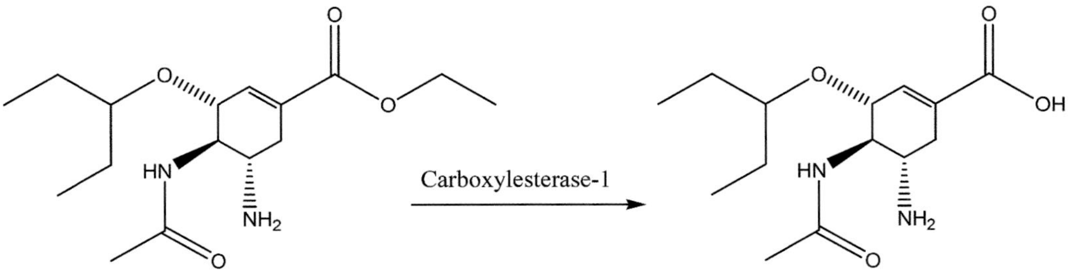

Carboxylesterase-1

Oseltamivir

Ro 64-0802

FIGURE 4.12 Bioactivation of oseltamivir.

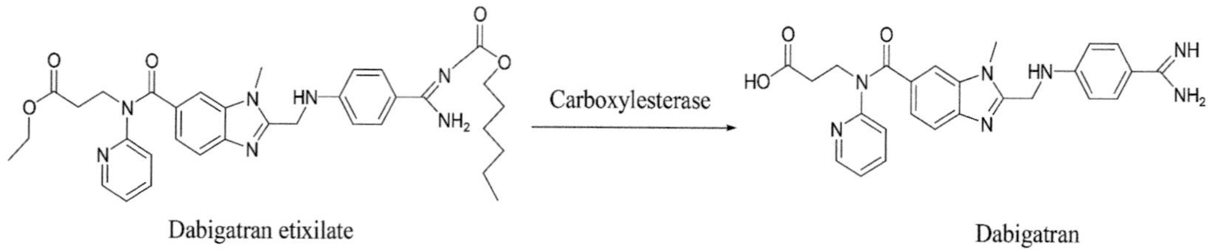

Dabigatran etixilate → Carboxylesterase → Dabigatran

FIGURE 4.13 Bioactivation of prodrug Dabigatran etixilate.

4.2.2.10 Furamidine

Furamidine is the active metabolite (DB 75) of Immtech's proprietary, orally bioavailable prodrug, pafuramidine (Huttunen et al. 2011). Furamidine has got the potential to inhibit DNA repair in cancerous cells that have been damaged by chemotherapy agents. Furamidine is poorly lipophilic drug, therefore DB 289 was designed as its prodrug to overcome this problem (Figure 4.14).

4.2.2.11 Norfloxacin

Norfloxacin is a second-generation fluoroquinolone. Its oral bioavailability is only 30%–40%. To overcome this disadvantage, dipalmitin prodrug of norfloxacin (Figure 4.15) was successfully synthesized which presented better lipophilicity, improved membrane permeability and bioavailability that resulted in remarkable improvement in activity as compared to norfloxacin at equimolar dose (Dhaneshwar et al. 2011).

4.2.2.12 A_{2B} Receptor Antagonist PSB-1115

Sulfonates that are highly polar compounds possess lower pKa value, i.e., less than 1. Due to their ionized state at body pH they are not per-orally bioavailable. Moreover, they fail to enter the

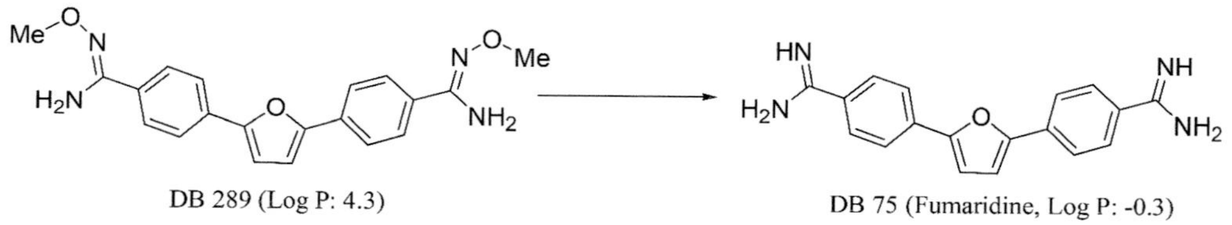

DB 289 (Log P: 4.3) → DB 75 (Fumaridine, Log P: -0.3)

FIGURE 4.14 Bioactivation of prodrug pafuramidine.

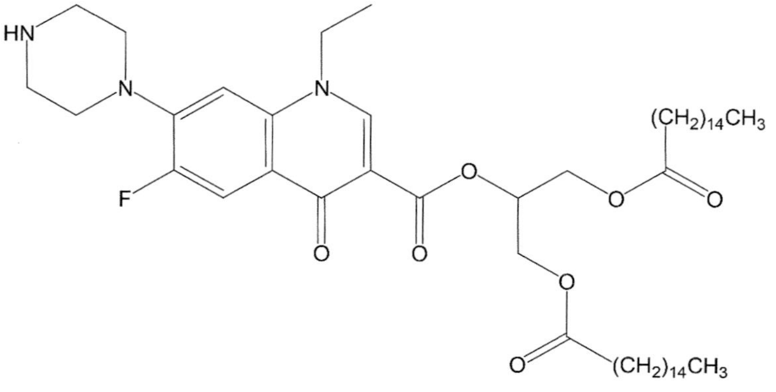

FIGURE 4.15 Dipalmitin prodrug of norfloxacin.

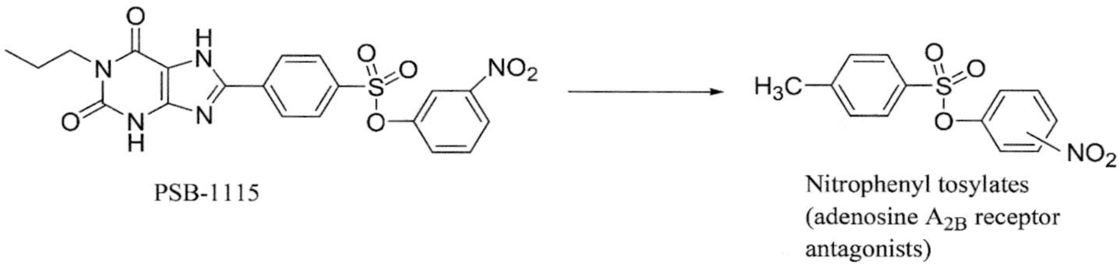

PSB-1115

Nitrophenyl tosylates (adenosine A_{2B} receptor antagonists)

FIGURE 4.16 Activation of A_{2B} receptor antagonist PSB-1115.

blood–brain barrier. Generally, sulfonamides or amides of sulfonic acids are highly stable and hence are unsuitable as prodrugs; sulfonic acid esters also are mostly unstable and because of this they are even used in substitution reactions as leaving groups (e.g., toluenesulfonates and methanesulfonates) (Yan and Müller 2004). Most stable class of sulfonic acid esters is that of nitrophenol esters; m-nitrophenyl sulfonates are found to be stable through a wide range of pH values from 1 to 10 as shown in Figure 4.16.

PSB-1115 (selective antagonist of adenosine A_{2B} receptor) was modified to its sulfonate ester prodrug with improved aqueous stability. It was found to be stable in artificial gastric acid and serum too. But on incubation with rat liver homogenate, it was readily hydrolyzed demonstrating an enzymatic hydrolysis pathway.

4.3 Prodrugs for Enhanced Topical Administration

As stated previously, prodrugs are aimed at enhancing the properties of active drug molecules, like solubility, permeability, chemical stability, metabolism and dosing frequency. Moreover, prodrugs may be employed to offer site-specific targeted delivery and also improve topical delivery. Prodrugs developed for ocular and dermal delivery are discussed below.

4.3.1 Timolol

Timolol is a β-adrenergic blocker used in glaucoma. *o*-Butyryl ester of timolol has been developed as a highly lipophilic prodrug of timolol which shows higher permeation rate across corneal epithelium. It has also been quantified that the dose of timolol given topically may possibly be lowered by minimum two times by means of this prodrug (Chang et al. 1987, Chien et al. 1991). It is activated to timolol in presence of esterases (Figure 4.17). Unfortunately, the reduced aqueous stability of this prodrug restricted its therapeutic effectiveness.

4.3.2 Ketorolac

Ketorolac, a non-selective cyclooxygenase-inhibitor, was converted into its ester prodrug by linking it with piperazinalkyl promoiety with a goal of improved topical penetration. As a consequence of the ionizable nitrogen atoms in the structure which have pKa values around 9 and the lipophilic alkyl moieties,

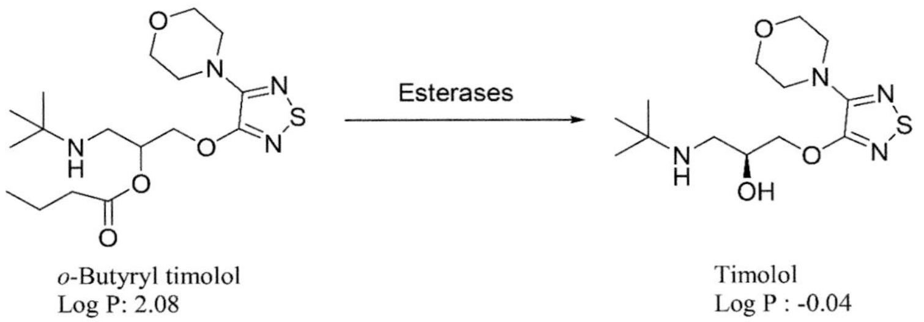

Esterases

o-Butyryl timolol
Log P: 2.08

Timolol
Log P : -0.04

FIGURE 4.17 Bioactivation of *o*-butyryl timolol to timolol by esterases.

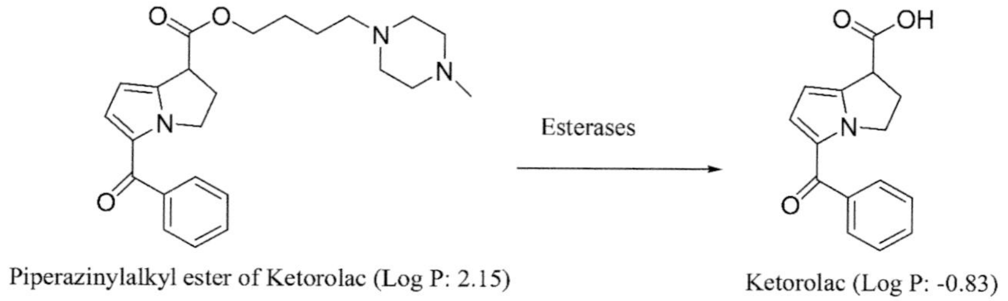

Piperazinylalkyl ester of Ketorolac (Log P: 2.15) — Esterases → Ketorolac (Log P: -0.83)

FIGURE 4.18 Bioactivation of piperazinylalkyl ester of ketorolac to ketorolac.

piperazinalkyl esters (Figure 4.18) can enable dissolution in aqueous as well as lipophilic phases of the skin (advantage over conventional aryl or alkyl esters) and thus can be regarded as a promising prodrug in the future (Qandil et al. 2008).

4.3.3 Latanoprost and Travoprost

Latanoprost acid and travoprost acid were derivatized into their isopropyl ester prodrugs named latanoprost and travoprost for improved lipophilicity and effortless entry into corneal epithelium where they are activated back to their respective active carboxylic acids. They are prostaglandin F2a (PGF2a) analogs used in the treatment of glaucoma (Figure 4.19).

They reduce intraocular pressure by enhancing the trabecular meshwork outflow and uveoscleral pathways (Bean and Camras 2008).

4.3.4 Tafluprost

Tafluprost (Saflutan) is a prodrug of a new synthetic prostaglandin F2μ (PGF2μ) analog. The isopropyl ester moiety of tafluprost is rapidly hydrolyzed to tafluprost acid in plasma and different tissues (for example the cornea) (Figure 4.20).

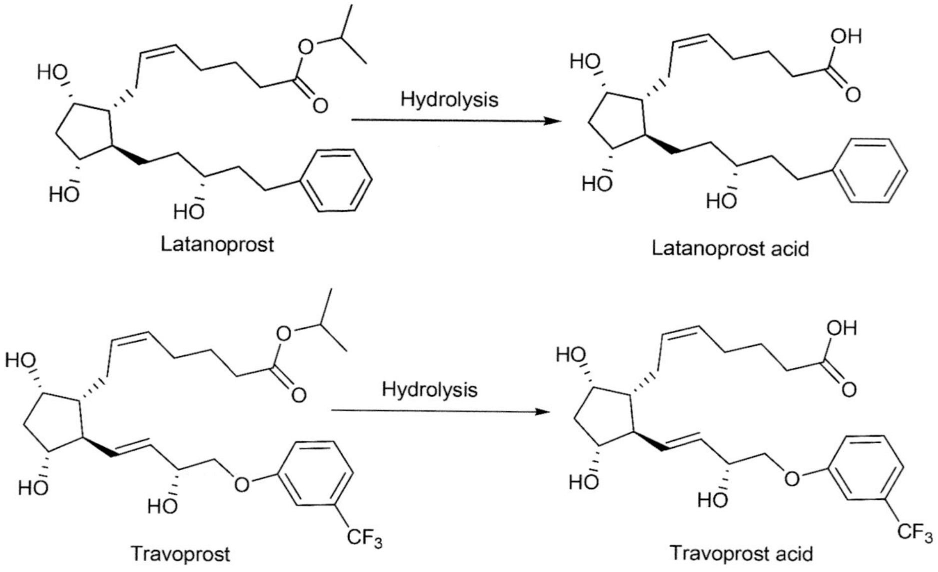

Latanoprost — Hydrolysis → Latanoprost acid

Travoprost — Hydrolysis → Travoprost acid

FIGURE 4.19 Bioactivation of latanoprost and travoprost.

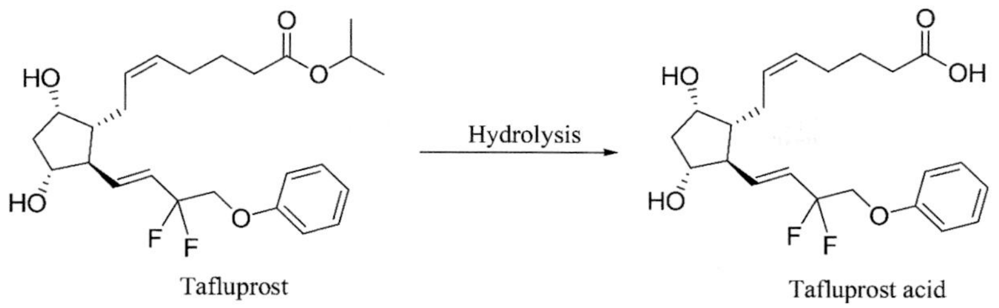

FIGURE 4.20 Hydrolysis of tafluprost to tafluprost acid.

4.3.5 Naproxen

Naproxen is one of the most potent nonsteroidal anti-inflammatory drugs (NSAIDs) which is generally used for treating the rheumatic diseases and other such painful disorders. However, naproxen displays bioavailability of merely 1%–2% following topical administration. Rautio et al. showed that piperazinylalkyl prodrugs of naproxen significantly improve the permeation of naproxen across human skin (Figure 4.21). The enhanced skin permeation noticed for these prodrugs may be accredited to their concurrently greater aqueous solubility as well as lipophilicity, which in combination are vital for topical penetration (Rautio et al. 2000) (Table 4.3).

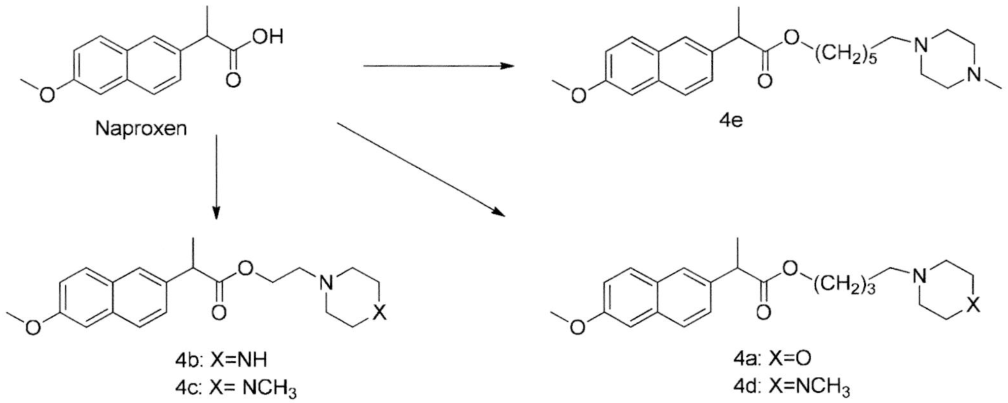

FIGURE 4.21 Prodrugs of naproxen.

TABLE 4.3

Aqueous Solubility (mean 6 ± S.D.; n = 2–4) and Apparent Partition Coefficient (log P, mean ± S.D.; n = 2–3) of Naproxen and Prodrugs

	Aqueous solubility (mM)		Log *P*app[a]	
Compound	**pH 5.0**	**pH 7.4**	**pH 5.0**	**pH 7.4**
Naproxen	0.4 ± 0.0	101.9 ± 1.3	2.38 ± 0.02	0.30 ± 0.03
4a	17.7 ± 0.5	0.2 ± 0.0	0.90 ± 0.08	2.60 ± 0.10
4b	18.2 ± 2.6	30.3 ± 7.1	0.20 ± 0.12	0.74 ± 0.02
4c	61.0 ± 6.8	32.6 ± 1.8	0.37 ± 0.03	2.29 ± 0.02
4d	54.6 ± 1.0	51.8 ± 2.6	1.13 ± 0.09	2.44 ± 0.09
4e	0.006 ± 0.001	0.006 ± 0.001	3.84 ± 0.05	3.92 ± 0.07

[a] Log *P*app is the apparent partition coefficient between phosphate buffer and 1-octanol at RT.

4.3.6 Tazarotene

Tazarotene is a prodrug as well as soft drug (undergoes oxidative deactivation) that is found in topical formulations used in the treatment of various conditions, such as psoriasis, acne, and sun damaged skin (photodamage) (Chandraratna 1996). It is bioconverted by esterases to give tazarotenic acid (Figure 4.22). Ester functionality resulted in better lipophilicity and preserved sufficient aqueous solubility that lead to better skin permeation (Dando and Wellington 2005).

Miscellaneous examples of prodrugs with enhanced lipophilicity along with their corresponding parent drugs are depicted in Table 4.4.

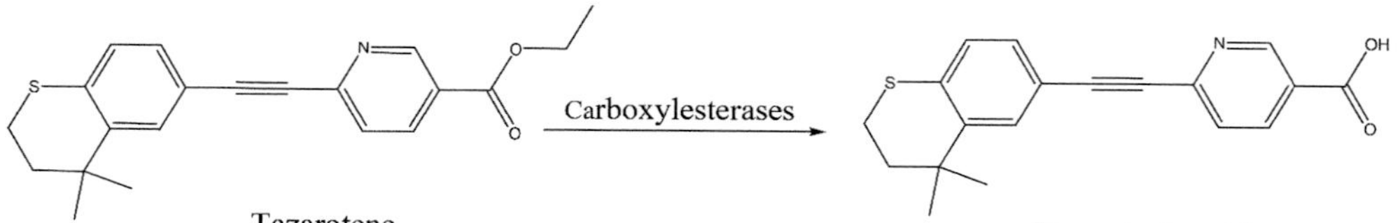

Carboxylesterases

Tazarotene

Tazarotenic acid

FIGURE 4.22 Bioactivation of tazarotene to tazarotenic acid.

TABLE 4.4

Prodrugs with Enhanced Lipophilicity

Sr.No.	Prodrug	Structure	Prodrug Approach Advantages
1	Enalpril (monoethyl ester of enalaprilat)	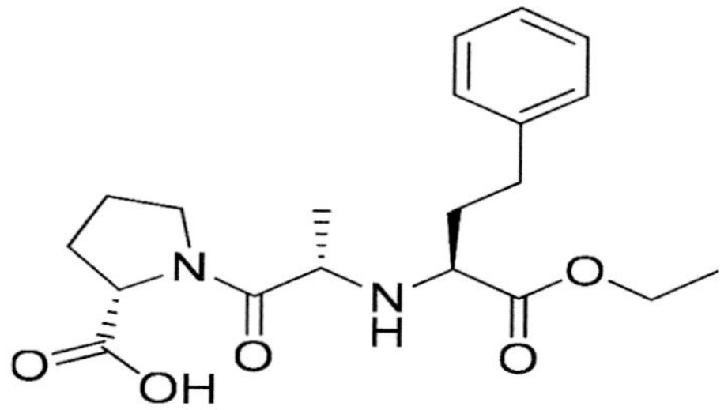	In liver it is bioconverted by esterases to enalaprilat (an angiotensin converting enzyme inhibitor). The oral bioavailability of enalaprilat in humans is 36%–44%. 53%–74% of the administered dose is absorbed.
2	Pivampicillin (pivaloyl methyl ester of ampicillin)	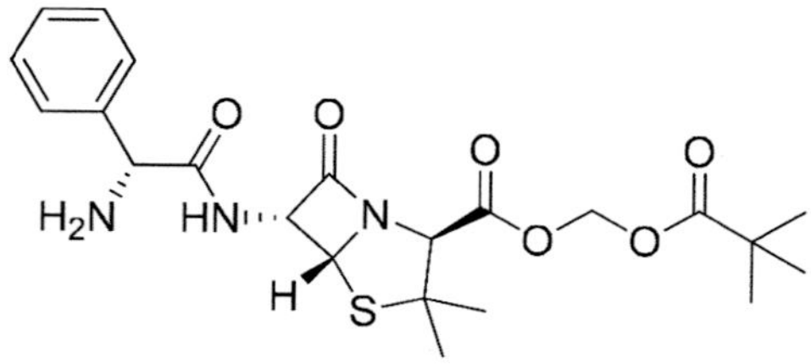	Bioactivated by esterases to ampicillin (*b*-lactam antibiotic). The oral bioavailability for ampicillin increased to 87%–94% for pivampicillin from 32%–55%.
3	Famciclovir (dimethyl ester of penciclovir)	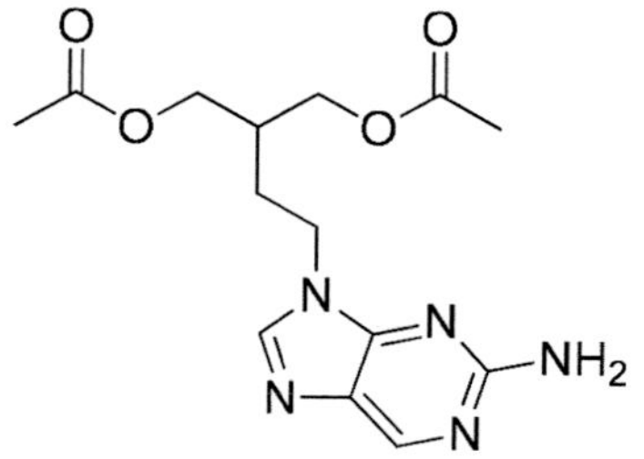	Activation mediated by aldehyde oxidase and esterases to penciclovir that arrests *Herpes* DNA synthesis.

(*Continued*)

TABLE 4.4 (*Continued*)

Prodrugs with Enhanced Lipophilicity

Sr.No.	Prodrug	Structure	Prodrug Approach Advantages
4	Abiraterone acetate(Abiraterone acetate (3β-acetate prodrug of abiraterone)	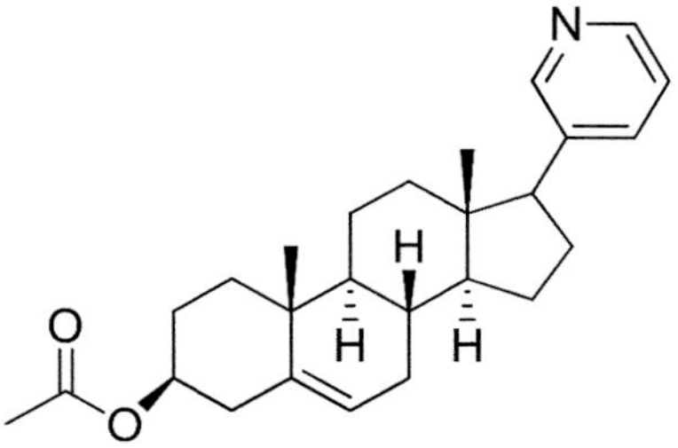	Bioconverted by esterases effective in hormone refractory prostate cancer (17-a-hydrolase/ C17,20 lyase).
5	Gabapentin encarbil (ester prodrug of gabapentin)	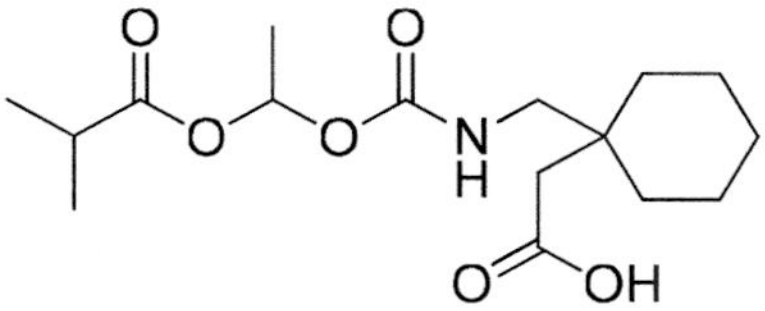	Bioactivated by esterases. New indication is in restless leg syndrome. Acts by modulating GABA and calcium channels.
6	Fenofibrate (Isopropyl ester prodrug of fenofibric acid)	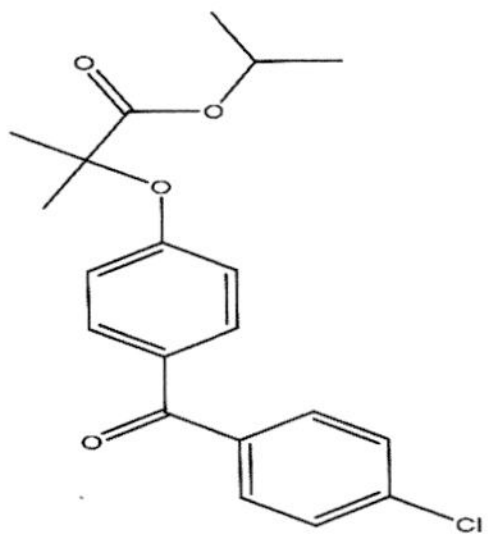	Bioactivated by esterases Hypercholesterolemia.
7	Ester prodrug of bexarotene	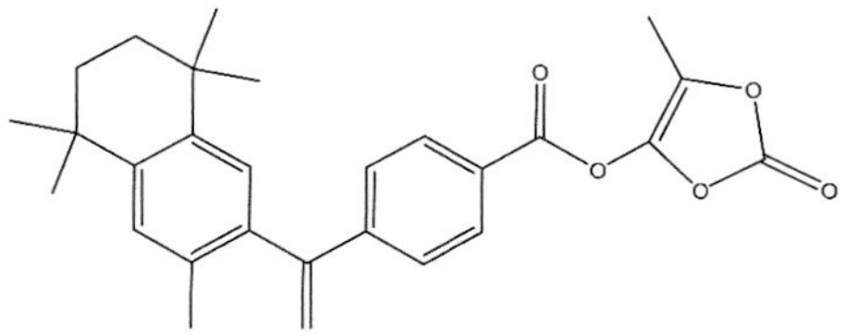	Bioactivated by esterases. Selectively delivery to intestinal lymphatics.

Source: Todd, P.A. and Heel, R.C., *Drugs*, 31, 198–248, 1986; Ehrnebo, M. et al., *J. Pharm. Biopharm.*, 7, 429–451, 1979; Jusko, W.J. and Lewis, G.P., *J. Pharm. Sci.*, 62, 69–76, 1973; Lee, J.B. et al., *J. Control. Release*, 286, 10–19, 2018.

4.4 Prodrugs for Masking Bitter Taste and Foul Odor

Undesirable taste is one of the reasons behind patient incompliance because even a small quantity of drug is sufficient to cause annoying taste. Older methods involving use of chemical agents like amino acid, flavoring agents and sweeteners are insufficient to modify bitterness of drugs which are especially given in large doses. Examples of some drugs with pronounced bitter taste are NSAIDs such as ibuprofen, COX inhibitors like etoricoxib and celecoxib, antibiotics like levofloxacin, penicillins, cefuroxime axetil, macrolide antibiotics, etc. (Ayenew et al. 2009). Efforts of researchers have been focused on developing new approaches which can efficiently mask foul taste of actives, among them prodrug approach is successful and has been used extensively. Most of research work is done on antibiotics with extreme sour taste. Flavor overlaying of drug via prodrug approach has been one of the successful strategies, few examples are discussed below.

Clindamycin's extreme bitterness precludes its suspension or solution formulation from being an acceptable oral dosage form. Further, the incidence of pain at the site of injection following intramuscular

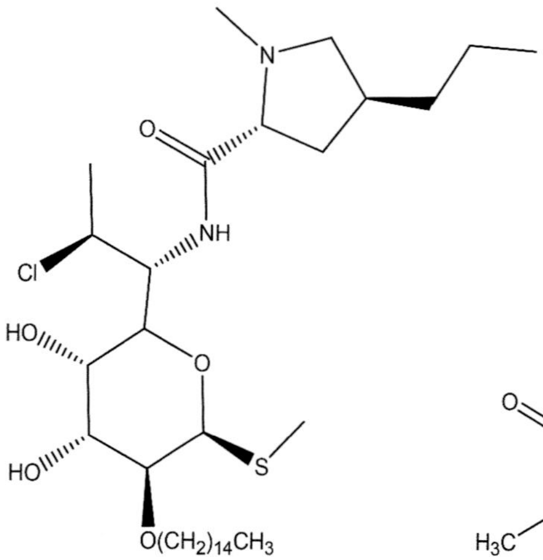

Clindamycin palmitate

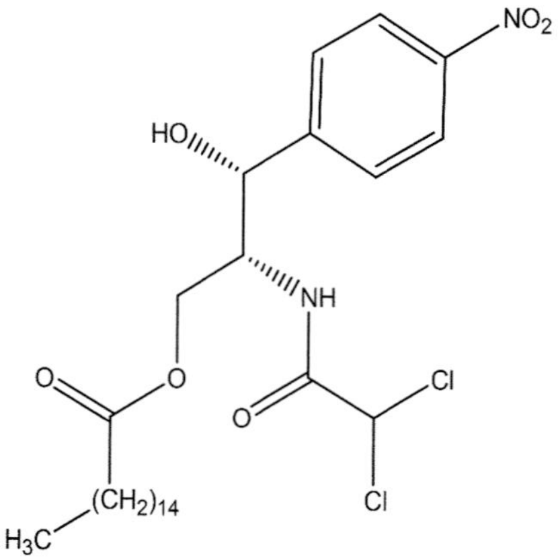

Chloramphenicol palmitate

FIGURE 4.23 Prodrugs with improved taste.

TABLE 4.5
Prodrugs with Improved Taste

Parent Drug	Prodrug with Improved Taste
Erythromycin	Alkyl ester
Triamcinolone	Diacetate ester
Lincomycin	Phosphate or alkyl ester 3,4,5-Trimethoxy benzoate
Sulfisoxazole	Acetyl ester
Clindamycin	Palmitate ester
N-acetyl cysteine	Palmitate ester
Chloramphenicol	Palmitate ester

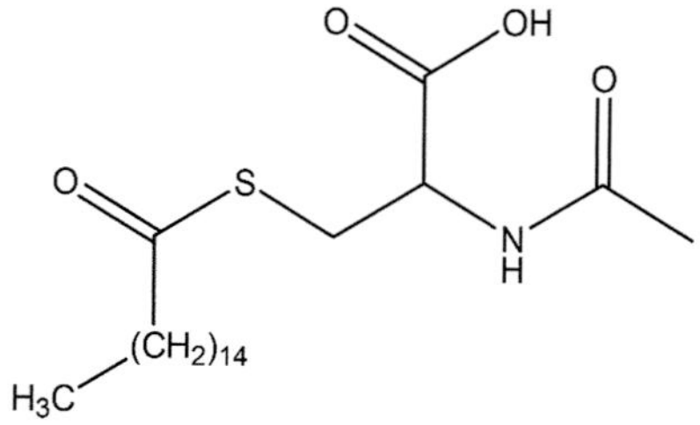

Thioester prodrug of N-acetyl cysteine

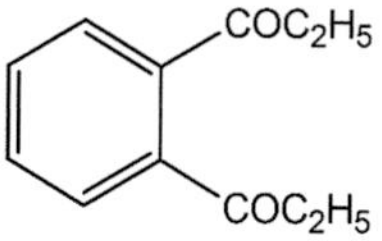

Pthalate ester of ethyl mercaptan

FIGURE 4.24 Prodrugs with improved odor.

injection is significant. These undesirable properties were overcome by prodrug design of clindamycin and chloramphenicol (Figure 4.23) conjugating it with palmitic acid. Di-carboxylic semi-ester derivatives of paracetamol were also used to design less bitter paracetamol prodrugs (Karaman 2014) (Table 4.5).

Masking of foul odor involves lowering of vapor pressure and thus boiling point (Chauhan 2017). Prodrug approach has been successfully employed in lowering vapor pressure of antileprotic agent ethyl mercaptan by preparing its phthalate ester (Figure 4.24).

Similarly, prodrug strategy used in synthesis of thioester prodrug of *N*-acetyl cysteine (Figure 4.24) for masking thiol group through thioester formation resulted in odorless prodrug (Bhilare et al. 2015).

4.5 Conclusion

Prodrugs have now become inherent members of the drug discovery and development process. Prodrug strategy can offer an attractive substitute for improving lipophilicity and thus eliminate various drug-related drawbacks like poor absorption, extensive first-pass metabolism, poor bioavailability, foul odor and taste as illustrated by various examples in this chapter. Apart from this, prodrugs with enhanced lipophilicity have also been used for the purpose of targeted delivery or site-specific delivery. Significance of prodrugs is evidenced by the fact that the growing number of new drug entities seeking approval each year are in fact in the form of a prodrug. It is anticipated that extensive application of various prodrug approaches right at the stage of drug development will definitely yield drug candidates with desired properties that would help in optimizing the delivery of a drug.

REFERENCES

Abrahamsson, Bertil, and Hans Lennernäs. 2005. Application of the biopharmaceutic classification system now and in the future. *Drug Bioavailability: Estimation of Solubility, Permeability, Absorption and Bioavailability* 18: 495–531.

Ali, Halah Hussein, and Ahmed Abbas Hussein. 2017. Oral solid self-nanoemulsifying drug delivery systems of candesartan citexetil: Formulation, characterization and in vitro drug release studies. *AAPS Open* 3: 6.

Aoki, Fred Y., and Karen E. Doucette. 2001. Oseltamivir: A clinical and pharmacological perspective. *Expert Opinion on Pharmacotherapy* 2:10, 1671–1683.

Ayenew, Zelalem, Vibha Puri, Lokesh Kumar, and Arvind K. Bansal. 2009. Trends in pharmaceutical taste masking technologies: A patent review. *Recent Patents on Drug Delivery & Formulation* 3:1, 26–39.

Bala, Vaskor, Shasha Rao, Peng Li, Shudong Wang, and Clive A. Prestidge. 2015. Lipophilic prodrugs of SN38: Synthesis and in vitro characterization toward oral chemotherapy. *Molecular Pharmaceutics* 13:1, 287–294.

Bardelmeijer, Heleen A., Olaf van Tellingen, Jan HM Schellens, and Jos H. Beijnen. 2008. The oral route for the administration of cytotoxic drugs: Strategies to increase the efficiency and consistency of drug delivery. *Investigational New Drugs* 18:3, 231–241.

Bardsley-Elliot, Anne, and Stuart Noble. 1999. Oseltamivir. *Drugs* 58:5, 851–860.

Baudy, Reinhardt B., John A. Butera, Magid A. Abou-Gharbia, Hong Chen, Boyd Harrison, Uday Jain, Ronald Magolda et al. 2009. Prodrugs of perzinfotel with improved oral bioavailability. *Journal of Medicinal Chemistry* 52: 771–778.

Bean, Gerald W., and Carl B. Camras. 2008. Commercially available prostaglandin analogs for the reduction of intraocular pressure: Similarities and differences. *Survey of Ophthalmology* 53:6, S69–S84.

Beaumont, Kevin, Robert Webster, Iain Gardner, and Kevin Dack. 2003. Design of ester prodrugs to enhance oral absorption of poorly permeable compounds: Challenges to the discovery scientist. *Current Drug Metabolism* 4:6, 461–485.

Benzaria, Samira, Hélène Pélicano, Richard Johnson, Georges Maury, Jean-Louis Imbach, Anne-Marie Aubertin, Georges Obert, and Gilles Gosselin. 1996. Synthesis, in vitro antiviral evaluation, and stability studies of Bis-(S-Acyl-2-Thioethyl) ester derivatives of 9-[2-(Phosphonomethoxy)Ethyl]adenine (PMEA) as potential PMEA prodrugs with improved oral bioavailability. *Journal of Medicinal Chemistry* 39:25, 4958–4965.

Bhilare, Neha, Dhaneshwar Suneela, Sinha Akanksha, Kandhare Amit, and Bodhankar Subhash. 2015. Novel thioester prodrug of N-acetylcysteine for odor masking and bio availability enhancement. *Current Drug Delivery* 4: 611–620.

Blech, Stefan, Thomas Ebner, Eva Ludwig-Schwellinger, Joachim Stangier, and Willy Roth. 2008. The metabolism and disposition of the oral direct thrombin inhibitor, dabigatran, in humans. *Drug Metabolism and Disposition* 36:2, 386–399.

Bodor, Nicholas, Kenneth B. Sloan, Takeru Higuchi, and Kunihiro Sasahara. 1977. Improved delivery through biological membranes. 4. Prodrugs of L-dopa. *Journal of Medicinal Chemistry* 20:11, 1435–1445.

Brahmankar, D. M., and Sunil Jaiswal. 2006. *Biopharmaceutics & Pharmacokinetics: A Treatise*. New Delhi, India: Vallabh Prakashan.

Castelli, Francesco, Maria Grazia Sarpietro, Flavio Rocco, Maurizio Ceruti, and Luigi Catte. 2007. Interaction of lipophilic gemcitabine prodrugs with biomembrane models studied by Langmuir–Blodgett technique. *Journal of Colloid and Interface Science* 313:1, 363–368.

Chan, O. Helen, and Barbra H. Stewart. 1996. Physicochemical and drug-delivery considerations for oral drug bioavailability. *Drug Discovery Today* 1:11, 461–473.

Chandraratna, R. A. 1996. Tazarotene—first of a new generation of receptor-selective retinoids. *British Journal of Dermatology* 135: 18–25.

Chang, Shih-chieh, H. Bundgaard, A. Buur, and V. H. Lee. 1987. Improved corneal penetration of timolol by prodrugs as a means to reduce systemic drug load. *Investigative Ophthalmology & Visual Science*, 28:3, 487–491.

Chauhan, Rohit. 2017. Taste masking: A unique approach for bitter drugs. *Journal of Stem Cell Biology and Transplantation* 1:2, 12.

Chien, Du-Shieng, Hitoshi Sasaki, Hans Bundgaard, Anders Buur, and Vincent H. L. Lee. 1991. Role of enzymatic lability in the corneal and conjunctival penetration of timolol ester prodrugs in the pigmented rabbit. *Pharmaceutical Research* 8:6, 728–733.

Dando, Toni M., and Keri Wellington. 2005. Topical tazarotene: A review of its use in the treatment of plaque psoriasis. *American Journal of Clinical Dermatology* 6:4, 255–272.

Dhaneshwar, Suneela, Kunal Tewari, Sonali Joshi, Dhanashree Godbole, and Pinaki Ghosh. 2011. Diglyceride prodrug strategy for enhancing the bioavailability of norfloxacin. *Chemistry and Physics of Lipids* 164:4, 307–313.

Dobrinska, M. R., W. Kukovetz, E. Beubler, H. Lorraine Leidy, H. J. Gomez, J. Demetriades, and J. A. Bolognese. 1982. Pharmacokinetics of the pivaloyloxyethyl (POE) ester of methyldopa, a new prodrug of methyldopa. *Journal of Pharmacokinetics and Biopharmaceutics* 10:6, 587–600.

Donghi, Monica, Barbara Attenni, Cristina Gardelli, Annalise Di Marco, Fabrizio Fiore, Claudio Giuliano, Ralph Laufer et al. 2009. Synthesis and evaluation of novel phosphoramidate prodrugs of 2′-methyl cytidine as inhibitors of hepatitis c virus NS5B polymerase. *Bioorganic & Medicinal Chemistry Letters* 19:5, 1392–1395.

Ehrnebo, Mats, Sten-Ove Nilsson, and Lars O. Boréus. 1979. Pharmacokinetics of ampicillin and its prodrugs bacampicillin and pivampicillin in man. *Journal of Pharmacokinetics and Biopharmaceutics* 7:5, 429–451.

Ettmayer, Peter, Gordon L. Amidon, Bernd Clement, and Bernard Testa. 2004. Lessons learned from marketed and investigational prodrugs. *Journal of Medicinal Chemistry* 47:10, 2393–2404.

Hansen, Laila Bach, Lona Louring Christrup, and Hans Bundgaard. 1992. Enhanced delivery of ketobemidone through porcine buccal mucosa in vitro via more lipophilic ester prodrugs. *International Journal of Pharmaceutics* 88: 237–242.

Hellriegel, Edward T., Thorir D. Bjornsson, and Walter W. Hauck. 1996. Interpatient variability in bioavailability is related to the extent of absorption: implications for bioavailability and bioequivalence studies. *Clinical Pharmacology & Therapeutics* 60:6, 601–607.

Hu, Longqin. 2004. Prodrugs: Effective solutions for solubility, permeability and targeting challenges. *IDrugs* 7:8, 736–742.

Huttunen, Kristiina M., and Jarkko Rautio. 2011. Prodrugs – An efficient way to breach delivery and targeting barriers. *Current Topics in Medicinal Chemistry* 11:18, 2265–2287.

Huttunen, Kristiina M., Anne Mannila, Krista Laine, Eeva Kemppainen, Jukka Leppanen, Jouko Vepsalainen, Tomi Jarvinen, and Jarkko Rautio. 2009. The first bioreversible prodrug of metformin with improved lipophilicity and enhanced intestinal absorption. *Journal of Medicinal Chemistry* 52:14, 4142–4148.

Huttunen, Kristiina M., Hannu Raunio, and Jarkko Rautio. 2011. Prodrugs – from serendipity to rational design. *Pharmacological Reviews* 63: 711–750.

Ichihashi, T., H. Kinoshita, and H. Yamada. 1991. Absorption and disposition of epithiosteroids in rats (1): Route of administration and plasma levels of epitiostanol. *Xenobiotica* 21:7, 865–872.

Jusko, William J., and George P. Lewis. 1973. Comparison of ampicillin and hetacillin pharmacokinetics in man. *Journal of Pharmaceutical Sciences* 62:1, 69–76.

Karaman, Rafik. 2014. Prodrugs for masking the bitter taste of drugs. In Ali Demir Sezer, ed., *Application of Nanotechnology in Drug Delivery*. Rijeka, Croatia: IntechOpen.

Landowski, Christopher P., Philip L. Lorenzi, Xueqin Song and Gordon L. Amidon. 2006. Nucleoside ester prodrug substrate specificity of liver carboxylesterase. *Journal of Pharmacology and Experimental Therapeutics* 316:2, 572–580.

Lee, Jong Bong, Atheer Zgair, Jed Malec, Tae Hwan Kim, Min Gi Kim, Joseph Ali, Chaolong Qin et al. 2018. Lipophilic activated ester prodrug approach for drug delivery to the intestinal lymphatic system. *Journal of Controlled Release* 286: 10–19.

Liederer, Bianca M., and Ronald T. Borchardt. 2006. Enzymes involved in the bioconversion of ester-based prodrugs. *Journal of Pharmaceutical Sciences* 95: 1177–1195.

McClellan, Karen J., and Karen L. Goa. 1998. Candesartan cilexetil. *Drugs* 56:5, 847–869.

McClellan, Karen, and Caroline M. Perry. 2001. Oseltamivir: A review of its use in influenza. *Drugs* 61:2, 263–283.

Noble, Stuart, and Karen L. Goa. 1999. Adefovir dipivoxil. *Drugs* 58: 479–487.

Peters, Godefridus J., Auke D. Adema, Irene V. Bijnsdorp, and Marit L. Sandvold. 2011. Lipophilic prodrugs and formulations of conventional (deoxy) nucleoside and fluoropyrimidine analogs in cancer. *Nucleosides, Nucleotides and Nucleic Acids* 30:12, 1168–1180.

Prichard, Mark N., Caroll B. Hartline, Emma A. Harden, Shannon L. Daily, James R. Beadle, Nadejda Valiaeva, Earl R. Kern, and Karl Y. Hostetler. 2008. Inhibition of herpesvirus replication by hexadecyloxypropyl esters of purine- and pyrimidine-based phosphonomethoxyethyl nucleoside phosphonates. *Antimicrobial Agents and Chemotherapy* 52: 4326–4330.

Qandil, Amjad, Soraya Al-Nabulsi, Bashar Al-Taani, and Bassam Tashtoush. 2008. Synthesis of piperazinylalkyl ester prodrugs of ketorolac and their in vitro evaluation for transdermal delivery. *Drug Development and Industrial Pharmacy* 34:10, 1054–1063.

Rautio, Jarkko, Hanna Kumpulainen, Tycho Heimbach, Reza Oliyai, Dooman Oh, Tomi Järvinen, and Jouko Savolainen. 2008. Prodrugs: Design and clinical applications. *Nature Reviews Drug Discovery* 7: 255.

Rautio, Jarkko, Tapio Nevalainena, Hannu Taipalea, Jouko Vepsalainen, Jukka Gynthera, Krista Laine, and Tomi Jarvinen. 2000. Piperazinylalkyl prodrugs of naproxen improve in vitro skin permeation. *European Journal of Pharmaceutical Sciences* 11: 157–163.

Shi, Deshi, Jian Yang, Dongfang Yang, Edward L. LeCluyse, Chris Black, Li You, Fatemeh Akhlaghi, and Bingfang Yan. 2006. Anti-influenza prodrug oseltamivir is activated by carboxylesterase human carboxylesterase 1, and the activation is inhibited by antiplatelet agent clopidogrel. *Journal of Pharmacology and Experimental Therapeutics* 319:3, 1477–1484.

Stangier, Joachim, Karin Rathgen, Hildegard Stähle, Dietmar Gansser, and Willy Roth. 2007. The pharmacokinetics, pharmacodynamics and tolerability of dabigatran etexilate, a new oral direct thrombin inhibitor, in healthy male subjects. *British Journal of Clinical Pharmacology* 64:3, 292–303.

Todd, Peter A., and Rennie C. Heel. 1986. Enalapril: A review of its pharmacodynamic and pharmacokinetic properties, and therapeutic use in hypertension and congestive heart failure. *Drugs* 31:3, 198–248.

Ulrika, Eriksson, Larryn W. Peterson, Boris A. Kashemirov, John M. Hilfinger, John C. Drach, Katherine Z. Borysko, Julie M. Breitenbach et al. 2008. Serine peptide phosphoester prodrugs of cyclic cidofovir: Synthesis, transport, and antiviral activity. *Molecular Pharmaceutics*, 5:4, 598–609.

Yan, Luo, and Christa E. Müller. 2004. Preparation, properties, reactions, and adenosine receptor affinities of sulfophenylxanthine nitrophenyl esters: Toward the development of sulfonic acid prodrugs with peroral bioavailability. *Journal of Medicinal Chemistry* 47:4, 1031–1043.

Lacteal does not contain basal membrane, and has a leaky constitution; hence, it permits infiltration of big colloid structures (200–800 mm), while blood capillaries contain intact basal membrane and tight junctions which do not allow permeation of big colloids (Swartz 2001). The assembly of lacteals together with the submucosal lymphatic vessels form the efferent trunks, followed by the thoracic duct, that joins the systemic blood flow at the point of left internal jugular and subclavian veins, thereby circumventing the hepatic blood flow and hepatic metabolism altogether (Bernier-Latmani and Petrova 2017). Incorporation of drugs/prodrugs into CM allows them to be lymphatically transported and to bypass the first-pass hepatic metabolism. Physicochemical properties that are proposed as a condition for lymphatic drug transport are log $P > 5$, and extensive solubility in TG > 50 mg/g; these conditions allow association of drug within the lipid core of CM. When it comes to prodrugs, in some cases fulfillment of these conditions is not necessary, due to the similar structure and nature of prodrugs to lipids in the body, and the ability of such prodrugs to incorporate into physiological lipid processing pathways (Shackleford et al. 2007). The lipidic prodrug approach that resulted in lymphatic transport includes a number of drugs; some examples are testosterone (Horst et al. 1976), L-dopa (Garzon-Aburbeh et al. 1986), and bexarotene (Lee et al. 2018). Drugs/prodrugs that undergo lymphatic transport can also be used for targeting lymphatic tissues in conditions such as particular autoimmune diseases and infections (Han et al. 2016; Porter et al. 2007), human immunodeficiency virus (HIV) (Bibby 1996; Pantaleo et al. 1993), tumor metastasis (Dart 2017).

5.2.2 Cholesterol Homeostasis

Cholesterol is the main precursor to all steroids produced in the body; it can be ingested as nutrient or synthesized de novo in the body. Cholesterol is crucial in maintaining cellular homeostasis (1), it serves as an integrated part of cell membranes and is enriched in lipid rafts (2), it makes up lipoproteins, which are responsible for lipid transport to tissues (3), plays an important role in intracellular signal transduction (4) and is a building block for all steroid hormones (5) (Ikonen 2008).

Cholesterol is mainly synthesized in the liver and then transported to cells through the systemic circulation in a form of a low-density lipoprotein (LDL) (Brown and Goldstein 1986); cholesterol that originates from nutrients is transferred from the gut to the liver as a part of large lipoproteins. LDL is entering cells by clathrin-mediated endocytosis, followed by a transfer to lysosomes via the endocytic pathway; in the lysosomes, LDL is hydrolyzed to single cholesterol molecules, which are transported to the cell membrane and different cell membrane–bound organelles (Brown and Goldstein 1986; Ikonen 2008). LDL is responsible for cholesterol transport from liver to the other tissues, whereas high-density lipoproteins (HDL) transfer cholesterol back to the liver. CM present ultra-low-density lipoproteins (Figure 5.1), produced in intestine; their role is to carry dietary lipids (TG, cholesterol) to different tissues. Very-low-density lipoproteins are synthesized in the liver, and are responsible for transferring TG to different tissues; the enzyme lipoprotein lipase can convert VLDL to LDL (Fredrickson et al. 1967).

Cholesterol is a precursor for all steroid hormones (sex hormones, glucocorticoids, mineralocorticoids), as well as bile salts. The pathogenesis of many malignant diseases is associated with higher/lower cholesterol levels. This variation between cholesterol uptake into normal vs cancerous cells/tissues can be used when designing steroid-based lipid prodrug. Cholesterol-drug conjugates can exploit the higher growth of cancerous cells, higher expression of LDL receptors and increased need for cholesterol. This approach can lead to more effective drug targeting, higher cellular drug uptake, and different pharmacokinetic profile (Radwan and Alanazi 2014).

The following sections describe different lipid-prodrug designs (FA, TG, PL, and steroid) and the ways they incorporate into lipid processing pathways, as well as their advantages and future uses.

5.3 Lipidic Prodrugs: Structures and Applications

Solubility obstacles that interfere with oral absorption of lipidic prodrugs are described in Section 5.1.1. Dietary lipids, on the other hand, do not face these limitations, and are entirely absorbed following ingestion; this points out to the fact that the physiological pathways included in digestion of lipid nutrients have the ability to overcome such solubility obstacles. Good knowledge of lipid metabolic pathways

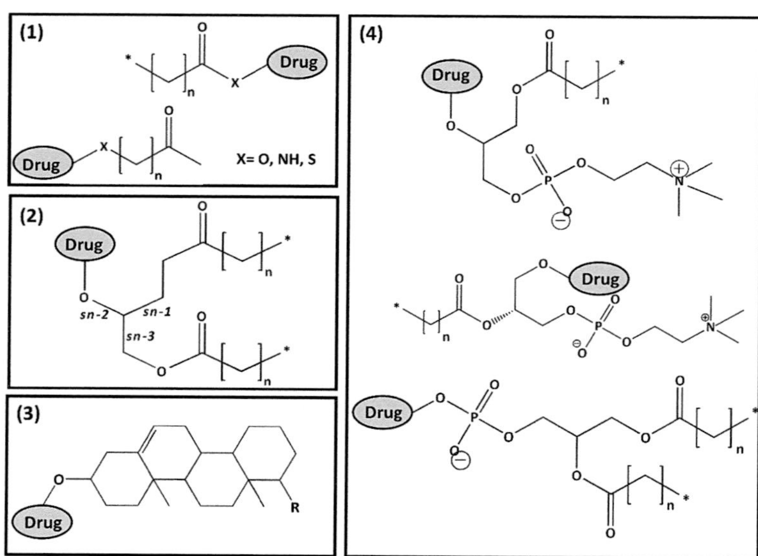

FIGURE 5.2 Types of lipid-drug conjugates: (1) FA-drug conjugates, where the drug moiety attached to the carboxyl group (top structure) or the ω-carbon (bottom structure) in the FA chain; (2) triglyceride-drug conjugate, typically connected to the *sn*-2 position; (3) steroid-drug conjugate with the drug connected to the hydroxyl functional group of the steroid ring (position 3); (4) phospholipid-drug conjugates, where the drug attached to the *sn*-2 position of the glycerol backbone (top structure), *sn*-1 position of the glycerol (middle structure) or to the phosphate group (bottom structure).

(Section 5.2) is a necessary prerequisite for designing lipid prodrugs; it is advantageous if lipidic prodrugs mimic endogenous lipids, and in that way are able to join lipid processing pathways. This allows lipidic prodrugs to deliver the parent drug across numerous barriers, and arrive at the site of action.

Lipidic prodrugs contain a drug covalently bound to a lipid moiety (FA, TG, steroid or PL), illustrated in Figure 5.2. Following sections describe chemical structure of lipidic prodrugs, detailed description of their applications and advantages in comparison to the parent drug alone and the future use of lipidic prodrugs.

5.3.1 FA-Drug Conjugates

FA-based prodrug approach is the most frequently employed strategy in lipidic prodrug approach. There are two possible strategies of attaching a parent drug to the FA: the drug is conjugated to the carboxyl group of the FA (1) or the carboxyl group is free and drug is conjugated to the modified ω-atom of the FA (2) (Figure 5.2). In cases when the carboxyl group of the FA is masked (by conjugation with the drug moiety), FA-drug conjugates are not able to join physiological lipid processing pathways, since the carboxyl group is responsible for FA recognition by the cell membrane transporters, as well as for binding between FAs and albumin; when the drug is conjugated in the ω-position, the carboxyl group is free to participate in the FA processing pathways. Interestingly, a recent study employed a thioether linker among the carboxyl group of the FA and the drug; in the cancerous tissue, the prodrug-containing thioether linker demonstrated higher sensitivity for redox heterogeneity in these tissues, in comparison to oxygen-containing groups (Luo et al. 2016).

FA-based prodrug approach demonstrated various advantages including improved chemical stability (Fernandez and Borgstrom 1990; Nakamura et al. 1975), enhanced biopharmaceutical profile, ability to join lymphatic transport, which allows prodrugs to surpass first-pass hepatic metabolism (Horst et al. 1976) and target the lymphatics (Bibby 1996), decreased irritation of the gastrointestinal tract (GIT), etc. (Mishima et al. 1990). Lymphatic transport is usually accomplished with prodrugs containing FAs with longer carbonic chain, whereas shorter and medium length FAs enter hepatic blood flow. Another reason that FA-drug conjugates may have limited lymphatic transport is considerable hydrolysis prior to absorption.

5.3.1.1 FA-Drug Conjugates Applications in Cancer

Unsaturated FAs (UFAs) contain at least a single unsaturated double bond. Depending upon the number of unsaturated double bonds we can have monounsaturated (e.g., oleic, elaidic acid) or polyunsaturated, where we differentiate ω-3 (eicosapentaenoic acid, docosahexaenoic) or ω-6 (linoleic acid, arachidonic), depending on the position of the unsaturated double bond. Due to their intrinsic cancer targeting features UFAs were conjugated to anticancer drugs aiming to enhance their pharmacological effect: paclitaxel-docosahexaenoic acid (Wolff et al. 2003), cytarabine-elaidic acid (Rizzieri et al. 2014) and gemcitabine-eladic acid (Stuurman et al. 2013) prodrugs. Paclitaxel-docosahexaenoic acid phase III clinical study demonstrated suboptimal results; one possible explanation is slow liberation of free paclitaxel at the activation site. In a recent study, novel paclitaxel FA-based prodrug was synthesized through conjugation with oleic acid through thioether and dithioether bond (Luo et al. 2016). As previously mentioned, thioether bond is sensitive to redox potential difference between cancer cells, unlike the ester bond, which served as a control. Paclitaxel-oleic acid conjugates were incorporated into redox-sensitive nanoparticles (NP), where the FA played an important role for self-assembly of the prodrug into NP. NPs had a core containing the conjugates and the shell composed of polyethylene glycol (PEG). Loading into PEG-labeled NP was very high (up to 57%), comparing to standard paclitaxel NP. Redox-sensitive prodrugs with one thioether bond showed higher drug liberation and *in vivo* pharmacological effect then the dithioether prodrug; the ester-linker conjugate, used as a control showed slow liberation of the parent drug and low efficacy (Luo et al. 2016). Special liker design and drug delivery system can significantly improve drug delivery; lipid-based conjugates and drug delivery systems can form hydrophobic interactions, stabilizing the drug loading and overall stability of the drug formulation. PEG also improved the stability of particles containing FA-paclitaxel prodrugs.

Saturated FA were also shown to be advantageous in the drug delivery of anticancer agents. SN-38 (7-ethyl-10-hydroxy camptothecin) is a powerful antineoplastic drug, active metabolite of irinotecan that inhibits DNA topoisomerase 1. Since it belongs to class 4 of the BCS, SN-38 has low permeability and low solubility (see Section 5.1.2), and presents an excellent candidate for using the lipidic prodrug approach (Bala et al. 2013). Conjugation of the C20 position of the SN-38 with undecanoic acid, resulted in improved permeability in the gut lumen, and liberation of the free drug in the blood (Bala et al. 2016a, 2016b). FA-SN-38 conjugate was incorporated into self-emulsifying delivery systems, which allowed oral drug administration and showed higher intestinal solubility; *in vivo* liberation of SN-38 from a prodrug complex demonstrated 100 times greater area under the curve (AUC) that the AUC of the oral SN-38 solution alone (Bala et al. 2016a).

A recent study using FA-based prodrugs in cancer employed cytarabine linked to the lauric acid (Liu et al. 2016). Cytarabine has low intestinal permeability, due to small lipophilicity, and poor metabolic stability. Conjugation with FA improved these shortcomings: intestinal permeability was improved, amino group of cytarabine was protected from deamination resulting in improved metabolic stability, and pharmacokinetic (PK) profile was significantly amended as well. Following oral administration, the free drug was released in a continuous manner and PK parameters were improved comparing to the cytarabine administration alone: drug bioavailability was 32 times better, peak plasma concentration (C_{max}) 5 times, and elimination half-life ($t_{1/2}$) 6.6 times higher. This study shows that designing a cytarabine-lauric acid conjugate is a more potent and reliable therapeutic approach for leukemia with improved bioavailability and metabolic stability (Liu et al. 2016, 2017).

5.3.1.2 FA-Drug Conjugates Applications in IBD

To date, the most common therapeutic approach for inflammatory bowel disease (IBD) are derivatives of 5-aminosalycilic acid (5-ASA). IBD is a life-long condition, and almost one-third of all patients are not responding to derivatives of 5-ASA; on the other hand, numerous side effects are associated with 5-ASA, making the FA-based prodrugs appealing approach for improving these aspects of treatment. In a recent study, 5-ASA derivative (sulfalasazine) was conjugated with two FA molecules (eicosapentaenoic acid and caprylic acid) through two ester bonds, aiming to provide improved targeting of diseased gut segments in IBD and decrease overall absorption of 5-ASA (Kandula et al. 2016). Double-lipid prodrug demonstrated high stability in the stomach and effective drug liberation, only once it is in the intestine. When compared to sulfasalazine, the double FA-prodrug showed improved retention in the ileum and colon, and higher safety, due to lower blood levels of 5-ASA. Pharmacological effect was shown to be similar to that of sulfasalazine; however, the two FA used for conjugation previously showed antimicrobial and anti-inflammatory effect, therefore further improving the effect of the prodrug (Hawthorne et al. 1992; Shimizu et al. 2003). This approach demonstrates that a choice of FA should also be carefully considered, since it can result in improved pharmacological efficacy.

5.3.1.3 FA-Drug Conjugates Applications for Hormone Therapy

FA-drug conjugates don't usually undergo lymphatic transport; however, in case of some hormones, careful design might govern FA-prodrug transport to the lymphatics. For instance, orally administered testosterone shows substantial first-pass hepatic metabolism, which results in extremely low testosterone levels in the blood and lack of efficacy (Nieschlag et al. 1975). Conjugation of testosterone to undecanoic acid increased the bioavailability of testosterone in humans to roughly 7% (Tauber et al. 1986). It was shown that the reason for this high increase was that testosterone undecanoate (TU) undergoes lymphatic transport and the metabolism in the liver is evaded; the prodrug enters the blood and free drug is liberated (Shackleford et al. 2003).

In conclusion, FA-drug conjugates have a number of different advantages including enhanced metabolic stability, improved biopharmaceutical profile due to higher lipophilicity of the conjugates in comparison to the parent drug, continuous drug release from the prodrug, and in some cases even lymphatic transport, all of which can largely influence the delivery of the free drug.

5.3.2 TG-Drug Conjugates

The structure of triglycerides (TG) consists of glyceride backbone and two FA molecules in position *sn*-1, *sn*-2 or *sn*-3 of the glyceride (Figure 5.2). Depending on which FA is replaced with a drug molecule, various TG-drug conjugates can be designed. Since TG compose majority of lipid nutrients, TG-based prodrug strategy relies on the physiological lipidic processing pathways to a great extent (Shackleford et al. 2007). This approach greatly influences the drug absorption, since it provides the opportunity to exploit intestinal lymphatic transport, which can result in lymphatic drug targeting (Loiseau et al. 1997; Loiseau et al. 1994), targeting to liver (for imagining purposes), brain (Bakan et al. 2000; Garzon-Aburbeh et al. 1986; Jacob et al. 1987) and other organs, as well as better oral drug delivery (Cullen 1984; Kumar and Billimoria 1978; Paris et al. 1980). The *sn*-2 position is the most common place of drug conjugation, since the *sn*-1 and *sn*-3 positioned FAs of the TG are hydrolyzed in the lumen of the intestine, leaving the *sn*-2 position monoglyceride free for absorption (Shackleford et al. 2007). However, the complex of lipase and colipase in the lumen will hydrolyze the FA in the *sn*-1 and *sn*-3 position merely if there are no steric strains between the prodrug and the enzymes. Bearing this in mind, *sn*-1 monoglycerides can be purposefully designed in a way to achieve lymphatic targeting (Sugihara et al. 1988).

5.3.2.1 TG-Drug Conjugates for Targeting Conditions with Involvement of the Lymphatic System

The TG-based prodrug approach is most commonly used lipidic prodrug strategy to achieve lymphatic transport; exploiting the physiological TG processing, this approach yielded in greatest success. Recently this approach was employed for mycophenolic acid (MPA), an immunosuppressant drug, conjugated to the glyceride through *sn*-2, resulting in 1,3-dipalmitoyl-2-mycophenoloyl glycerol (2-MPA-TG) (Han et al. 2014, 2015, 2016). By exploiting the TG processing pathways such as TG deacylation-reacylation pathway, 2-MPA-TG can undergo lymphatic transport following oral administration. Comparing the intraduodenal administration of free MPA and the prodrug to rats, an 80 times higher lymphatic transport of moieties related to MPA and 100 times higher MPA level inside lymphocytes were shown (Han et al. 2014). Liberation of the 2-MPA-monoglyceride from the conjugate allowed the re-esterification in the enterocyte. Mesenteric lymph nodes contained 28 times higher levels of MPA in a group that was administered with a prodrug, in comparison to the control that received the free MPA, and drug concentration was sustained for 8h, demonstrating the continuous liberation of MPA from the re-esterified TG-prodrug. In addition to this, FA-drug conjugates of MPA were studied as well; however, it was shown that the sufficient lymphatic transport was not achieved due to low metabolic stability and low absorption (Han et al. 2014).

Additional study was conducted in greyhound dogs: 2-MPA-TG considerably increased lymphatic transport, almost 290 times higher than the administration of the free MPA (Han et al. 2016). The amount of free drug in the lymph-residing lymphocytes and lymph nodes was up to 6 times and 21 times greater, respectively (Han et al. 2016). Comparison with the rat study showed that the lymphatic transport was greater in dogs, where approximately 3-fold higher MPA dose was found in the lymph (Han et al. 2014, 2016). This study on free drug levels in the lymph nodes and lymph-residing lymphocytes demonstrates the ability of TG-prodrugs to target the diseased lymphatic tissues by entering the TG deacylation-reacylation pathways, hence increasing the MPA immunomodulatory effects.

Lymph-residing lymphocytes make up 90% of all lymphocytes, whereas 50% belong to intestinal lymphatic tissues. TG-drug conjugates have the ability to incorporate TG processing pathways and undergo lymphatic transport therefore circumventing the liver on the way to the blood circulation. It is advantageous to use such approach in conditions that involve lymphatic system (i.e., if the targets are lymphocytes).

5.3.2.2 TG-Drug Conjugates Applications for Hormone Therapy

In addition to FA-based prodrug approach, TG-drug conjugates were also employed for overcoming extensive high first-pass hepatic metabolism of testosterone (Amory et al. 2003). A study in rabbits was conducted comparing orally administered testosterone-triglyceride conjugate (TTC) and TU with 4 and 8 mg/kg. Pharmacokinetic study resulted in 2–3 times greater testosterone AUC of TTC in comparison to TU following oral administration, demonstrating enhanced absorption after TTC administration, most likely due to higher extent of lymphatic transport.

An additional study demonstrates that a change in design, for instance incorporation of linker in the *sn*-2 position between testosterone and TG yields in higher stability (Hu et al. 2016). Linker was a 5-carbon chain with a self-immolative linker introduced to improve systemic drug release, and an alkyl group in the β position of the ester bond was introduced to interfere (steric hindrance) with the GIT enzymes. Testosterone oral bioavailability was 90 times higher than that following TU administration. Careful consideration of branched and/or self-immolative linkers can highly influence the stability in the GIT.

5.3.3 Steroid-Drug Conjugates

The most commonly used place of conjugation for steroid-drug conjugates is the hydroxyl group in position three of the four-ring steroid structure (Radwan and Alanazi 2014) (Figure 5.2). Steroid-based prodrug approach can be a useful tool for targeting cholesterol rich tissues in diseases with accumulation of cholesterol (e.g., ovarian cancer), infertility, etc.

5.3.3.1 Steroid-Drug Conjugates Applications for Cancer Therapy

Resistance to anticancer agents is a rising problem, hence an innovative approach needs to be used, for instance drugs in an enzyme-instructed self-assembly (EISA). EISA uses enzymes to create assemblies of small drug molecules on the surface or within the cell, specifically targeting growing cancerous tissues. Phosphotyrosine-cholesterol conjugates were created in an attempt to treat platinum-resistant ovarian cancer cells (Wang et al. 2016). Cholesterol is present on both inside and outside of the cell surface, reacts with proteins that influence essential functions in the cells, and it can play a role in the self-assembly of nanostructures. Cholesterol conjugate of D-phosphotyrosine was involved in production of EISA, which resulted in binding to the surface/inside of the platinum-resistant cancer cells and triggering a chain of events resulting in cell death signaling. This conjugate was proven to be more selective and pharmacologically active than cisplatin. In this case direct conjugation between cholesterol and D-phosphotyrosine is favored, since a linker based on phenylalanine between the lipid carrier and the drug resulted in lower cytotoxicity. In this study cholesterol-drug conjugate was not used for oral drug delivery, but it can provide an important basis for using this approach in oral drug delivery in the future.

Cholesterol is needed for speedy cancer cell growth and a number of LDL receptors are overexpressed on the cells, as mentioned in Section 5.2.2. This can be used as a tool to direct cholesterol-drug conjugates to the specific cancerous tissues and their incorporation into LDL. An example of such approach was shown on 5-fluorouracil (5FU), where a 5FU-cholesterol conjugate was incorporated within the LDL; the LDL particles were then bound to the LDL receptors and taken up by cancerous cells, resulting in an increased efficacy comparing to the free 5FU in an *in vivo* system (Radwan and Alanazi 2014).

5.3.3.2 Steroid-Drug Conjugates Applications for Antiviral Therapy

Zidovudine is anti-AIDS agent and drug for hepatitis, characterized by low brain permeability, due to mechanisms of resistance for antiviral agents, such as efflux transporters. Ursodeoxycholic acid (UDCA) is a bile acid, used for conjugation with zidovudine, due to its ability to permeate into the brain. It contains three hydroxyl groups for conjugation with a drug (Dalpiaz et al. 2012, 2014). The resulting conjugate demonstrated higher antiviral effect than zidovudine alone, due to decreased liberation of the free drug in the plasma, improved CNS permeation, and overcoming the mechanism of resistance to antiviral agents (Dalpiaz et al. 2012).

5.3.4 PL-Drug Conjugates

PLs originate from dietary and endogenous (bile salts) sources; they travel through intestinal lumen and prior to absorption get hydrolyzed by the stereospecific enzyme phospholipase A_2 (PLA_2), which liberates the *sn*-2 positioned FA, and yields two moieties ready for absorption: an *sn*-1 lysophospholipid (LPL) and a FA. LPL is reacylated to PL within the enterocyte by lysophosphatidylcholine acyltransferase; this PL is now attached to the surface of the lipoprotein.

The drug in the PL-drug conjugates can be connected to the phosphate group, to the *sn*-2 or to the *sn*-1 position of the glyceride (Figure 5.2). When linking the drug to the phosphate group the most common scenario is conjugation between the drug hydroxyl functional group and the PL phosphate. This is employed when hydrolysis by the enzyme PLA_2 is desired; the prodrug is absorbed and enters PL physiological processing (Alexander et al. 2005; Kucera et al. 2001). On the contrary, conjugation through the *sn*-2 position exploits the action of PLA_2, using it as a prodrug-activating enzyme (Dahan et al. 2007, 2017a; Markovic et al. 2019a; Pan et al. 2012). Conjugation in this case occurs between the *sn*-2 hydroxyl group of the PL and amino/hydroxyl group of the drug moiety; it was shown that a linker of particular length between the glyceride backbone and the drug is advantageous and can alter the drug release/prodrug activation significantly.

5.3.4.1 PL-Drug Conjugates Application for Alzheimer's Disease

PL-prodrug where a drug is attached to the *sn*-2 position of the PL have similar surface properties and aggregation behavior to endogenous PLs (Kurz and Scriba 2000). PL-prodrug approach was employed for valproic acid (DP-VPA), aiming to pass the BBB, and permeate into CNS intact; levels of PLA_2 enzyme are higher during epileptic seizure, which would permit liberation of the valproic acid in the targeted site (epileptic focus) (Dahan et al. 2008; Isoherranen et al. 2003; Labiner 2002). Valproic acid is associated with severe side effects; DP-VPA has the potential to lower the systemic exposure to valproic acid, thus the dose could be reduced, and side effects improved. It has been shown that following oral ingestion of DP-VPA in dogs, 10%–18% of the dose undergoes intestinal lymphatic transport, showing just how important the structural similarity between the PL-prodrug and endogenous PLs is (Shackleford et al. 2007). Additional study in rats was conducted with DP-VPA in two different formulations (long and medium chain TGs) following oral absorption in fed and fasted state (Dahan et al. 2008); 5% from the overall dose of DP-VPA was transferred via the lymphatics, and the overall systemic bioavailability was 8.8%; the percentage of absorbed DP-VPA in correlation with lymphatic transport was found to be 60% determined by direct measurement of DP-VPA in the lymph. Oral absorption of DP-VPA in long-chain TG formulation was three times higher than the DP-VPA in medium chain TG formulation; long-chain TGs are involved in chylomicron synthesis and lymphatic transport (Dahan and Hoffman 2005, 2006), hence these results demonstrate the significance of chylomicrons and intestinal lymphatic transport for absorption of DP-VPAs. Additional confirmation of lymphatic transport was an increase (3-fold) in the absorption in fed state vs fasted state (Dahan et al. 2008). PLA_2 knockout mice vs control mice were studies to determine the effect of PLA_2-mediated hydrolysis of DP-VPA. However, similar plasma profiles were found for both mice groups; the DP-VPA includes direct conjugation (without linker) among the PL and the drug, and this prodrug design was shown to be resistant to PLA_2-mediated activation. It is evident that DP-VPA travels through the intestinal lumen, permeates through the gut wall, and passes into the enterocytes intact; at this point it becomes a part of CMs, and undergoes lymphatic transport (Dahan et al. 2008).

5.3.4.2 Potential Uses of PL-Drug Conjugates for Inflammatory Conditions

A number of inflammatory conditions and various tumors demonstrated overexpression of secretory PLA_2 (e.g., atherosclerosis (Goncalves et al. 2012), rheumatoid arthritis (Pruzanski et al. 1985) and colon cancer (Tribler et al. 2007)). In IBD significant overexpression was shown in the inflamed regions of the intestine (Haapamaki et al. 1997, 1999; Minami et al. 1994; Peterson et al. 1996). We propose a strategy for exploiting the overexpression of PLA_2 via PL-prodrugs. Following oral administration, the PL-prodrug can pass through the gut lumen, where activation would occur, once the prodrug arrives to the places with overexpression of PLA_2; the active drug would be liberated from the prodrug complex in the specific site of inflammation. This strategy could be used in various inflammatory disease with overexpression of PLA_2.

Steric hindrance within the PL and the drug results in suboptimal PLA_2-mediated activation of the PL-drug conjugates, due to interference with the active site of the enzyme. Incorporating a linker between the drug and the PL moiety can increase the distance between them, and help overcome this hindrance. PL-indomethacin and PL-diclofenac prodrugs containing carbonic linkers of different length between the drug and the lipid were considered and optimized using *in silico* simulations grounded on thermodynamic integration method and weighted histogram analysis method (WHAM)/umbrella sampling (Dahan et al. 2016, 2017b). *In vitro* studies on PLA_2-mediated liberation from the PL-drug complex demonstrated excellent correlation with *in silico* predictions (Dahan et al. 2008, 2016, 2017a, 2017b). For PL-diclofenac, 6-carbon linker was demonstrated as optimal linker for this prodrug, resulting in highest degree of PLA_2-mediated activation (Dahan et al. 2017a, 2017b). The optimal PL-indomethacin prodrug contained a linker with 5 -CH_2 units (DP-155). *In vivo* absorption studies in rats after oral administration of DP-155 were conducted; no absorption of intact conjugate was revealed, but liberation of the free indomethacin from the complex took place within the gut lumen by the action of PLA_2, demonstrating controlled-release drug profile in the systemic circulation (Dahan et al. 2007). This is evident from $t_{1/2}$ of free indomethacin (10h), and $t_{1/2}$ of DP-155 (24h), which points out that presence of free indomethacin is due to PLA_2-mediated liberation from the DP-155 in the intestinal lumen (Dahan et al. 2007). Prodrug containing shorter linker length (2-carbon linker, DP-157) resulted in 10-fold lower drug release. Optimal linker length is to be evaluated on a case-by-case basis, since the drug size and/or volume can vary, and can impact the steric hindrance.

The important takeaway from these studies is the fact that linker length between the drug and the PL dictates the rate of PLA_2-mediated activation and the overall occurrence of the liberated active drug in the diseased site.

5.3.4.3 PL-Drug Conjugates Applications for Antiviral Therapy

Peptidophospholipid prodrugs were synthesized by conjugating oligopeptides with antiviral activity to the *sn*-2 PL position (Rosseto and Hajdu 2014). Hydrolysis in the *sn*-2 position produced free oligopeptides, corresponding to the rate of hydrolysis of endogenous PLs. Prerequisite for PLA_2 activation was α-methylene group of the oligopeptides, and peptidophospholipids which lacked this functional group were not activated by PLA_2 (Bonsen et al. 1972); this group may be considered a linker that permits binding between the conjugate and the active site of the enzyme (Rosseto and Hajdu 2014).

Drug moieties may be connected to the PLs' phosphate group, as mentioned in Section 5.3.4. A number of nucleoside drugs with antiviral/antineoplastic effect, were conjugated in this way. The physicochemical properties (low lipophilicity) of nucleoside drugs cause low absorption and poor pharmacokinetic profile. Additionally, lower transporter expression causing the worse drug uptake is among the key reasons for nucleoside resistance (i.e., deoxycytidine kinase (dCK) loss of expression and lower activation of the drug, human equilibrative nucleoside transporter 1 (hENT1), higher efflux by P-glycoprotein (P-gp) and cytidine deaminase/nuclear exonucleases/ribonucleotide reductase loss of function), which becomes an increasingly big problem for cancer/antiviral therapy (Alexander et al. 2016; Zaro 2015). PL-nucleoside conjugates have the ability to enhance passive permeability and avoid transport resistance barriers (Lambert 2000; Zaro 2015). By decreasing the MDR-mediated drug efflux from the cell, the drug accumulates within the cell and achieves appropriate pharmacological effect (Alexander et al. 2005). PL-nucleoside prodrugs have the ability to improve the activity of nucleoside kinase activity by releasing the drug in a monophosphate form within the cell, and by doing so circumvent dependence on dCK-mediated activation (Alexander et al. 2016; Zaro 2015).

5.3.4.4 PL-Drug Conjugates Applications for Cancer Therapy

PL-based prodrug approach was shown to be very useful tool for cancer therapy, and it represents a rather novel tool for drug targeting to the cancer cells. In a recent study, an antiangiogenesis agent, fumagillin was conjugated to PL in the *sn*-2 position, aiming to reduce poor chemical stability in the blood and photoinstability of fumagilin. NP drug delivery system was used for targeting the prodrug to the angiogenic endothelium expressing $\alpha_v\beta_3$-integrin, and allowed passive transfer to the targeted cell (Pan et al. 2012; Zhou et al. 2012). Fumagilin chemical structure contains two epoxide rings and conjugated decatetraenedioic tail, responsible for its lack of chemical stability and photoinstability, respectively; during prodrug synthesis, a modification in the tail resulted in a 7-carbon acyl linker in the PLs' *sn*-2 position. Loading of the prodrug into the NPs aided the endocytosis (Lanza et al. 2002) and the passive transport of the PL-fumagilin prodrug to the target cell (Partlow et al. 2008; Soman et al. 2009); NP binds to the target cell creating a hemifusion complex among the cell membrane and NP, and PL-fumagilin conjugate is transferred into the cell (contact-facilitated drug delivery mechanism). Within the cell, the PL-fumagilin prodrugs are cleaved by the local PLA_2 enzymes, and free fumagilin is liberated from the *sn*-2 position of the prodrug. Animal studies showed reduction of angiogenesis with fumagilin NP in comparison to control NP (Pan et al. 2012). This study clearly demonstrates the significance of a linker/spacer among the PL and the drug, in this case a linker is a part of the fumagilin structure, which enables the prodrug to be activated by PLA_2. In addition to this, combined strategy of lipid-prodrug approach and specialized NP can considerably improve chemical stability (but leaving the active functional groups intact), drug loading and targeting to the site of action.

A contact-facilitated drug delivery mechanism using PL-prodrug approach was also used in a different study for treatment of multiple myeloma, where the MYC-MAX dimerization inhibitor was conjugated to the PL moiety (Soodgupta et al. 2015). PL-prodrug of MYC-MAX inhibitors managed to overcome difficulties in protein-protein and protein-DNA interaction inhibition, that the free drug has, and increased efficacy toward myeloma cell cultures. Mice model with myeloid leukemia was used for testing integrin-targeted NP loaded with PL-prodrug. Prodrug-rich NP demonstrated considerably longer survival of mice when compared to controlled targeted/ untargeted NP (Soodgupta et al. 2015). Without the delivery system (NP), both the PL-prodrug and the Myc-inhibitor 1 failed to produce any pharmacological effect. More recently, prodrugs of fumagilin and docetaxel integrated in the $\alpha_v\beta_3$-targeted micellar nanotherapy resulted in anti-angiogenesis effect and ability to alleviate asthma in rats. Airway hyper-responsiveness and bigger microvascularity in asthma model were significantly decreased using both treatments (Lanza et al. 2017). Targeted drug delivery system and the use of PL-conjugates in such cases is of very important.

Gemcitabine-prodrug was synthesized, and the free drug was conjugated thought he phosphate group of the PL. PL-gemcitabine was designed to improve drugs' PK profile, to avoid multiple resistance pathways, stop protein kinase C (PKC), and enhance permeation to CNS (Alexander et al. 2003, 2005). As opposed to free gemcitabine, the PL-prodrug avoided hENT1 transporter, were not substrates for MDR-1 efflux transporter, and dCK action was not required.

Overall, PL-based prodrugs have the ability to integrate the physiological PL metabolic pathways, go through lymphatic transport, take advantage of PLA_2 overexpression in certain conditions as a prodrug-activating enzyme, or achieve controlled-release profile of the parent drug. Novel computational simulation can lead the future PL-drug design and structure optimization, allowing better and faster drug development.

5.4 Discussion

A lipid-prodrug approach can improve many aspects of oral drug delivery by modifying the physicochemical features of the drug itself (e.g., increasing the drug lipophilicity improves intestinal drug permeability). One of the main goals for designing a lipid prodrug is joining the metabolic lipid processing pathways, and in that way passing the biological barriers and delivering the drug to the site of action. This chapter reviews advantages, obstacles, applications, and promising future uses of the lipid-prodrug approach.

In some cases, FA-, TG- or PL-drug conjugates have the ability to join intestinal lymphatic transport, which allows the drugs to surpass first-pass hepatic metabolism; by avoiding this pre-systemic metabolism, higher percentage of dose enters the systemic blood. This design is particularly valuable in conditions where lymphatic targeting is needed (Dahan et al. 2008; Han and Amidon 2000; Han et al. 2015).

Lipidic prodrugs can influence the safety profile of the parent drug and minimize adverse effects. TG-based prodrug approach was employed for many nonsteroidal anti-inflammatory drugs (NSAIDS) drugs, where the drug conjugation to the *sn*-2 position of the TG resulted in less GIT irritation, showing a better safety profile (Kumar and Billimoria 1978).

The choice of lipid carrier is very important, especially in cases when the lipid itself has innate pharmacological activity. Following prodrug hydrolysis, two moieties are present, a drug and a carrier, both contributing to therapeutic effect (Kandula et al. 2016). Lipidic prodrug approach is also successful in improving drugs' chemical and metabolic stability (Borkar et al. 2015, 2016, 2017; Pan et al. 2012; Zhang et al. 2015). However, in some cases the PL-prodrugs are intended and designed to exploit the overexpression of PLA_2 in cancer or inflammatory disease, and use it as a prodrug-activating enzyme (Arouri et al. 2013). Since the PLA_2 is not specific for a certain type of FA, the design of the PL-prodrug can be planned ahead in order to ensure activation by this enzyme. In this case, the overexpression of PLA_2 in the inflamed/cancerous tissues can release the parent drug in the diseased tissues, accomplishing drug targeting to the site of action, at the same time avoiding the non-diseases regions, a strategy that could be successfully employed in diseases such as IBD (Dahan et al. 2017a, 2017b; Haapamaki et al. 1997, 1999; Markovic et al. 2019a).

As previously mentioned, there is an increasing use of linkers between the lipid carrier and the drug, which highly influences prodrug features such as drug liberation from the prodrug complex, drug targeting by exploiting enzymes involved in lipid processing and drug stability in the intestinal lumen. Self-immolative linkers can also aid in drug activation by enzymes, where the linker is removed spontaneously, and the free drug is liberated in the disease tissue.

A combination of a lipid-prodrug approach and different formulation strategies can largely enhance drug delivery and targeting (Arouri and Mouritsen 2011; Bala et al. 2013, 2016a; Gong et al. 2011; Kuznetsova et al. 2012; Pan et al. 2012; Zhou et al. 2012). Lipid carrier can also improve drug loading into formulation systems by forming hydrophobic bonds between the lipid carrier and regions of delivery system; by doing so, leaking of the drug can be avoided, stability of the delivery system can be improved (Arouri and Mouritsen 2011; Sun et al. 2016). In some cases, the delivery systems are not needed, since the lipid-drug conjugates have the ability to create self-assembled nanoparticles, that result in successful drug delivery (Gong et al. 2011; Sagnella et al. 2011).

However, before considering the lipidic prodrug design, there are certain aspects of this approach that need to be taken into account. Lipid-drug conjugates are inactive as such, and liberation of the free drug is necessary to achieve pharmacological effect; mechanisms of prodrug activation need to be considered prior to prodrug design. Good understanding of the barriers that both the drug and the prodrug come across on their way to the site of action is needed. Estimates of the rate of drug release need to be considered prior to prodrug synthesis (some chemical bonds between the lipid carrier and the drug degrade faster than the others) in order to ensure desired PK (Duhem et al. 2014; Irby et al. 2017).

Prodrug development can be long, expensive and challenging; this is why in order to design an optimal prodrug, modern *in silico* approaches are becoming increasingly employed. Prodrug optimization prior to actual synthesis of the lipid-conjugate can significantly decrease experimentation time, which was demonstrated for some PL-drug conjugates (Dahan et al. 2016, 2017b; Markovic et al. 2019a, 2019b).

5.5 Conclusion

This chapter presents a review of the lipidic prodrug approach, their applications, future potentials and challenges. Drug conjugation with FA, TG, steroid or PL can improve oral bioavailability, alter the PK profile of the parent drug, allow delivery/targeting to the lymphatics, and enable site-specific drug delivery. Careful design of lipid-drug conjugates personalized for particular disease may introduce an innovative therapeutic approach for the therapy of challenging conditions. Lipidic prodrug approach is anticipated to grow in the near future.

ACKNOWLEDGMENTS

This work is a part of Ms. Milica Markovic's PhD dissertation.

ABBREVIATIONS

2-MG	2-monoglyceride
2-MPA-TG	1,3-dipalmitoyl-2-mycophenoloyl glycerol
5-ASA	5-aminosalycilic acid
5-FU	5-fluorouracil
ADMET	absorption, distribution, metabolism, excretion and toxicity
AUC	area under the curve
BBB	blood-brain barrier
BCS	Biopharmaceutical Classification System
CM	chylomicron
CNS	central nervous system
CYP	Cytochrome P450
dCK	deoxycytidine kinase
DP-155	PL-indomethacin prodrug with 5-carbon linker
DP-157	PL-indomethacin prodrug with 2-carbon linker
DP-VPA	PL-conjugate of valproic acid
EISA	enzyme-instructed-self-assembly
FA	fatty acid
G3P	glycerol-3-phosphate
GIT	gastrointestinal tract
HDL	high-density lipoprotein
hENT1	human equilibrative nucleoside transporter 1
HIV	human immunodeficiency virus
IBD	inflammatory bowel disease
LDL	low-density lipoprotein
LPL	lysophospholipid
MAX	bHLH ZIP protein
MDR	multiple drug resistance
MG	monoglyceride
MPA	mycophenolic acid

6

The Progress of Prodrugs in Drug Solubility

Guilherme Felipe dos Santos Fernandes, Igor Muccilo Prokopczyk, Chung Man Chin, and Jean Leandro dos Santos

CONTENTS

6.1 Introduction

Prodrug design has been widely used as a successful approach to achieve selectivity, reduce toxicity, and improve pharmaceutical, pharmacokinetic, and pharmacodynamics features of drugs for some time. It has been estimated that approximately 10% of all the drugs in the market are prodrugs, implying that a prodrug strategy is a powerful approach for discovering novel drugs (Rautio et al. 2008). In the period between 2008 and 2017, the US Food & Drug Administration (FDA) approved at least 30 prodrugs, which corresponds to an average of 3 prodrugs approved each year during the period (Rautio et al. 2018). Historically, in the nineteenth century, serendipity allowed the discovery of several prodrugs, such as codeine (**1**), methenamine (**2**), phenacetin (**3**), and protonsil (**4**) (Brune and Hinz 2004; Huttunen et al. 2011). However, the term "prodrug" was first used by Adrian Albert in 1958 to describe the compounds whose biological activity was dependent on previous biotransformation (Albert 1958).

Prodrugs are compounds that undergo chemical and/or enzymatic bioconversion *in vivo*, releasing the active parent drug responsible for promoting the biological effect (Figure 6.1) (Chung et al. 2008; Jornada et al. 2015). Prodrugs have been classified into the following types: (a) classical prodrugs; (b) bioprecursors; (c) mutual prodrugs or co-drugs; and (d) selective prodrugs. Classical prodrugs are a result of linkage between a certain drug and a transport moiety which does not exhibit any pharmacological activity by itself. In general, the choice of a suitable carrier moiety promotes the improvement in the pharmacological activity by increasing the bioavailability of the drug, reducing its toxicity, prolonging its action, and increasing its selectivity. The examples of classical prodrugs include enalapril (**5**), pivampicillin (**6**), oseltamivir (**7**), dipivefrin (**8**), tazarotene (**9**), fosfluconazole (**10**), propofol phosphate (**11**), and others. Bioprecursors are another kind of prodrugs without a carrier moiety, which undergo *in vivo* metabolic activation, and generate on their own, a metabolite responsible for the pharmacological effect. The examples of bioprecursors include acyclovir (**12**), losartan (**13**), omeprazole (**14**), sulindac (**15**), and

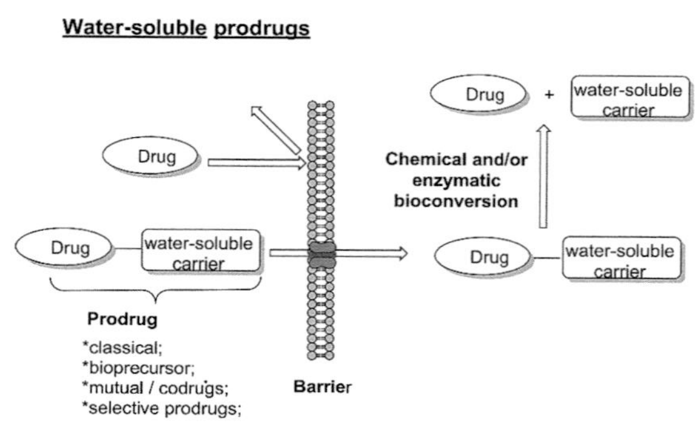

FIGURE 6.1 Prodrug approach to improve water-solubility.

others. Mutual prodrugs or co-drugs are the types of prodrugs that contain a temporary linkage between two active compounds/drugs, and which, after undergoing *in vivo* biotransformation, release the parental drugs that exert a synergic effect through different mechanisms of action (Das et al. 2010). Sultamicillin (**16**) is an example of mutual prodrugs/co-drugs, which is formed by a linkage between ampicillin (**17**) and sulbactam (**18**). Selective prodrugs, on the other hand, contain carriers capable of selectively releasing the drug at the site of action, minimizing adverse effects and reducing toxicity. The main carrier moieties used by selective prodrugs are macromolecules (e.g., antibodies, proteins, and glycoproteins) and synthetics (e.g., polymers, liposomes, etc.). The use of antibodies allows the preparation of selective bioconjugates, which release the drug inside the target cell post endocytosis. One approach, referred to as the Antibody Directed Enzyme Prodrug Therapy (ADEPT), has demonstrated potential to serve as a promising therapy for treating solid cancers (Sharma and Bagshawe 2017).

The prodrug approach offers enormous applicability as it overcomes several barriers in their applications, including poor aqueous solubility, local irritation after application, inadequate permeability and bioavailability, chemical instability, fast presystemic metabolism, and inappropriate physicochemical characteristics of the drug, among others (Walther et al. 2017).

Solubility is a key molecular property for a drug. In order to achieve the absorption and action of a drug, it must be able to be solubilized in the biological fluids. Despite its importance, poor solubility remains to be a common obstacle encountered in drug development. It has been estimated that approximately 40% of the drug candidates identified in High-Throughput Screening (HTS) campaigns exhibit low aqueous solubility (<10 μM). Moreover, approximately 37% of the top 200 oral drug products in Great Britain, Japan, Spain, and the United States exhibit solubility less than 0.1 mg/mL, demanding strategies to overcome this issue (Takagi et al. 2006).

Currently, the strategies commonly used to improve drug solubility include: (a) co-solvents; (b) surfactants; (c) inclusion in cyclodextrin complexes; (d) lipid-based systems; (e) co-crystals; (f) polymorphism; (g) amorphous solid dispersion; (h) particle size reduction; (i) formation of salts; (j) solubilizing moieties; and (k) the prodrug approach (Williams et al. 2013). At certain times, the use of strategies such as salt formation, solubilizing agents, and particle size reduction do not result in achieving the desired levels of solubility. In addition, certain surfactants used in parenteral drug formulations may induce toxic effects and anaphylactic reactions (Williams et al. 2013). The solubilizing moieties may be categorized into reversible and irreversible. In the case of reversible moieties, the compound is a prodrug. The main solubilizing moieties used are (salts of) carboxylic acid, amines, sulfonates, phosphates, sugar derivatives, polyethylene glycol, etc.

In this chapter, the application of prodrugs as a promising approach to improve drug solubility will be described, along with highlighting the progress in utilizing this strategy which generated several approved drugs worldwide.

6.2 Ester Prodrugs

Ester is an organic functional group that has been widely explored in prodrug designing due to its several favorable features, including being amenable to hydrolysis both *in vitro* and *in vivo*, a certain level of resistance to hydrolysis favoring intact absorption, and controlled kinetic release modulated through steric and electronic effects. *In vivo*, the ester group may be hydrolyzed by esterases. The products of this hydrolysis (alcohol and carboxylic acid derivatives) are polar in nature and water-soluble, allowing renal elimination. In humans, two carboxylesterases (hCE1 and hCE2) have been identified to be involved in ester hydrolysis. Both the enzymes are present in the liver, although the levels of hCE1 are higher to those of hCE2. On the other hand, only hCE2 is expressed in the intestine. Moreover, the hCE1 enzyme exhibits a preference for ester containing a large, bulky acyl group and a small alcohol group, while hCE2 has been demonstrated to prefer small acyl groups and a large alcohol group. The examples of drugs metabolized by hCE1 are enalapril (**5**), oseltamivir (**7**), ramipril (**19**), meperidine (**20**), simvastatin (**21**), lovastatin (**22**), fenofibrate (**23**), and methylphenidate (**24**), while those metabolized by hCE2 are prasugrel (**25**), acetylsalicylic acid (**26**), oxybutynin (**27**), tenofovir disoproxil (**28**), and adefovir dipivoxil (**29**), among others (Figure 6.2) (Casey Laizure et al. 2013).

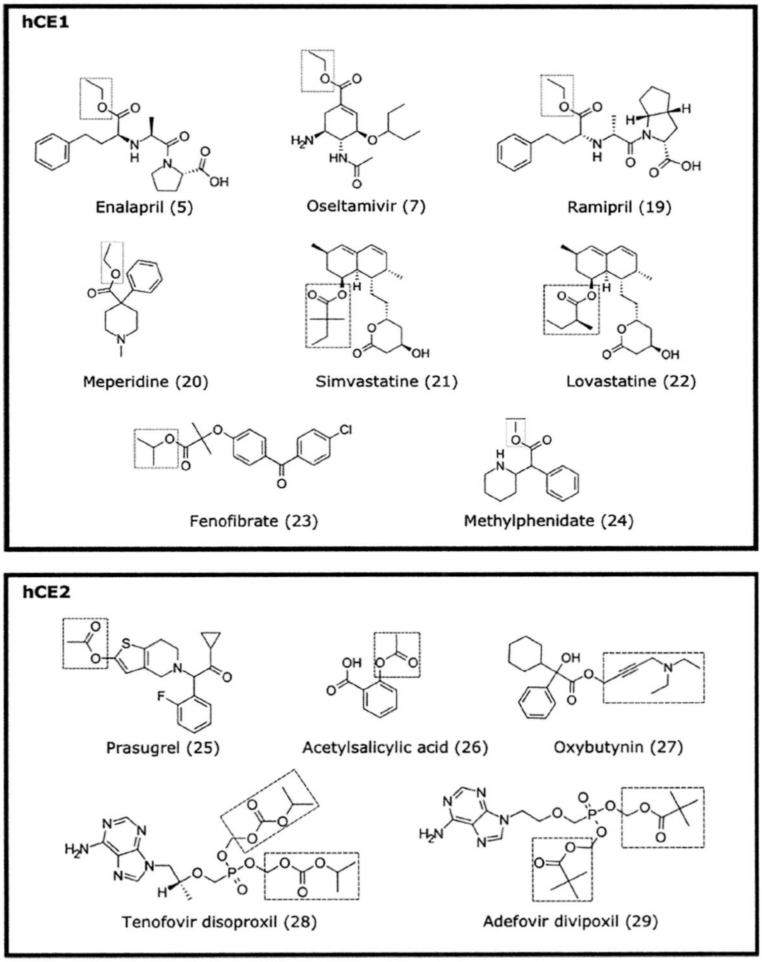

FIGURE 6.2 Ester prodrugs hydrolyzed by hCE1 and hCE2.

Most of the well-known examples of prodrugs important for improving drug solubility are present in the class of taxoids. Taxol is a diterpene that exhibits anticancer activity, which is present in the bark of the *Taxus brevifolia* and is capable of promoting the formation and stabilization of tubulin polymers. As a result of its low water-solubility, it was initially formulated as a solution in 50% dehydrated alcohol and 50% Cremophor EL; however, challenges encountered in the formulation and administration of this drug demanded novel alternatives such as prodrugs (Adams et al. 1993). In this context, different kinds of taxoid prodrugs aiming to increase water-solubility have been identified in the literature.

Prodrugs formed of docetaxel-glucopyranosides were prepared through chemo-enzymatic reactions using lactase, β-galactosidase, and β-xylosidase. These prodrugs exhibited water-solubility that increased 39- to 52-fold compared to that of docetaxel (**30**). In addition, cytotoxicity of the docetaxel-sugar prodrugs (**31–33**) was observed to be reduced in KC and MCF-7 cells. The incubation of prodrugs with KB human cancer cells revealed a relative resistance to hydrolysis, as the conversion to docetaxel ranged between 45% and 88% only (Table 6.1) (Shimoda and Kubota 2011).

Paclitaxel (**34**) prodrugs which contain an ester of malic acid (**35**, **37–38**) were reported to cause a 20-fold increase in water-solubility in comparison to the parent drug. The sodium salt (**36**) of this prodrug exhibited a threefold increase in water-solubility in comparison to the malic acid derivative. Both the prodrug and its salt exhibited effects similar to those of paclitaxel *in vitro*. However, the sodium salt was relatively more active than the parent drug in the *in vivo* p388 tumor model, demonstrated by an increase observed in the long-term survival of animals (Table 6.1) (Damen et al. 2000).

Isotaxel (**39**) is a water-soluble prodrug of paclitaxel that exhibits a water-solubility 1800-fold superior to that of paclitaxel (0.00025 mg·mL^{-1}). This 2′-*O*-benzoyl isoform undergoes a pH-dependent O–N acyl migration without generating any auxiliary products or byproducts. The kinetic study of O–N benzoyl migration at pH 7.4 revealed a $t_{1/2}$ value of 15.1 ± 1.3 min, which was described as adequate for systemic distribution (Hayashi et al. 2003). The studies on the structure-activity relationship have demonstrated that most of the water-soluble paclitaxel prodrugs presented derivatization at 2′-OH and 7-OH positions, which are the two most liable positions for the addition of hydrophilic and/or charged solubilizing moieties (Table 6.1) (Fang and Liang 2005).

Another anticancer compound with low water-solubility is etoposide (**40**). This drug is a topoisomerase inhibitor with a variable pharmacokinetic profile due to its limited water-solubility. In order to improve its water-solubility, a series of etoposide prodrugs were synthesized. These synthesized compounds (**41–44**) exhibited a 23- to 120-fold increase in solubility in comparison to etoposide, especially the malic acid derivative in salt form. In the case of the sodium salts, it was possible to consider the use of lyophilized powder for i.v. administration without the use of any surfactant such as Tween 80. The plasmatic half-life for all the prodrugs ranged from 3 to 24 h when the compounds were evaluated against a panel of human cancer cell lines (K562, KB, KB-R, A549, and U251). The prodrugs exhibited activity similar to etoposide, although certain derivatives exhibited increased activity in comparison to etoposide against the KB-R (1-resistant KB) resistant cells (Table 6.1) (Chen and Du 2013).

Propofol (**45**), an intravenous anesthetic with a rapid onset and short-term effect, also exhibits low water-solubility. It is common to use emulsion (Diprivan®); however, the disadvantages of its usage include injection-site pain, increased bacterial infection, interference with fat metabolism, and "propofol infusion syndrome," which is an obstacle that demands the development of soluble prodrugs. The reaction of propofol with 1,4-dioxane-2,6-dione leads to a water-soluble prodrug known as sodium 2-(2-(2,6-diisopropylphenoxy)-2-oxoethoxy) acetate (HX0921) (**46**). This compound has demonstrated fast onset, short duration of action, and fast recuperation in comparison to the parent drug, and also the other prodrugs such as fospropofol (**47**) (Table 6.1) (Zhang et al. 2019).

At certain instances, polar groups may enhance not just the water-solubility of the drug, they may even contribute to improvement in the bioavailability and biomembrane passage of the drug. One example

TABLE 6.1

Esters Prodrugs

Prodrug	Parent Drug	Prodrug Strategy	References
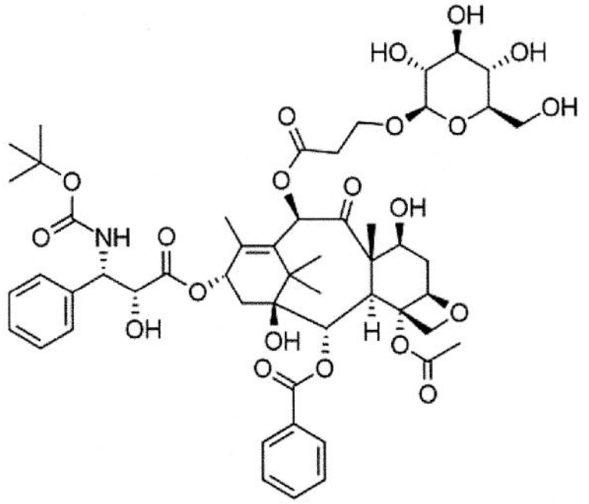 (31)	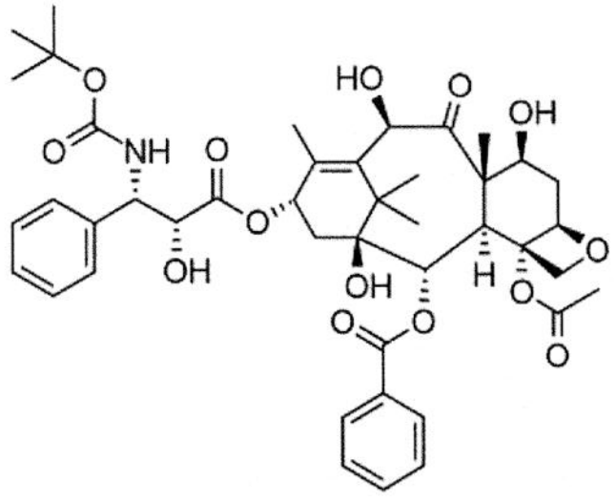 Docetaxel (30)	Improve water-solubility 52-fold	Shimoda and Kubota (2011)
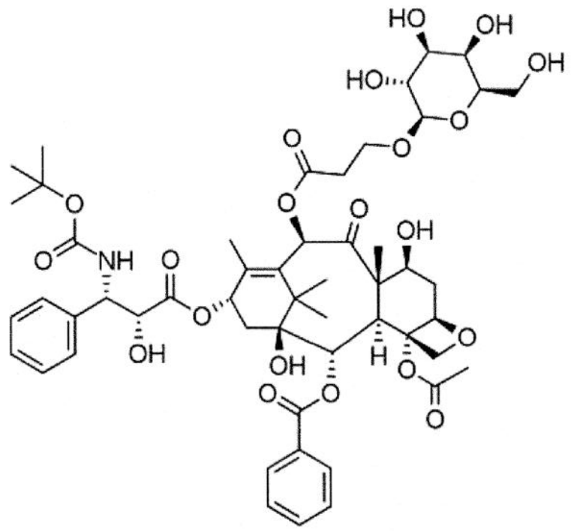 (32)		Improve water-solubility 49-fold	Shimoda and Kubota (2011)

(Continued)

TABLE 6.1 (*Continued*)

Esters Prodrugs

Prodrug	Parent Drug	Prodrug Strategy	References
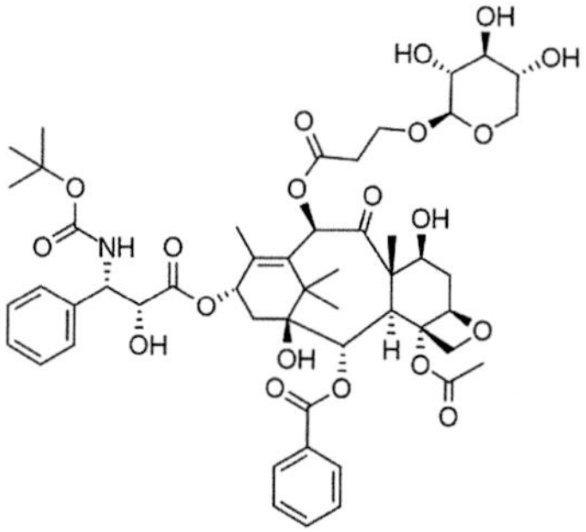 (33)		Improve water-solubility 39-fold	Shimoda and Kubota (2011)
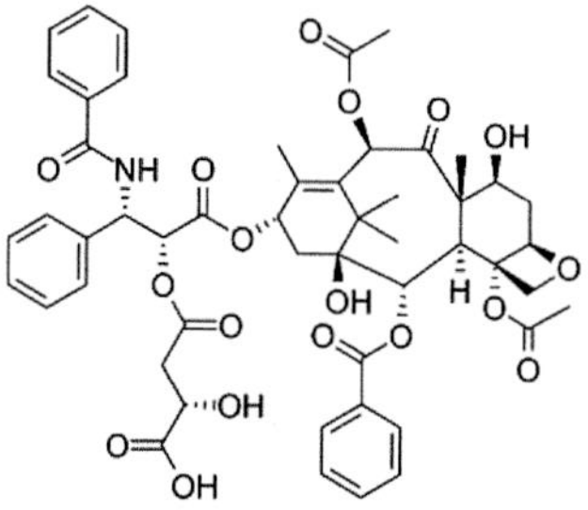 (35)	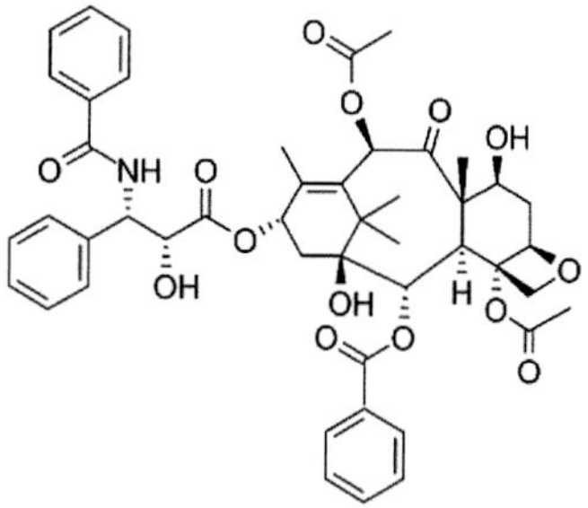 Paclitaxel (34)	Improve water-solubility 20-fold	Damen et al. (2000)

(Continued)

TABLE 6.1 (*Continued*)

Esters Prodrugs

Prodrug	Parent Drug	Prodrug Strategy	References
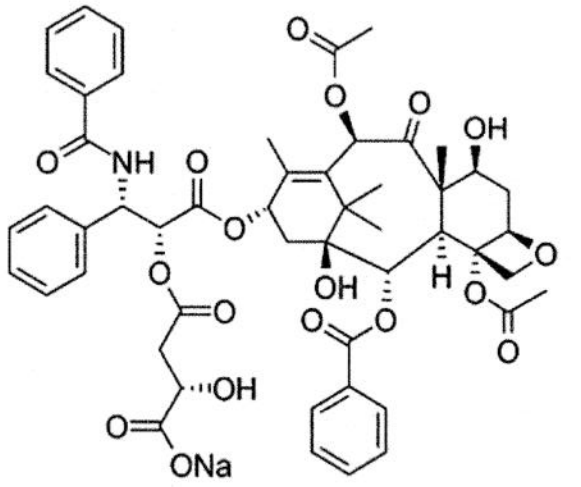 (36)		Improve water-solubility 3-fold	Damen et al. (2000)
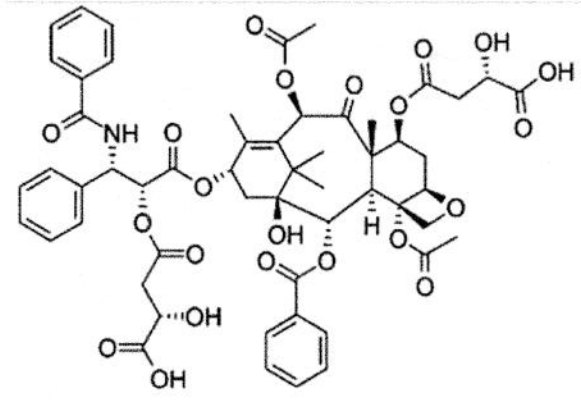 (37)		Improve water-solubility 20-fold	Damen et al. (2000)
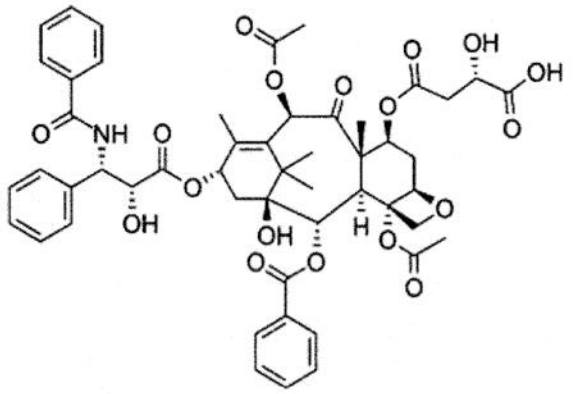 (38)		Improve water-solubility 20-fold	Damen et al. (2000)

(*Continued*)

TABLE 6.1 (*Continued*)

Esters Prodrugs

Prodrug	Parent Drug	Prodrug Strategy	References
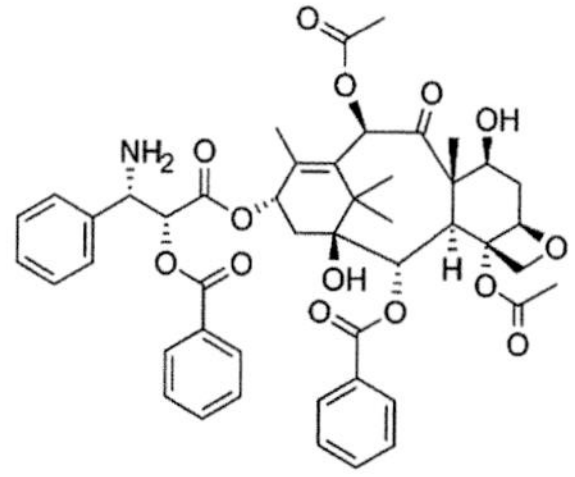 Isotaxel (39)		pH-dependent O-N acyl migration Improve water-solubility	Hayashi et al. (2003)
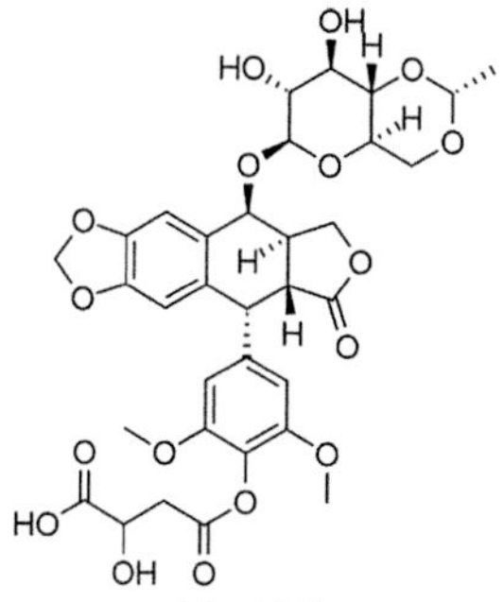 6Ia (41)	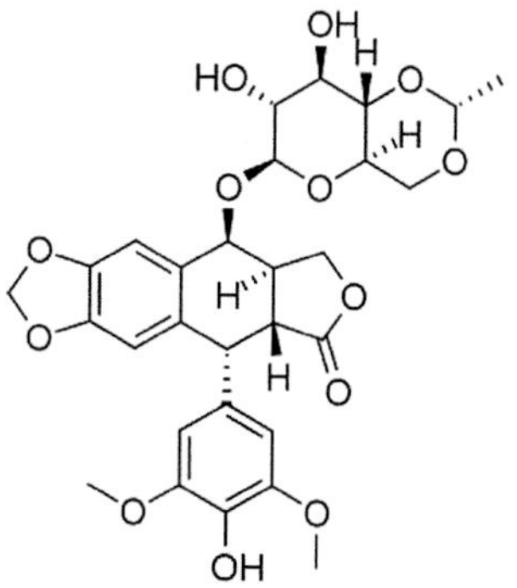 Etoposide (40)	Acid form Improve water-solubility 62-fold	Chen and Du (2013)

(Continued)

TABLE 6.1 (*Continued*)

Esters Prodrugs

Prodrug	Parent Drug	Prodrug Strategy	References
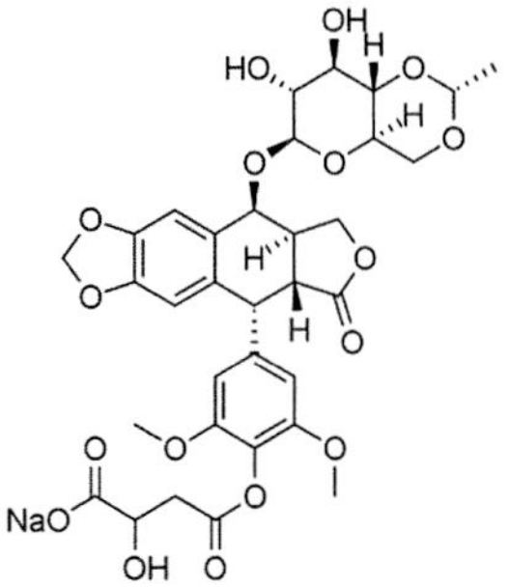 6IIa (42)		Sodium salt Improve water-solubility 107-fold	Chen and Du (2013)
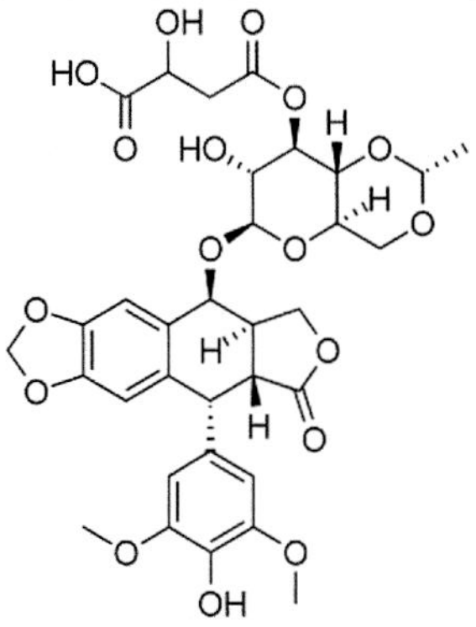 7Ia (43)		Acid form Improve water-solubility 76-fold	Chen and Du (2013)

(Continued)

TABLE 6.1 (*Continued*)
Esters Prodrugs

Prodrug	Parent Drug	Prodrug Strategy	References
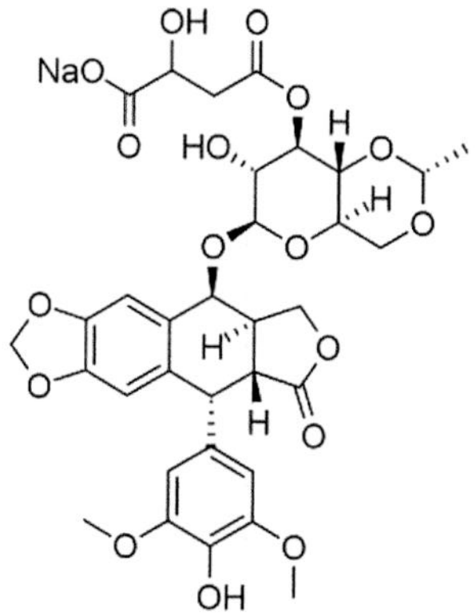 NaO OH O O HO H H O O O O O H O H O O O OH 7IIa (44)		Sodium salt Improve water-solubility 120-fold	Chen and Du (2013)
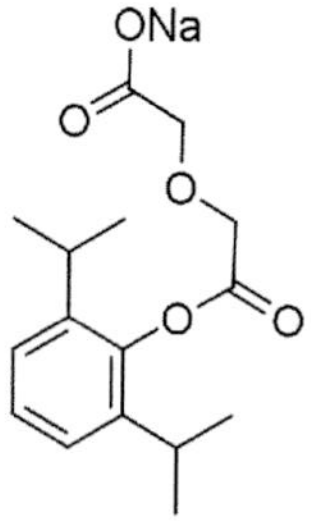 ONa O O O O HX0921 (46)	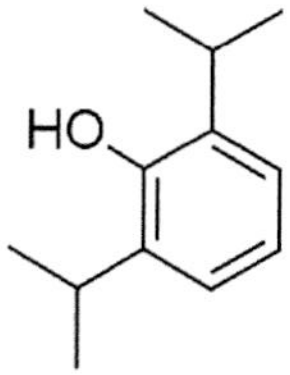 HO Propofol (45)	Improved water-solubility Fast onset, short duration of action and fast recuperation	Zhang et al. (2019)

(*Continued*)

TABLE 6.1 (*Continued*)

Esters Prodrugs

Prodrug	Parent Drug	Prodrug Strategy	References
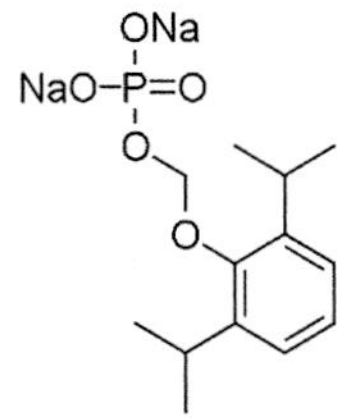 Fospropofol (47)		Improve water-solubility	Zhang et al. (2019)
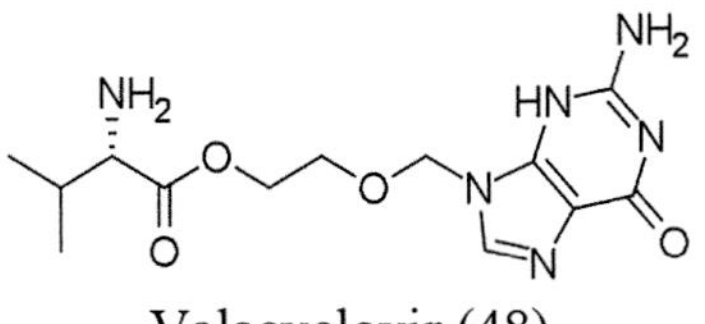 Valacyclovir (48)	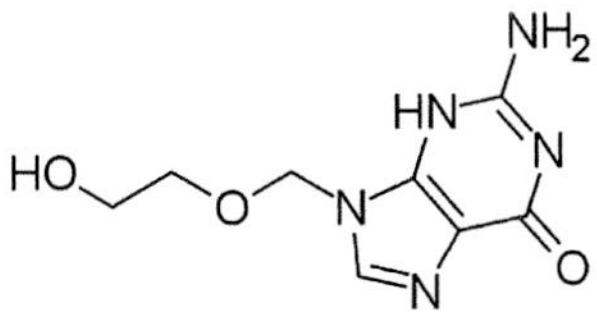 Acyclovir (12)	Improve bioavailability (55%) Biomembrane passage	Sinko and Balimane (1998), MacDougall and Guglielmo (2004)
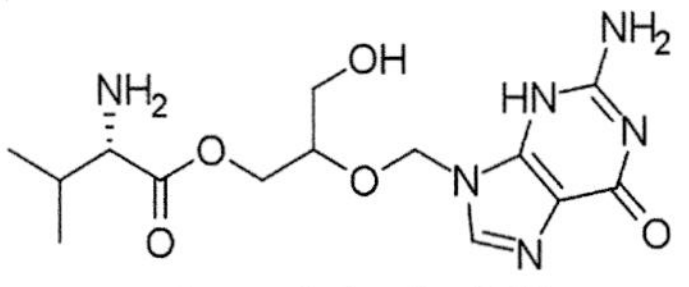 valganciclovir (50)	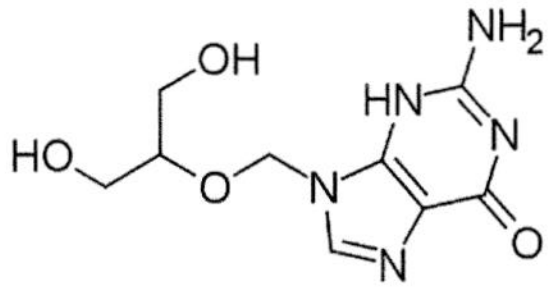 Ganciclovir (49)	Improve bioavailability (60%) Biomembrane passage	Stockmann et al. (2015)
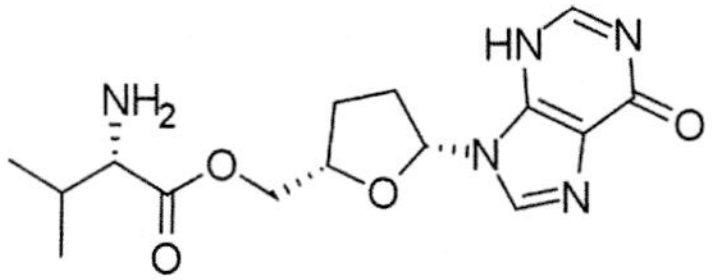 Didanosine valine (52)	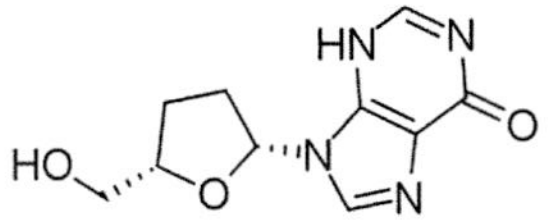 Didanosine (51)	Improve water-solubility 1,3-fold	Krečmerová (2017)

(*Continued*)

TABLE 6.1 (*Continued*)

Esters Prodrugs

Prodrug	Parent Drug	Prodrug Strategy	References
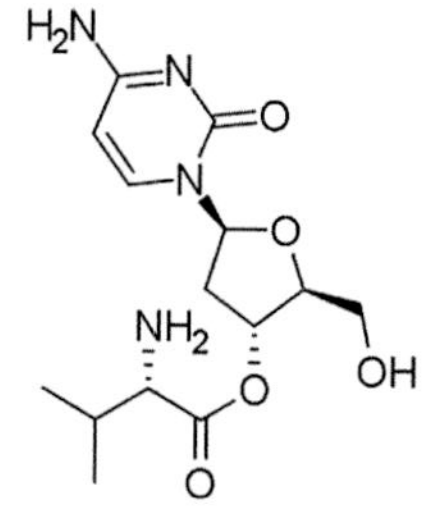 Valtorcitabine (54)	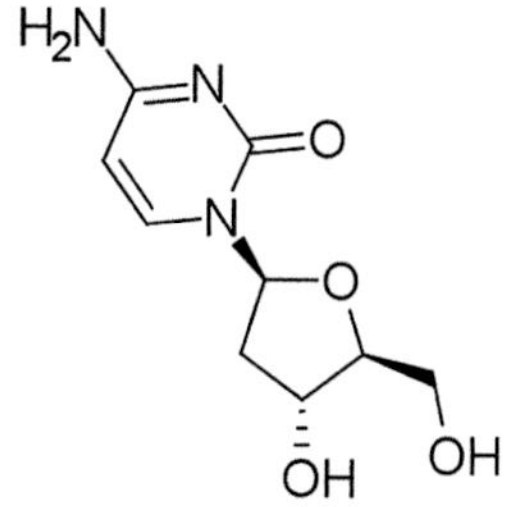 L-deoxycytidine (53)	Improve water-solubility 120-fold	Krečmerová (2017)
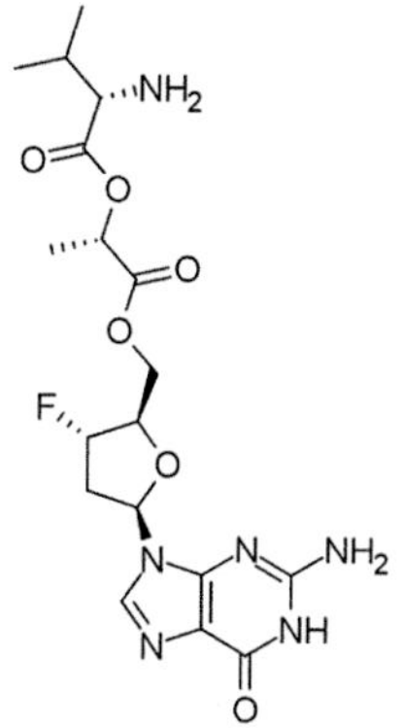 Lagociclovir valactate (56)	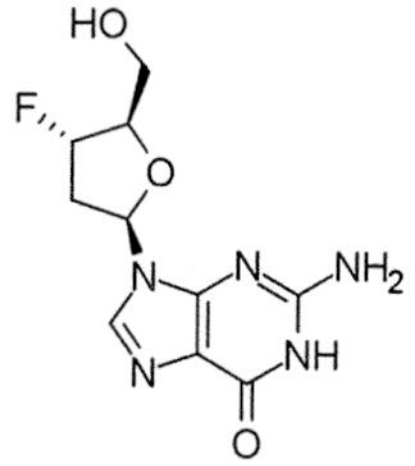 Lagociclovir (55)		Krečmerová (2017)
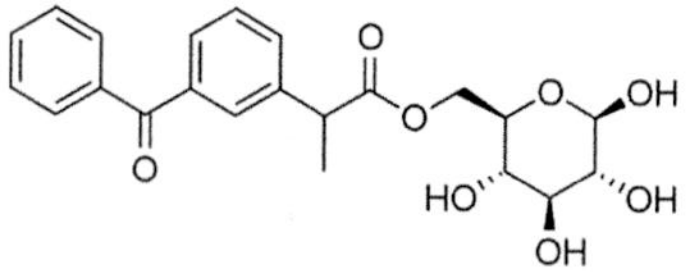 Ketoprofen-glucose (58)	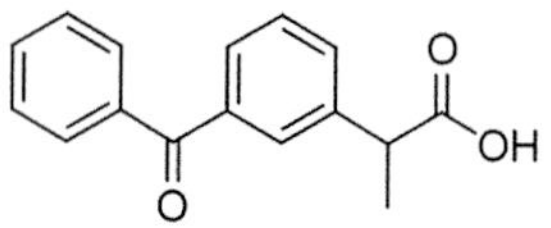 Ketoprofen (57)	Enhance the brain uptake Use GLUT-1 transporter	Gynther et al. (2009)

(*Continued*)

TABLE 6.1 (*Continued*)

Esters Prodrugs

Prodrug	Parent Drug	Prodrug Strategy	References
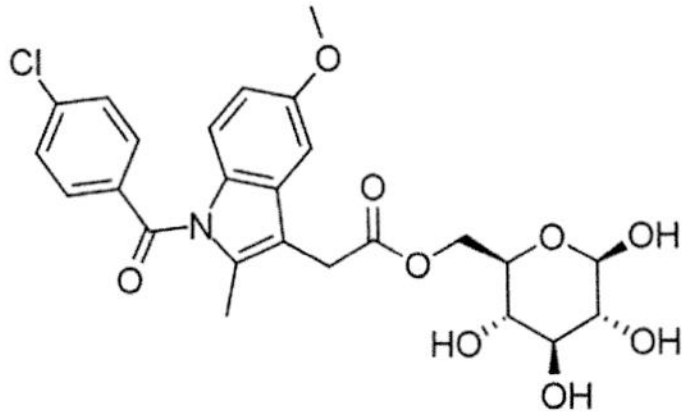 Indomethacin-glucose (60)	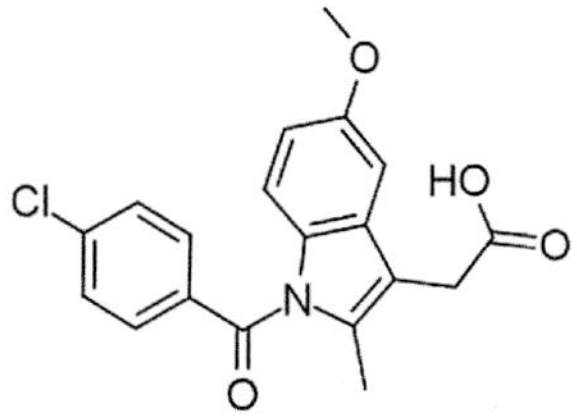 Indomethacin (59)	Enhance the brain uptake Use GLUT-1 transporter	Gynther et al. (2009)
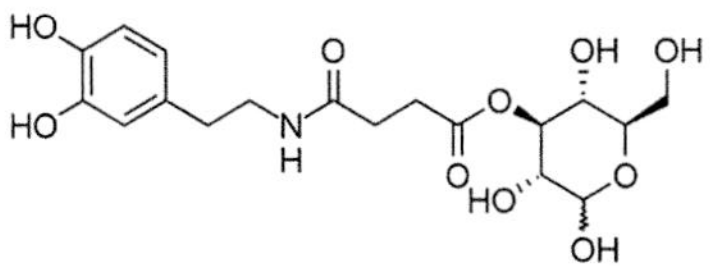 L-Dopa- succinyl -glucose (62)	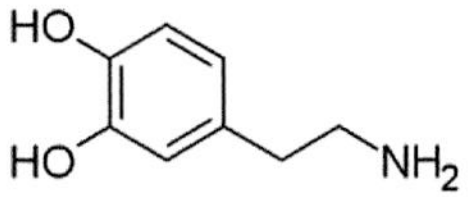 Dopamine (61)	Enhance bioavailability	Dalpiaz et al. (2007)
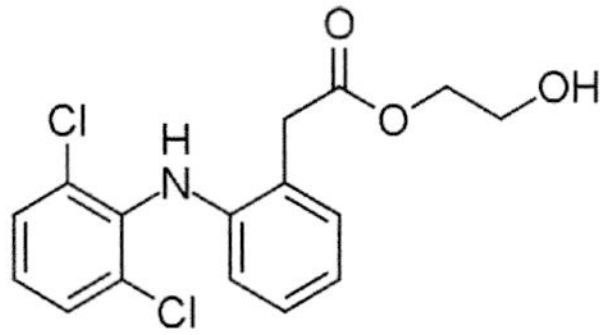 Glycerol diclofenac (64)	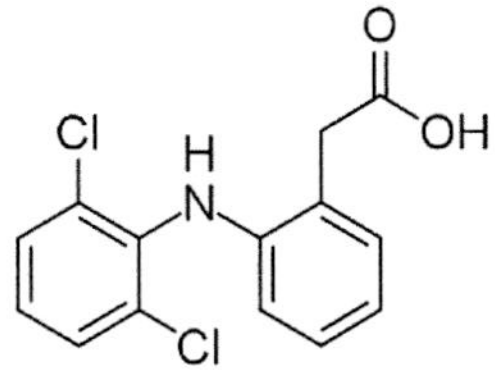 Diclofenac (63)	Improve water-solubility 22-fold	Lobo et al. (2014)
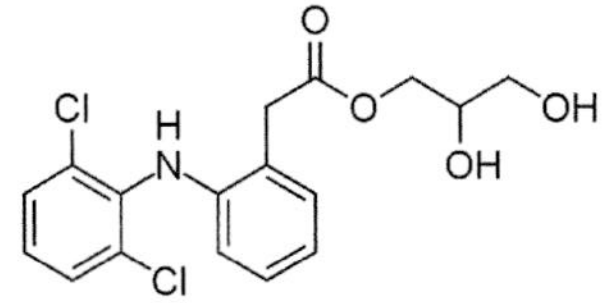 Ethylene glycol diclofenac (65)		Improve water-solubility 162-fold	Lobo et al. (2014)

of this is the antiviral prodrug valacyclovir (**48**). This drug, marketed as Valtrex®, is used to treat herpes virus infection, including herpes zoster, genital herpes (HSV-2), and herpes labialis (HSV-1). After metabolization, this prodrug is converted to acyclovir (**12**). Structurally, the membrane transporter PepT1 recognizes the presence of valine in the prodrug, allowing the absorption of the prodrug with a bioavailability of ~55%, a value superior to that observed for acyclovir (**12**) (~10%–20%) (Sinko and Balimane 1998; MacDougall and Guglielmo 2004). The X-ray structure of PepT1 was elucidated in 2011, and this transporter may be used to improve the oral bioavailability of certain prodrugs. PepT1 is expressed in the small intestine and corneal epithelium, contributing to and exhibiting broad substrate specificity (Newstead et al. 2011). Another example of a prodrug exploring the PepT1 transport system is valganciclovir (**50**) (Valcyte®). This prodrug is indicated for the treatment of cytomegalovirus (CMV) retinitis in adult patients with acquired immunodeficiency syndrome (AIDS), and also as prophylaxis for the CMV disease in adult and pediatric recipients in solid organ transplant. After oral administration, the bioavailability of valganciclovir (**50**) was observed to be approximately 60%, a value better than that observed for its parent drug ganciclovir (**49**) (5%) (Stockmann et al. 2015). The use of valine as a substrate for PepT1 was also explored for other antiviral compounds, such as didanosine (**51**), lagociclovir (**53**), valtocitabine (**55**), etc. (Table 6.1) (Krečmerová 2017).

In addition to causing improvement in water-solubility, the polar groups may also be used to improve the target specificity for certain tissues such as the central nervous system (CNS). In drug designing, one of the most common challenges is to cross the blood-brain barrier (BBB). The relative impermeability of the BBB may be comprehended, in parts, by its specialized tight junctions that decrease the permeability of organic compounds among the capillary endothelial cells. In case of certain molecules, such as glucose, specialized transport systems allow the flux of polar compounds into the brain. GLUT-1 transporter mediates the uptake of glucose and other hexoses from blood, as well as their transport to CNS. Therefore, the prodrugs containing glucose as a carrier may be recognized by the GLUT-1 transporter, thereby improving the uptake of these drugs. Ketoprofen (**57**) and indomethacin (**59**) are the non-steroidal anti-inflammatory drugs (NSAIDs), which have limited access to the CNS. Studies on binding have revealed that the prodrugs containing glucose as a carrier bind to the GLUT-1 receptor and enhance the availability of these NSAIDs in the brain. The brain uptake was determined in situ by using the brain reperfusion technique, in which the ketoprofen-glucose prodrug (**58**) (used at 150 μM) exhibited brain uptake at an average rate of 1.33 pmol/mg/min, while the indomethacin-glucose prodrug (**60**) exhibited a value of 1.99 pmol/mg/min for the brain uptake. However, despite bioconversion, these prodrugs were reported to undergo extensive enzymatic hydrolysis in the liver (Table 6.1) (Gynther et al. 2009). Another example of the application of this approach was its use to enhance the bioavailability of dopamine prodrugs in Parkinson's disease. L-Dopa (**62**) was converted, *in vivo*, to dopamine (**61**) by the action of enzyme dopa decarboxylase; however, the peripheral bioconversion caused unwanted adverse effects. Subsequently, a prodrug formed of dopamine linked to D-glucose through a succinyl moiety as a spacer was synthesized and evaluated. This water-soluble prodrug was able to interact with GLUT-1, while dopamine (**61**) (without conjugation) could not (Table 6.1) (Dalpiaz et al. 2007).

Diclofenac-ester prodrugs were designed to provide enhanced transdermal delivery. Interestingly, it was observed that hydrophilic prodrugs containing alcohols, such as glycerol (**64**) and ethylene glycol (**65**), exhibited higher aqueous solubility, low partition coefficient (logP value), and higher fluxes across the skin, in comparison to diclofenac (**63**). The water-solubility values obtained for diclofenac, glycerol prodrug, and ethylene glycol prodrug were 0.0034, 0.0747, and 0.551 $\mu mol \cdot mL^{-1}$, respectively. The half-lives of bioconversion of the aforementioned two prodrugs into diclofenac in plasma were observed to be 1.66 and 11.4 min, respectively (Table 6.1) (Lobo et al. 2014).

6.3 Phosphate Esters

Phosphate esters have been used widely to design water-soluble prodrugs as the dianionic phosphate group in their structures is able to increase water-solubility by several orders of magnitude. Chemically, the phosphate prodrugs are stable and convenient to prepare through organic synthesis. If required for the oral route, prodrugs containing phosphate ester decrease or eliminate the requirement of solubilizing agents, thereby offering a possibility to design more convenient regimens for high-dose compounds, enhancing patient's compliance (Pradere et al. 2014; Schultz 2003).

After oral administration, the phosphate prodrug metabolism is initiated at the intestinal brush border by the action of alkaline phosphatase, a metalloenzyme that catalyzes the removal of phosphate groups. This enzyme is widely distributed across different tissues, being expressed mainly in the liver. In general, phosphate prodrugs are cleaved rapidly by the alkaline phosphatases, releasing the parent drug (Fawley and Gourlay 2016; Yang et al. 2011).

There are several examples of drugs that utilize the phosphate ester group for improving water-solubility. Fospropofol (**47**), for example, is a water-soluble prodrug of the hypnotic and sedative drug, propofol. Several disadvantages of propofol (**45**) emulsion, such as hyperlipidemia, pain at the injection site, and contamination with microorganisms, justify the search for novel water-soluble derivatives. The presence of the phosphonooxymethyl group in the fospropofol prodrug (**47**) increases the aqueous solubility of the parent drug from 150 $\mu g{\cdot}mL^{-1}$ to 500 $mg{\cdot}mL^{-1}$ (Table 6.2) (Kumpulainen et al. 2008; Dinis-Oliveira 2018). Another prodrug, referred to as propofol phosphate (**11**), was also developed with the aim of increasing water-solubility. When used in laboratory animals, this phosphate prodrug achieved anesthetic and sedative levels similar to those of the parent drug propofol (**45**) (Banaszczyk et al. 2002). Clinical studies have revealed that the prodrug fospropofol (**47**) provides a maximum plasma concentration (C_{max}) faster than that provided by propofol phosphate. In humans, after the i.v. administration of 10 $mg{\cdot}kg^{-1}$ of fospropofol (**47**), it was possible to observe rapid metabolism, with the C_{max} value for propofol being achieved just after 8 min. The presence of –OCH2– spacer exposed the phosphate group in fospropofol (**47**) to be recognized and cleaved by alkaline phosphatase, in a manner different from the phosphate group linked directly to the steric hindrance-causing phenol group in case of propofol phosphate (**11**) (Table 6.2) (Rautio et al. 2018).

Another example is fosfluconazole (**10**), the prodrug prepared from the antifungal drug fluconazole. The low water-solubility of fluconazole (**66**) is a pharmaceutical inconvenience. In clinical practice, this low solubility (4 $mg{\cdot}mL^{-1}$) demands the use of high volumes, thereby hindering the management of hospitalized patients under critical conditions. In contrast, the prodrug fosfluconazole (**10**) exhibits higher solubility in water (>100 $mg{\cdot}mL^{-1}$), facilitating intravenous administration. Inside the body, fosfluconazole (**10**) exhibits rapid bioconversion into fluconazole (**66**), with less than 4% of prodrug remaining intact and being excreted in urine (Table 6.2) (Sobue et al. 2004; Aoyama et al. 2012).

Fosphenytoin (**67**) is another prodrug containing phosphate ester functional group. The presence of this functional group increases the water-solubility of phenytoin (**68**) from 20–25 $\mu g{\cdot}mL^{-1}$ to 140 $mg{\cdot}mL^{-1}$ for fosphenytoin (**67**). This improvement in water-solubility allows the preparation of formulations to be used for the intravenous route. After infusion, fosphenytoin (**67**) is rapidly bioconverted to phenytoin (**68**) by alkaline phosphatase, exhibiting a half-life ranging from 7 to 15 min (Table 6.2) (Browne et al. 1996; Browne 1997).

As the phosphate ester group induces improvement in water-solubility, it is common for this group to be used in designing prodrugs for which the route of administration is intravenous. However, it is also possible to use this approach for the other routes of administration. The use of phosphate prodrugs for oral administration is promising for insoluble drugs, mainly for the ones administered at high doses in cases where the absorption is dissolution-rate limited due to low solubility and when there are medical requirements for higher doses. Estramustine phosphate (**69**) is one example of a phosphate ester prodrug used for oral route administration since the mid-1970s. In humans, it is possible to use the prodrug orally at doses up to 1400 $mg{\cdot}day^{-1}$ (in 2–3 divided doses) to treat the metastatic and/or progressive prostate cancer (Hoisaeter and Bakke 1983). Different from estramustine (**70**), whose water-solubility is close to 1 $\mu g{\cdot}mL^{-1}$, the prodrug estramustine phosphate (**69**) exhibits a water-solubility of 50 $mg{\cdot}mL^{-1}$. After oral administration, approximately 75% of the estramustine phosphate (**69**) prodrug remains intact at the time of absorption, although the prodrug undergoes fast presystemic dephosphorylation resulting in the release of estramustine (**70**) (posteriorly oxidized to generate the active cytotoxic isomer of estromustine (**70**)). The half-life of estramustine phosphate (**69**) was observed to be around 1.27 h, while the active metabolite estromustine (**70**) remains for longer periods ranging from 8 to 22.7 h (Table 6.2) (Perry and McTavish 1995).

TABLE 6.2
Phosphate Prodrugs

Prodrug	Parent Drug	Prodrug Strategy	References
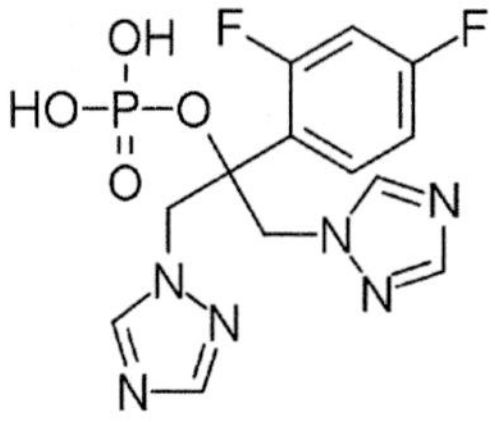 Fosfluconaxole (10)	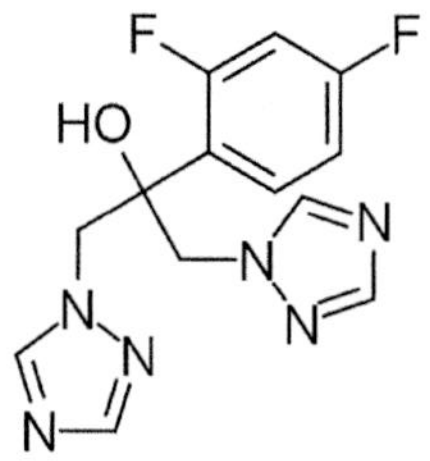 Fluconaxole (66)	Improve water-solubility > 25-fold	Sobue et al. (2004), Aoyama et al. (2012)
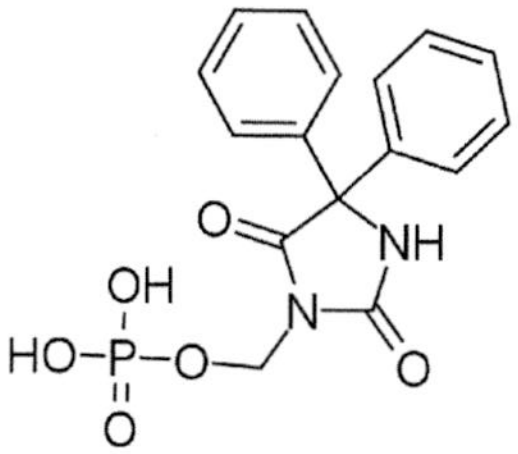 Fosphenytoin (67)	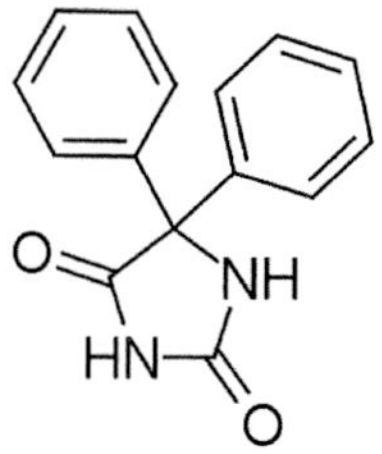 Phenytoin (68)	Improve water-solubility 7000-fold	Browne et al. (1996), Browne (1997)
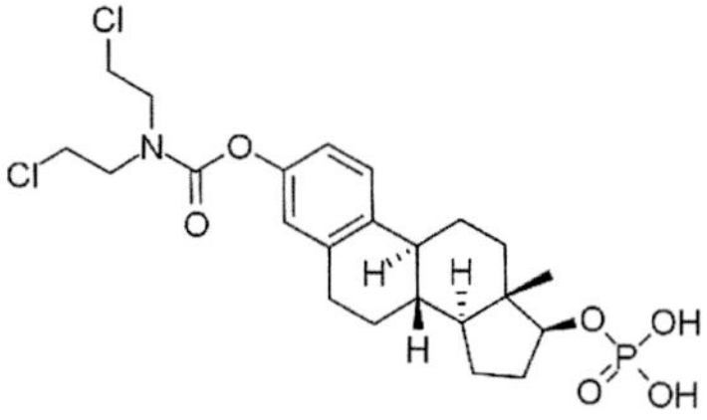 Estramustine-phosphate (69)	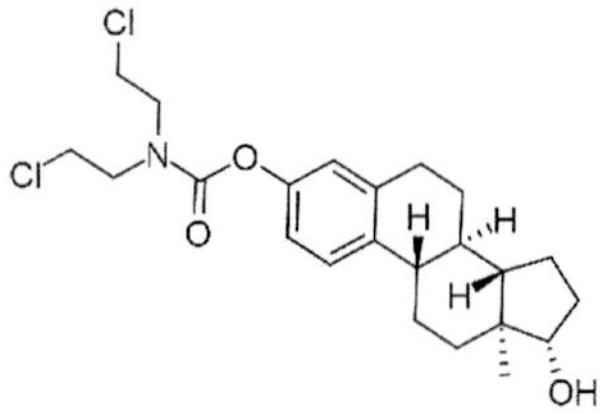 Estramustine (70)	Improve water-solubility 50000-fold	Perry and McTavish (1995)

(Continued)

TABLE 6.2 (*Continued*)

Phosphate Prodrugs

Prodrug	Parent Drug	Prodrug Strategy	References
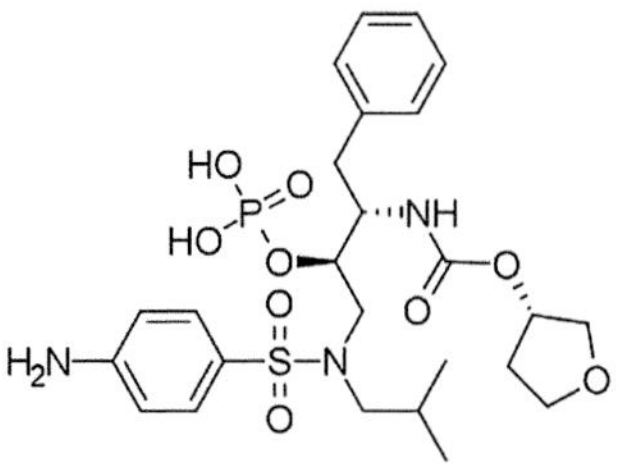 Fosamprenavir (71)	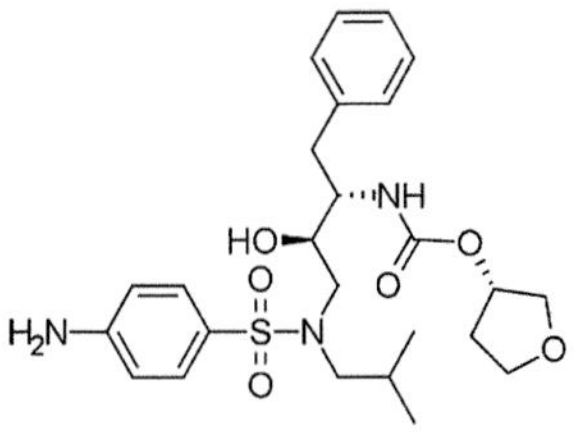 Amprenavir (72)	Improve water-solubility 10-fold	Hester et al. (2006), Subbaiah et al. (2017)
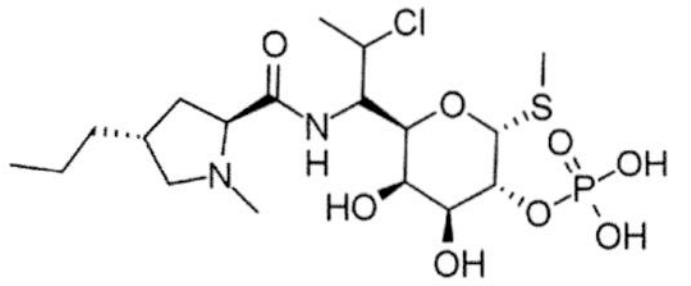 Clindamycin phosphate (73)	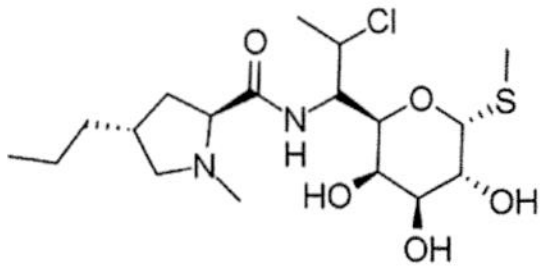 Clindamycin (74)	Improve water-solubility 1,3-fold	DeHaan et al. (1973), Hugo et al. (1977)
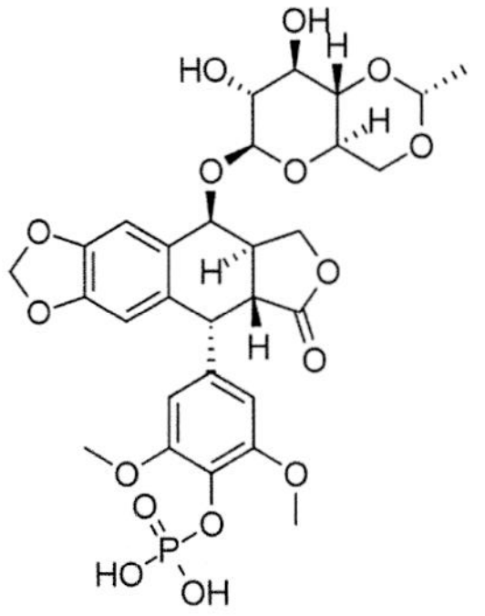 Etoposide phosphate (75)	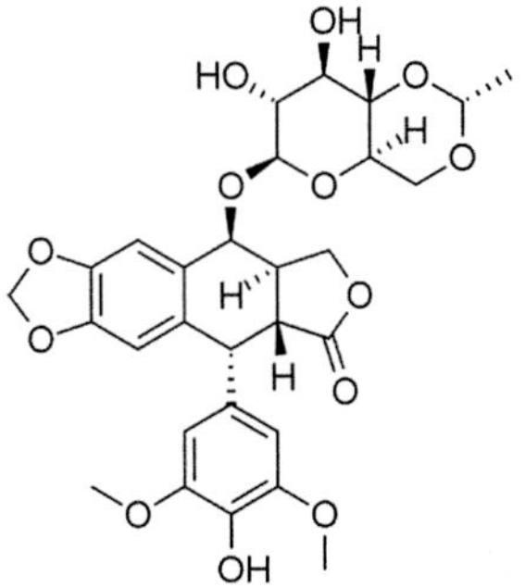 Etoposide (24)	Improve water-solubility 120-fold	Witterland et al. (1996), Budman (1996), Schacter (1996)

(*Continued*)

TABLE 6.2 (*Continued*)
Phosphate Prodrugs

Prodrug	Parent Drug	Prodrug Strategy	References
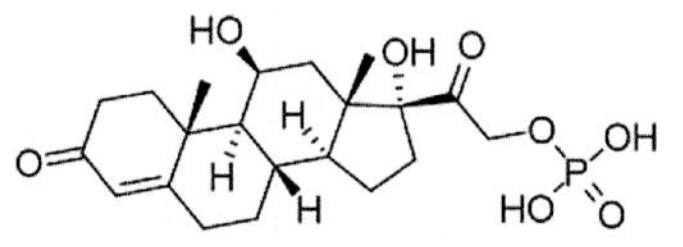 Hydrocortisone phosphate (76)	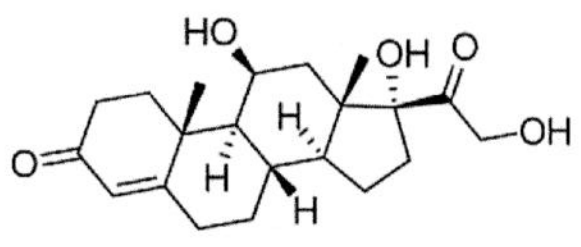 Hydrocortisone (77)		Botte (1968)
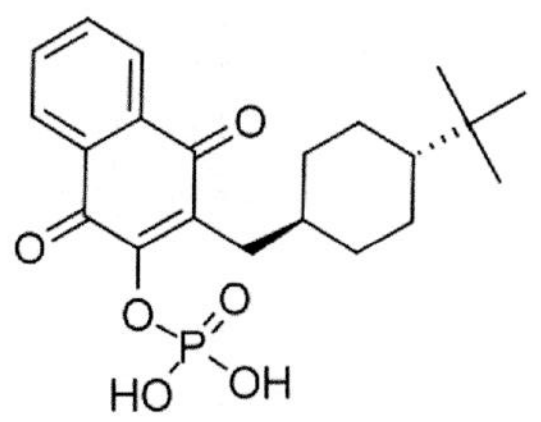 Bupavarquone phosphate (78)	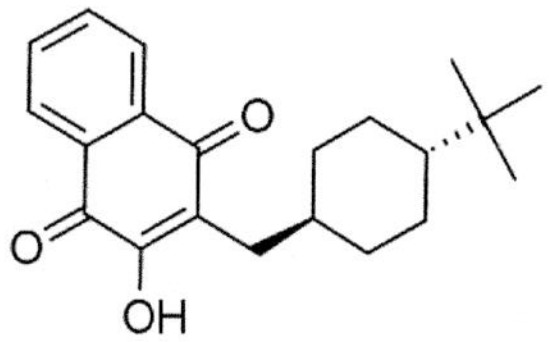 Bupavarquone (79)		Mäntylä et al. (2004a, 2004b)

Another example of a phosphate ester prodrug formulated for oral administration is fosamprenavir (**71**). The phosphate monoester of amprenavir is an inhibitor of HIV protease that exhibits 10-fold increased aqueous solubility compared to its parent drug. The low solubility of amprenavir (**72**) (0.04 mg·mL^{-1} at 25°C) demanded the use of solubilizing agents, such as polyethylene glycol 400 and propylene glycol, which became unnecessary for its prodrug. The favorable physicochemical profile of the calcium salt of fosamprenavir (**71**) allowed the development of formulations that were further simplified, enhancing patient compliance to the treatment and reducing the pill burden associated with amprenavir (**72**). After oral administration, fosamprenavir (**71**) could be bioconverted to amprenavir (**72**) by alkaline phosphatase in the intestinal epithelium (Table 6.2) (Hester et al. 2006; Subbaiah et al. 2017).

Several other examples of phosphate ester being used to improve the water-solubility of drugs have been described in the literature, for example, clindamycin phosphate (**73**) (DeHaan et al. 1973; Hugo et al. 1977), etoposide phosphate (**75**) ((Witterland et al. 1996; Budman 1996; Schacter 1996), hydrocortisone-phosphate (**76**) (Botte 1968), buparvaquone phosphate (**78**) (Table 6.2) (Mäntylä et al. 2004a, 2004b).

6.4 Amide Prodrugs

The enzymes named amidases hydrolyze the amide prodrugs *in vivo*. In general, the mechanism of this hydrolysis is similar to that observed in the case of the ester functional group. The main difference between the hydrolysis of the amide and the ester functional groups is that the amide functional group exhibits greater resistance for the chemical hydrolysis compared to the ester functional group. The property of amides to be hydrolyzed slower than esters is being explored for designing drugs having longer half-lives compared to ester prodrugs (Weber et al. 2013). Several examples of the use of amides for improving the water-solubility of drugs are available in the literature.

The antiviral drug, acyclovir (**12**), exhibits limited solubility (2.5 mg·mL^{-1} at 37°C) and low bioavailability (10%–20%). Since the 4′-hydroxyl group exhibits superior reactivity compared to the 2-amino group, it is more common to find ester prodrugs of acyclovir (**12**) in comparison to the amide prodrugs (**80**) of acyclovir (**12**). In order to explore the hydrolytic effect of dipeptidyl-peptidase IV (DPPIV), an amide prodrug (**80**) was synthesized from acyclovir (**12**). The enzyme DPPIV exhibits a peptidase catalytic activity, removing dipeptides from the N-terminus of the substrate, which contains proline (or alanine) at the penultimate position. An improvement in water-solubility of 17-fold was observed for the amide prodrug (**80**) compared to acyclovir (**12**). The amide prodrug was observed to be stable in PBS for up to 24 h, with a small percentage bioconverting into acyclovir (**12**). In human and bovine serum, the bioconversion to the parent drug occurred within 1 h. Interestingly, when a similar assay was performed in the presence of vildagliptin (25 μM), which is a potent inhibitor of dipeptidyl-peptidase, it was observed that up to 24 h, greater than 80% of the prodrug remained intact, suggesting that the metabolism was performed by the DPPIV enzyme. The amide prodrug (**80**) also exhibited *in vitro* anti-herpetic activity (Table 6.3) (Diez-Torrubia et al. 2013).

Taurine (2-aminoethanesulfonic acid) is a non-essential beta-amino acid most abundant in mammals, and present in several tissues including the CNS where it is involved in regulating several pathways and providing a neuroprotective effect. Taurine, GABA, and glycine serve as inhibitory neurotransmitters in the CNS. In addition, certain studies have suggested that reduced levels of taurine may aggravate epilepsy in certain patients. The prodrugs of valproic acid (**81**) and taurine were designed to obtain better effective and safe antiepileptic compounds. Valproic acid (**81**) is a first-line drug that is able to treat several epileptic seizure types, such as myoclonus, absence, generalized, and partial seizures; however, limitations due to severe adverse effects, such as teratogenicity and hepatotoxicity, restrict the clinical use of this drug. Taurine increases the water-solubility of compounds by decreasing the partition coefficient (log *P* value). The cLogP value for valproic acid (**81**) was 2.6, while the cLog P value for the amide prodrug was 1.1. Despite this, access to CNS is available to this prodrug due to the presence of taurine transporters (Foos and Wu 2002). The valproic acid-taurine prodrug (**82**) exhibited anticonvulsant activity with reduced teratogenicity, demonstrating that, in addition to the increased water-solubility, the adverse effects of the parent drug may also be modulated by using the prodrug approach (Table 6.3) (Isoherranen et al. 2003).

TABLE 6.3
Amide Prodrugs

Prodrug	Parent Drug	Prodrug Strategy	References
Acyclovir amide (80)	Acyclovir (12)	Improved water-solubility 17-fold stable in PBS	Diez-Torrubia et al. (2013)
Valproic acid-taurine (82)	Valproic acid (81)	Improved water-solubility reduced teratogenicity	Foos and Wu (2002), Isoherranen et al. (2003)
Ibuprofen taurine (84)	Ibuprofen (83)	Improve water-solubility 50-fold No gastrotoxicity	Vizioli et al. (2009), Chung et al. (2012)

(*Continued*)

TABLE 6.3 (*Continued*)
Amide Prodrugs

Prodrug	Parent Drug	Prodrug Strategy	References
Mesalazine taurine (85)	Mesalazine (86)	Reduced absorption on superior intestine	Jung et al. (2003)
Phe-Ala-primaquine (90)	Primaquine (89)	Less cytotoxic Not induce hemolysis in G6PD-deficient red blood cells	Davanço et al. (2014)
TXY-436 (92)	PC190723 (91)	Improve water-solubility 100-fold Ionizable group at pH 7.4	Kaul et al. (2013a, 2013b)
TXY-541 (93)		Improve water-solubility 143-fold Ionizable group at pH 7.4	Kaul et al. (2013a, 2013b)

Beyond the CNS action, taurine has also been reported to exert anti-inflammatory effects. Taurine plays an important role in the inflammation process associated with oxidative stress, because of its ability to react with the hypochlorous acid produced by neutrophil-myeloperoxidase, leading to the formation of the anti-inflammatory compound taurine chloramine (TauCl) (Marcinkiewicz and Kontny 2014). The prodrugs of nonsteroidal anti-inflammatory drugs (NSAIDs)-taurine were synthesized in order to improve the water-solubility of the parent NSAIDs. The water-solubility of ibuprofen (**83**) was determined to be 0.021 mg·mL^{-1} (20°C), while the ibuprofen-taurine prodrug (**84**) exhibited a 50-fold increase in the water-solubility (Table 6.3). Moreover, after oral administration, the prodrug demonstrated anti-inflammatory activity when evaluated using carrageenan-induced paw edema in rats, without causing gastrotoxicity. While ibuprofen-induced ulcerative gastric lesions, the prodrug did not exhibit this adverse effect (Vizioli et al. 2009; Chung et al. 2012).

In certain cases, the improvement in the water-solubility may be explored to provide tissue selectivity. One example is 5-aminosalicyl-taurine prodrug (**85**), which was designed as a colon-specific prodrug of 5-aminosalicylic acid (mesalazine) (**86**) with an aim to act against the inflammatory conditions, such as Crohn's disease and ulcerative colitis. To date, mesalazine (**86**) remains one of the main therapeutic arsenals for treating bowel inflammation (Bosquesi et al. 2011). However, after oral administration, this drug is rapidly absorbed in the upper intestine, reaching colon at low levels. The prodrug of mesalazine-taurine (**85**) was observed to exhibit reduced absorption in the upper intestine, thereby reaching the colon at higher levels compared to the parent drug mesalazine (**86**). Considering the anti-inflammatory effect of mesalazine (**86**), the prodrug could act synergically, reducing the inflammation. In order to characterize the hydrolytic profile of the prodrug, the compound was incubated in buffer solutions with pH 1.2 and 6.8 along with the homogenates of tissue (contents of the stomach and/or small intestine of rats). No hydrolysis of the prodrug was observed under these conditions. However, when the mesalazine-taurine prodrug (**85**) was incubated with cecal and the colonic content of rats for 8 h, the levels of mesalazine (**86**) resulting from the bioconversion of the prodrug were observed to be 45% and 20%, respectively (Table 6.3) (Jung et al. 2003).

Another example of a water-soluble prodrug that targets a specific site/tissue is the sulfamethoxazole prodrug (**87**), which is capable of selectively delivering the antimicrobial drug to the kidneys (Figure 6.3). The clogP value of sulfamethoxazole (**88**) is 0.36, while that for its prodrug is –1.09, implying that the prodrug exhibits higher water-solubility in comparison to the parent drug. After oral administration, the enzyme *N*-acylamino acid deacylase performs initial deacetylation of the glutamic subunit. This enzyme is present in high concentration in the kidneys. After the first step, a second bioconversion is performed by the enzyme gamma-glutamyl transpeptidase, which is also present in the kidneys. These sequences of cleavages allow the selective release of the drug in the kidneys, while simultaneously improving the water-solubility of the parent drug (Orlowski et al. 1980; Murakami et al. 1998).

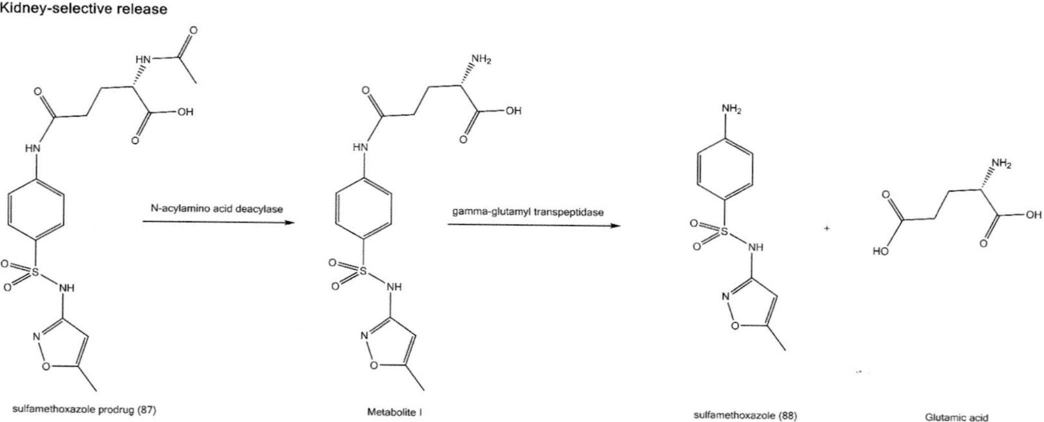

FIGURE 6.3 Bioconversion of sulfamethoxazole prodrug.

The antimalarial primaquine (PQ) (**89**) that exhibits low water-solubility has certain advantages over the other antimalarial drugs, including activity against the gametocytes of all species of *Plasmodium* and effective against latent liver forms involved in relapsing malaria caused by *Plasmodium vivax* and *Plasmodium ovale*. PQ (**89**) has also demonstrated activity against the chloroquine-resistant *P. falciparum* strains. In addition to this spectrum of action, PQ (**89**) presents certain limitations, such as the induction of severe hemolytic anemia in the patients with deficiency of glucose-6-phosphate dehydrogenase (G6PD) and the presence of inappropriate pharmacokinetic parameters due to its short half-life. The rapid deamination of PQ (**89**) during metabolism leads to the generation of the inactive metabolite carboxyprimaquine (Mihaly et al. 1984). The prodrugs approach might be able to conveniently solve the major limitations of PQ (**89**) by increasing its water-solubility and modifying its metabolic profile (Table 6.3).

PQ-Peptides prodrugs were expected to exhibit better water-solubility, achieving favorable consequences in its ADMET profile. The logP experimental data for phenylalanine-alanine-PQ prodrug (**90**) (named Phe-Ala-PQ), obtained using n-octanol/water, generated a value of 0.94 for this prodrug, while the value obtained for PQ (**89**) was 2.1. PQ (**89**) released from the prodrug Phe-Ala-PQ (**90**) demonstrated antimalarial activity, reducing the number of oocytes and sporozoites by 100% at a dose of 1.9 mg/kg. In addition, this prodrug was less cytotoxic in comparison to PQ (**89**) when tested on several cell lines, and also did not induce hemolysis in the G6PD-deficient red blood cells (Table 6.3) (Davanço et al. 2014).

In drug designing, the selection of the carrier determines the success in improving the water-solubility of the drug. Certain polar groups, preferentially the ones containing ionizable groups, offer a high increase in water-solubility. In case of a benzamide derivative, namely, PC190723 (**91**), with low water-solubility, which exhibits antimicrobial activity due to the inhibition of the FtsZ protein, the inclusion of a carrier containing ionizable groups improved the solubility by 100- to 143-fold. While the parent compound PC190723 (**91**) exhibited no activity against the methicillin-susceptible *Staphylococcus aureus* and methicillin-resistant *Staphylococcus aureus*, the more water-soluble prodrugs TXY-436 (**92**) and TXY-541 (**93**) exhibited such activity, thereby increasing the survival rate among the treated animals up to 100% (Table 6.3) (Kaul et al. 2013a, 2013b).

6.5 Carbonate and Carbamate Prodrugs

Carbonates and carbamates have been reported to be more stable than esters and less stable in comparison to amides. Although no specific enzyme is available for their hydrolysis, carbonates are generally degraded by esterases to release the parent drugs (Karaman 2013). Carbamate prodrugs also require esterases to release the parent drug, generating phenol or alcohol along with carbamic acid. The chemical instability of the latter leads to its break down into carbon dioxide and the respective amines. It has been reported that carbamates of primary amines are able to break down into alcohol and isocyanate upon treatment with bases (Ghosh and Brindisi 2015).

The examples of carbonate and carbamates used for the improvement of water-solubility are available in the literature (Karaman 2013; Ghosh and Brindisi 2015). The gastric lesions induced by the long-term use of acetylsalicylic acid (ASA) (**26**) have been reported widely in the literature. ASA carbonate prodrug (**94**) was reported to exhibit a water-solubility value of 6.7 mg·mL^{-1} at 37°C in 0.1 N HCl solution, while in the same conditions, ASA (**26**) exhibited a water-solubility value of 5.3 mg·mL^{-1}. The half-life of ASA carbonate prodrug (**94**) was reported to be 41 h at pH 7.4 (Table 6.4) (Dittert et al. 1968).

In early clinical trials, pharmacokinetic and solubility issues were observed with the anticancer compound CHS828 (**95**). In order to solve the issue of low water-solubility (0.5 μg·mL^{-1}), several prodrugs were synthesized using a carbonate functional group that contained a tetraethylene-glycol subunit. One of these prodrugs, namely, EB1627 (**96**), exhibited a 120-fold increase in the water-solubility in comparison to the parent compound CHS828 (**95**). This improvement in water-solubility allowed the administration of this prodrug through the intravenous route (Table 6.4) (Binderup et al. 2005).

Further higher water-solubility-exhibiting taxoid prodrugs, with water-solubility ranging from 0.8 to 1.1 mg·mL^{-1}, have been reported in the literature. Carbonate prodrugs (**97**) have been reported to cause

TABLE 6.4
Carbonate and Carbamate Prodrugs

Prodrug	Parent Drug	Prodrug Strategy	References
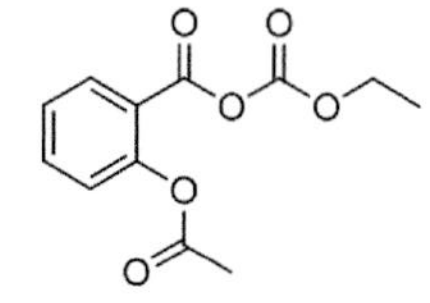 ASA carbonate (94)	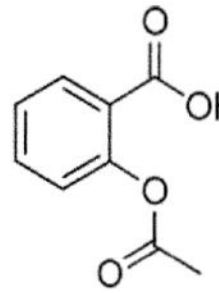 Acetylsalicylic acid (10)	Improve water-solubility 1,3-fold	Dittert et al. (1968)
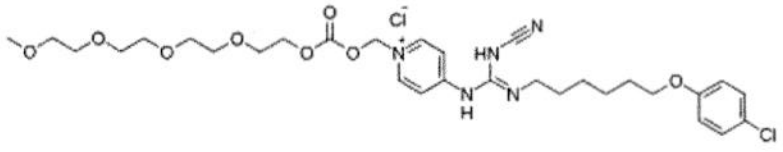 EB1627 (96)	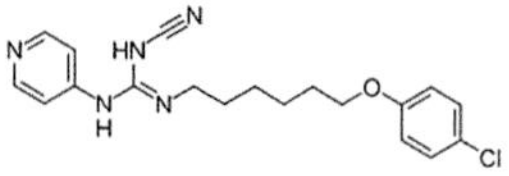 CHS828 (95)	Improve water-solubility 120-fold	Binderup et al. (2005)
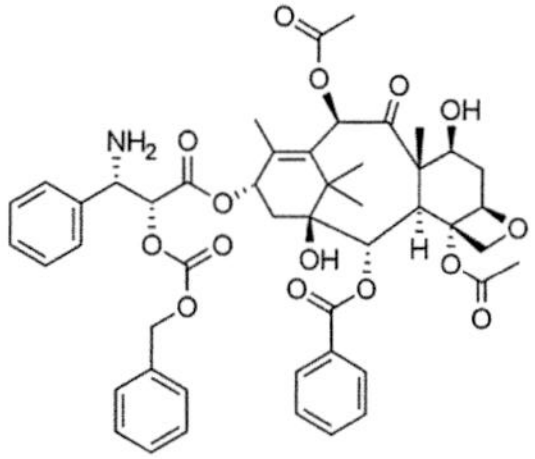 2′-O-Benzyloxycarbonyl-3′N-debenzoylpaclitaxel (97)	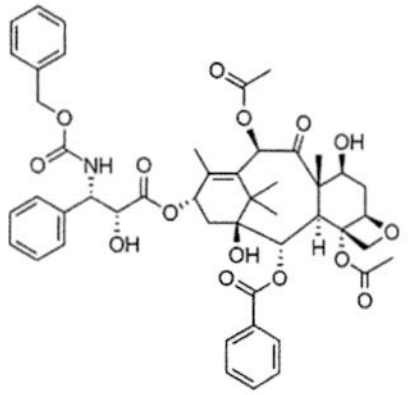 3′-N-Benzyloxycarbonyl-3′-N-debenzoylpaclitaxel (98)	Improve water-solubility 4000-fold	Skwarczynski et al. (2005)
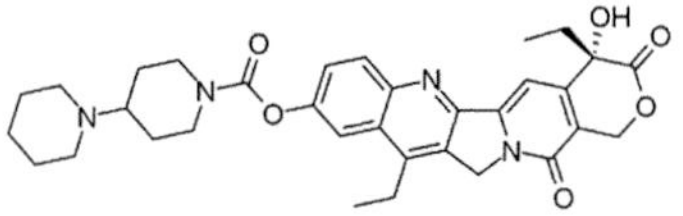 Irinotecan (99)	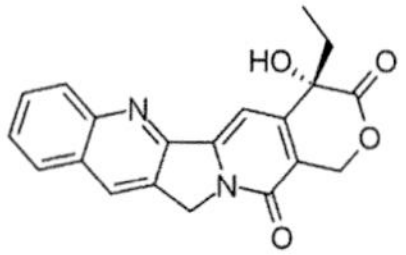 Camptothecin (100)	Protonation of tertiary amino group in the piperidine ring at low pHs Improve water-solubility 1000-fold	Slatter et al. (2000), Sanghani, et al. (2004)

a 4000-fold increase in water-solubility in comparison to the parent drug (**98**). The parent-drug release from the carbonate prodrugs was achieved after a pH-dependent rearrangement of atoms within the molecule, involving O–N intramolecular acyl migration. Interestingly, no enzyme activity or additional functional auxiliaries were necessarily required to promote the bioconversion of these prodrugs to the parent drug (Table 6.4) (Skwarczynski et al. 2005).

Irinotecan (**99**) is a topoisomerase I inhibitor that is used as an anticancer drug. Structurally, irinotecan is a dipiperidino carbamate prodrug of camptothecin (**100**), which contains a 1,4′-dipiperidinyl promoiety that increases water-solubility to 20 mg·mL^{-1} (at pH 3–4) from a value of 2–3 μg·mL^{-1} for camptothecin (**100**). A part of this water-solubility improvement is caused by the protonation of the tertiary amino group present in the piperidine ring at low pH values. Human carboxylesterases, especially those present in the liver and in tumors, are responsible for the bioconversion of the prodrug to the parent drug camptothecin (**100**) through the hydrolysis of the carbamate functional group. *In vivo*, irinotecan (**99**) and camptothecin (**100**) co-exist in a pH-dependent equilibrium between the carboxyl state and lactone. However, only lactone is the active form. Despite the complex pharmacokinetics, after the intravenous administration of irinotecan, the T_{max} value for camptothecin (**100**) was observed to be 2.3 h (Table 6.4) (Slatter et al. 2000; Sanghani et al. 2004).

Isavuconazole (**101**) is a triazole antifungal compound, which causes the inhibition of enzyme 14α-lanosterol demethylase, disrupting the fungal ergosterol biosynthesis. The drug has demonstrated a broad antifungal spectrum and is used mainly to treat the invasive aspergillosis and mucormycosis (Miceli and Kauffman 2015). The low water-solubility of this drug is a drawback that limits its clinical use; therefore, the prodrug approach was used to develop water-soluble acyloxyalkyl triazonium salt-based isavuconazonium sulfate prodrug (**102**). This prodrug exhibited high water-solubility (>100 mg·mL^{-1}), allowing its administration through the intravenous route. The prodrug release is dependent on the route of administration used. For instance, after intravenous infusion, isavuconazonium is initially metabolized by plasmatic esterases (mainly butyrylcholinesterase), releasing a benzyl alcohol derivative. In the next step, the alcohol functional group attacks the electrophilic carbonyl functional group, leading to a rapid and spontaneous intramolecular cyclization that induces the release of isavuconazole (**101**) (Figure 6.4). On the contrary, after the oral administration, the prodrug undergoes a rapid presystemic bioconversion without detectable levels in the

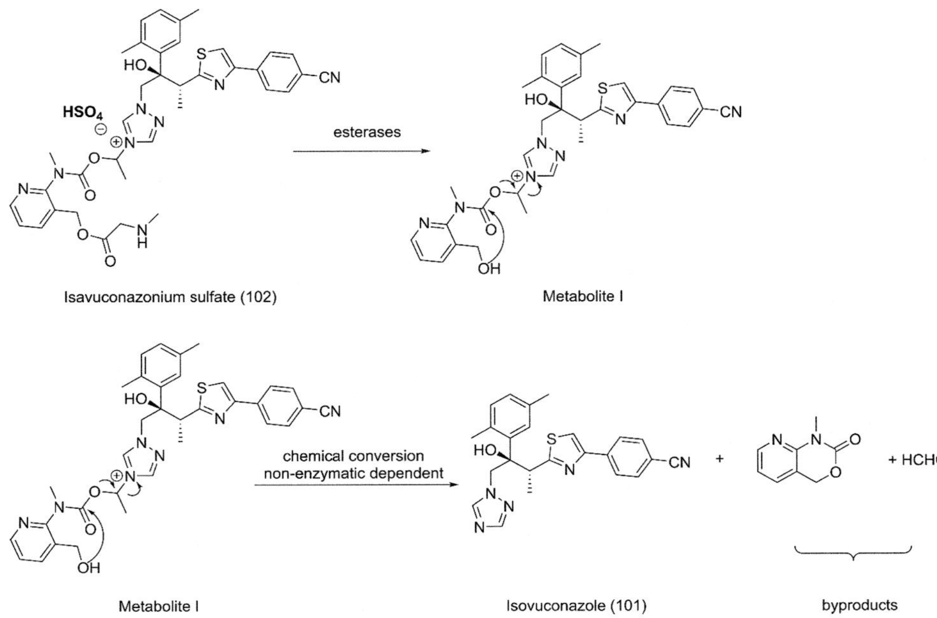

FIGURE 6.4 Bioactivation of isavuconazonium sulfate prodrug (**102**).

TABLE 6.5
Other Prodrugs

Prodrug	Parent Drug	Prodrug Strategy	References
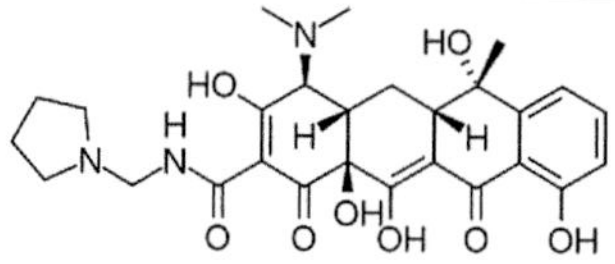 Rolitetracycline (103)	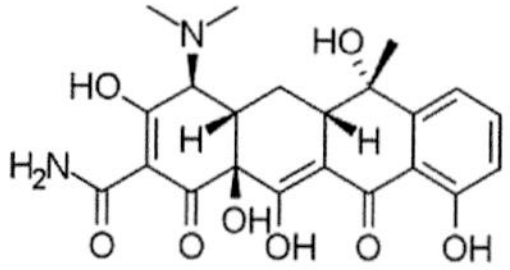 Tetracycline (104)	Improve water-solubility I.V. administration	Agwuh and MacGowan (2006), Karaman (2013)
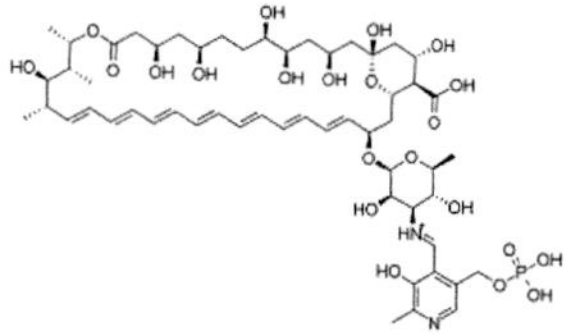 AmB pyridoxal phosphate (107)	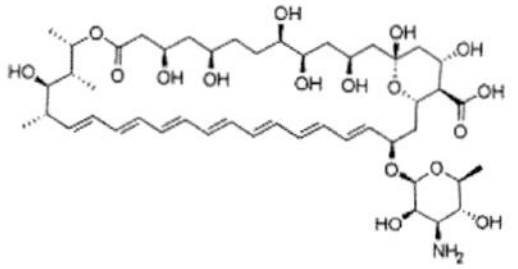 AmB (105)	Improve water-solubility	Day et al. (2011)
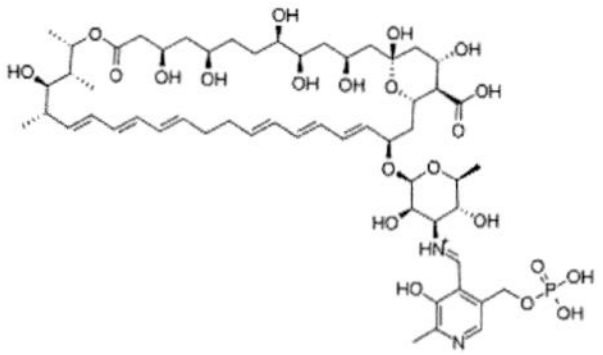 Nystatin pyridoxal phosphate (108)	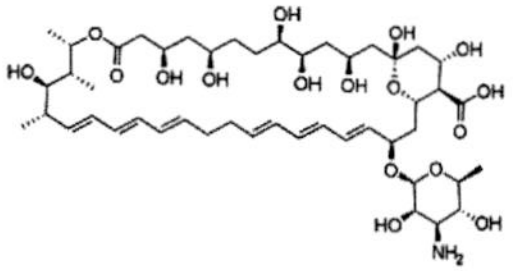 Nystatin (106)	Improve water-solubility	Day et al. (2011)

(Continued)

TABLE 6.5 (*Continued*)

Other Prodrugs

Prodrug	Parent Drug	Prodrug Strategy	References
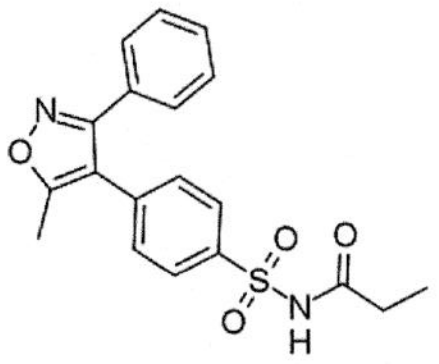 Parecoxib (109)	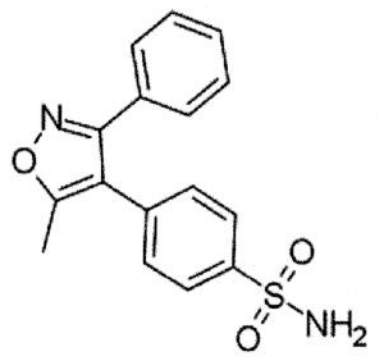 Valdecoxib (110)	Improve water-solubility	Cheer and Goa (2001)
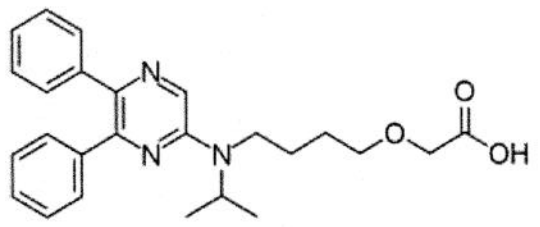 ACT333679 (112)	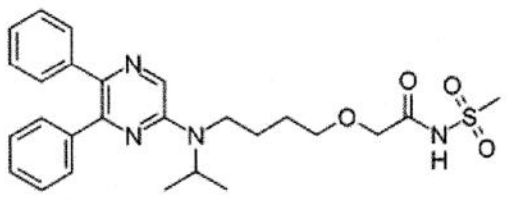 Selexipag (111)	Prolonged action Reduced side effects	Asaki et al. (2015)

plasma. In addition, this prodrug exhibits a favorable pharmacokinetic profile, with a bioavailability value of approximately 98% and low inter-individual variability (Ohwada et al. 2003; Schmitt-Hoffmann et al. 2006).

6.6 Other Prodrugs

Several other chemical functions/strategies may be explored to improve the water-solubility of compounds. These include the use of imines, oximes, enamines, ethers, N-Mannich bases, azo conjugates, acylsulfonamides, sulfonamides, N-acyloxyalkylamines, phosphoramidates, phosphorodiamidates, ProTide phosphoramidates/phosphonamidates, carbonyloxymethyls, alkoxyalkyl monoesters, dithiodiethanol, cyclosal, and S-acetylthioethanol, among others.

Rolitetracycline (**103**) is the water-soluble N-Mannich base prodrug of tetracycline (**104**), which is available for intravenous administration. Chemically, N-Mannich bases are prepared conveniently through reactions involving an aldehyde, an ethanol, and an NH-acid compound. *In vivo*, chemical hydrolysis converts the prodrug into the active drug (Table 6.5) (Agwuh and MacGowan 2006; Karaman 2013).

Prodrugs containing an imine function may also be explored for increasing the water-solubility of compounds. Amphotericin B (**105**) and nystatin (**106**) are two antifungal drugs exhibiting limited water-solubility. The prodrugs containing pyridoxal phosphate (**107** and **108**), as a solubilizing subunit, attached to the aforementioned antifungal drugs through an imine functional group were synthesized, and their water solubilities were evaluated. Both the amphotericin B and nystatin prodrugs (**107** and **108**) exhibited water-solubility value as good as 100 mg/mL (Table 6.5) (Day et al. 2011).

The acid nature of the N–H bond in *N*-acylsulfonamides allows the production of water-soluble salts designed as prodrugs. One example of this is the NSAIDs COX-2-selective parecoxib (**109**), which is the water-soluble prodrug of valdecoxib (**110**). The enhanced water-solubility of parecoxib (**109**) allows its administration through the intravenous route (Cheer and Goa 2001). Another example belonging to the same class of compounds is the non-prostanoid prostacyclin receptor agonist named selexipag (**111**). This prodrug is used in the treatment of pulmonary hypertension and provides a prolonged action with reduced side effects. *In vivo*, subsequent to its metabolism by hepatic esterase CES1, selexipag (**111**) is gradually bioconverted into the active metabolite ACT333679 (**112**), which contains carboxylic acid (Table 6.5) (Asaki et al. 2015).

6.7 Conclusions

One of the major challenges encountered by the researchers during drug designing is the inadequacy of a particular physicochemical property of the prototype, usually the low water-solubility. The drugs that are administered through the oral route encounter low water-solubility, drug dissolution, and absorption, leading to an erratic pharmacokinetic profile of the drug. In order to solve these issues, the prodrug approach has been developed as a powerful tool to improve the water-solubility of the drugs and to modulate their pharmacokinetic properties. The use of polar carriers, especially those containing ionizable functional groups, increases the water-solubility of the parent drug. One such prodrug approach widely explored is the conjugation of a drug with amino acids or peptides, which results in the creation of a charge at physiological pH values that enhances the solubility of the parent drug. Moreover, the salt of the prodrug is also a possible alternative that would provide even further improvement in the water-solubility of the parent drug. At certain times, the use of amino acids and peptides as carriers not only increases the water-solubility but also enhances the bioavailability once certain transporters are able to recognize some particular residues as substrates improving the uptake of the prodrug.

Another aspect to be considered is the functional group used to link the carrier with the drug. It is well established that differences among the rate of hydrolysis are related directly to the chemical/enzymatic susceptibility of the chemical functional group used in the prodrug design. In this chapter, a few of these chemical functional groups, including esters, ester phosphates, amides, carbonates, carbamates, and others, were discussed.

The prodrug approach may be used at different stages of drug design and development. The sooner the awareness regarding the applications of the prodrug strategy is realized, the greater will be the probability of avoiding the elimination of promising prototypes during the early stages of drug development.

REFERENCES

Adams, J. D., Flora, K. P., Goldspiel, B. R., et al. 1993. Taxol: A History of Pharmaceutical Development and Current Pharmaceutical Concerns. *Journal of the National Cancer Institute. Monographs*: 141–47.

Agwuh, K. N., and MacGowan, A. 2006. Pharmacokinetics and Pharmacodynamics of the Tetracyclines Including Glycylcyclines. *Journal of Antimicrobial Chemotherapy* 58: 256–65.

Albert, A. 1958. Chemical Aspects of Selective Toxicity. *Nature* 182: 421–23.

Aoyama, T., Hirata, K., Hirata, R., et al. 2012. Population Pharmacokinetics of Fluconazole after Administration of Fosfluconazole and Fluconazole in Critically Ill Patients. *Journal of Clinical Pharmacy and Therapeutics* 37: 356–63.

Asaki, T., Kuwano, K., Morrison, K., et al. 2015. Selexipag: An Oral and Selective IP Prostacyclin Receptor Agonist for the Treatment of Pulmonary Arterial Hypertension. *Journal of Medicinal Chemistry* 58: 7128–37.

Banaszczyk, M. G., Carlo, A. T., Millan, V., et al. 2002. Propofol Phosphate, a Water-Soluble Propofol Prodrug: In Vivo Evaluation. *Anesthesia and Analgesia* 95: 1285–92.

Binderup, E., Björkling, F., Hjarnaa, P. V., et al. 2005. EB1627: A Soluble Prodrug of the Potent Anticancer Cyanoguanidine CHS828. *Bioorganic & Medicinal Chemistry Letters* 15: 2491–94.

Bosquesi, P. L., Melo, T. R. F., Vizioli, E. O., dos Santos, J. L., and Chung. M. C., 2011. Anti-Inflammatory Drug Design Using a Molecular Hybridization Approach. *Pharmaceuticals* 4: 1450–74.

Botte, V., 1968. The Hydrocortisone 21-Phosphate as Substrate for Alkaline Phosphatase. Histochemical Remarks. *Histochemie. Histochemistry. Histochimie* 16: 195–96.

Browne, T. R., 1997. Fosphenytoin (Cerebyx). *Clinical Neuropharmacology* 20: 1–12.

Browne, T. R., Kugler, A. R., and Eldon, M. A. 1996. Pharmacology and Pharmacokinetics of Fosphenytoin. *Neurology* 46: S3–7.

Brune, K., and Hinz, B. 2004. The Discovery and Development of Antiinflammatory Drugs. *Arthritis & Rheumatism* 50: 2391–99.

Budman, D. R. 1996. Early Studies of Etoposide Phosphate, a Water-Soluble Prodrug. *Seminars in Oncology* 23: 8–14.

Casey Laizure, S., Herring, V., Hu, Z., Witbrodt, K., and Parker, R. B. 2013. The Role of Human Carboxylesterases in Drug Metabolism: Have We Overlooked Their Importance? *Pharmacotherapy: The Journal of Human Pharmacology and Drug Therapy* 33: 210–22.

Cheer, S. M., and Goa, K. L. 2001. Parecoxib (Parecoxib Sodium). *Drugs* 61: 1133–41.

Chen, J., and Du, W. 2013. Synthesis and Evaluation of Water-Soluble Etoposide Esters of Malic Acid as Prodrugs. *Medicinal Chemistry* 9: 740–47.

Chung, M. C., Ferreira, E. I., Santos, J. L., et al. 2008. Prodrugs for the Treatment of Neglected Diseases. *Molecules* 13: 616–77.

Chung, M. C., Malatesta, P., Bosquesi, P. L., et al. 2012. Advances in Drug Design Based on the Amino Acid Approach: Taurine Analogues for the Treatment of CNS Diseases. *Pharmaceuticals* 5: 1128–46.

Dalpiaz, A., Filosa, R., de Caprariis, P., et al. 2007. Molecular Mechanism Involved in the Transport of a Prodrug Dopamine Glycosyl Conjugate. *International Journal of Pharmaceutics* 336: 133–39.

Damen, E., Peter, W. P., Wiegerinck, H. G., Braamer, L., et al. 2000. Paclitaxel Esters of Malic Acid as Prodrugs with Improved Water Solubility. *Bioorganic & Medicinal Chemistry* 8: 427–32.

Das, N., Dhanawat, M., Dash, B., Nagarwal, R. C., and Shrivastava, S. K. 2010. Codrug: An Efficient Approach for Drug Optimization. *European Journal of Pharmaceutical Sciences* 41: 571–88.

Davanço, M. G., Aguiar, A. C. C., dos Santos, L. A., et al. 2014. Evaluation of Antimalarial Activity and Toxicity of a New Primaquine Prodrug. *PLoS One* 9: 1–10.

Day, T. P., Sil D., Shukla, N. M., et al. 2011. Imbuing Aqueous Solubility to Amphotericin B and Nystatin with a Vitamin. *Molecular Pharmaceutics* 8: 297–301.

DeHaan, R. M., Metzler, C. M., Schellenberg, D., and Vandenbosch, W. D. 1973. Pharmacokinetic Studies of Clindamycin Phosphate. *Journal of Clinical Pharmacology* 13: 190–209.

Diez-Torrubia, A., Cabrera, S., de Castro, S., et al. 2013. Novel Water-Soluble Prodrugs of Acyclovir Cleavable by the Dipeptidyl-Peptidase IV (DPP IV/CD26) Enzyme. *European Journal of Medicinal Chemistry* 70: 456–68.

Dinis-Oliveira, R. J. 2018. Metabolic Profiles of Propofol and Fospropofol: Clinical and Forensic Interpretative Aspects. *BioMed Research International* 2018: 1–16.

Dittert, L. W., Caldwell, H. C., Ellison, T., et al. 1968. Carbonate Ester Prodrugs of Salicylic Acid. Synthesis, Solubility Characteristics, *in vitro* Enzymatic Hydrolysis Rates, and Blood Levels of Total Salicylate Following Oral Administration to Dogs. *Journal of Pharmaceutical Sciences* 57: 828–31.

Fang, W. S., and Liang, X. T. 2005. Recent Progress in Structure Activity Relationship and Mechanistic Studies of Taxol Analogues. *Mini-Reviews in Medicinal Chemistry* 5: 1–12.

Fawley, J., and Gourlay, D. M. 2016. Intestinal Alkaline Phosphatase: A Summary of Its Role in Clinical Disease. *Journal of Surgical Research* 202: 225–34.

Foos, T. M., and Wu, J. Y. 2002. The Role of Taurine in the Central Nervous System and the Modulation of Intracellular Calcium Homeostasis. *Neurochemical Research* 27: 21–26.

Ghosh, A. K., and Brindisi, M. 2015. Organic Carbamates in Drug Design and Medicinal Chemistry. *Journal of Medicinal Chemistry* 58: 2895–2940.

Gynther, M., Ropponen, J., Laine, K., et al. 2009. Glucose Promoiety Enables Glucose Transporter Mediated Brain Uptake of Ketoprofen and Indomethacin Prodrugs in Rats. *Journal of Medicinal Chemistry* 52: 3348–53.

Hayashi, Y., Skwarczynski, M., Hamada, Y., et al. 2003. A Novel Approach of Water-Soluble Paclitaxel Prodrug with No Auxiliary and No Byproduct: Design and Synthesis of Isotaxel. *Journal of Medicinal Chemistry* 46: 3782–84.

Hester, E. K., Chandler, H. V., and Sims, K. M. 2006. Fosamprenavir: Drug Development for Adherence. *Annals of Pharmacotherapy* 40: 1301–10.

Hoisaeter, P. A., and Bakke, A. 1983. Estramustine Phosphate (Estracyt): Experimental and Clinical Studies in Europe. *Seminars in Oncology* 10: 27–33.

Hugo, H., Dornbusch, K., and Sterner, G. 1977. Studies on the Clinical Efficacy, Serum Levels and Side Effects of Clindamycin Phosphate Administered Intravenously. *Scandinavian Journal of Infectious Diseases* 9: 221–26.

Huttunen, K. M., Raunio, H., and Rautio, J. 2011. Prodrugs—From Serendipity to Rational Design. *Pharmacological Reviews* 63: 750–71.

Isoherranen, N., Yagen, B., Spiegelstein, O., et al. 2003. Anticonvulsant Activity, Teratogenicity and Pharmacokinetics of Novel Valproyltaurinamide Derivatives in Mice. *British Journal of Pharmacology* 139: 755–64.

Jornada, D. H., dos Santos Fernandes, G. F., Chiba, D. E., et al. 2015. The Prodrug Approach: A Successful Tool for Improving Drug Solubility. *Molecules* 21: 1–31.

Jung, Y. J., Kim, H. H., Kong, H. S., and Kim, Y. M. 2003. Synthesis and Properties of 5-Aminosalicyl-Taurine as a Colon-Specific Prodrug of 5-Aminosalicylic Acid. *Archives of Pharmacal Research* 26: 264–69.

Karaman, R. 2013. Prodrugs Design Based on Inter- and Intramolecular Chemical Processes. *Chemical Biology & Drug Design* 82: 643–68.

Kaul, M., Mark, L., Zhang, Y., et al. 2013a. An FtsZ-Targeting Prodrug with Oral Antistaphylococcal Efficacy *In Vivo*. *Antimicrobial Agents and Chemotherapy* 57: 5860–69.

Kaul, M., Mark, L., Zhang, Y., et al. 2013b. Pharmacokinetics and *in Vivo* Antistaphylococcal Efficacy of TXY541, a 1-Methylpiperidine-4-Carboxamide Prodrug of PC190723. *Biochemical Pharmacology* 86: 1699–707.

Krečmerová, M. 2017. Amino Acid Ester Prodrugs of Nucleoside and Nucleotide Antivirals. *Mini-Reviews in Medicinal Chemistry* 17: 818–33.

Kumpulainen, H., Järvinen, T., Mannila, A., et al. 2008. Synthesis, in Vitro and in Vivo Characterization of Novel Ethyl Dioxy Phosphate Prodrug of Propofol. *European Journal of Pharmaceutical Sciences* 34: 110–17.

Lobo, S., Li, H., Farhan, N., and Yan, G. 2014. Evaluation of Diclofenac Prodrugs for Enhancing Transdermal Delivery. *Drug Development and Industrial Pharmacy* 40: 425–32.

MacDougall, C., and Guglielmo, B. J. 2004. Pharmacokinetics of Valaciclovir. *Journal of Antimicrobial Chemotherapy* 53: 899–901.

Mäntylä, A., Garnier, T., Rautio, J., et al. 2004a. Synthesis, *in vitro* Evaluation, and Antileishmanial Activity of Water-Soluble Prodrugs of Buparvaquone. *Journal of Medicinal Chemistry* 47: 188–95.

Mäntylä, A., Rautio, J., Nevalainen, T., et al. 2004b. Design, Synthesis and in Vitro Evaluation of Novel Water-Soluble Prodrugs of Buparvaquone. *European Journal of Pharmaceutical Sciences* 23: 151–58.

Marcinkiewicz, J., and Kontny, E. 2014. Taurine and Inflammatory Diseases. *Amino Acids* 46: 7–20.

Miceli, M. H., and Kauffman, C. A. 2015. Isavuconazole: A New Broad-Spectrum Triazole Antifungal Agent. *Clinical Infectious Diseases* 61: 1558–65.

Mihaly, G. W., Ward, S. A., Edwards, G., Orme, M. L., and Breckenridge, A. M. 1984. Pharmacokinetics of Primaquine in Man: Identification of the Carboxylic Acid Derivative as a Major Plasma Metabolite. *British Journal of Clinical Pharmacology* 17: 441–46.

Murakami, T., Kohno, K., Yumoto, R., Higashi, Y., and Yata, N. 1998. *N*-Acetyl-L-Gamma-Glutamyl Derivatives of p-Nitroaniline, Sulphamethoxazole and Sulphamethizole for Kidney-Specific Drug Delivery in Rats. *The Journal of Pharmacy and Pharmacology* 50: 459–65.

Newstead, S., Drew, D., Cameron, A. D., et al. 2011. Crystal Structure of a Prokaryotic Homologue of the Mammalian Oligopeptide-Proton Symporters, PepT1 and PepT2. *The EMBO Journal* 30: 417–26.

Ohwada, J., Tsukazaki, M., Hayase, T., et al. 2003. Design, Synthesis and Antifungal Activity of a Novel Water Soluble Prodrug of Antifungal Triazole. *Bioorganic & Medicinal Chemistry Letters* 13: 191–96.

Orlowski, M., Mizoguchi, H., and Wilk, S. 1980. *N*-Acyl-Gamma-Glutamyl Derivatives of Sulfamethoxazole as Models of Kidney-Selective Prodrugs. *Journal of Pharmacology and Experimental Therapeutics* 212: 167–72.

Perry, C. M., and McTavish, D. 1995. Estramustine Phosphate Sodium. *Drugs & Aging* 7: 49–74.

Pradere, U., Garnier-Amblard, E. C., Coats, S. J., Amblard, F., and Schinazi, R. F. 2014. Synthesis of Nucleoside Phosphate and Phosphonate Prodrugs. *Chemical Reviews* 114: 9154–218.

Rautio, J., Kumpulainen, H., Heimbach, T., et al. 2008. Prodrugs: Design and Clinical Applications. *Nature Reviews Drug Discovery* 7: 255–70.

Rautio, J., Meanwell, N. A., Di, L., and Hageman, M. J. 2018. The Expanding Role of Prodrugs in Contemporary Drug Design and Development. *Nature Reviews Drug Discovery* 17: 559–87.

Sanghani, S. P., Quinney, S. K., Fredenburg, T. F., et al. 2004. Hydrolysis of Irinotecan and Its Oxidative Metabolites, 7-Ethyl-10-[4-N-(5-Aminopentanoic Acid)-1-Piperidino] Carbonyloxycamptothecin and 7-Ethyl-10-[4-(1-Piperidino)-1-Amino]-Carbonyloxycamptothecin, by Human Carboxylesterases CES1A1, CES2, and a Newly E. *Drug Metabolism and Disposition: The Biological Fate of Chemicals* 32: 505–11.

Schacter, L. 1996. Etoposide Phosphate: What, Why, Where, and How? *Seminars in Oncology* 23: 1–7.

Schmitt-Hoffmann, A., Roos, B., Heep, M., et al. 2006. Single-Ascending-Dose Pharmacokinetics and Safety of the Novel Broad-Spectrum Antifungal Triazole BAL4815 after Intravenous Infusions (50, 100, and 200 Milligrams) and Oral Administrations (100, 200, and 400 Milligrams) of Its Prodrug, BAL8557, in Healthy. *Antimicrobial Agents and Chemotherapy* 50: 279–85.

Schultz, C. 2003. Prodrugs of Biologically Active Phosphate Esters. *Bioorganic & Medicinal Chemistry* 11: 885–98.

Sharma, S. K., and Bagshawe, K. D. 2017. Antibody Directed Enzyme Prodrug Therapy (ADEPT): Trials and Tribulations. *Advanced Drug Delivery Reviews* 118: 2–7.

Shimoda, K., and Kubota, N. 2011. Chemo-Enzymatic Synthesis of Ester-Linked Docetaxel-Monosaccharide Conjugates as Water-Soluble Prodrugs. *Molecules* 16: 6769–77.

Sinko, P. J., and Balimane, P. V. 1998. Carrier-Mediated Intestinal Absorption of Valacyclovir, the L-Valyl Ester Prodrug of Acyclovir: 1. Interactions with Peptides, Organic Anions and Organic Cations in Rats. *Biopharmaceutics & Drug Disposition* 19: 209–17.

Skwarczynski, M., Sohma, Y., Noguchi, M., et al. 2005. No Auxiliary, No Byproduct Strategy for Water-Soluble Prodrugs of Taxoids: Scope and Limitation of O-N Intramolecular Acyl and Acyloxy Migration Reactions. *Journal of Medicinal Chemistry* 48: 2655–66.

Slatter, J. G., Schaaf, L. J., Sams, J. P., et al. 2000. Pharmacokinetics, Metabolism, and Excretion of Irinotecan (CPT-11) Following I.V. Infusion of [(14)C]CPT-11 in Cancer Patients. *Drug Metabolism and Disposition* 28: 423–33.

Sobue, S., Sekiguchi, K., Shimatani, K., and Tan, K. 2004. Pharmacokinetics and Safety of Fosfluconazole after Single Intravenous Bolus Injection in Healthy Male Japanese Volunteers. *Journal of Clinical Pharmacology* 44: 284–92.

Stockmann, C., Roberts, J. K., Knackstedt, E. D., Spigarelli, M. G., and Sherwin, C. M. T. 2015. Clinical Pharmacokinetics and Pharmacodynamics of Ganciclovir and Valganciclovir in Children with Cytomegalovirus Infection. *Expert Opinion on Drug Metabolism & Toxicology* 11: 205–19.

Subbaiah, M. A. M., Meanwell, N. A., and Kadow, J. F. 2017. Design Strategies in the Prodrugs of HIV-1 Protease Inhibitors to Improve the Pharmaceutical Properties. *European Journal of Medicinal Chemistry* 139: 865–83.

Takagi, T., Ramachandran, C., Bermejo, M., et al. 2006. A Provisional Biopharmaceutical Classification of the Top 200 Oral Drug Products in the United States, Great Britain, Spain, and Japan. *Molecular Pharmaceutics* 3: 631–43.

Vizioli, E. O., Chung, M. C., Menegon, R. F., Blau, L., dos Santos, J. L., Longo, M. C. Novel Compounds Derived from Taurine, Process of Their Preparation and Pharmaceutical Compositions Containing These. WO 2009/124371 A2 (2009), issued 2009.

Walther, R., Rautio, J., and Zelikin, A. N. 2017. Prodrugs in Medicinal Chemistry and Enzyme Prodrug Therapies. *Advanced Drug Delivery Reviews* 118: 65–77.

Weber, B. W., Kimani, S. W., Varsani, A., et al. 2013. The Mechanism of the Amidases: Mutating the Glutamate Adjacent to the Catalytic Triad Inactivates the Enzyme Due to Substrate Mispositioning. *The Journal of Biological Chemistry* 288: 28514–23.

Williams, H. D., Trevaskis, N. L., Charman, S. A., et al. 2013. Strategies to Address Low Drug Solubility in Discovery and Development. *Pharmacological Reviews* 65: 315–499.

Witterland, A. H., Koks, C. H., and Beijnen, J. H. 1996. Etoposide Phosphate, the Water Soluble Prodrug of Etoposide. *Pharmacy World & Science: PWS* 18: 163–70.

Yang, Y, Aloysius, H., Inoyama, D., Chen, Y., and Hu, L. 2011. Enzyme-Mediated Hydrolytic Activation of Prodrugs. *Acta Pharmaceutica Sinica B* 1: 143–59.

Zhang, W., Yang, J., Fan, J., et al. 2019. An Improved Water-Soluble Prodrug of Propofol with High Molecular Utilization and Rapid Onset of Action. *European Journal of Pharmaceutical Sciences* 127: 9–13.

pumps (most prominently those of P-glycoprotein/P-gp, multidrug resistance-associated proteins and breast cancer resistance protein) cannot cross the BBB from the blood and/or maintain therapeutic concentrations within the organ (Deli, 2011).

Certain physicochemical properties of CNS drugs, such as lipophilicity, can be strategically manipulated by prodrug design to aid BBB uptake in the absence of specific transport mechanisms. Lipophilicity is most frequently characterized by the logarithm of octanol/water "P" partition coefficient for neutral molecules, logP or clogP, where c denotes calculated. The logarithm of D_{pH} distribution coefficient at the given pH should be used for weak electrolytes. Lipophilicity is a profoundly important descriptor of drug-likeness, and perhaps the most important parameter in terms of CNS entry by passive transport. Unfortunately, it cannot be increased indefinitely to enhance transport through the lipoidal BBB. In their seminal work, Hansch et al. (1967) showed that there is a parabolic relationship between logP and potency, proposing that the optimal logP range for BBB transport is around 1.5–2. Hann (2011) has also introduced the fitting term "molecular obesity" referring to the often-dogmatic tendency to increase lipophilicity in hope to increase membrane transport. This in turn contributes to molecular "promiscuity" (polypharmacology), causing unwanted side effects and toxicity, bringing about concerns regarding therapeutic safety. Moreover, increased lipophilicity may also increase plasma protein binding, which works against the concentration gradient that is the force behind simple diffusion of an unbound drug in the blood across the BBB (Fick's Law). Increased lipophilicity may also increase the likelihood of recognition by efflux pumps.

Another fundamental physicochemical property in the context of CNS drug delivery from the circulation is the molecular weight (MW) or molecular volume. It is generally accepted that agents with MW> 500–600 Da cannot diffuse through the BBB; however, using the molecular volume and flexibility instead of only the MW may give a much better prediction for drug permeability. This size exclusion limit also is probably unfounded for highly flexible molecules. Nevertheless, the rate of diffusion is inversely proportional to the square of molecular size, according to Graham's law. Altogether, high lipophilicity and large MW/volume, thus, "molecular obesity" (Hann, 2011) are quite unfavorable for passive transport across the BBB.

An additional impediment is that most drug molecules are weak electrolytes; they carry functional groups that, depending on their pKa (the negative logarithm of "Ka" acid dissociation constant), are more or less ionized under physiological conditions. Many of these functional groups are pivotal to binding to receptors or other biological macromolecules. For example, aliphatic amino groups are very common basic functional groups in drugs, and they are mostly ionized at physiological pH, diminishing thereby BBB uptake from the blood. Similarly, a typical acidic functional group in drugs is the carboxyl group that also shows a pH-dependent ionization. It has been proposed that agents with a pKa < 4 for acids or a pKa > 10 for bases do not cross the BBB by passive diffusion in therapeutic level (Fischer et al., 1998), as practically complete ionizations of these agents occur at physiological pH. It has also been recommended that the pKa should be between 7.5 and 10.5 for basic CNS drugs (Pajouhesh and Lenz, 2005). In any case, the famous "pH partition theory" (Shore et al., 1957), originally proposed for absorption through the GI tract, can provide a guide for predicting transport of weak electrolyte drugs through biological membranes, such as the BBB. It assumes that only neutral species of drugs can partition into the membranes. On the other hand, a certain degree of aqueous solubility is absolutely necessary for multiple reasons, and ionization is pivotal in this regard. Ideally, an optimal "lipophilic-hydrophilic balance" is the goal. For example, an oral drug (solute) has to be dissolved in the aqueous environment of the gastrointestinal tract and partition into the lipid membrane of the gut wall in order to get into the blood (i.e., for absorption). The great advantage of prodrug design is that ionizable functional groups can be transiently masked for BBB transport by strategically selected bioreversible chemical modifications to make the prodrugs neutral at physiological conditions, and also to balance water solubility with membrane affinity. The above mentioned ionizable functional groups, just like many other functional groups in drug molecules, can also participate in hydrogen bonding. They can function both as donors and acceptors, although nonionizable polar groups (such as carbonyl) in drug molecules are acceptors only. Intermolecular hydrogen bonding with water is a very important contributor to aqueous solubility, but once the total number of hydrogen bonds reaches around 8–10, the BBB uptake is presumed negligible (Fischer et al., 1998; Pajouhesh and Lenz, 2005; Wager et al., 2010). Therefore, beside the previously mentioned metabolic stability problems with peptides, hydrogen bonding adds another layer of difficulties to peptide delivery into the CNS (Prokai and Prokai-Tatrai, 2003). It has also been shown that there is a correlation between hydrogen bond count and P-gp recognition (Desai et al., 2012).

Altogether, a general conclusion can be drawn: adequate lipophilicity, small molecular volume, lack of sufficient ionization at physiological pH, and limited hydrogen bonding are favorable for BBB penetration from the circulation. The list of factors affecting free diffusion through this physical, transport and enzymatic barrier is, however, far from being complete and rather case specific (Fischer et al., 1998; Pajouhesh and Lenz, 2005; Rankovic, 2015). Nevertheless, some of the critical physicochemical characteristics to improve/optimize ADME (absorption, distribution, metabolism, and excretion) properties of CNS agents to achieve CNS-targeting can be successfully modified by prodrug designs, as illustrated below.

7.3 Prodrugs for CNS Drug Delivery by Passive Transport

As most CNS maladies (such as affective disorders, neurodegenerative diseases) require chronic treatment, targeting drugs into the site of action is essential not only to ensure efficacy but also therapeutic safety. Off-target peripheral impact of centrally acting medications may prevent the clinical usefulness of the otherwise promising CNS agents. Targeted pharmacotherapy with prodrugs can be achieved with site-directed delivery through, e.g., receptor-mediated and gene-directed enzyme prodrug systems, or site-specific facile bioactivation of the prodrug to the pharmacological active parent agent by enzymes preferentially expressed in the CNS (Prokai et al., 2015). It is surely not easy to create prodrugs that have easy access to the CNS and rely on CNS-specific enzymes for prodrug bioactivation to achieve true brain targeting (Prokai-Tatrai et al., 2008a; Prokai et al., 2015). In certain cases, we can take advantage of the specific physiological conditions, such as the acidic environment in tumors, for targeted prodrug delivery. Another criterion that must be met for a successful CNS-prodrug approach is the retention of the drug liberated from the prodrug construct to achieve and maintain therapeutic concentration.

In any case, for prodrug-mediated CNS drug targeting a high ratio of bioconversion rates in brain relative to plasma is desired. Anderson (2007) has provided a great example to illustrate this by using a poorly brain-permeable CNS agent that is also recognized by efflux pump transporters. 2′,3′-Dideoxyinosine has been approved to treat NeuroAIDS. Metabolic stabilities of its various alky and aryl ester prodrugs with those of its halogenated analogs were compared. *In vitro* half-lives in plasma and brain homogenate showed that the prodrugs had significantly shorter half-lives in plasma (<1 min) than in brain homogenate (1–20 min), essentially making the prodrugs unsuitable for CNS drug delivery (Anderson, 2007).

Prodrugs are synthesized by bioreversible, and thus, transient chemical modifications of the parent drugs taking advantage of functional groups, as "synthetic handles" present in the drug molecules (Prokai-Tatrai and Prokai, 2009). Strategic selections of the drug's functional group(s) and the auxiliary promoiety(ies) are mission-critical to improve whatever shortcomings (pharmacokinetics, pharmaceutical, etc.) the parent drug have in the context of drug delivery across the BBB. In the prodrug construct, the promoiety (promoieties) is (are) covalently attached to the drug either directly (e.g., by simple esterification) or through a linker (e.g., an N-hydroxyalkyl group) resulting in bipartite and tripartite prodrugs, respectively. This is the classical/traditional or carrier-linked prodrug approach that mostly includes ester- and amide-types of prodrugs (Prokai-Tatrai and Prokai, 2003, 2009; Pavan et al., 2008). It is important, however, to distinguish these simple prodrugs from another type of prodrugs, the so-called bioprecursor prodrugs, in light of recent advances of their use for CNS drug targeting (Prokai et al., 2015; Prokai-Tatrai et al., 2018). Of note, a new classification of prodrugs (Type I, Type II, and associated subtypes) based on their cellular sites of the final bioactivation step to the active drug has also been introduced (Wu, 2009).

Bioprecursor prodrugs do not have auxiliary promoiety(ies) but the drug molecule per se is modified bioreversibly. This type of prodrugs mostly undergoes Phase I metabolism (such as oxidation, reduction, decarboxylation, etc.) to regenerate the active agent from the prodrug. One of the advantages of bioprecursor prodrugs over classical prodrugs is that promoiety-related toxicity or innate pharmacological effects are eliminated. The already mentioned prodrugs relying on LAT1 for transport into the brain are actually bioprecursor prodrugs (Fahn et al., 2004; Yaksh et al., 2017). It is worth noting that, to avoid utilizing the saturable LAT1 for the brain delivery of amino acid-type chemotherapeutic agents, efforts have been put forward to design prodrugs capable of reaching the brain by the nonsaturable passive diffusion (Killian and Chikhale, 2000).

7.3.1 Small-Molecule Prodrugs for CNS Drug Delivery

Let us illustrate the applicability of the prodrug strategy for true brain targeting with work from our laboratory that resulted in the preclinical development of estrogen-derived bioprecursor prodrugs for CNS-selective estrogen therapy (Prokai et al., 2003, 2015; Prokai-Tatrai et al., 2018). Figure 7.2a shows these agents; the first one is 10β,17β-dihydroxyestra-1,4-dien-3-one (DHED), prodrug of the main human estrogen 17β-estradiol (E2). DHED is synthesized from E2 via a one-step stereoselective oxidation of the phenolic A-ring (Prokai et al., 2015). Similarly, from the isomeric 17α-estradiol (αE2), 10β,17α-dihydroxyestra-1,4-dien-3-one (αDHED) bioprecursor prodrug was prepared using microwave-assisted oxidation (Prokai-Tatrai et al., 2007, 2018), while 10β-hydroxyestra-1,4-dien-3,17-dione (HEDD) was derived from another human estrogen (estrone, E1) (Prokai et al., 2003). The recognition of these steroidal *para*-quinols as first-in-class bioprecursor prodrugs (Prokai et al., 2003) originates from our discovery of a novel antioxidant cycle for estrogens as direct free-radical scavenging phenolic antioxidants (Prokai et al., 2013). The antioxidant activity of the hormones is a critical contributor to the nongenomic component of estrogens' powerful neuroprotective actions (Prokai-Tatrai et al., 2008b, 2017). As the structures of DHED, αDHED and HEDD show (Figure 7.2a), there is no auxiliary promoiety in the prodrugs because the phenolic A-rings of the steroids were oxidized to *para*-quinols. The bioactivation of this unique type of bioprecursor prodrugs in the CNS is essentially the opposite process of their oxidative chemical synthesis, and is carried out through an NADP(H)-dependent catalytic process by short-chain dehydrogenase/reductase selectively expressed in the CNS (Figure 7.2b). This reductive bioactivation is expected to avoid toxic metabolites commonly seen with oxidative metabolisms (Rivera-Portalatin et al., 2007). The distinguishing feature of prodrugs shown in Figure 7.2a is that they provide unprecedented CNS-targeting of the corresponding estrogen to treat the estrogen-deprived or diseased brain without exposing the rest of the body to the hormones (Prokai et al., 2015; Merchenthaler et al., 2016; Tschiffely et al., 2016, 2018; Prokai-Tatrai and Prokai, 2017; Prokai-Tatrai et al., 2018). This site-directed delivery of estrogens is gender-independent (Tschiffely et al., 2016, 2018) and, even after chronic exposure to the prodrugs, no increase in circulating estrogens can be measured (Prokai et al., 2015). Representative experimental outcomes are given in Figure 7.3. The profoundly beneficial effect of E2 produced from the orally bioavailable DHED on lowering tail-skin temperature in a hot flush model of menopause is shown in Figure 7.3a (Prokai et al., 2015; Merchenthaler et al., 2016). Due to lack of appreciable oral bioavailability of E2, the clinically used synthetic estrogen ethynyl estradiol (EE) was used here as a positive control. The excellent oral

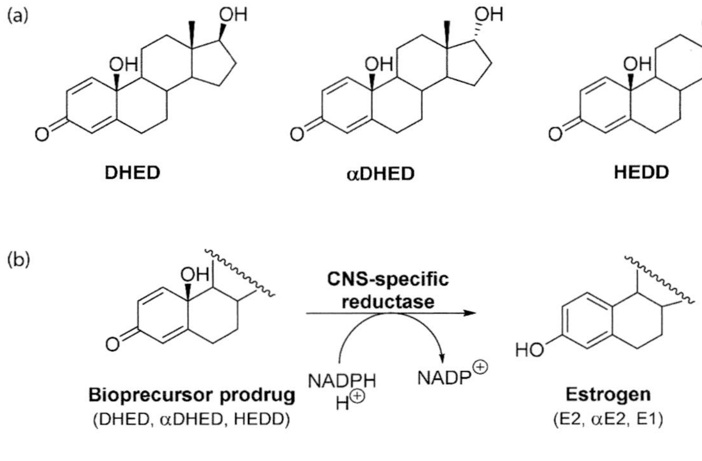

FIGURE 7.2 (a) Chemical structures of bioprecursor prodrugs of estrogens: 10β,17β-dihydroxyestra-1,4-dien-3-one (DHED) for E2; 10β,17α-dihydroxyestra-1,4-dien-3-one (αDHED) for αE2; and 10β-hydroxyestra-1,4-dien-3,17-dione for E1. (b) Schematic illustration of CNS-specific reductive bioactivation of bioprecursor prodrugs shown in panel (a) via an NADP(H)-dependent short-chain dehydrogenase/reductase to the corresponding estrogen (E2, αE2 or E1).

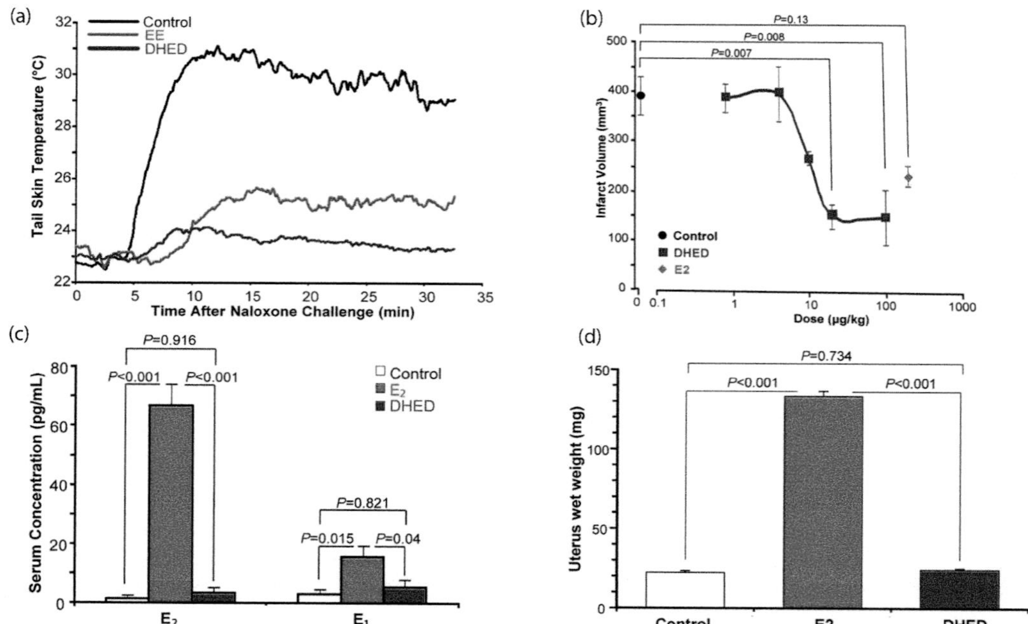

FIGURE 7.3 DHED treatment elicits estrogen-responsive neuropharmacological effects without peripheral E2 exposure. (a) Tail-skin temperature changes in an OVX rat hot flush model after oral administrations of DHED (30 μg/kg). The orally active strong synthetic estrogen EE (200 μg/kg) was used as positive control. (b) DHED treatment elicits dose-dependent neuroprotection in a rat model of ischemic stroke. OVX rats were treated with various doses of DHED (subcutaneously, s.c.), 1 h before tMCAO followed by 24-h reperfusion. A nearly 10-fold higher concentration of E2 is needed to achieve the same neuroprotection with direct administration of E2 than with that of DHED. (c) Serum E2 and E1 concentrations and (d) Wet uterine weight of OVX Fischer-344 rats continuously treated with E2 and DHED, respectively, at 4 μg/day by s.c. ALZET osmotic pumps for 48 days (averages ± SEM, n = 4–5 per treatment group). *P* values were determined by ANOVA using post hoc SNK multiple comparison test. (From Prokai, L. et al., *Sci. Transl. Med.*, 7, 297ra113, 2015. Reprinted with permission from AAAS.)

bioavailability of DHED is very important in the context of translational research. The dose-dependent neuroprotective effect of E2 (Prokai et al., 2015) when formed from the prodrug in the brain can be seen in Figure 7.3b. Owing to the CNS-specific E2 delivery by DHED and the prodrug's significantly improved physicochemical properties for brain uptake from the circulation relative to those of E2 (e.g., clogP of 1.67 versus 4.01 for the highly lipophilic parent E2), it is not surprising that approximately 10-fold higher E2 concentration was needed upon direct E2 administration to manifest the same degree of neuroprotection in a rat model of stroke. This model is based on transient middle cerebral artery occlusion (tMCAO) followed by reperfusion. Post-stroke treatment with DHED also brought about profound neuroprotection with associated functional recovery (Prokai et al., 2015). It is important to note that with DHED, αDHED and HEDD (Figure 7.2a) approximately a 200-fold "delipidization" of the highly lipophilic corresponding parent drugs occurred instead of the traditional "lipidization" to increase membrane permeability by prodrug designs. Concomitantly, a significant increase in water solubility of the prodrugs relative to the practically water-insoluble hormones also occurred. "Delipidization" via a simple prodrug design was also undertaken when SB-3CT, 2-[[(4-phenoxyphenyl)sulfonyl]methyl]-thiirane, a potent and selective matrix metalloproteinase-2 and -9 inhibitor, was converted into its phosphate ester prodrug to increase water solubility for probing the drug's utility in a model of severe traumatic brain injury (Lee et al., 2015).

In another set of experiments (Prokai et al., 2015), ovariectomized (OVX) animals without appreciable level of circulating E2 and E1 received continuous DHED treatment for over 6 weeks, yet no increase in blood estrogen levels (Figure 7.3c) could be established as quantified by a liquid chromatography-coupled tandem mass spectrometry (LC-MS/MS)-based assay (Szarka et al., 2013). Accordingly, the profound estrogenic effect on the uterus associated with exogenously added estrogens was also absent (Figure 7.3d). The distinguishing lack of the uterotropic effect is another independent confirmation of the site-specific E2 delivery by DHED, as the uterus is one of the most sensitive organs to circulating estrogens. The uterine wet weight increase due to fluid imbibition is a frequently used marker for estrogenicity (uterotropic assay). In contrast, direct E2 treatment, as expected, produced a profound increase in circulating estrogen levels and, thus, in uterine wet weight (Figure 7.3c and d, respectively, red panels). Altogether, our data on the patented DHED approach (Prokai et al., 2006, 2007) show a previously unseen brain-targeted delivery of E2 with a small-molecule bioprecursor prodrug having practically an ideal clogP (=1.61) and MW (=288 Da) for passive transport across the BBB, for confining E2's beneficial effects into the CNS for the treatment of centrally regulated and estrogen-responsive maladies (Prokai et al., 2015; Merchenthaler et al., 2016; Tschiffely et al., 2016, 2018; Prokai-Tatrai and Prokai, 2017; Prokai-Tatrai et al., 2018; Rajsombath et al., 2019; Yan et al., 2019). Similar degree of brain targeting of αE2 with αDHED (Prokai-Tatrai et al., 2018) and E1 with HEDD (Prokai et al., 2003) has also been achieved.

Simple prodrugs of E2, such as the clinically used E2-valerate, -acetate and -benzoate cannot achieve brain targeting, as bioactivation of these ester prodrugs by ubiquitous hydrolases/esterases occurs all over the body. Similarly, when the 17-N,N-dimethylaminobutyl water-soluble ester prodrug of E2 was administered intranasally to bypass the BBB, even though the prodrug produced a high level of E2 in the CSF, a very large concentration of E2 (ng/mL level) was also present in the plasma (Al-Ghananeem et al., 2003). A series of alkyl ester prodrugs of L-Dopa has also been synthesized for nasal drug delivery (Kao et al., 2000; Lee et al., 2014), but they turned out not to be more beneficial than L-Dopa itself. One of the major limitations with the most commonly used simple esterase-triggered prodrugs is the lack of target specificity owing to nonspecific hydrolysis of the prodrugs by the abundantly expressed esterases in the body. Therefore, it is very challenging to create an ester prodrug that is sufficiently stable in plasma but rapidly liberates the active agent in the brain.

For several decades, it was assumed that the CDS uniquely provides differential brain delivery of drugs (Prokai-Tatrai et al., 2012). This concept utilizes 1,4-dihydropyridine-based promoieties that are easily oxidized in the brain to the corresponding pyridiniums. Mostly ester- and carbamate-type of prodrugs/CDSs have been prepared. The oxidized CDS/prodrug is presumed to stay inside the organ owing to its permanent charge. In the brain, hydrolase then removes the pyridinium promoiety to release the parent drug at the site of action (Prokai et al., 2000). It has always been speculated that the liberation of the parent drug from the CDS in the circulation would be negligible, because the oxidized, quaternary CDS would rapidly be eliminated from the periphery. The CDS concept was applied to practically any and all available compounds with CNS therapeutic potentials. It remains to be seen whether any of these CDSs were recognized by efflux pumps.

The best-known application of this concept has probably been the E2-CDS for estrogen neurotherapy. While this prodrug, indeed, provides a sustained E2 release within the brain, its administration also produces a profound and sustained E2 level in the blood (Prokai-Tatrai et al., 2012). It has unequivocally been shown by an LC-MS/MS assay (Szarka et al., 2013) that, even after a single dose of E2-CDS incorporated in a cyclodextrin and termed Estredox, it took more than 8 days to eliminate the prodrug-derived E2 from the circulation (Prokai-Tatrai et al., 2012). This finding was in agreement with what was reported much earlier by another research group (Sarkar et al., 1989). It has been shown that, for example, 48-h post-Estredox treatment around 350 pg/mL E2 was present in the blood of OVX animals compared to approximately 13 pg/mL in the control group. Consequently, the average wet uterine weight was around 64 mg; a 3-fold increase relative to the control group. Therefore, it is not surprising, that the uterus of Estredox-treated animals had approximately 3 ng/g E2 concentration, while in the control group the E2 level was around 500 pg/g (Prokai-Tatrai et al., 2012). In contrast, even after chronic treatment with DHED (Figure 7.3c and d) and related prodrugs (Prokai et al., 2003; Prokai-Tatrai et al., 2018), no increase in circulating E2 and markers of peripheral estrogenicity (such as the already mentioned uterotropic effect) could be established (Prokai et al., 2003, 2015; Merchenthaler et al., 2016; Tschiffely et al., 2016; Prokai-Tatrai et al., 2018). Altogether, the available data clearly show that E2-CDS/Estredox performs poorly compared to the chemically stable, orally bioavailable and significantly more innovative DHED approach in terms of CNS-targeting of E2.

The CDS redox concept was also attempted for brain-enhanced delivery of oximes. Oximes, such as pralidoxime (1-methylpyridine-6-carbaldehyde oxime), acting as nucleophiles, are important components of therapeutic interventions against organophosphate poisoning to quickly regenerate the phosphorylated serine residue of acetylcholinase (Eddleston et al., 2002). While organophosphates are capable of crossing the BBB, pralidoxime has a quaternary structure. Therefore, its neutral dihydropyridine-based bioprecursor prodrug (N-methyl-1,6-dihydropyridine-2-carbaldoxime hydrochloride) was synthesized by reducing the pyridinium to dihydropyridine (Prokai et al., 2000). This prodrug then, on the analogy of the CDS approach, was expected to oxidize in the brain to the quaternary parent drug. The prodrug, however, performed marginally; one possible reason could be insufficient brain targeting with the inherently unstable dihydropyridine. Chemical tractability issues are undoubtedly serious drawbacks of the CDS approach. Because of this, another redox prodrug concept with "lock-in" mechanism into the brain may be more amenable for practical CNS-targeting (Ishikura et al., 1995).

Alzheimer's disease (AD) is the most common form of dementia in elderly people, and without curative or curtailing therapeutic modalities. Acetylcholinesterase (AChE) inhibitors show some benefit in improving cognition in early cases of AD. The problem with AChE inhibitors is that due to their limited brain uptake they cause adverse peripheral side effects. Taking the reversible acetylcholine (ACh) inhibitor donezapil (2,3-dihydro-5,6-dimethoxy-2-[[1-(phenylmethyl)-4-piperidinyl]methyl]-1H-inden-1-one hydrochloride) as a template, Azzouz et al. (2018) have synthesized N-benzylpyridinium derivatives of this drug. Several aryl ketone-derived N-benzylpyridinium analogs showed high AChE inhibitory activity *in vitro*. These pyridinium analogs are prodrug-amenable, as their corresponding neutral dihydropyridine bioprecursor prodrugs are expected to deliver and retain the drug in the brain, on the analogy of the CDS approach discussed previously (Prokai et al., 2000). *In vivo* substantiation of the design has yet to come. While N-benzyl-1,4-dihydropyridines are more stable chemically than the N-methyl counterpart used in the CDS; chemical tractability and formulation problems may halt the otherwise interesting approach. Other bioprecursor prodrug-amenable pyridinium analogs of CNS agents have also been reported before (Prokai-Tatrai et al., 2002, 2005, 2013).

Besides the already discussed prodrug approaches for the treatment of NeuroAIDS (Anderson 2007; Dalpiaz and Pavan, 2008), novel prodrugs have recently been created for the same therapeutic indication (Nedelcovych et al., 2017b). Studies of affected patients have showed a disrupted glutamate metabolism in the CNS. Therefore, next to viremic control, normalizing glutamate homeostasis by the use of glutamine antagonists has therapeutic potential for the treatment of NeuroAIDS. To avoid peripheral toxicity, the brain uptake of these agents, however, must be enhanced. To do that, the research group has created lipophilic prodrugs of the glutamine antagonist 6-diazo-5-oxo-L-norleucine. A series of (pivaloyloxy)alkoxy-carbonyl prodrugs has been synthesized and a promising lead was selected after *in vitro* and *in vivo* studies in swine. The lead prodrug, isopropyl 6-diazo-5-oxo-2-(((phenyl(pivaloyloxy) methoxy)-carbonyl)amino)hexanoate, provided a significantly increased CSF-to-plasma and brain-to-plasma ratio of the parent drug, compared to the direct administration of 6-diazo-5-oxo-L-norleucine. The same group has also reported (Nedelcovych et al., 2017a) that intranasal administration of a lipophilic ester prodrug of the hydrophilic 2-(phosphonomethyl)pentanedioic acid, a potent and selective glutamate carboxypeptidase-II inhibitor for the treatment of the diseased CNS, has doubled the parent drug's concentration in the CSF of primates relative to direct administration of 2-(phosphonomethyl) pentanedioic acid.

Similarly to other NSAIDs, dexibuprofen, the S(+)-isomer of ibuprofen, has low bioavailability to the brain due to its structural features (carboxylic acid) that also result in high plasma protein binding, as well as recognition by efflux pumps. Efforts have been made to overcome problems associated with this NSAID's limited brain uptake by prodrug designs that either utilized carrier-mediated (Yue et al., 2018) or passive transport for drug delivery. The latter includes ester-type of prodrugs that were prepared with ethanolamine and N-methylethanolamine, taking advantage of the carboxylic group of the parent drug (Zhang et al., 2012; Li et al., 2015). *In vitro, in vivo* and mechanistic studies showed that these prodrugs reached the brain by simple diffusion where they were hydrolyzed to dexibuprofen, most probably by cholinesterase due to the structural similarity of ethanolamine to choline. Recently, new N-benzylamide-type of conjugates have also been synthesized for a series of NSAIDs for their potential brain-enhanced delivery (Eden et al., 2019).

Another acidic drug, the thyromimetic sobetirome, has a great potential for remyelination therapy in multiple sclerosis. Its ethanolamine-ester prodrug has been reported (Placzek et al., 2016). This prodrug has favorably produced enhanced brain and lower blood levels of sobetirome, relative to the direct administration of the parent drug. The methylacetamide prodrug of sobetirome has also produced enhanced brain delivery, and was efficacious in a genetic mouse model of demyelination (Hartley et al., 2019). It remains to be seen if these prodrugs reached the brain by diffusion only or active transports have also contributed to this process. Noteworthy, that the dimethyl diester prodrug (or pro-prodrug) of fumaric acid (Tecfidera®) is a US Food and Drug Administration (FDA)-approved drug for the management of relapsing multiple sclerosis (FDA, 2019). The monomethyl ester prodrug of fumaric acid, however, has very recently received tentative approval by the FDA for the very same therapeutic indication by promising higher potency than Tecfidera® (mdmag, 2019). Interestingly, the feasibility of the transdermal delivery (where the formidable obstacle is the stratum corneum) of Tecfidera® for CNS therapy has also been probed (Ameen and Michniak-Kohn, 2017).

7.3.2 Peptide Prodrugs for CNS Drug Delivery

Peptide prodrug delivery into the CNS remains an important and intriguing field as many peptides have unfulfilled neurotherapeutic potentials for a wide variety of maladies affecting the CNS and currently without efficacious pharmacological interventions. A book entitled *Peptide Transport and Delivery into the Central Nervous System*, part of the Progress in Drug Research series (Prokai and Prokai-Tatrai, 2003), is dedicated to the subject, and interested readers are referred to this book to gain more insight into the field that is not possible to offer here due to space limitations.

CNS peptide delivery, and here only small- and medium-sized peptides are considered (Prokai and Prokai-Tatrai, 2011), is even more complicated than that of comparable size of nonpeptidic solutes because of the general enzymatic-liability and chemical nature of peptides. Many peptides with free N- and C-termini are degraded by exopeptidases usually within a few minutes (Prokai and Prokai-Tatrai, 2003). Therefore, providing necessary metabolic stability of these peptides in blood by prodrug design is essential.

An excellent example of neuropeptides with tremendous potential to treat certain CNS maladies is thyrotropin-releasing hormone (TRH) with the sequence pGlu-His-Pro-NH_2 (Prokai, 2002; Fröhlich and Wahl, 2019). This tripeptide is highly hydrophilic (clogP $\backsim -3.5$) and extremely susceptible to enzymatic degradation; thus, it exhibits very short biological half-lives (Prokai-Tatrai and Prokai 2009). Many of TRH's central beneficial effects, such as analeptic- and antidepressant-like effects, robust modulation of ACh synthesis and release, are shared with the metabolically more stable TRH-like peptides having pGlu-XX-Pro-NH_2 sequence (XX denotes an L-amino acid residue or mimetic) (Prokai-Tatrai and Prokai 2009; Bílek et al., 2011; Kobayashi et al., 2018). Therefore, initial efforts for TRH's CNS-targeting rather used the metabolically stable and centrally acting [Leu^2]TRH by essentially tailoring the CDS approach for peptides (Prokai et al., 1999). As the commonly applied 1,4-dihydropyridine-based promoiety (Prokai et al., 2000) would not furnish the necessary lipophilicity to the prodrug construct for free diffusion across the BBB, further "lipidization" was done by using the cholesteryl ester of the progenitor sequence of the TRH analog, as shown in Figure 7.4a. Critical was in the design the use of Gln-Leu-Pro-Gly progenitor sequence. This was proposed by Prokai on the analogy of TRH's biosynthesis (Prokai et al., 1994). The progenitor sequence has two important advantages over the direct use of the parent peptide; it provides a suitable synthetic handle at the N-terminus for covalent attachment of appropriate promoieties, and the C-terminal amino acid allows for an easy attachment of highly lipophilic alcohols (such as cholesterol or adamantanol) to ensure the necessary increase in membrane affinity for diffusion across the BBB. It remains to be seen whether a bulky ester at the C-terminus would disguise the peptide against recognition by degrading peptidase, as it would require the use of metabolically unstable peptides, such as TRH itself.

Another important aspect of the design is the use of a specific enzyme-sensitive linker between the dihydropyridine and progenitor sequence. This enzyme-sensitive (dipeptidyl or single amino acid) linker (Prokai-Tatrai et al., 2008a) is based on the inherent feature of prolyl oligopeptidase (POP, a.k.a. post-proline cleaving enzyme) (Polgar and Szeltner, 2008). POP selectively hydrolyzes a peptide bond at the carboxyl side of internal proline in small and medium-sized peptides. The idea behind the use of a

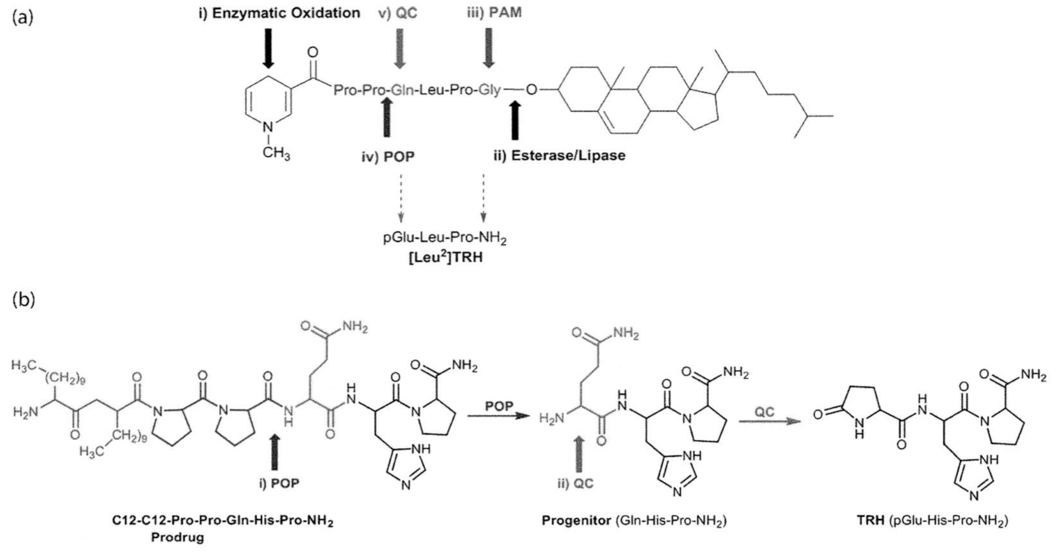

FIGURE 7.4 Schematic illustration of the prodrug concepts developed for brain-enhanced delivery of (a) [Leu^2]TRH utilizing Gln-Leu-Pro-Gly progenitor sequence and a series of enzymatic metabolisms (i to iv) for liberating [Leu^2]TRH in the brain. After (i) enzymatic oxidation the pyridinium construct in the brain, (ii) releases the bulky cholesteryl ester group by hydrolase; (iii) followed by α-amidation on the C-terminus by peptidyl-glycine alpha amidating monooxygenase (PAM). Next, (iv) prolyl oligopeptidase (POP) releases the progenitor sequence, that (v) glutaminyl cyclase (QC) converts to [Leu^2]TRH. (Adapted from Prokai-Tatrai, K. and Prokai, L., *Molecules*, 14, 633–654, 2009. Under the terms of Creative Commons Attribution License.) (b) Brain-enhanced delivery of TRH by its C12-C12-Pro-Pro-Gln-His-Pro-NH_2 prodrug extended from the N-terminus of Gln-His-Pro-NH_2 progenitor with a pair of 2-aminododecanoyl (C12) residues (magenta) and Pro-Pro as a POP-sensitive linker (blue). The two-step bioactivation leading to the liberation of TRH preferentially occurs in the brain (i) by POP (blue), (ii) followed by QC (red). (Adapted from Prokai-Tatrai, K. et al., *Pharmaceutics* 11, 349, 2019. Under the terms of Creative Commons Attribution License.)

POP-sensitive linker in the prodrug construct was two-folded; since highest activity of POP was found in the brain (Myöhänen et al., 2008), utilizing this enzyme for removing the promoiety should occur preferentially within the brain, i.e., leading to brain-enhanced delivery of the peptide. Additionally, direct attachment of the dihydropyridine to the N-terminus would not have resulted in useful prodrugs owing to the general low amidase activity in the CNS. This has been unequivocally shown with constructs having no POP-sensitive linker not only for TRH-like peptides (Prokai et al., 1994; Prokai-Tatrai and Prokai, 2009), but also for a metabolically stable leucine-enkephalin analog, Tyr-D-Ala-Gly-Phe-D-Leu (DADLE) (Prokai-Tatrai et al., 1996, 2008a).

The prodrug construct based on these design principles for [Leu^2]TRH is quite complex, as shown in Figure 7.4a. It requires a series of sequential metabolic steps to liberate the neuropeptide within the brain. Specifically, the expectation was that, after the intact prodrug reaches the brain, the dihydropyridine will be oxidized to pyridinium, which restricts the oxidized peptides from effluxing into the circulatory system. Then, hydrolysis of the lipophilic cholesteryl ester occurs, followed by α-amidation on the C-terminus by peptidyl-glycine alpha amidating monooxygenase (PAM). This enzyme catalyzes the formation of C-terminal peptide amides from Gly-extended precursors (Stevens et al., 2005). The next pivotal step is the liberation of the progenitor sequence by POP, followed by metabolism of Gln to pGlu by glutaminyl cyclase (QC), generating then the TRH analog within the brain. When tested *in vivo*, prodrugs having various POP-sensitive linkers (Ala, Pro, or the combinations thereof) and cholesteryl ester at the C-terminus (Figure 7.4a) produced centrally-mediated analeptic effect, indicating a successful delivery of the TRH-like peptide into the brain (Prokai et al., 1999). The most efficacious prodrug had Pro-Pro as the POP-sensitive linker. Administration of this prodrug also produced a significant and sustained increase in extracellular ACh release, when measured after *in vivo* intracranial microdialysis sampling

(Prokai and Zharikova, 2002). Similarly, when the same design principle was applied to brain delivery of the metabolically stable enkephalin analog DADLE, the most efficacious prodrug in terms of producing analgesia in rats (monitored by the tail-flick latency measurements) carried an identical POP-sensitive linker (Pro-Pro) (Prokai-Tatrai et al., 1996, 2008a). The obtained *in vivo* efficacy data showed excellent correlations with the measured *in vitro* half-lives of the oxidized prodrugs.

Despite these attractive *in vivo* data, this complex "molecular packaging" CDS approach for small neuropeptides (Prokai et al., 1994, 1999; Prokai-Tatrai et al., 2008a; Prokai-Tatrai and Prokai, 2009, 2011) is theoretically interesting, yet without realistic translational value owing to persistent problems associated with dihydropyridines as promoieties (Prokai-Tatrai et al., 2012). Therefore, the CDS approach is an excellent example for a prodrug-mediated CNS drug delivery system whose translational value has been reduced owing to overlooked nontargeting issues (such as formulation). Noteworthy, that Tsuzuki et al. (1991) probed the brain delivery of Tyr-D-Ala-Gly-Phe-Leu, ([D-Ala^2]Leu-enkephalin), a metabolically liable Leu-enkephalin by derivatization with adamantyl moiety to increase the lipophilicity of the peptide. Modification at the N-terminus abolished activity most probably owing to the metabolic stability of the amide bond between the promoiety and [D-Ala^2]Leu-enkephalin (Prokai-Tatrai et al., 2008a). Systemic administration of the adamantyl ester prodrug on the other hand produced analgesia. The bulky alcohol apparently had protective effect against degradation of the peptide by proteolytic enzymes.

Simple alkyl ester prodrugs were also synthesized for successful brain-enhanced delivery of another TRH-like peptide, [Glu^2]TRH (Prokai-Tatrai et al., 2003; Prokai-Tatrai and Prokai, 2011). Membrane affinities of these ester prodrugs as predictors of the transport across the BBB were compared using immobilized artificial membrane chromatography (IAMC) (Braddy et al., 2002). Lipophilic prodrugs with half-lives around 20-25 min in mouse brain homogenate yielded a significant enhancement in the CNS delivery, as monitored by the parent peptide's profound analeptic effect (Prokai-Tatrai and Prokai, 2009). The n-hexyl ester afforded the most promising prodrugs for brain delivery of [Glu^2]TRH that has otherwise poor access to the CNS from the circulation.

A novel promising prodrug approach has been introduced very recently for brain-enhanced delivery of TRH itself (Prokai-Tatrai et al., 2019). This approach is devoid of the above addressed limitations of the dihydropyridine-based CDS approach. As shown in Figure 7.4b, the promoiety of this prodrug is the result of synergistic combination of lipoamino acid(s) (LAA) and a linker cleaved by POP for the preferential brain delivery of the peptide. The prototype of this design, C12-C12-Pro-Pro-Gln-His-Pro-NH_2, carries two 2-aminododecanoyl (C12) residues as LAA, and a diprolyl POP-sensitive linker attached to the N-terminus of the Gln-His-Pro-NH_2 progenitor (Figure 7.4b). The LAA was included into the prodrug construct to augment not only lipophilicity but also to introduce amphiphilicity to facilitate the prodrug's interaction with cell membranes (Tóth et al., 1999). Essential in this design is the previously utilized POP-sensitive linker (Prokai-Tatrai et al., 1996, 2008a; Prokai et al., 1999; Prokai-Tatrai and Prokai, 2009) for the liberation of TRH after its preferential delivery into the brain (Figure 7.4b) owing to the highest POP activity measured in this organ (Myöhänen et al., 2008). *In vitro* metabolic stability studies were conducted in rodent plasma and brain homogenate with or without the presence of a POP inhibitor (KYP-2047). The prodrug showed the desired high ratio of bioconversion rates in brain relative to plasma for brain-enhanced delivery, and the POP inhibitor completely prevented prodrug metabolism in the brain homogenate. *In vivo* substantiations of this approach have been conducted using typical neuropharmacodynamic and neurochemical effects of TRH with or without the use of the brain-permeable KYP-2047 (Prokai-Tatrai et al., 2019). It is believed that this is the first successful example of noninvasive delivery of the metabolically unstable and poorly BBB-permeable TRH by a prodrug that reaches the brain by diffusion across the BBB, and then metabolizes to TRH by enzymes preferentially expressed in the brain.

7.4 Conclusion

We hope this chapter gives an insight into the challenges associated with drug delivery into the central nervous system. Here, we focused on drug delivery based on prodrugs that are capable of traversing the BBB from the circulation by simple diffusion. The BBB, as a gatekeeper, represents multiple blockades for drug delivery. However, for a certain set of small-molecule CNS agents, the prodrug approach allows for transient and strategic manipulations of physicochemical properties to make these agents "BBB friendly" and, thus, these prodrugs can deliver the parent drugs into the brain. Important recent developments in this regard were highlighted. Novel types of prodrugs and formulation approaches have been and continued to be developed, therefore, the future of brain-targeting prodrug technologies remains exciting yet not without challenges.

REFERENCES

Alavijeh, M.S., Chishty, M., Qaiser, M.Z., and Palmer, A.M. 2005. Drug metabolism and pharmacokinetics, the blood-brain barrier, and central nervous system drug discovery. *NeuroRx®* 2:554–571.

Al-Ghananeem, A.M., Traboulsi, A.A., Dittert, L.W., and Hussain, A.W. 2003. Targeted brain delivery of 17ß-estradiol via nasally administered water-soluble prodrugs. *AAPS PharmSciTech* 3 (1):40–47.

Ameen, D., and Michniak-Kohn, B. 2017.Transdermal delivery of dimethyl fumarate for Alzheimer's disease: Effect of penetration enhancers. *International Journal of Pharmaceutics* 529:465–473.

Anderson, B.A. 2007. Prodrug approaches for drug delivery to the brain. In *Prodrugs: Challenges and Rewards. Part 1*, eds. V.J. Stella, R.T. Borchardt, M.J. Hageman, R. Oliyai, H. Maag, and J.W. Tilley, 573–651. New York: AAPS Press, Springer.

Azzouz. R., Peauger, L., Gembus, V., et al. 2018. Novel donepezil-like N-benzylpyridinium salt derivatives as AChE inhibitors and their corresponding dihydropyridine "bio-oxidizable" prodrugs: Synthesis, biological evaluation and structure-activity relationship. *European Journal of Medicinal Chemistry* 145:165–190.

Bellavance, M.-A., Blanchette, M., and Fortin, D. 2008. Recent advances in blood–brain barrier disruption as a CNS delivery strategy. *AAPS Journal* 10: 166–177.

Bickerton, G.R., Paolini, G.V., Besnard, J., Muresan, S., and Hopkins. A.L. 2012. Quantifying the chemical beauty of drugs. *Nature Chemistry* 4:90–98.

Bílek, R., Bičíková, M., and Šafařík, L. 2011. TRH-like peptides. *Physiological Research* 60:207–215.

Braddy, A.C., Janáky, T., and Prokai, L. 2002. Immobilized artificial membrane chromatography coupled with atmospheric pressure ionization mass spectrometry. *Journal of Chromatography A* 966:81–87.

Cimler, R., Maresova, P., Kuhnova, J., and Kuca, K. 2019. Predictions of Alzheimer's disease treatment and care costs in European countries. *PLoS One* 14(1):e0210958.

D'Agata, F., Ruffinatti, F.A., Boschi, S., et al. 2017. Magnetic nanoparticles in the central nervous system: Targeting principles, applications and safety issues. *Molecules* 23:9.

Dalpiaz, A., and Pavan, B. 2018. Nose-to-brain delivery of antiviral drugs: A way to overcome their active efflux? *Pharmaceutics* 10:39.

Dalpiaz, A., Paganetto G., and Pavan, B. 2012. Zidovudine and ursodeoxycholic acid conjugation: Design of a new prodrug potentially able to bypass the active efflux transport systems of the central nervous system. *Molecular Pharmaceutics* 9:957–968.

Dalvi, S., On, N., Nguyen, H., Pogorzelec, M., Miller, D.W., and Hatch, G.M. 2014. The blood brain barrier—Regulation of fatty acid and drug transport. In *Neurochemistry*, ed. T. Heinbockel, Chapter 1. Rijeka, Croatia: IntechOpen.

Daneman, R., and Prat, A. 2015. The blood–brain barrier. *Cold Spring Harbor Perspectives in Biology* 7:a02041.

Deli, M.A. 2011. Drug transport and the blood-brain barrier. In *Solubility, Delivery, and ADME Problems of Drugs and Drug-Candidates*, eds. K. Tihanyi, and M. Vastag, 144–165. Sharjah: Bentham Science Publisher Ltd.

Demeule, M., Regina, A., Jodoin, J., et al. 2002. Drug transport to the brain: Key roles for the efflux pump P-glycoprotein in the blood-brain barrier. *Vascular Pharmacology* 38:339–348.

Desai, P. V., Raub, T. J., and Blanco, M.-J. 2012. How hydrogen bonds impact P-glycoprotein transport and permeability. *Bioorganic & Medicinal Chemistry Letters* 22:6540–6548.

Eddleston, M., Szinicz, L., Eyer, P., and Buckley, N. 2002. Oximes in acute organophosphate pesticide poisoning: A systematic review of clinical trials. *QJM An International Journal of Medicine* 95:275–283.

Eden, B.D., Rice, A.J., Lovett, T.D., et al. 2019. Microwave-assisted synthesis and in vitro stability of N-benzylamide nonsteroidal anti-inflammatory drug conjugates for CNS delivery. *Bioorganic & Medicinal Chemistry Letters* 29:1487–1491.

Engelhardt, B., and Sorokin, L. 2009. The blood–brain and the blood–cerebrospinal fluid barriers: Function and dysfunction. *Seminars in Immunopathology* 31:497–511.

Fahn, S., Oakes, D., Shoulson, I., Kieburtz, K., Rudolph, A., Lang, A., Olanow, C.W., Tanner, C., Marek, K., The Parkinson Study Group. 2004. Levodopa and the progression of Parkinson's disease. *The New England Journal of Medicine* 351:2498–2508.

Fan, C.H., Lin, C.Y., Liu, H.L., and Yeh, C.K. 2017. Ultrasound targeted CNS gene delivery for Parkinson's disease treatment. *Journal of Controlled Release* 261:246–262.

FDA. 2019. NDA 204063 Tecfidera (dimethyl fumerate) approved labeling text. https://www.accessdata.fda.gov/drugsatfda_docs/label/2016/204063s014lbl.pdf (accessed July 5, 2019).

Fernández, C., Nieto, O., Fontenla, J.A., Rivas, E., de Ceballos, M.L., and Fernández-Mayoralas, A. 2003. Synthesis of glycosyl derivatives as dopamine prodrugs: Interaction with glucose carrier GLUT-1. *Organic & Biomolecular Chemistry* 1:767–771.

Fischer, H., Gottschlich, R., and Seelig, A. 1998. Blood-brain barrier permeation molecular parameters governing passive diffusion. *The Journal of Membrane Biology* 165:201–211.

Fröhlich, E., and Wahl, R. 2019. The forgotten effects of thyrotropin-releasing hormone: Metabolic functions and medical applications. *Frontiers Neuroendocrinology* 52:29–43.

Gänger, S., and Schindowski, K. 2018. Tailoring formulations for intranasal nose-to-brain delivery: A review on architecture, physicochemical characteristics and mucociliary clearance of the nasal olfactory mucosa. *Pharmaceutics* 10:116.

Gomez-Zepeda, D., Taghi, M., Smirnova, M., et al. 2019. LC-MS/MS-based quantification of efflux transporter proteins at the BBB. *Journal of Pharmaceutical and Biomedical Analysis* 164:496–508.

Gooch, C.L., Pracht, E., and Borenstein, A.R. 2017. The burden of neurological disease in the United States: A summary report and call to action. *Annals in Neurology* 81:479–484.

Gribkoff, V.K., and Kaczmarek, L.K. 2017. The need for new approaches in CNS drug discovery: Why drugs have failed, and what can be done to improve outcomes. *Neuropharmacology* 120:11–19.

Gupta, S., Kesarla, R., and Omri, A. 2019. Approaches for CNS delivery of drugs—Nose to brain targeting of antiretroviral agents as a potential attempt for complete elimination of major reservoir site of HIV to aid AIDS treatment. *Expert Opinion on Drug Delivery* 16:287–300.

Haddad, F., Sawalha, M., Khawaja, Y., Najjar, A., and Karaman, R. 2018. Dopamine and Levodopa prodrugs for the treatment of Parkinson's disease. *Molecules* 23:40.

Haduch, A., and Daniel, W.A. 2018. The engagement of brain cytochrome P450 in the metabolism of endogenous neuroactive substrates: A possible role in mental disorders. *Drug Metabolism Reviews* 50:415–429.

Hann, M.M. 2011. Molecular obesity, potency and other addictions in drug discovery. *Medicinal Chemistry Communications* 2:349–355.

Hansch, C., Steward, A.R., Anderson. S.M., and Bentley, D. 1967. The parabolic dependence of drug action upon lipophilic character as revealed by a study of hypnotics. *Journal of Medicinal Chemistry* 11:1–11.

Hartley, M.D., Banerji, T., Tagge, I.J., et al. 2019. Myelin repair stimulated by CNS-selective thyroid hormone action. *JCI Insight* 4(8):e126329.

Hawkins, B.T., and Davis, T.P. 2005. The blood-brain barrier/neurovascular unit in health and disease. *Pharmacological Reviews* 57:173–185.

Hersh, D.S., Wadajkar, A.S., Roberts, N.B., et al. 2016. Evolving drug delivery strategies to overcome the blood brain barrier. *Current Pharmaceutical Design* 22:1177–1193.

Huttunen, J., Gynther, M., and Huttune, K.M. 2018. Targeted efflux transporter inhibitors—A solution to improve poor cellular accumulation of anti-cancer agents. *International Journal of Pharmaceutics* 550:278–289.

Ishikura, T. Senou, T., Ishihara, H., Kato, T., and Ito, T. 1995. Drug delivery to the brain. DOPA prodrugs based on a ring-closure reaction to quaternary thiazolium compounds. *International Journal of Pharmaceutics* 116:5–63.

Kalari, K.R., Thompson, K.J., Nair, A.A., et al. 2016. BBBomics-human blood brain barrier transcriptomics hub. *Frontiers in Neuroscience* 10:71.

Kao, H.D., Traboulsi, A., Itoh, S., Dittert, L., and Hussain, A. 2000. Enhancement of the systemic and CNS specific delivery of L-dopa by the nasal administration of its water-soluble prodrugs. *Pharmaceutical Research* 17:978–984.

Kevadiya, B.D., Ottemann, B.M., Thomas, M.B., et al. 2019. Neurotheranostics as personalized medicines. *Advanced Drug Delivery Reviews*, 148:252–289.

Khan, M.S., and Roberts, M.R. 2018. Challenges and innovations of drug delivery in older age. *Advanced Drug Delivery Reviews* 135:3–38.

Killian, D.M., and Chikhale, P.J. 2000. A bioreversible prodrug approach designed to shift mechanism of brain uptake for amino-acid-containing anticancer agents. *Journal of Neurochemistry* 76:966–974.

Kim, J., Ahn, S.I., and Kim, Y.T. 2019. Nanotherapeutics engineered to cross the blood-brain barrier for advanced drug delivery to the central nervous system. *Journal of Industrial and Engineering Chemistry* 73:8–18.

Kobayashi, N., Sato, N., Fujimura, Y., et al. 2018. Discovery of the orally effective thyrotropin-releasing hormone mimetic: 1-{N-[(4S,5S)-(5-Methyl-2-oxooxazolidine-4-yl)carbonyl]-3-(thiazol-4-yl)-l-alanyl}-(2R)-2-methylpyrrolidine trihydrate (rovatirelin hydrate). *ACS Omega* 3:13647–13666.

Lawther, B.K., Kumar, S., and Krovvidi, H. 2011. Blood–brain barrier. *Continuing Education in Anaesthesia, Critical Care & Pain* 11:128–132.

Lee, M., Chen, Z., Tomlinson, B.N., et al. 2015. Water-soluble MMP-9 inhibitor reduces lesion volume after severe traumatic brain injury. *ACS Chemical Neuroscience* 6:1658–1664.

Lee, Y.H., Kim, K.H., Yoon, I.K., et al. 2014. Pharmacokinetic evaluation of formulated levodopa methyl ester nasal delivery systems. *European Journal of Drug Metabolism and Pharmacokinetics* 39:237–242.

Li, Y., Zhou, Y., Jiang, J., et al. 2015. Mechanism of brain targeting by dexibuprofen prodrugs modified with ethanolamine-related structures. *Journal of Cerebral Blood Flow & Metabolism* 35:1985–1994.

Manfredini, S., Pavan, B., Vertuani, S., et al. 2002. Design, synthesis and activity of ascorbic acid prodrugs of nipecotic, kynurenic and diclophenamic acids, liable to increase neurotropic activity. *Journal of Medicinal Chemistry* 45:3559–3562.

McNamara, P.J., and Leggas, M. 2009. Drug distribution, In *Pharmacology, Principles and Practice*, eds. M. Hacker, W. Messer, and K. Bachmann, 113–129. London, UK: Academic Press.

mdmag. 2019. Monomethyl fumarate challenges multiple sclerosis. https://www.mdmag.com/medical-news/monomethyl-fumarate-challenges-multiple-sclerosis (accessed July 7, 2019).

Merchenthaler, I., Lane, M., Sabni, G., et al. 2016. Treatment with an orally bioavailable prodrug of 17β-estradiol alleviates hot flushes without hormonal effects in the periphery. *Scientific Reports* 6:30721.

Miller, D.S. 2015. Regulation of ABC transporters at the blood–brain barrier. *Clinical Pharmacology & Therapeutics* 97:395–403.

Minocha, M. Khurana, V., Dhananjay, Qin., B., Pal, D., and Mitra, A.K. 2012. Enhanced brain accumulation of pazopanib by modulating P-gp and Bcrp1 mediated efflux with canertinib or erlotinib. *International Journal of Pharmaceutics* 436:127–134.

Myöhänen, T.T., Venäläinen, J.I., García-Horsman, J.A., Piltonen, M., and Männistö, P.T. 2008. Distribution of prolyl oligopeptidase in the mouse whole-body sections and peripheral tissues. *Histochemistry and Cell Biology* 130:993–1003.

Nedelcovych, M., Dash, R.P., Tenora, L., et al. 2017a. Enhanced brain delivery of 2-(phosphonomethyl) pentanedioic acid following intranasal administration of its γ-substituted ester prodrugs. *Molecular Pharmaceutics* 14:3248–3257.

Nedelcovych, M.T., Tenora, L., Kim, B.-H., et al. 2017b. N-(Pivaloyloxy)alkoxy-carbonyl prodrugs of the glutamine antagonist 6-diazo-5-oxo-l-norleucine (DON) as a potential treatment for HIV associated neurocognitive disorders. *Journal of Medicinal Chemistry* 60:7186–7198.

Ohtsuki, S., Hirayama, M., Ito, S., Uchida, Y., Tachikawa, M., and Terasaki. T. 2014. Quantitative targeted proteomics for understanding the blood–brain barrier: Towards pharmacoproteomics. *Expert Reviews in Proteomics* 11:303–313.

Pajouhesh, H., and Lenz, G.R. 2005. Medicinal chemical properties of successful central nervous system drugs. *NeuroRx®* 2:541–553.

Pan, Y., and Nicolazzo, J.A. 2018. Impact of aging, Alzheimer's disease and Parkinson's disease on the blood-brain barrier transport of therapeutics. *Advanced Drug Delivery Reviews* 135:62–74.

Pangalos, M.N., Schechter, L.E., and Hurk, O. 2007. Drug development for CNS disorders: Strategies for balancing risk and reducing attrition. *Nature Reviews Drug Discovery* 6:521–532.

Parodi, A., Rudzinska, M., Deviatkin, A.A., Soond, S.M., Baldin, A.V., and Zamyatnin, Jr., A.A. 2019. Established and emerging strategies for drug delivery across the blood-brain barrier in brain cancer. *Pharmaceutics* 11:245.

Pathak, V. 2018. Nasal delivery—A promising route of drug delivery to the brain: Scientific considerations. *Drug Development & Delivery* 18:63–68.

Pavan, B., Dalpiaz, A., Ciliberti, N., Biondi, C., Manfredini, S., and Vertuani, S. 2008. Progress in drug delivery to the central nervous system by the prodrug approach. *Molecules* 13:1035–1065.

Pezron, I., Tirucherai, G.S., Duvvuri, S., and Mitra, A.K. 2002. Prodrug strategies in nasal drug delivery. *Expert Opinion on Therapeutic Patents* 12:331–340.

Placzek, A.P., Ferrara, S.J., Hartley, M.D., Sanford-Crane, H.S., Meinig, J.M., and Scanlan, T.S. 2016. Sobetirome prodrug esters with enhanced blood–brain barrier permeability. *Bioorganic & Medicinal Chemistry* 24:5842–5854.

Polgar, L., and Szeltner, Z. 2008. Structure, function and biological relevance of prolyl oligopeptidase. *Current Protein & Peptide Science* 9:96–107.

Prokai, L. 2002. Central nervous system effects of thyrotropin-releasing hormone and its analogues: Opportunities and perspectives for drug discovery and development. In *Progress of Drug Research*, ed., E. Jucker, Vol. 59, 133–170. Basel, CH: Birkhauser.

Prokai, L., and Prokai-Tatrai, K. 2003. *Peptide Transport and Delivery Into the Central Nervous System*, Vol. 61. 1–245. Basel, CH: Birkhauser.

Prokai, L., and Zharikova, A.D. 2002. Neuropharmacodynamic evaluation of the centrally active thyrotropin-releasing hormone analogue [Leu^2]TRH and its chemical brain-targeting system. *Brain Research* 952:268–274.

Prokai, L., Nguyen, V., Szarka, S., et al. 2015. The prodrug DHED selectively delivers 17β-estradiol to the brain for treating estrogen-responsive disorders. *Science Translational Medicine* 7(297):297ra113.

Prokai, L., Ouyang, X., Wu, W-M., and Bodor, N. 1994. Chemical delivery system to transport a pyroglutamyl peptide amide to the central nervous system. *Journal of the American Chemical Society* 111:2643–2644.

Prokai, L., Prokai-Tatrai, K., and Bodor, N. 2000. Targeting drugs to the brain by redox chemical delivery systems. *Medicinal Research Reviews* 20:367–416.

Prokai, L., Prokai-Tatrai, K., and Simpkins, J.W. 2006. Steroidal quinols and their use for antioxidant therapy. US Patent 7026306.

Prokai, L., Prokai-Tatrai, K., and Simpkins, JW. 2007. Steroidal quinols and their use for estrogen replacement therapy. US Patent 7300926.

Prokai, L., Prokai-Tatrai, K., Ouyang, X., et al. 1999. Metabolism-based brain-targeting system for a thyrotropin-releasing hormone analogue. *Journal of Medicinal Chemistry* 42:4563–4571.

Prokai, L., Prokai-Tatrai, K., Perjesi, P., et al. 2003. Quinol-based cyclic antioxidant mechanism in estrogen neuroprotection. *Proceedings of the National Academy Sciences USA* 100:11741–11746.

Prokai, L., Prokai-Tatrai, K., Zharikova, A.D., Nguyen, V., and Stevens, S.M., Jr. 2004. Centrally-acting and metabolically stable thyrotropin-releasing hormone analogues by replacement of histidine with substituted pyridinium. *Journal of Medicinal Chemistry* 47:6025–6033.

Prokai, L., Rivera-Portalatin, N.M., and Prokai-Tatrai, K. 2013. Quantitative structure–activity relationships predicting the antioxidant potency of 17β-estradiol-related polycyclic phenols to inhibit lipid peroxidation. *International Journal of Molecular Sciences* 14:1443–1454.

Prokai-Tatrai, K., and Prokai, L. 2003. Modifying peptide properties by prodrug design for enhanced transport into the CNS. In *Peptide Transport and Delivery into the Central Nervous System*, eds. L. Prokai, and K. Prokai-Tatrai, Vol. 61, 189–220. Basel, CH: Birkhauser.

Prokai-Tatrai, K., and Prokai, L. 2009. Prodrugs of thyrotropin-releasing hormone and related peptides as central nervous system agents. *Molecules* 14:633–654.

Prokai-Tatrai, K., and Prokai, L. 2011. Prodrug design for brain delivery of small-and medium-sized neuropeptides. In *Neuropeptides. Methods in Molecular Biology (Methods and Protocols)*, ed. A. Merighi, 313–336. Totowa, NJ: Humana Press.

Prokai-Tatrai, K., and Prokai, L. 2017. 17β-Estradiol as a neuroprotective agent. In *Sex Hormones in Neurodegenerative Processes and Diseases*, ed. G. Drevenšek, 21–39, Rijeka, Croatia: IntechOpen.

Prokai-Tatrai, K., De La Cruz, D.L., Nguyen, V., Ross, B.R., Toth, I., and Prokai, L. 2019. Brain delivery of thyrotropin-releasing hormone via a novel prodrug approach. *Pharmaceutics* 11:349.

Prokai-Tatrai, K., Kim, H.S., and Prokai, L. 2008a. The utility of oligopeptidase in brain-targeting delivery of an enkephalin analogue by prodrug design. *Open Medicinal Chemistry Journal* 20:97–100.

Prokai-Tatrai, K., Nguyen, V., and Prokai, L. 2018. 10β,17α-Dihydroxyestra-1,4-dien-3-one: A bioprecursor prodrug preferentially producing 17α-estradiol in the brain for targeted neurotherapy. *ACS Chemical Neuroscience* 9:2528–2533.

Prokai-Tatrai, K., Nguyen, V., Szarka, S., Konya, K., and Prokai, L. 2013. Design and exploratory neuropharmacological evaluation of novel thyrotropin-releasing hormone analogs and their brain-targeting bioprecursor prodrugs. *Pharmaceutics* 5:318–328.

Prokai-Tatrai, K., Nguyen, V., Zharikova, A.D., Braddy, A.C., Stevens, S.M., and Prokai, L. 2003. Prodrugs to enhance central nervous system effects of the TRH-like peptide pGlu-Glu-Pro-NH_2. *Bioorganic & Medicinal Chemistry Letters* 24:1011–1014.

Prokai-Tatrai, K., Perjesi, P., Rivera-Portalatin, N.M., Simpkins, J.W., and Prokai L. 2008b. Mechanistic investigations on the antioxidant action of a neuroprotective estrogen derivative. *Steroids* 73:280–288.

Prokai-Tatrai, K., Perjesi, P., Zharikova, A.D., Li, X., and Prokai, L. 2002. Design, synthesis, and biological evaluation of novel, centrally-acting thyrotropin-releasing hormone analogues. *Bioorganic & Medicinal Chemistry Letters* 12:2171–2174.

Prokai-Tatrai, K., Prokai, L., and Bodor, N. 1996. Brain-targeting delivery of a leucine-enkephalin analogue by retrometabolic design. *Journal of Medicinal Chemistry* 22:4775–4782.

Prokai-Tatrai, K., Rivera-Portalatin, N.M., Rauniyar, N., and Prokai, L. 2007. A facile microwave-assisted synthesis of p-quinols by lead (IV) acetate oxidation. *Letters in Organic Chemistry* 4:265–267.

Prokai-Tatrai, K., Szarka, S., Nguyen, V., et al. 2012. "All in the mind"? Brain-targeting chemical delivery system of 17β-estradiol (Estredox) produces significant uterotrophic side effect. *Pharmaceutica Analytica Acta* S7:002.

Prokai-Tatrai, K., Teixido, M., Nguyen, V., Zharikova, A.D., and Prokai, L. 2005. A pyridinium-substituted analog of the TRH-like tripeptide pGlu-Glu-Pro-NH_2 and its prodrugs as central nervous system agents. *Medicinal Chemistry* 1:141–152.

Puris, E., Gynther, M., Huttunen, J., Auriola, S., and Huttune, K.M. 2019. L-type amino acid transporter 1 utilizing prodrugs of ferulic acid revealed structural features supporting the design of prodrugs for brain delivery. *European Journal of Pharmaceutical Sciences* 129:99–109.

Rajsombath, M.M., Nam, A.Y., Ericsson, M., and Nuber, S. 2019. Female sex and brain-selective estrogen benefit α-synuclein tetramerization and the PD-like motor syndrome in 3K transgenic mice. *Journal of Neuroscience* 39:7628–7640.

Rankovic, Z. 2015. CNS drug design: Balancing physicochemical properties for optimal brain exposure. *Journal of Medicinal Chemistry* 58:2584–2608.

Rivera-Portalatin, N.M., Vera-Serrano, J.L., Prokai-Tatrai, K., and Prokai, L. 2007. Comparison of estrogen-derived ortho-quinone and para-quinol concerning induction of oxidative stress. *Journal of Steroid Biochemistry and Molecular Biology* 105:71–75.

Sarkar, D.K., Friedman, D.J., Yen, S.S.C., and Frautschy, S.A. 1989. Chronic inhibition of hypothalamic-pituitary-ovarian axis and body weight gain by brain-directed delivery of estradioi-17β in female rats. *Neuroendocrinology* 50:204–210.

Servick, K. 2019. Another major drug candidate targeting the brain plaques of Alzheimer's disease has failed: What's left? *Science* 10. https://www.sciencemag.org/news/2019/03/another-major-drug-candidate-targeting-brain-plaques-alzheimer-s-disease-has-failed (accessed July 2, 2019).

Shore, P.A., Brodie, B.B., and Hogben, C.A.M. 1957. The gastric secretion of drugs: A pH partition hypothesis. *Journal of Pharmacology and Experimental Therapeutics* 119:361–369.

Stein, D.G. 2015. Embracing failure: What the Phase III progesterone studies can teach about TBI clinical trials. *Brain Injury* 29:1259–1272.

Stevens, Jr., S.M., Prokai-Tatrai, K., and Prokai, L. 2005. Screening of combinatorial libraries for substrate preference by mass spectrometry. *Analytical Chemistry* 77:698–701.

Szarka, S., Nguyen, V., Prokai, L., and Prokai-Tatrai, K. 2013. Separation of dansylated 17β-estradiol, 17α-estradiol and estrone on a single HPLC column for simultaneous quantitation by LC-MS/MS. *Analytical and Bioanalytical Chemistry* 405:3399–3406.

Tóth, I., Malkinson, J.P., Flinn, N.S., et al. 1999. Novel lipoamino acid- and liposaccharide-based system for peptide delivery: Application for oral administration of tumor-selective somatostatin analogues. *Journal of Medicinal Chemistry* 42:4010–4013.

Tschiffely, A.E., Schuh, R.A., Prokai-Tatrai, K., Ottinger, M.A., and Prokai, L. 2018. An exploratory investigation of brain-selective estrogen treatment in males using a mouse model of Alzheimer's disease. *Hormones and Behaviors* 98:16–21.

Tschiffely, A.E., Schuh, R.A., Prokai-Tatrai, K., Prokai, L., and Ottinger, M.A. 2016. A comparative evaluation of treatments with 17β-estradiol and its brain-selective prodrug in a double-transgenic mouse model of Alzheimer's disease. *Hormones and Behaviors* 83:39–44.

Tsuzuki, N., Hama, T., Hibi, T., Konishi, R., Futak, S., and Kitagawa, K. 1991. Adamantane as a brain-directed drug carrier for poorly absorbed drug: Antinociceptive effects of [D-Ala^2]Leu-enkephalin derivatives conjugated with the 1-adamantane moiety. *Biochemical Pharmacology* 41:R5–R8.

Wager, T.T., Chandrasekaran, R.Y., Hou, X., et al. 2010. Defining desirable central nervous system drug space through the alignment of molecular properties, in vitro ADME, and safety attributes. *ACS Chemical Neuroscience* 1:420–434.

Wang, X., Li, J., Xu, C., et al. 2014. Scopine as a novel brain-targeting moiety enhances the brain uptake of chlorambucil. *Bioconjugate Chemistry* 19:2046–2054.

Wen, P.Y., Chang, S.M., Lambornet, K.R., et al. 2014. Phase I/II study of erlotinib and temsirolimus for patients with recurrent malignant gliomas: North American brain tumor consortium trial 04-02. *Neurooncology* 16:567–578.

Wu, K.-M. 2009. A new classification of prodrugs: Regulatory perspective. *Pharmaceuticals* 2:77–81.

Yaksh, T.L., Schwarcz, R., and Snodgrass, H.R. 2017. Characterization of the effects of L-4-chlorokynurenine on nociception in rodents. *Journal of Pain* 18:1184–1196.

Yamazaki, Y., and Kanekiyo, T. 2017. Blood-brain barrier dysfunction and the pathogenesis of Alzheimer's disease. *International Journal of Molecular Sciences* 18:1965.

Yan, W., Wu, B.S., Luo. Q., and Xu, Y. 2019. Treatment with a brain-selective prodrug of 17β-estradiol improves cognitive function in Alzheimer's disease mice by regulating klf5-NF-κB pathway. *Naunyn-Schmiedeberg's Archives of Pharmacology* 392:879–886.

Yue, Q., Peng, Y., Zhao, Y., et al. 2018. Dual-targeting for brain-specific drug delivery: Synthesis and biological evaluation. *Drug Delivery* 25:426–434.

Zhang, X., Liu, X., Gong, T., Sun, X., and Zhang, Z.R. 2012. In vitro and in vivo investigation of dexibuprofen derivatives for CNS delivery. *Acta Pharmacologica Sinica* 33:279–288.

Zhao, J., Zhang, L., Peng, Y., et. al. 2018. GLUT1-mediated venlafaxine-thiamine disulfide system-glucose conjugates with "lock-in" function for central nervous system delivery. *Chemical Biology & Drug Design* 91:707–716.

Zuchero, Y.J.Y., Chen, X., Bien-Ly, N., et al. 2016. Discovery of novel blood-brain barrier targets to enhance brain uptake of therapeutic antibodies. *Neuron* 89:70–82.

8

Prodrugs for Cancer Treatment

William L. Stone, Victoria E. Palau, and K. Krishnan

CONTENTS

8.1 Introduction: Precision Oncology and Precision Prodrugs

Cancer management is no longer based solely on the anatomical site of the tumor and is rapidly evolving to take into account the individual differences giving rise to a tumor's set of phenotypes, i.e., precision oncology (NCI dictionary 2019). Nevertheless, most cancer treatments still rely on a "one size fits all," or a "few sizes" fits all approach. Advances in histopathology image analyses combined with "omics" technologies, e.g., bioinformatics, genomics, and proteomics, are rapidly changing cancer management (Stone et al. 2017; Yu et al. 2017a, 2017b). In parallel with advances in precision oncology, there have been major advances in the rational design of drugs/prodrugs that can exploit alterations in oncogenic signal transduction pathways resulting from genetic alterations in oncogenes and tumor suppressor genes. This approach has been termed "targeted chemotherapy." This review will focus on the use of cancer esterases as prodrugs targets with particular emphasis on oxidized protein hydrolase (EC:3.4.19.1) and prodrugs based on redox cancer therapy. We suggest that oxidative-stress biomarkers would be useful in identifying highly aggressive cancers and also be the basis for targeted pro-oxidant chemotherapy when combined with a biomarker for oxidized protein hydrolase (OPH).

8.2 Prodrugs in Targeted Chemotherapy: Converting Conventional Cancer Drugs to Prodrugs

A prodrug is an inactive and less toxic agent than the bioactivated agent derived from the prodrug. Typically, the prodrug is bioactivated by an enzyme such as an esterase. The use of prodrugs to treat cancer is a widely accepted pharmacologic strategy but has yet to achieve widespread clinical utilization (Giang et al. 2014; Spiegelberg et al. 2019; Malekshah et al. 2016). Many conventional chemotherapeutic drugs such as cisplatin, chlorambucil, and 5-fluorouracil (5-FU) can be converted into prodrugs (Giang et al. 2014; Pathak et al. 2017; Malet-Martino et al. 2002). The goals of converting a conventional drug into a prodrug include: (a) creating a prodrug that could be delivered by oral administration when the conventional drug can only be administered intravenously; (b) increasing the delivery of the drug to tumor cells; (c) increasing bioavailability, and (d) decreasing the toxic side effects associated with all conventional chemotherapeutic agents.

Despite its well-documented toxicity, the fluoropyrimidine 5-FU remains one of the most widely used chemotherapeutic agents for treating solid tumors (Sara et al. 2018). 5-FU is typically delivered intravenously and this, along with the goal of diminishing toxicity, has been key impetuses for developing orally administered 5-FU prodrugs (Malet-Martino and Martino 2002). These efforts have proven very fruitful: capecitabine is an oral 5-FU prodrug and its bioactivation to 5-FU is partially catalyzed by thymidine phosphorylase (TP) which is an enzyme found to be elevated in many tumors (Siddiqui et al. 2019). The tumor localization of TP promotes the selective bioactivation of capecitabine and, thereby, can help diminish systemic toxicity (Siddiqui et al. 2019).

8.3 Endogenous Enzyme Bioactivation of Rationally-Designed Cancer Prodrugs

As indicated in Figure 8.1, a key step for cancer prodrug development is identifying a suitable bioactivating target enzyme. Ideally, this could be an enzyme that is selectively overexpressed in a tumor, as detailed above for capecitabine and thymidine phosphorylase. Rooseboom et al. (2004) have written an excellent and comprehensive review on the enzyme-catalyzed bioactivation of anticancer prodrugs. This article includes a detailed list of potential endogenous bioactivating enzymes, both by class and by specific enzymes such as aldehyde oxidase, glutathione S-transferase and carboxylesterase.

Rapid advances in bioinformatics provide the opportunity to accelerate treatment options for cancer by focusing on the rational design, and subsequent organic synthesis, of prodrugs likely to be good substrates for a bioactivating target enzyme (Brown and Bishop 2017). As detailed below and outlined in Figure 8.1, this approach is particularly useful if the three-dimensional (3D) structure of the candidate bioactivating target enzyme is known along with ligands bound to the active site. Nevertheless, structural bioinformatics has advanced to the point where the amino acid sequence of an enzyme is often sufficient to predict the 3D structure of a protein (a ".pdb" file) which can then guide subsequent rationale drug/prodrug development. Moreover, there are powerful and free online tools for accomplishing this goal, as will be illustrated for oxidized protein hydrolase (OPH).

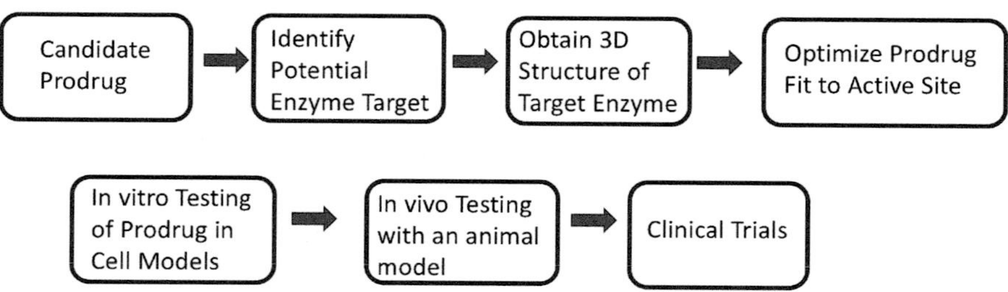

FIGURE 8.1 Example workflow for the rational design and development of a candidate prodrug.

8.4 Directed Enzyme Prodrug Therapy (DEPT)

An alternative strategy to the use of endogenous activating enzymes is "directed enzyme prodrug therapy" or DEPT. The goal of DEPT is to artificially introduce a bioactivating enzyme into a tumor that will specifically activate its cognate prodrug. A variety of very innovative techniques have been proposed for achieving this goal (Xu and McLeod 2001). One such strategy is that of "suicide gene therapy," also termed "gene-directed enzyme prodrug therapy" or GDEPT (Malekshah et al. 2016). In this approach, tumor cells are transfected with a gene expressing a bioactivating enzyme specific for a chemotherapeutic prodrug. For this technique to be effective, the bioactivating enzyme should be: (a) selectively expressed in the tumor; (b) have a high affinity (low Km) and high Vmax for the prodrug; (c) the active drug should have a long half-life. An activated cytotoxic drug with a long half-life can facilitate the "bystander killing effect" and kill nearby tumor cells not having taken up the bioactivating gene. An added advantage of GDEPT is that the expression of the bioactivating enzyme can potentially be regulated by tumor-specific promotors (Malekshah et al. 2016). The technical barriers for implementing GDEPT are formidable, and there are currently no commercially available GDEPT products (Zhang et al. 2015). An alternative strategy is antibody-directed enzyme prodrug therapy or ADEPT where the goal is to link a monoclonal antibody to a tumor antigen with a bioactivator for its cognate prodrug (Xu and McLeod 2001).

Another DEPT variation, called inhibitor-DEPT or IDEPT, has been developed, and "proof of concept" *in vitro* experiments have been completed with the prostate cancer cell line LNCaP (Martin et al. 2014). In this technique, a suicide enzyme, yCDtriple, is delivered to LNCaP cells expressing prostate-specific membrane antigen (PSMA) which is a biomarker for prostate cancer. yCDtriple has a hydrolytic deamination activity with the ability to convert 5-fluorocytosine (5-FC) to 5-FU. 5-FC is the nontoxic prodrug in this strategy, and its activation to 5-FU inhibits DNA synthesis and induces apoptosis. PSMA, also known as glutamate carboxypeptidase II (GCPII), is a well-characterized transmembrane protein that indirectly activates the AKT/PI3K protein kinase pathway (more on this below). Both the expression level of PSMA and the activation of AKT/PI3K protein kinase oncogenic pathway are correlated with aggressive prostate cancer (Kaittanis et al. 2018). Martin et al. (2014) point out that 5-FU can freely diffuse through cell membranes and would, therefore, exhibit a bystander killing effect when potentially used against a heterogenous prostate cancer tumor with only some tumor cells overexpressing PSMA.

8.5 Systems Medicine-Directed Prodrug Therapy (SMDPT) and the Power of Proteogenomics

The molecular characteristics of a tumor at a given anatomical site can vary from individual to individual even when the tumor shows similar histopathology and stage. Nevertheless, the majority of genetic alterations giving rise to oncogenic phenotypes reside in a relatively small set of oncogenes (about 100) and tumor suppressor genes (about 10–20) controlling functions such as cellular growth, differentiation, cancer invasiveness, and angiogenesis. Precision oncology is based on the assumption that the molecular characteristics of a given tumor can provide biomarkers to guide treatment. In the context of prodrug development, precision oncology would involve treating cancers with prodrugs guided by conventional histopathology images as well as the molecular characteristics of a given tumor. Biomarkers that reflect genetic alterations occurring in critical oncogenic signaling pathways represent ideal guiding factors for prodrug development.

Systems medicine attempts to integrate information from conventional medicine and omic technologies to provide individualized treatment and health care. Although only in its infancy, a systems medicine approach can be applied to prodrug/drug development Surprisingly, this overall approach does not yet have a descriptive term, so we propose "systems medicine-directed prodrug therapy" or SMDPT. Genomics, transcriptomics, and proteomics are the primary omic platforms now contributing to SMDPT (Jiang and Zhou 2005). SMDPT can contribute to cancer prodrug/drug therapy by: (1) identifying specific oncogenic molecular alterations in a patient's cancer thereby providing insight into the best available chemotherapy; (2) providing insight into the development of new prodrugs; (3) identifying omic biomarkers for monitoring treatment effectiveness and prognosis.

There are reasons to be optimistic about SMDPT since precision oncology is rapidly overcoming the daunting task of integrating information from electronic health records, histopathology images, and various "omic" platforms. Moreover, the complex software for analyzing these disparate data streams is rapidly progressing to the point of being clinically useful. For example, artificial intelligence (AI) methods have recently been employed to help predict lung cancer prognosis based on histopathology images combined with data input from a tumor's proteome and transcriptome, i.e., proteogenomics (Yu et al. 2017a). The ability of AI to generate prognostic predictions for lung cancer was markedly improved by combining histopathology image analysis data with the omic data compared with either data stream alone (Yu et al. 2017a).

8.6 Circulating Tumor-DNA (ctDNA) in Precision Oncology

While it is not always possible to obtain a surgical tumor biopsy for genomic analysis, circulating tumor-DNA (ctDNA) can be obtained by a "liquid biopsy" from a non-invasive 5–10 mL blood draw. ctDNA are small fragments of DNA that are shed from cancer cells into the bloodstream and provide tumor-specific genomic information (Heitzer et al. 2019). In addition to adult cancers, it is now recognized that ctDNA can also provide a useful genomic characterization of pediatric solid tumors (Klega et al. 2018). The analysis of ctDNA by next-generation sequencing represents a major advance in precision oncology since it holds the promise of early detection, which is a key factor driving successful treatment for many cancers. Moreover, the ability to obtain serial ctDNA samples opens the door to monitoring cancer treatment effectiveness and tumor evolution (Heitzer et al. 2019). Nevertheless, the ctDNA test is now expense, and even a 10 mL blood sample may not be sufficient to detect very early asymptomatic cancer (Fiala and Diamandis 2018).

8.7 Cancer Esterases as Prodrug Targets

The primary focus of this chapter is on the use of cancer cell esterases as prodrug bioactivators with special emphasis on OPH as a target for redox-directed chemotherapy (Wondrak 2009). Esterases cleave esters (RCO-OR′) into an alcohol (R′OH) and a carboxylate anion (RCOO-). The chemical synthesis of prodrug esters is relatively straight-forward, and esters can generally diffuse cross biomembranes.

8.8 Early History of Cancer Esterases

Nonspecific esterase activities in normal tissues and tumors have traditionally been visualized in native-polyacrylamide gel (n-PAGE) bands by use of alpha-naphthyl acetate, or similar substrates such as alpha-naphthyl butyrate. Using an animal model, Kreusser (1966) found the nonspecific esterase gel banding pattern in a normal tissue type was fairly unique to that tissue type. In marked contrast, tumors did not exhibit the gel pattern of the parent tissue type and tended to have similar nonspecific esterase banding patterns regardless of their tissue origin. These results were among the first to suggest that cancer esterases could be important in oncology.

In contrast to nonspecific esterases, esterproteases, like OPH, are more specifically visualized in gels by using alpha-naphthyl esters of peptides such as L-alanine or L-methionine. Kirkeby and Moe (1988) found that some esterproteases n-PAGE bands colocalized with nonspecific esterases. Yamazaki et al. (1995) were among the first to observe that some esterases are overexpressed in cancer cells compared to the corresponding normal tissues. These investigators suggested that endogenous esterases in cancer cells could be exploited to activate anticancer prodrug esters but never identified any of proteins giving rise to the esterase activity and never developed a prototype ester prodrug.

8.9 Cancer, the Serine Hydrolase Superfamily and Activity-Based Proteomics

Esterases are a class of enzymes belonging to the serine hydrolase superfamily which also includes proteases, peptidases, lipases, amidases, and thioesterases. Activity-based proteomics has been used to study the activity profiles of both membrane-bound and secreted serine hydrolases with respect to cancer cell invasiveness (Jessani et al. 2002; Liu et al. 1999). In this approach, all the serine nucleophilic active sites in a given cell line are first labeled with a rhodamine-coupled-fluorophosphonate (FP) also covalently tagged with biotin (Liu et al. 1999). The FP-labeled proteins are subsequently isolated using biotinylated-avidin affinity purification followed by SDS-PAGE. The FP-labeled polypeptides are then visualized with a flatbed fluorescence scanner. Mass spectrometry was used to identify the polypeptide bands in the SDS-PAGE gels. Membrane-bound, secreted, and soluble intracellular compartments were analyzed from a variety of breast and melanoma cell lines. The results were striking: most of the serine hydrolases in invasive cancer cell lines were downregulated compared to the esterase activities in the non-invasive cancer cell lines. Quite remarkably, however, the most invasive cancer cell lines exhibited a unique set of up-regulated secreted and membrane serine hydrolases (Jessani et al. 2002). These researchers concluded that the proteomic phenotype of aggressive cancers might be similar since they tend to revert to a "common pluripotent embryonic state." This conclusion, in the year 2002, is reminiscent of that mentioned above by Kreusser (1966). OPH, also called acylamino-acid-releasing enzyme or AARE, is usually considered an intracellular soluble protein. Jessani et al. (2002) found, however, that OPH in all three of the compartments they analyzed, i.e., membrane proteins, secreted proteins and intracellular soluble proteins (more on this below).

8.10 Oxidized Protein Hydrolase (OPH) is Elevated in Tumorigenic Prostate Cancer Cell Lines with High Levels of Intrinsic Oxidative Stress

OPH has recently been proposed as a target for the development of anticancer drugs (Palmieri et al. 2011; Bergamo et al. 2013; McGoldrick et al. 2014a, 2014b). Histological data in the Human Protein Atlas shows that OPH is strongly expressed in melanoma, thyroid, and prostate cancers (Atlas 2015). McGoldrick et al. (2014b) studied the esterases differentially expressed in tumorigenic and nontumorigenic prostate epithelial cells. A unique activity-based proteomics methodology was developed for characterizing cellular OPH esterase (McGoldrick 2013; McGoldrick et al. 2014a, 2014b). In this method, cellular lysates, with the same amount of protein or cells, are first subjected to native-polyacrylamide gel electrophoresis (n-PAGE) followed by treatment with alpha-naphthyl N-acetylalaninate/diazonium salts to visualize the bands of OPH esterase activity. Alpha-naphthyl N-acetylalaninate is a selective substrate for OPH (Kobayashi and Smith 1987). Nanospray liquid chromatography followed by tandem mass spectrometry of the excised diazonium-visualized band (digested with trypsin) confirmed the presence of OPH (McGoldrick et al. 2014b). The OPH esterase activity band was found to disappear after treatment with a serine protease inhibitor (diisopropylfluorophosphate) or by immunoprecipitation with anti-OPH antibody (McGoldrick 2013; McGoldrick et al. 2014b). Collectively, these data show that all the esterase activity observed with the N-acetylalaninate substrate is due to OPH, i.e., there are no other esterase bands in the cell lysates utilizing the OPH specific substrate. Moreover, these investigators found that N-acetyl alanyl esters are not substrates for porcine liver carboxylesterase which is typically used in pharmacology to study the hydrolysis of ester drugs (McGoldrick et al. 2014b).

McGoldrick et al. (2014b) found that tumorigenic human prostate cancer cell lines have a higher level of OPH activity (as measured by gel densitometry) than a normal nontumorigenic prostate epithelial cell line. Tumorigenic prostate cells are also known to have a significantly higher level of intrinsic oxidative stress than non-tumorigenic prostate cells (Kumar et al. 2008; McGoldrick et al. 2014b). Collectively, these data suggest that the increased cellular oxidative stress found in tumorigenic prostate cells may require a compensatory increase in OPH expression for cancer cells to maintain viability by promoting the degradation of toxic oxidized proteins. As detailed below, increased oxidative stress in aggressive cancers appears to be a very generalized phenomenon and very relevant to the design of redox-directed prodrug therapy. In addition to prostate cell lines, several breast cancer cell lines, melanoma cancer cell lines, and bladder cancer cell lines also overexpress OPH. More research will be necessary to determine if overexpression of OPH a general characteristic of aggressive cancer cells. Given the potential importance of utilizing OPH as a prodrug target, we will briefly review its known functions and relationship with oxidative stress.

8.11 What is Known About the Functions of OPH?

OPH is a serine protease with three known functions: (1) an exopeptidase activity that unblocks N-acetyl peptides with a selective preference for N-acetyl alanyl peptides (Kobayashi and Smith 1987); (2) an endopeptidase activity toward oxidized and glycated proteins (Harmat et al. 2011; Nakai et al. 2012; Shimizu et al. 2004, 2009; Kikugawa 2004); and (3) an ability to associate with aggresomes, which are pericentrosomal accumulations of misfolded or oxidized protein (Shimizu et al. 2004). Moreover, work by Shimizu et al. (2004) suggests that under conditions of oxidative stress the proteasome and OPH work coordinately to clear cells of oxidized (carbonylated) proteins that might otherwise be cytotoxic. Drugs that inhibit the proteasome (e.g., bortezomib) are thought to have anticancer effects by permitting the accumulation of toxic misfolded/oxidized proteins which induce apoptosis. Increased reactive oxygen species (ROS) production has been observed during the initiation of apoptosis in cancer cells treated with bortezomib (Bergamo et al. 2013). Nevertheless, a comprehensive physiological understanding of OPH and its potential roles in cancer biology remains elusive. The acetylation of the N-terminal alpha-amine group of proteins is, however, the most common post-translational modification in eukaryotic proteins yet, little is known about the biological role of N-alpha-terminal acetylation, and even less is known about the role of enzymes like OPH that catalyze the hydrolysis of an N-terminally acetylated peptide to release an N-acetylamino acid.

8.12 Redox-Directed Prodrug Therapy (RDPT) for Cancer

Most effective agents now used to kill cancer cells do so by directly or indirectly generating reactive oxygen species (ROS), which subsequently cause oxidative-stress-induced apoptosis (Watson 2012; Chandel and Tuveson 2014). It has been suggested, for example, that 5-FU induces oxidative-stress apoptosis in cancer cells and that antioxidants could decrease this cytotoxic effect (Fu et al. 2014). Similarly, oxidative stress generated by cisplatin appears to be critical to its anti-neoplastic effect (Yu et al. 2018). Despite mounting evidence, there have been minimal efforts to rationally optimize the efficacy of pro-oxidant drugs/prodrugs or identify biomarkers that would suggest which cancer patients would be likely responders.

8.13 Oxidative Stress and Cancer Redox Therapy

Quite remarkably, many cancer cells, including tumorigenic prostate cancer cells, exhibit a high level of intrinsic oxidative stress (Kumar et al. 2008; Nogueira et al. 2008). This is not a metabolic quirk but is required to maintain aggressive cancer phenotypes such as angiogenesis, invasive migration tumor metastasis and resistance to chemotherapy (Kumar et al. 2008; Fiaschi and Chiarugi 2012; Glasauer and Chandel 2014). Oxidative stress can result from an overproduction of ROS and pioneering work by Irani et al. (1997) has shown that ROS plays a key role as second messengers in promoting oncogenic phenotypes. As indicated in Figure 8.2, the response of cancer cells to ROS is, however, biphasic: while high levels of ROS promote cancer phenotypes a small additional increment in oxidative stress can overwhelm antioxidant defense mechanisms resulting in oxidative apoptosis (Nogueira et al. 2008). The high level of intrinsic oxidative stress in aggressive

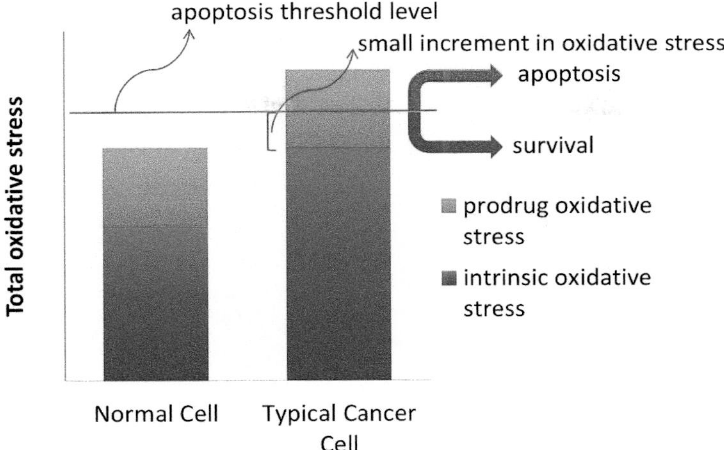

FIGURE 8.2 Cancer cells usually have a higher level of intrinsic oxidative stress (blue) than normal cells. A small increment in oxidative stress (red) from a pro-oxidant prodrug can increase the total level of oxidative stress beyond the threshold level needed to induce oxidative apoptosis in cancer cells but not normal cells.

cancer cells can be exploited by pro-oxidant agents, causing only a small additional increment oxidative stress while not affecting normal cells (Figure 8.2).

Oxidative apoptosis has been termed the "Achilles heel" of cancer (Nogueira et al. 2008). Glutathione (GSH) is the primary intracellular water-soluble antioxidant, and therapeutic agents causing a decrease its concentration are effective at selectively inducing oxidative-stress apoptosis in cancers with high levels of intrinsic oxidative stress (Traverso et al. 2013). It is important to note that intracellular GSH depletion need not be "complete" to induce oxidative apoptosis in cancer cells already having a high intrinsic (basal) level of oxidative stress (Figure 8.2), i.e., only a small additional increment in oxidative stress may be sufficient to reach the threshold needed to induce oxidative stress apoptosis.

8.14 Cancer, Akt Activation/Phosphorylation and Oxidative-Stress Apoptosis

One underlying cause for both increased cancer aggressiveness and increased intrinsic oxidative stress is the activation (i.e., phosphorylation) of the proto-oncogene Akt kinase via the phosphoinositide-3-kinase (PI3K) pathway. Clinical studies have shown, for example, that increased prostate tumor Akt phosphorylation is correlated with a higher Gleason score and a poor clinical outcome (Malik et al. 2002; Ayala et al. 2004). As shown in Figure 8.3 (simplified), Akt activation by phosphorylation increases intracellular ROS production by stimulating mitochondrial (MIT) metabolic activity while simultaneously decreasing the expression of key ROS detoxifying enzymes thereby increasing cellular sensitivity to oxidative-stress apoptosis (Nogueira et al. 2008). Akt activation has also been implicated in the activation of NADPH oxidase 4 activity (Edderkaoui et al. 2011) as well as an increased tumor expression of NADPH oxidase 4 protein (Govindarajan et al. 2007). NADPH oxidase is a pro-oxidant enzyme capable of generating large amounts of ROS (Govindarajan et al. 2007).

Akt activation is commonly caused by mutations in components of its signaling cascade and results in cancer cell lines with an increased ability to escape normal apoptosis and proliferate. As noted above, this is a "double-edged sword" since Akt activation also causes an increased production of ROS as well as an increased sensitivity to pro-oxidant drugs causing oxidative apoptosis (Dolado and Nebreda 2008). Akt activation has been extensively studied in a variety of cancer lines and animal models (Majumder and Sellers 2005). Most significantly, *in vivo* work by Li et al. (2007) in a xenograft model has shown that conditional Akt activation (i.e., inducible) in LNCaP human prostate cancer cells plays a causative role in promoting a transition in this cell's normal androgen-dependent growth to androgen-independent growth.

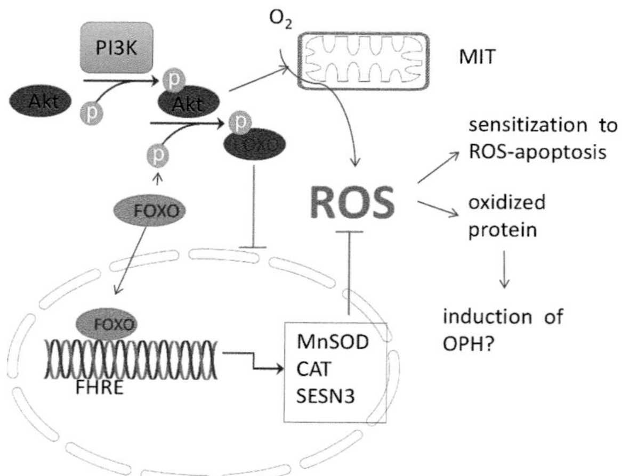

FIGURE 8.3 Activated Akt increases oxidative stress and may induce oxidized protein hydrolase (OPH) caused by increased levels of oxidized protein. Akt activation increases mitochondrial (MIT) oxygen consumption and phosphorylates forkhead box O (FOXO), thereby inhibiting its role as a transcriptional activator by blocking its sequestration in the cell nucleus. Non-phosphorylated FOXO binds to the FHRE (forkhead response element) transcription factor-binding site in the cell nucleus and induces the expression and the expression of key antioxidant enzymes, i.e., manganese superoxide dismutase (MnSOD), catalase (CAT) and sestrin3 (SESN3) Phosphorylating FOXO decreases the expression of MnSOD, CAT, and SESN3. The net effect of Akt activation is an increased production of reactive oxygen species (ROS) and oxidized protein.

8.15 Rational Redox-Directed Prodrug Cancer Therapy (RDPT)

We reasoned that a prodrug ester could be designed that would deplete intracellular GSH levels in cancer cells when activated by the esterase activity of OPH and subsequently cause oxidative-stress-induced apoptosis (Stone et al. 2014). As mentioned above, OPH has an esterase activity specific for N-acetyl-L-alanyl esters. We, therefore, proposed a novel N-acetyl-L-alanyl prodrug ester (**NPAA**, Figure 8.4) that would release a quinone methide (QM) upon potential hydrolysis by OPH. QM can covalently bind GSH,

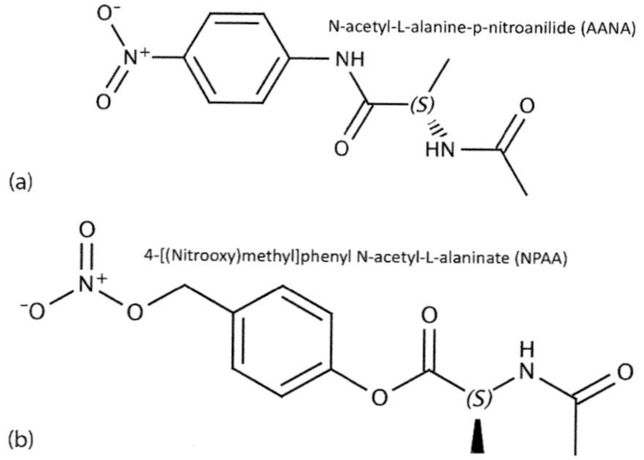

(a) N-acetyl-L-alanine-p-nitroanilide (AANA)

(b) 4-[(Nitrooxy)methyl]phenyl N-acetyl-L-alaninate (NPAA)

FIGURE 8.4 (a) The structure of the *N*-acetyl-L-alanine-p-nitroanilide (AANA), which is a specific substrate for OPH. (b) The structure of 4-[(nitrooxy)methyl]phenyl *N*-acetyl-L-alaninate (NPAA) a potential anticancer prodrug similar in structure to both AANA and nitric oxide-donating aspirin. The "S" in the structure indicates a chiral carbon center with the "S" configuration. In silico protein-ligand docking suggests that both AANA and NPAA bind to the active site of human OPH models derived from protein structure homology-modeling servers (see text).

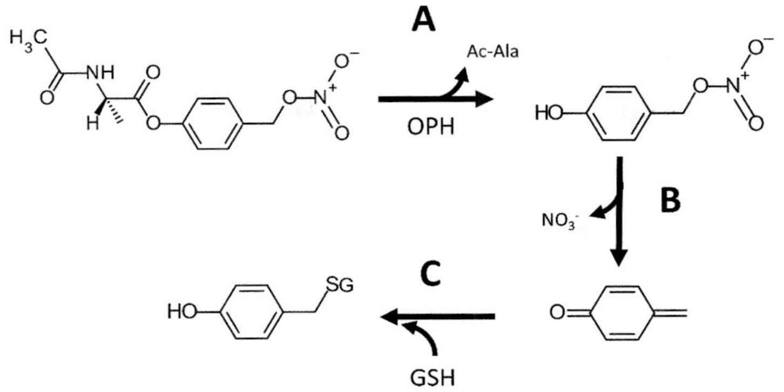

FIGURE 8.5 Mechanism of N-acetylalaninate prodrug (NPAA) activation by OPH and subsequent depletion of glutathione: (A) The ester bond of the prodrug (NPAA) is cleaved by the esterase activity of OPH releasing acetylalaninate and a (4-hydroxyphenyl)methyl nitrate intermediate; (B) The intermediate quickly undergoes elimination releasing NO_3^- and forming a reactive quinone methide (QM); (C) The QM rapidly reacts with the thiol group of reduced GSH in a Michael addition leaving GSH unavailable to participate in protective cellular redox reactions.

thereby decreasing its intracellular level (Stone et al. 2014; McGoldrick et al. 2014b) (see Figure 8.5). NPAA has a structure somewhat similar to nitric oxide-donating aspirin, which is a promising anticancer prodrug that also releases a QM upon hydrolysis and is less toxic than aspirin itself (Sun et al. 2009; Rigas 2007; Gao and Williams 2012; Song et al. 2018).

Before expending considerable effort on the organic synthesis of NPAA, we used protein-ligand docking software to determine if NPAA could bind to the active site of a 3D model of OPH. As indicated in Figure 8.1, this follows the general work-flow for the rational design of any prodrug. As a reference, we also performed protein-ligand docking with AANA since this compound is a substrate used in the colorimetric assay for OPH. If both NPAA and AANA bind to the active site of OPH, this will lend support for NPAA being a potential cancer prodrug. The structure of OPH has not yet been determined by either NMR or x-ray crystallography. Knowing the 3D structure of a protein drug target opens the door to the modern cornucopia of bioinformatic tools increasingly used in rational drug design (Nero et al. 2018). Rational drug design can help refine the specificity of a lead prodrug/drug by guiding its chemical synthesis and chemical modification to give an optimized fit in terms of shape, hydrogen bonds and other non-covalent interactions with a druggable ligand-binding site in the target protein.

8.16 Predicting the Structure of OPH as a Step in Rational Redox Prodrug Design

We utilized the I-TASSER (zhanglab.ccmb.med.umich.edu/I-TASSER/) and the Phyre-2 (www.sbg.bio.ic.ac.uk/~phyre2/) structure prediction web servers to calculate the potential 3D structure of OPH and to provide functional predictions. I-TASSER is free to use and is ranked as one the best server for 3D predictions with only the primary amino acid sequence as the required input (Zhang 2008, 2009; Yang and Zhang 2015). Similarly, the (Kelley et al. 2015) Phyre-2 server is free to use, is highly ranked and also has a very intuitive user interface.

The I-TASSER server picked the crystal structure of *Pyrococcus horikoshii* acylaminoacyl peptidase (4HXE.pdb) and the crystal structure of S9 peptidase (active form) from *Deinococcus radiodurans* R1 (5YZN.pdb) as the best structural analogs of human OPH. The consensus gene ontology (GO) terms used to describe the predicted molecular functions of OPH are "the hydrolysis of proteins into smaller polypeptides and/or amino acids by cleavage of their peptide bonds" and "catalysis of the hydrolysis of peptide bonds in a polypeptide chain by a catalytic mechanism that involves a catalytic triad consisting of a serine nucleophile that is activated by a proton relay involving an acidic residue (e.g., aspartate or

TABLE 8.1

Evaluation of Structural Accuracy by Ramachandran Plot Analyses

	% Residues in Ramachandran Regions		
Model Server	**Favored**	**Allowed**	**Outliers**
Initial values from the protein structure prediction web servers			
I-TASSER	63.4	20.2	16.4
Phyre-2	82.0	11.6	6.4
After YASARA energy minimization			
I-TASSER	82.6	12.8	4.6
Phyre-2	86.6	10.7	2.7

glutamate) and a basic residue (usually histidine)." These functions are remarkably consistent with the known functions of OPH (https://www.uniprot.org/uniprot/P13798). The Phyre2 server picked C5OLJA.pdb and C6EOTG.pdb as the best structural analogs of OPH. Both C5OLJA.pdb and C6EOTG.pdb are dipeptidyl peptidases with hydrolase activities. Quite interestingly, the Phyre2 server predicted that OPH could have a transmembrane helix spanning residues 585–615. This prediction is consistent with the finding of Jessani et al. (2002) showing that OPH can be found in the cell membrane fraction as well as the cytosolic and secreted fractions. It should also be noted that the 585–590 region of OPH has a region with either an active site amino acid, a conserved amino acid or an amino acid in contact with the OPH substrate (see more on this below).

Based on the ligands bound to the proteins used as structural homology templates, I-TASSER provides a residue-specific ligand-binding probability (p) for each of the 732 residues in OPH. Ser587 and His707 had among the highest probabilities (0.979 and 0.976 on a scale of 0 to 1) of being involved with ligand-binding. Identifying the residues involved in ligand-binding is an important step in rational drug design (Yang et al. 2013). The active site of OPH is known to have a "classic" Ser587/His707/Asp675 charge relay system at its active site which is typical of most serine proteases (https://www.uniprot.org/uniprot/P13798) (Ekici et al. 2008). It is interesting that I-TASSER did not identify Asp675 ($p = 0.05$) as a high probability ligand-binding residue but did identify His588 ($p = 0.95$) as such. Some "unconventional" serine proteases also utilize a Ser/His/His triad (Ekici et al. 2008).

The best 3D models of OPH provided by I-TASSER and Phyre2 were submitted to the MOLPROBITY web server (http://molprobity.biochem.duke.edu/) to provide an independent evaluation of structural accuracy by Ramachandran plot analyses (Lovell et al. 2003; Chen et al. 2010). As indicated in Table 8.1, 16.4% of the residues in the OPH I-TASSER model were in the "outlier" regions of the Ramachandran plot with 83.6% in favored or allowed regions. For the OPH-Phyre2 model, 6.4% were in the outlier region, while 93.6% were in favored or allowed regions. Typically, less than 2% of the residues in high-quality 3D protein structures are outliers. To improve structural accuracy the OPH models were submitted to the free online YASARA energy minimization server which explicitly takes into account a solvent shell surrounding the protein (Krieger et al. 2009) (http://www.yasara.org/minimizationserver.htm). The YASARA energy minimized I-TASSER OPH model (OPH -YASARA-I-TASSER.pdb) showed a remarkable improvement in the Ramachandran plot analysis, with only 4.6% of residues in outlier regions. Similarly, the percent outliers for the YASARA energy minimized Phyre2-OPH model (OPH -YASARA-Phyre2.pdb) decreased to 2.7%.

8.17 Protein-Ligand Docking Results with AANA and NPAA

Both energy minimized model structures of OPH were submitted to the I-TASSER BSP-SLIM "blind" molecular docking server which is optimized for low-resolution protein structures (Lee and Zhang 2012) with the two organic structures shown in Figure 8.4 as the "ligands." The BSP-SLIM server does

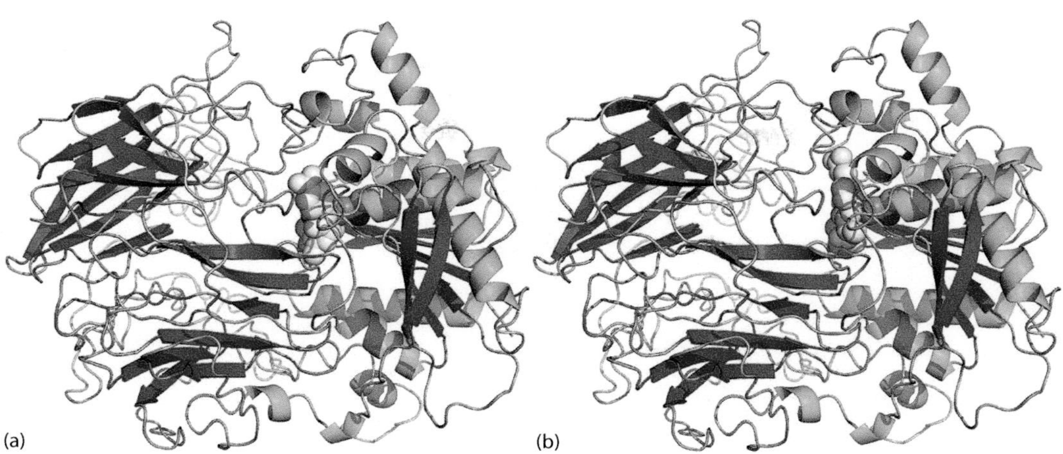

FIGURE 8.6 Predicted structural models for OPH-ligand complexes. (a) The binding of the AANA (pale-cyan) to OPH -YASARA-I-TASSER.pdb (see text). (b) The binding of NPAA (pale-cyan). The prodrug NPAA binds to the same active site region as the known substrate, AANA. The protein chain is colored by secondary structure and Ser587, His707, and Asp675 at the active site are colored by atom type. Only Ser587 is visible at the immediate right of the bound ligands.

TABLE 8.2

Protein-ligand Interface Analysis

Model	Ligand	Residues in Contact with Ligand
OPH -YASARA-I-TASSER.pdb	AANA	Gly586, Ser587, Arg677, His707
OPH -YASARA-I-TASSER.pdb	NPAA	Gly508, Gly509, Ser587, Asp624, Ile625, Cys629, Arg677, His707
OPH -YASARA-Phyre2.pdb	AANA	Glu146, Gly509, Pro510, His511, His588, Phe635
OPH -YASARA-Phyre2.pdb	NPAA	Glu146, Gly354, Pro510, His511, His588, Glu632, Phe635, Arg677, His707

Note: The conserved amino acid residues for all OPH models are Gly585, Asp675, Pro24, Gly508, Gly509, Gly529, Asn536, Gly539, Gly544, Asp562, Ser587, Gly589, Gly590, Pro656, His707.

not assume that the ligand-binding site is known and performs a search that covers the entire surface of the target protein. The calculated protein-ligands structural complexes for the OPH -YASARA-I-TASSER.pdb model with AANA and NPAA as the ligands are shown in Figure 8.6 and summarized in Table 8.2. Figure 8.6 was generated PyMOL (https://pymol.org/2/), which can also provide an analysis of evolutionarily conserved residues in the target protein along with a list of residues in contact with the ligand. PyMOL is a very flexible software package and provides high-resolution images. Evolutionarily conserved residues are likely to occur at or near the active site of an enzyme and/or have functional significance (Ramanathan and Agarwal 2011). As indicated in Table 8.2 for the OPH -YASARA-I-TASSER.pdb model, both AANA and NPAA were found to be in contact with active site residues Ser587 and His707. Moreover, as shown in Figure 8.6, the calculated docking site of AANA and NPAA is in close proximity to Ser587/His707/Asp675 residues at the active site of the OPH -YASARA-I-TASSER.pdb model. The results for the NPAA docking to OPH -YASARA-Phyre2.pdb were likewise positive with this ligand being in contact with His707.

Collectively, the data in Table 8.2 and Figure 8.6 support the binding of prodrug NPAA at the active site of OPH and therefore the subsequent synthesis and *in vitro* testing. The subsequent *in vitro* testing confirmed that NPAA crosses the plasma membrane (see Figure 8.7), depletes intracellular GSH when specifically activated by OPH and subsequently induces oxidative apoptosis in

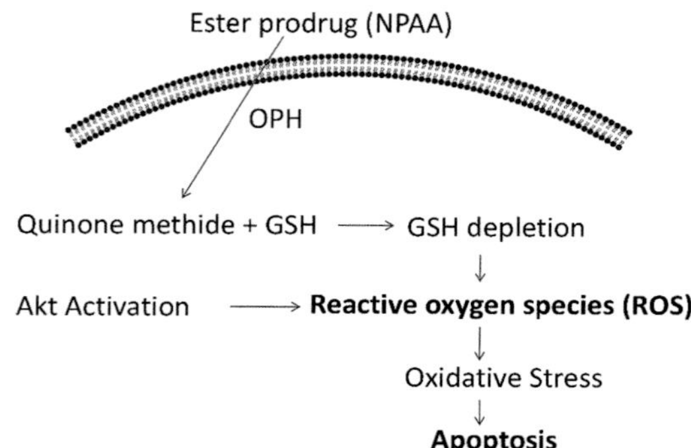

FIGURE 8.7 Cancer cells with increased OPH activity and activated Akt are ideal targets for GSH depleting prodrug esters.

tumorigenic prostate cancer cell lines with high levels of oxidative stress and/or high expression of OPH (McGoldrick et al. 2014a).

8.18 Future Research with Prodrug NPAA and Prostate Cancer

As indicated in Figure 8.1, the step before clinical trials would be demonstrating *in vivo* efficacy with an animal model. The optimal approach would be utilizing a human cancer cell line already well characterized by *in vitro* experiments. This human cell line could then be used in a mouse xenograft model such as the hairless male athymic nude mice (Hsd:Athymic Nude-*Foxn1nu/nu*). For prostate cancer, a variety of commercially available human cell lines (LNCaP, DU145, and PC3) are available that express red fluorescent protein (RFP). This permits *in vivo* fluorescent imaging for the non- invasive detection and quantification of tumor growth and metastases. This powerful methodology would enable a longitudinal assessment of the chemotherapeutic effect of the **NPAA** prodrug by monitoring tumor growth and metastases *in vivo* (Hoffman 2005).

8.19 Conclusions

There is a clear need to develop biomarkers that can distinguish indolent from aggressive forms of cancer as well as prodrug chemotherapy specifically targeted to aggressive forms of cancer. In this chapter, we posit that oxidative-stress biomarkers could serve this purpose and simultaneously provide the basis for targeted pro-oxidant chemotherapy when combined with a biomarker for oxidized protein hydrolase (OPH). One underlying cause for both increased cancer aggressiveness and increased intrinsic oxidative stress is the activation (i.e., phosphorylation) of the proto-oncogene Akt kinase via the phosphoinositide-3-kinase (PI3K) pathway. Clinical studies have shown that increased prostate tumor Akt phosphorylation is correlated with a higher Gleason score and a poor clinical outcome (Malik et al. 2002; Ayala et al. 2004). Clinical pharmacology studies will be needed to test the hypothesis that the NPAA prodrug detail above would be particularly effective in prostate cancer subjects in which a tumor biopsy demonstrated immunohistochemical evidence for a high level of phospho-Akt and OPH (see Figure 8.7).

ACKNOWLEDGMENTS

This research was supported in part by the National Institutes of Health grant C06RR0306551, the East Tennessee State University Robert W. Summers Pediatric Research Endowment and ETSU RDC grant 20-001M (WLS).

REFERENCES

Ayala, G., T. Thompson, G. Yang, A. Frolov, R. Li, P. Scardino, M. Ohori, T. Wheeler, and W. Harper. 2004. High levels of phosphorylated form of Akt-1 in prostate cancer and non-neoplastic prostate tissues are strong predictors of biochemical recurrence. *Clin Cancer Res* 10 (19): 6572–6578. https://doi.org/10.1158/1078-0432.CCR-04-0477.

Bergamo, P., E. Cocca, R. Palumbo, M. Gogliettino, M. Rossi, and G. Palmieri. 2013. RedOx status, proteasome and APEH: Insights into anticancer mechanisms of t10,c12-conjugated linoleic acid isomer on A375 melanoma cells. *PLoS One* 8 (11): e80900. https://doi.org/10.1371/journal.pone.0080900.

Brown, D. K., and Ö. T. Bishop. 2017. Role of structural bioinformatics in drug discovery by computational SNP analysis: Analyzing variation at the protein level. *Glob Heart* 12 (2): 151–161. https://doi.org/10.1016/j.gheart.2017.01.009.

Chandel, N. S., and D. A. Tuveson. 2014. The promise and perils of antioxidants for cancer patients. *N Engl J Med* 371 (2): 177–178. https://doi.org/10.1056/NEJMcibr1405701.

Chen, V. B., W. B. Arendall, J. J. Headd, D. A. Keedy, R. M. Immormino, G. J. Kapral, L. W. Murray, J. S. Richardson, and D. C. Richardson. 2010. MolProbity: All-atom structure validation for macromolecular crystallography. *Acta Crystallographica A* 66 (Pt 1): 12–21. https://doi.org/10.1107/S0907444909042073.

Dolado, I., and A. R. Nebreda. 2008. AKT and oxidative stress team up to kill cancer cells. *Cancer Cell* 14 (6): 427–429. https://doi.org/10.1016/j.ccr.2008.11.006.

Edderkaoui, M., C. Nitsche, L. Zheng, S. J. Pandol, I. Gukovsky, and A. S. Gukovskaya. 2011. NADPH oxidase activation in pancreatic cancer cells is mediated through Akt-dependent up-regulation of p22phox. *J Biol Chem* 286 (10): 7779–7787. https://doi.org/10.1074/jbc. M110.200063.

Ekici, O. D., M. Paetzel, and R. E. Dalbey. 2008. Unconventional serine proteases: Variations on the catalytic Ser/His/Asp triad configuration. *Protein Sci* 17 (12): 2023–2037. https://doi.org/10.1110/ps.035436.108.

Fiala, C., and E. P. Diamandis. 2018. Utility of circulating tumor DNA in cancer diagnostics with emphasis on early detection. *BMC Med* 16 (1): 166. https://doi.org/10.1186/s12916-018-1157-9.

Fiaschi, T., and P. Chiarugi. 2012. Oxidative stress, tumor microenvironment, and metabolic reprogramming: A diabolic liaison. *Int J Cell Biol* 2012: 762825. https://doi.org/10.1155/2012/762825.

Fu, Y., G. Yang, F. Zhu, C. Peng, W. Li, H. Li, H. G. Kim, A. M. Bode, and Z. Dong. 2014. Antioxidants decrease the apoptotic effect of 5-Fu in colon cancer by regulating Src-dependent caspase-7 phosphorylation. *Cell Death Dis* 5: e983. https://doi.org/10.1038/cddis.2013.509.

Gao, L., and J. L. Williams. 2012. Nitric oxide-donating aspirin induces G2/M phase cell cycle arrest in human cancer cells by regulating phase transition proteins. *Int J Oncol* 41 (1): 325–330. https://doi.org/10.3892/ijo.2012.1455.

Giang, I., E. L. Boland, and G. M. Poon. 2014. Prodrug applications for targeted cancer therapy. *AAPS J* 16 (5): 899–913. https://doi.org/10.1208/s12248-014-9638-z.

Glasauer, A., and N. S. Chandel. 2014. Targeting antioxidants for cancer therapy. *Biochem Pharmacol* 92 (1): 90–101. https://doi.org/10.1016/j.bcp.2014.07.017.

Govindarajan, B., J. E. Sligh, B. J. Vincent, M. Li, J. A. Canter, B. J. Nickoloff, R. J. Rodenburg et al. 2007. Overexpression of Akt converts radial growth melanoma to vertical growth melanoma. *J Clin Invest* 117 (3): 719–729.

Harmat, V., K. Domokos, D. K. Menyhard, A. Pallo, Z. Szeltner, I. Szamosi, T. Beke-Somfai, G. Naray-Szabo, and L. Polgar. 2011. Structure and catalysis of acylaminoacyl peptidase: Closed and open subunits of a dimer oligopeptidase. *J Biol Chem* 286 (3): 1987–1998.

Heitzer, E., I. S. Haque, C. E. S. Roberts, and M. R. Speicher. 2019. Current and future perspectives of liquid biopsies in genomics-driven oncology. *Nat Rev Genet* 20 (2): 71–88. https://doi.org/10.1038/s41576-018-0071-5.

Hoffman, R. M. 2005. In vivo cell biology of cancer cells visualized with fluorescent proteins. *Curr Top Dev Biol* 70: 121–144. https://doi.org/10.1016/S0070-2153(05)70006-5.

The Human Protein Atlas. 2015. The Human Protein Atlas, antibody staining in cancers. http://www.proteinatlas.org/ENSG00000164062-APEH/cancer.

Irani, K., Y. Xia, J. L. Zweier, S. J. Sollott, C. J. Der, E. R. Fearon, M. Sundaresan, T. Finkel, and P. J. Goldschmidt-Clermont. 1997. Mitogenic signaling mediated by oxidants in Ras-transformed fibroblasts. *Science* 275 (5306): 1649–1652.

Jessani, N., Y. Liu, M. Humphrey, and B. F. Cravatt. 2002. Enzyme activity profiles of the secreted and membrane proteome that depict cancer cell invasiveness. *Proceedings of the National Academy of Sciences of the United States of America* 99 (16): 10335–10340. https://doi.org/10.1073/pnas.162187599.

Jiang, Z., and Y. Zhou. 2005. Using bioinformatics for drug target identification from the genome. *Am J Pharmacogenomics* 5 (6): 387–396.

Kaittanis, C., C. Andreou, H. Hieronymus, N. Mao, C. A. Foss, M. Eiber, G. Weirich et al. 2018. Prostate-specific membrane antigen cleavage of vitamin B9 stimulates oncogenic signaling through metabotropic glutamate receptors. *J Exp Med* 215 (1): 159–175. https://doi.org/10.1084/jem.20171052.

Kelley, L. A., S. Mezulis, C. M. Yates, M. N. Wass, and M. J. Sternberg. 2015. The Phyre2 web portal for protein modeling, prediction and analysis. *Nat Protoc* 10 (6): 845–858. https://doi.org/10.1038/nprot.2015.053.

Kikugawa, K. 2004. Defense of living body against oxidative damage. *Yakugaku zasshi* 124 (10): 653–666.

Kirkeby, S., and D. Moe. 1988. A comparison between activities for non-specific esterases and esterproteases. *Acta Histochem* 83 (1): 11–19.

Klega, K., A. Imamovic-Tuco, G. Ha, A. N. Clapp, S. Meyer, A. Ward, C. Clinton et al. 2018. Detection of somatic structural variants enables quantification and characterization of circulating tumor DNA in children with solid tumors. *JCO Precis Oncol* 2018. https://doi.org/10.1200/PO.17.00285.

Kobayashi, K., and J. A. Smith. 1987. Acyl-peptide hydrolase from rat liver. Characterization of enzyme reaction. *J Biol Chem* 262 (24): 11435–11445.

Kreusser, E. H. 1966. Nonspecific esterases in normal and neoplastic tissues of the Syrian hamster: A zymogram study. *Cancer Res* 26 (10): 2181–2185.

Krieger, E., K. Joo, J. Lee, S. Raman, J. Thompson, M. Tyka, D. Baker, and K. Karplus. 2009. Improving physical realism, stereochemistry, and side-chain accuracy in homology modeling: Four approaches that performed well in CASP8. *Proteins* 77 (Suppl 9): 114–122.

Kumar, B., S. Koul, L. Khandrika, R. B. Meacham, and H. K. Koul. 2008. Oxidative stress is inherent in prostate cancer cells and is required for aggressive phenotype. *Cancer Res* 68 (6): 1777–1785. https://doi.org/10.1158/0008-5472.CAN-07-5259.

Lee, H. S., and Y. Zhang. 2012. BSP-SLIM: A blind low-resolution ligand-protein docking approach using predicted protein structures. *Proteins* 80 (1): 93–110. https://doi.org/10.1002/prot.23165; 10.1002/prot.23165.

Li, B., A. Sun, H. Youn, Y. Hong, P. F. Terranova, J. B. Thrasher, P. Xu, and D. Spencer. 2007. Conditional Akt activation promotes androgen-independent progression of prostate cancer. *Carcinogenesis* 28 (3): 572–583. https://doi.org/10.1093/carcin/bgl193.

Liu, Y., M. P. Patricelli, and B. F. Cravatt. 1999. Activity-based protein profiling: The serine hydrolases. *Proceedings of the National Academy of Sciences of the United States of America* 96 (26): 14694–14699.

Lovell, S. C., I. W. Davis, W. B. Arendall, 3rd, P. I. de Bakker, J. M. Word, M. G. Prisant, J. S. Richardson, and D. C. Richardson. 2003. Structure validation by Calpha geometry: Phi, psi and Cbeta deviation. *Proteins* 50 (3): 437–450.

Majumder, P. K., and W. R. Sellers. 2005. Akt-regulated pathways in prostate cancer. *Oncogene* 24 (50): 7465–7474. https://doi.org/10.1038/sj.onc.1209096.

Malekshah, O. M., X. Chen, A. Nomani, S. Sarkar, and A. Hatefi. 2016. Enzyme/prodrug systems for cancer gene therapy. *Curr Pharmacol Rep* 2 (6): 299–308. https://doi.org/10.1007/s40495-016-0073-y.

Malet-Martino, M., P. Jolimaitre, and R. Martino. 2002. The prodrugs of 5-fluorouracil. *Curr Med Chem Anticancer Agents* 2 (2): 267–310.

Malet-Martino, M., and R. Martino. 2002. Clinical studies of three oral prodrugs of 5-fluorouracil (capecitabine, UFT, S-1): A review. *Oncologist* 7 (4): 288–323.

Malik, S. N., M. Brattain, P. M. Ghosh, D. A. Troyer, T. Prihoda, R. Bedolla, and J. I. Kreisberg. 2002. Immunohistochemical demonstration of phospho-Akt in high Gleason grade prostate cancer. *Clin Cancer Res* 8 (4): 1168–1171.

Martin, S. E., T. Ganguly, G. R. Munske, M. D. Fulton, M. R. Hopkins, C. E. Berkman, and M. E. Black. 2014. Development of inhibitor-directed enzyme prodrug therapy (IDEPT) for prostate cancer. *Bioconjug Chem* 25 (10): 1752–1760. https://doi.org/10.1021/bc500362n.

McGoldrick, C. A. 2013. Novel Ester Substrates for the Detection and Treatment of Prostate Cancer. Dissertation/Thesis, East Tennessee State University. http://dc.etsu.edu/etd/2308/.

McGoldrick, C. A., Y. L. Jiang, M. Brannon, K. Krishnan, and W. L. Stone. 2014a. In vitro evaluation of novel N-acetylalaninate prodrugs that selectively induce apoptosis in prostate cancer cells. *BMC Cancer* 14: 675. https://doi.org/10.1186/1471-2407-14-675.

McGoldrick, C. A., Y. L. Jiang, V. Paromov, M. Brannon, K. Krishnan, and W. L. Stone. 2014b. Identification of oxidized protein hydrolase as a potential prodrug target in prostate cancer. *BMC Cancer* 14: 77. https://doi.org/10.1186/1471-2407-14-77.

Nakai, A., Y. Yamauchi, S. Sumi, and K. Tanaka. 2012. Role of acylamino acid-releasing enzyme/oxidized protein hydrolase in sustaining homeostasis of the cytoplasmic antioxidative system. *Planta* 236: 427–436. https://doi.org/10.1007/s00425-012-1614-1.

NCI dictionary, 2019. Precision medicine. https://www.cancer.gov/publications/dictionaries/cancer-terms/def/precision-medicine.

Nero, T. L., M. W. Parker, and C. J. Morton. 2018. Protein structure and computational drug discovery. *Biochem Soc Trans* 46 (5): 1367–1379. https://doi.org/10.1042/BST20180202.

Nogueira, V., Y. Park, C. C. Chen, P. Z. Xu, M. L. Chen, I. Tonic, T. Unterman, and N. Hay. 2008. Akt determines replicative senescence and oxidative or oncogenic premature senescence and sensitizes cells to oxidative apoptosis. *Cancer Cell* 14 (6): 458–470. https://doi.org/10.1016/j.ccr.2008.11.003.

Palmieri, G., P. Bergamo, A. Luini, M. Ruvo, M. Gogliettino, E. Langella, M. Saviano, R. N. Hegde, A. Sandomenico, and M. Rossi. 2011. Acylpeptide hydrolase inhibition as targeted strategy to induce proteasomal down-regulation. *PLoS One* 6 (10): e25888. https://doi.org/10.1371/journal.pone.0025888.

Pathak, R. K., R. Wen, N. Kolishetti, and S. Dhar. 2017. A prodrug of two approved drugs, cisplatin and chlorambucil, for chemo war against cancer. *Mol Cancer Ther* 16 (4): 625–636. https://doi.org/10.1158/1535-7163.MCT-16-0445.

Ramanathan, A., and P. K. Agarwal. 2011. Evolutionarily conserved linkage between enzyme fold, flexibility, and catalysis. *PLoS Biol* 9 (11): e1001193. https://doi.org/10.1371/journal.pbio.1001193.

Rigas, B. 2007. Novel agents for cancer prevention based on nitric oxide. *Biochem Soc Trans* 35 (Pt 5): 1364–1368.

Rooseboom, M., J. N. Commandeur, and N. P. Vermeulen. 2004. Enzyme-catalyzed activation of anticancer prodrugs. *Pharmacol Rev* 56 (1): 53–102.

Sara, J. D., J. Kaur, R. Khodadadi, M. Rehman, R. Lobo, S. Chakrabarti, J. Herrmann, A. Lerman, and A. Grothey. 2018. 5-fluorouracil and cardiotoxicity: A review. *Ther Adv Med Oncol* 10: 1758835918780140.

Shimizu, K., M. Ikegami-Kawai, and T. Takahashi. 2009. Increased oxidized protein hydrolase activity in serum and urine of diabetic rat models. *Biol Pharm Bull* 32 (9): 1632–1635.

Shimizu, K., Y. Kiuchi, K. Ando, M. Hayakawa, and K. Kikugawa. 2004. Coordination of oxidized protein hydrolase and the proteasome in the clearance of cytotoxic denatured proteins. *Biochem Biophys Res Commun* 324 (1): 140–146. https://doi.org/10.1016/j.bbrc.2004.08.231.

Siddiqui, N. S., A. Godara, M. M. Byrne, and M. W. Saif. 2019. Capecitabine for the treatment of pancreatic cancer. *Expert Opin Pharmacother* 20 (4): 399–409. https://doi.org/10.1080/14656566.2018.1560422.

Song, J. M., P. Upadhyaya, and F. Kassie. 2018. Nitric oxide-donating aspirin (NO-Aspirin) suppresses lung tumorigenesis in vitro and in vivo and these effects are associated with modulation of the EGFR signaling pathway. *Carcinogenesis* 39 (7): 911–920. https://doi.org/10.1093/carcin/bgy049.

Spiegelberg, L., R. Houben, R. Niemans, D. de Ruysscher, A. Yaromina, J. Theys, C. P. Guise et al. 2019. Hypoxia-activated prodrugs and (lack of) clinical progress: The need for hypoxia-based biomarker patient selection in phase III clinical trials. *Clin Transl Radiat Oncol* 15: 62–69. https://doi.org/10.1016/j.ctro.2019.01.005.

Stone, W. L., K. J. Klopfenstein, M. J. Hajianpour, M. I. Popescu, C. M. Cook, and K. Krishnan. 2017. Childhood cancers and systems medicine. *Front Biosci (Landmark Ed)* 22: 1148–1161.

Stone, W. L., L. J. Yu, C. McGoldrick, M. Brannon, and K. Krishnan. 2014. The design, synthesis and in vitro evaluation of a novel pro-oxidant anticancer prodrug substrate targeted to acylamino-acid releasing enzyme. In *Free Radicals: The Role of Antioxidants and Pro-oxidants in Cancer Development*, edited by William L. Stone, 189. New York: Nova Biomedical.

Sun, Y., J. Chen, and B. Rigas. 2009. Chemopreventive agents induce oxidative stress in cancer cells leading to COX-2 overexpression and COX-2-independent cell death. *Carcinogenesis* 30 (1): 93–100. https://doi.org/10.1093/carcin/bgn242.

Traverso, N., R. Ricciarelli, M. Nitti, B. Marengo, A. L. Furfaro, M. A. Pronzato, U. M. Marinari, and C. Domenicotti. 2013. Role of glutathione in cancer progression and chemoresistance. *Oxid Med Cell Longev* 2013: 972913. https://doi.org/10.1155/2013/972913.

Watson, J. 2012. Oxidants, antioxidants and the current incurability of metastatic cancers. *Open Biol.* doi: 10.1098/rsob.120144.

Wondrak, G. T. 2009. Redox-directed cancer therapeutics: Molecular mechanisms and opportunities. *Antioxid Redox Signal* 11 (12): 3013–3069. https://doi.org/10.1089/ars.2009.2541.

Xu, G., and H. L. McLeod. 2001. Strategies for enzyme/prodrug cancer therapy. *Clin Cancer Res* 7 (11): 3314–3324.

Yamazaki, Y., Y. Ogawa, A. S. Afify, Y. Kageyama, T. Okada, H. Okuno, Y. Yoshii, and T. Nose. 1995. Difference between cancer cells and the corresponding normal tissue in view of stereoselective hydrolysis of synthetic esters. *Biochim Biophys Acta* 1243 (3): 300–308.

Yang, J., A. Roy, and Y. Zhang. 2013. Protein-ligand binding site recognition using complementary binding-specific substructure comparison and sequence profile alignment. *Bioinformatics* 29 (20): 2588–2595. https://doi.org/10.1093/bioinformatics/btt447.

Yang, J., and Y. Zhang. 2015. Protein structure and function prediction using I-TASSER. *Curr Protoc Bioinformatics* 52: 5–8 https://doi.org/10.1002/0471250953.bi0508s52.

Yu, K. H., G. J. Berry, D. L. Rubin, C. Ré, R. B. Altman, and M. Snyder. 2017a. Association of omics features with histopathology patterns in lung adenocarcinoma. *Cell Syst* 5 (6): 620–627.e3. https://doi.org/10.1016/j.cels.2017.10.014.

Yu, K. H., M. R. Fitzpatrick, L. Pappas, W. Chan, J. Kung, and M. Snyder. 2017b. Omics AnalySIs System for PRecision Oncology (OASISPRO): A web-based omics analysis tool for clinical phenotype prediction. *Bioinformatics*. https://doi.org/10.1093/bioinformatics/btx572.

Yu, W., Y. Chen, J. Dubrulle, F. Stossi, V. Putluri, A. Sreekumar, N. Putluri, D. Baluya, S. Y. Lai, and V. C. Sandulache. 2018. Cisplatin generates oxidative stress which is accompanied by rapid shifts in central carbon metabolism. *Sci Rep* 8 (1): 4306. https://doi.org/10.1038/s41598-018-22640-y.

Zhang, J., V. Kale, and M. Chen. 2015. Gene-directed enzyme prodrug therapy. *AAPS J* 17 (1): 102–110. https://doi.org/10.1208/s12248-014-9675-7.

Zhang, Y. 2008. I-TASSER server for protein 3D structure prediction. *BMC Bioinformatics* 9: 40. https://doi.org/10.1186/1471-2105-9-40.

Zhang, Y. 2009. I-TASSER: Fully automated protein structure prediction in CASP8. *Proteins 77* (Suppl 9): 100–113. https://doi.org/10.1002/prot.22588.

9

Prodrugs in Cancer Nanomedicine and Therapy

Vikas Pandey, Tanweer Haider, Vishal Gour, Vandana Soni, and Prem N. Gupta

CONTENTS

9.1 Introduction

The efficacy and selectivity of cancer therapy have been improved by the use of targeting concept through the unique markers overexpressed on the cancer cells as compared to normal tissues. The therapeutics delivery of drug through the various drug carriers is a promising and pioneering move toward the improvement in cancer treatment. The use of prodrugs for the targeted therapy is providing great diversity in the cancer therapy, in terms of activation chemistry, selection of target, size, and prodrug physicochemical nature. Prodrugs are the conjugate between the drug and a polymer, which help to carry the drug to desired site and then releasing it to show its action. Prodrugs may be less active or inactive derivatives of parent drug molecules, which may undergo chemical or enzymatic transformation to convert back into its parent active form of drug. The prodrugs can be grouped in two different way, one as a bioprecursor prodrug, which utilizes chemical or metabolic alterations converting prodrug into an active form in the cancer cells; the second is carrier-linked prodrugs where a drug in active form is covalently linked to some carrier that gets released on reaching the target site. Various prodrugs have been extensively found to be useful in the targeted therapy of anticancer drugs to cancerous cells (Giang et al., 2014b). The various chemotherapeutic agents including paclitaxel, doxorubicin, cisplatin, camptothecan have been converted into prodrugs which can be used for targeting and treatment of particular type of cancer. Hong et al., developed the prodrug-based pH-sensitive nanomedicine for the lung cancer therapy comprising the combination of doxorubicin and curcumin. They developed the nanoparticles encapsulating the dual drug to overcome the multidrug resistance and toxic side effects of doxorubicin. A U11 peptide, a uPA receptor, conjugated doxorubicin prodrug was synthesized to be used as materials to produce nanoparticles bearing the curcumin for the treatment of lung cancer (Hong et al., 2019).

9.2 Targeted Actively Drugs

The cancer-targeted prodrugs use the diverse range of chemistry to convert the prodrugs into its active cytotoxic parts. They may get converted by either passively or actively which depends upon the mechanism the prodrug follows. The passive activation uses the local physicochemical changes, like pH, hypoxia or physiologic alteration of tumor tissue, like overexpression of surface receptors, which are not or less present in normal cells/tissues. The active activation of prodrugs uses the specific activation chemistry.

9.2.1 Passively Activated Conversion of Cancer-Targeted Prodrugs

Passive activation of prodrug involves the activation through the endogenous enzymes. Various enzymes are found and unusually upregulated in cancer cells, like proteases. These different proteases enzymes are overexpressed in tumor cells which are contributing and leading to an metastatic phenotype. They can be targeted through the incorporation of proper substrate in the prodrug. Most common examples include lysosomal proteases (like legumain and cathepsins) and proteases enzymes of extracellular matrix (ECM) which include urokinase-type plasminogen activator (uPA) and matrix metalloproteases (MMPs). Targeting is accomplished through the incorporation of a peptide linker which is sequence-specific in nature as a "trigger" moiety. This would help in preventing the free diffusion of prodrug into cells and upon cleavage the cytotoxic agent gets released (Choi et al., 2012).

Also, various nonproteolytic targets, like cell-surface receptors, which are overexpressed on cancer cells are also used as a target by prodrug conjugates. Numerous receptors may undergo endocytosis process to transport substrates, thus acting as precise portals for the entry into cells. Targeting ligands range widely in size and chemistry. Short peptides and folic acid, due to ease of availability and synthesis, are the most common ligands used for targeted prodrug conjugates (Low and Antony, 2004). In case of intracellular targets, DT-diaphorase enzyme is a cytosolic enzyme. This enzyme mediates the two-electron reductase of quinone substrates. In different tumors, like colorectal carcinoma, breast carcinoma, Non-Small Cell Lung Cancer (NSCLC), and liver cancer the level of DT-diaphorase are increased. DT-diaphorase enzyme can activate the mitomycin C. An alkylating agent RH1, activate aziridine-based mustards in cancer cells through the bioreduction of attached quinine (Hargreaves et al., 2000). Telomerase is an intracellular cancer target which is normally repressed enzyme but is active in certain cancer like pancreatic cancers. A thymidine analog prodrug acycloguanosyl 5′-thymidyltriphosphate gets hydrolyzed through telomerase to acyclovir diphosphate, which is the active form of drug acyclovir (Polvani et al., 2011). Others alternative approaches used for targeting specific cellular targets is the use of tumor microenvironment (TME) to target aberrant.

9.2.2 Active Conversion of Cancer-Targeted Prodrugs

Exogenous enzymes are also used as a target for cancer cells for activating the particular administered prodrug to a cytotoxic moiety. The example of such type of application is cytosine deaminase, the most commonly used enzyme. Recombinant cytosine deaminase which is cloned from a yeast, fungal, or bacterial converts the prodrug 5-fluorocytosine (5-FC) into 5-fluorouracil (5-FU). The antimetabolite prodrugs acyclovir and ganciclovir which are nucleoside analogs are activated to their respective active

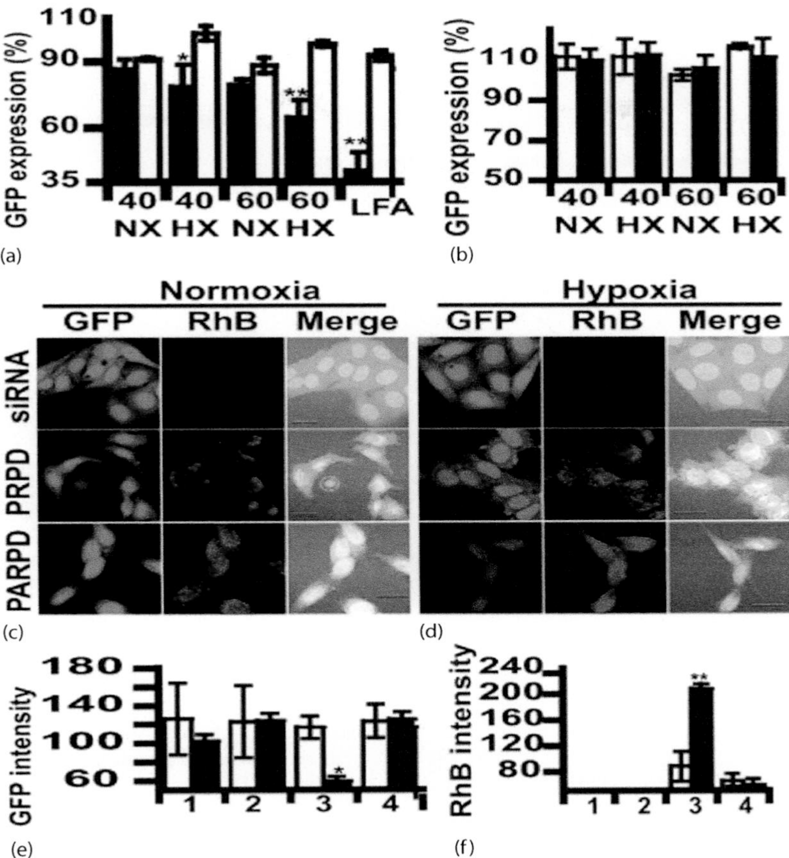

FIGURE 9.3 GFP downregulation in vitro. GFP downregulation was evaluated both by flow cytometry (a, b) and microscopy (c–f) after transfection under normoxic (NX) or hypoxic (HX) conditions. Relative geometric mean fluorescence from FACS analysis of HeLa/GFP cells transfected with (a) PEG-Azo-PEI-DOPE (PAPD)/siRNA complexes, (b) PEG-PEI-DOPE (PPD)/siRNA complexes in the presence of 10% FBS. Polyplexes were prepared at N/P ratios of 40 and 60 with antiGFP siRNA (black bars) or scrambled siRNA (white bars). Lipofectamine 2000 (LFA) was used as a positive control, $*p<0.05$, $**p<0.01$ compared with scrambled siRNA complexes. CLSM images of HeLa/GFP cells transfected with Rhodamine B labeled copolymers PEG-Azo-Rhodamine-PEI-DOPE (PARPD), PEG-Rhodamine-PEI-DOPE (PRPD) and GFP siRNA under normoxia (c) and hypoxia (d). Mean pixel intensities of GFP (e) and Rhodamine B (f) after transfection under normoxia (white bars) and hypoxia (black bars); 1: PBS, 2: free siRNA, 3: PARPD, 4: PRPD. $p<0.05$, $** p<0.01$ compared with normoxia. (Copyrighted, reused from Perche, F. et al., *Angew. Chem. Int. Ed. Engl.*, 53, 3362–3366, 2014. With permission.)

the silencing of hypoxia-activated green fluorescent protein *in vitro* along with the downregulation in GFP-expressing tumors (Figure 9.3) (Perche et al., 2014).

AZO derivative is a novel type of hypoxia-responsive moiety and developed a series of hypoxia-sensitive fluorescent probes for *in-vivo* imaging (Alimoradi et al., 2016). The AZO group is hypoxia-responsive for the efficient targeting and killing of tumor cells and is considered as the key factor in the successful treatment of tumor. The AZO-based derivatives are reduced by one electron reduction under the hypoxic condition and further stepwise reduction, i.e., hydrazine and then finally amines formation takes place. Liu et al., prepared the self-assembled micelles in which an AZO bond, which is crosslinked with conjugated PEG–hexanethiol (PEG–C6) with combretastatin A-4 (CA4) to form PEG–C6–AZO–CA4 amphiphilic molecule. The prepared PEG–C6–AZO–CA4 molecules get self-assembled into micelles encapsulating the doxorubicin. Under the hypoxic condition, CA4, and doxorubicin are

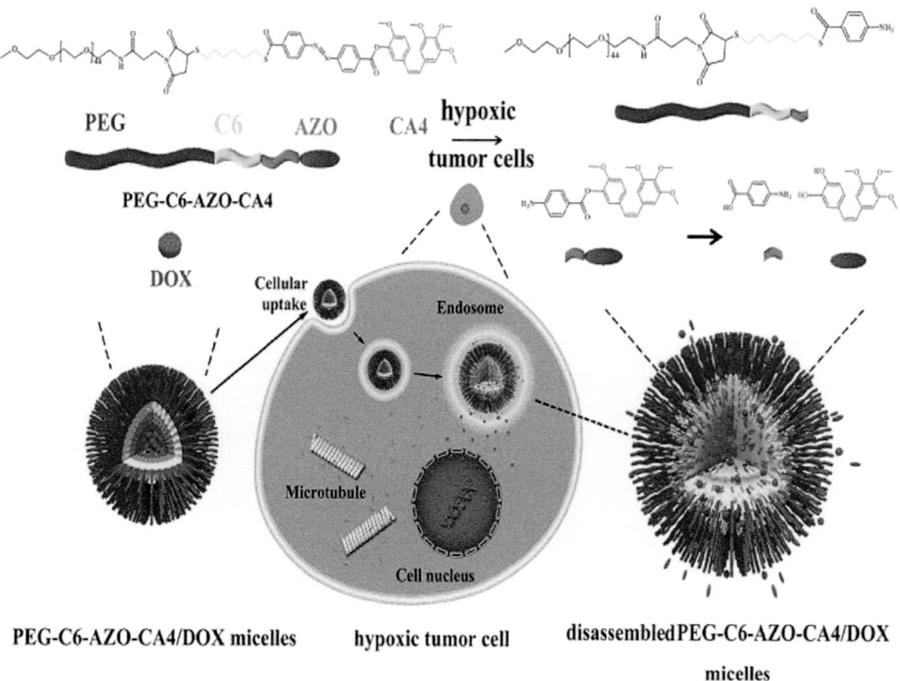

FIGURE 9.4 The chemical structure of PEG–C6–AZO–CA4 molecule and schematic illustration of PEG–C6–AZO–CA4/DOX micelles as codelivery platform for CA4 and DOX inhibiting the tumor growth in a combined way. Under hypoxic tumor cells, the PEG–C6–AZO–CA4/DOX micelles release CA4 and DOX rapidly. Released CA4 inhibits tubulin polymerization and the released DOX enters into the nucleus to kill cancer cells. (Copyrighted, reused from Liu, H.M. et al., *RSC Adv.*, 5, 20848–20857, 2015. With permission.)

released rapidly. Figure 9.4 describes the scheme and the cytotoxic mechanism of prepared PEG–C6–AZO–CA4 molecules containing micelles (Liu et al., 2015).

Glucose Oxidase (GOx) acts as a catalyst against β-d-glucose and GOx's underlying biocompatible and biodegradable properties have generated great attention in cancer management. GOx catalysis can successfully eliminate glucose and oxygen supplies, which results in acidic condition, low oxygen level and hydrogen peroxide in TME (Mukerabigwi et al., 2019). Such attractive catalytic properties make GOx as a possible therapeutic agent for handling cancer. Combining the properties of GOx with photosensitizers, enzymes, certain hypoxia-active prodrugs provide a good option for targeted drug delivery for cancer treatment (Fu et al., 2019). Zhang et al. (2018b) used the PEG-modified stealth liposomal for the delivery of GOx and AQ4N to tumors, thus developed an inventive cancer treatment approach. Zhang and co-worker prepared the GOx and TPZ bearing hyaluronic acid (HA) tethered calcium carbonate ($CaCO_3$) nanoparticles which show acidic decomposition. The GOx-motivated oxidation reaction results in the glucose cutoff in tumors along with the induction of tumorous abnormality which includes increased acidity and aggravated hypoxia. Amplification in acidic conditions leads to $CaCO_3$ decomposition, which offers spatial control over the embedded TPZ liberation within the tumor and temporal control on timely chemotherapy initiation. This helps in matching the hypoxia amplification and providing the synergistic benefit of the chemotherapy and starvation therapy (Zhang et al., 2018a).

Various other approaches have also been developed for the synthesis of hypoxia-activated prodrug/delivery system used in the treatment and diagnosis of cancer. Winn et al., prepared the hypoxia-targeted bioreductively activatable prodrug conjugated of phenstatin (anticancer potent inhibitor of tubulin polymerization). They synthesized nitroimidazole, nitrobenzyl, nitrothienyl, and nitrofuranyl prodrugs of phenstatin incorporating mono-methyl, nor-methyl, and gem-dimethyl variants which are attached to nitro compounds. The results suggested that in the hypoxia condition, the gem-dimethyl nitrofuran and gem-dimethyl nitrothiophene undergo enzymatic cleavage in the company of NADPH cytochrome P-450 oxidoreductase. The prodrug has better anticancer effect than parent phenstatin (Zhang et al., 2018a). Hypoxia-sensitive phosphorescent metal complexes transition is now budding as innovative tumor imaging moiety with high specificity and sensitivity. In cancer diagnosis, the hypoxia-targeted probes have been widely used, i.e., nitroididazole, PdII/PtII porphyrins, Cu(II)-diacetylabis (N4 methylithiosicarzone), etc. By keeping the center point of metal transition complexes, Zheng et al. (2015) prepared micelles based nanoprobe with thepoly(N-vinylpyrrolidone)-conjugated iridium (III) complex (Ir-PVP) and poly(e-caprolactone)-b-PVP, and assessed the hypoxia response performance in monolayer cell (human neuroblastoma SH-SY5Y) cultures *in vitro*. The cells treated with prepared micelles nanoprobe under hypoxic condition show intense phosphorescent emission signal whereas under normoxia conditions the signal was quite weak (Wang et al., 2015).

9.4.2 pH-Activated Prodrugs in Cancer Therapy

The tumor cells have a low pH of 6.2–6.8 in comparison to normal healthy cells (pH of 7.2–7.4) (Montcourrier et al., 1997). The major causes of the low pH ECM of the tumor are pH lower clearance of acid from hypoxic cells, over lactic acid production in ECM, Na^+-HCO_3^- co-transporter and Cl^-/HCO_3^- contribution, etc. (Denny, 2001; Kato et al., 2013). The prodrug also activated in endosomal in the low pH environment (Denny, 2004). HMR 1826 (N-[4-Glucuronyl-3-nitrobenzyl-oxycarbonyl] doxorubicin), nontoxic glucuronide prodrug of doxorubicin, is activated by β–glucuronidase in low pH condition (Mürdter et al., 2002). TP300 is a prodrug of CH0793076 (hexacyclic camptothecin analog) converted pH depended in active CH0793076 showed higher antitumor activity in various human xenograft models (Ohwada et al., 2009).

Ruthenium drugs are considered to be potent anticancer drugs but due to their less selectivity, the efforts have been made to synthesize the prodrug. Hufziger et al., synthesize the low pH-activated prodrug of ruthenium i.e., ruthenium dihydroxybipyridine complexes which activated in low pH and induced by blue light, by using pH-sensitive ligand i.e., 6,6′-dihydroxy-2,2′-bipyridine (66′bpy$(OH)_2$), to generate $[Ru(bpy)_2(66'(bpy(OH)_2)]^{2+}$ and found that the cytotoxic effect of Ru(bpy$)_2$(66′(bpy$(OH)_2)]^{2+}$ is more potent than the other ruthenium complexes (Hufziger et al., 2014). Similarly, Qu et al., synthesized the five different types of the type $([(N,N)_2Ru(PL)]^{2+,}$ where N,N is different compounds) ruthenium complexes Metallo-prodrugs by using photolabile ligands (4,4′-dimethyl-6,6′-dihydroxybipyridine) which activated in low pH condition and triggered by light. The cytotoxic study was performed at different breast cancer cells and found that the complex with N,N = 2,3-dihydro-[1,4]dioxino[2,3-f][1,10]phenanthroline has IC_{50}values as low as 4 μM with blue light (Qu et al., 2017).

The acidic pH of lysosomes (4–5) and endosomes (5–6) are found to be applicable and useful in development of pH-responsive drug delivery systems, which are designed and developed in such a way to only release the loaded drugs after endocytosis. The pH-responsive polymers may have carboxylic acid or amino groups which may be based on the protonation-induced change in polymer hydrophobicity. As the pH changes from 10 to 4, the diameter of PEG-b-poly(L-lysine)-b-poly(L-phenylalanine) (PEG-PLL-PLP) micelles increases from 15 to 60 nm, thus helping in the drug release. As the aqueous medium pH gets raised, the reversibly hydrogels possessing carboxylic acid groups swells.

The pH-responsive polymer-prodrugs are also developed for chemotherapeutic uses. The hydrazone linkage, linked for pH-responsive polymer-prodrug, may undergo hydrolysis under the low pH condition (Cheng et al., 2014). Due to the rapid hydrolysis rate in acidic pH, hydrazone linkage is often applied to the conjugation of drugs in the polymer backbone, intended for site-specific drug delivery, thus reducing systemic toxicity (Sonawane et al., 2017). This strategy of pH-responsive drug delivery system is of current interest for pharmaceutical scientists in the cancer chemotherapy. A variety of copolymer-drug conjugated like DOX-pullulan (Lu et al., 2009), N-(2-hydroxypropyl) methacrylamide (PHPMA)-DOX conjugates (Etrych et al., 2014), PHPMA pirarubicin conjugates (Nakamura et al., 2014), modified HA-DOX conjugates (Gurav et al., 2016), chitosan-hydrazone-mPEG nanoparticles with prednisone (Kumar et al., 2017), methoxy PEG and poly(β-benzyl-L-aspartate)-doxorubicin conjugates micelles

(Gao and Lo, 2018), etc., via the hydrazone bond, were prepared for the pH-responsive drug delivery. Wang et al., prepared the self-assembled polymer micelles prodrug, in which P(2-(methacryloyloxy)-ethyl phosphorylcholine)-b-P(2-methoxy-2-oxoethyl methacrylate) polymer was synthesized and doxorubicin were linked with polymer through a pH-responsive hydrazone bond. The *in vitro* drug release studies suggested that doxorubicin released more at pH 5.0 than pH 7.4 and the micellar system has great *in vitro* anticancer potency (Wang et al., 2013). Similarly, Yu et al., prepared the triple-layered siRNA and alkylated cisplatin prodrug (Pt(IV)-OC) loaded pH-responsive micelleplex by using the PEG-block-poly(aminolated glycidyl methacrylate)-block-poly(2-(diisopropyl amino) ethyl methacrylate) (PEG-b-PAGA-b-PDPA) triblock copolymers for targeting the NF-Kappa B of metastatic breast cancer treatment. The self-assembling properties of polymer and endosomal degradation of micelleplex inside cells are demonstrated in Figure 9.5 (Yu et al., 2016). Wan et al. (2018) prepared the pH-responsive polymeric prodrug nanoparticles developed by the conjugation of 6-TG (antimetabolite drug) to dialdehyde sodium alginate Schiff base linkage which are found to be applicable in the acute myelocytic leukemia chemotherapy.

Furthermore, Poly-L-histidine (polyHis) is used for the development of pH-responsive drug delivery system and used as core forming material of polymer micelles. The polyhisionization break the hydrophobic contact within the core and help in the release of drugs (Kim et al., 2008, 2009). Lee et al., prepared the polyHis-polyPEGdiblock copolymers and is needed for polymeric micelles development. The developed micelles are having the property to respond to local pH changes in body and taking advantage of the fact that most solid tumors have an acidic extracellular pH, applications are expected to be used for pH-sensitive micelles for solid tumor treatment (Lee et al., 2003). Lee et al., extended their work and prepared polymeric micelles flower-like assembly of PLA and polyHis blocks in the core and PEG block as the shell and PLA-*b*-PEG-*b*-polyHis micelles providing drug release from the micelles triggered through the small pH change, i.e., pH 7.2–6.5 (Lee et al., 2007). The copolymer polyHis-graft-PLL is also used in the delivery of gene by the using polyplex particles, which has electrostatic interactions of plasmid DNA. Lee et al., prepared the hybrid nanoparticles by using the chlorin e6 conjugated polyHis-HA copolymers for the diagnostic and therapeutic application in breast cancer. In the acidic media, the size of nanoparticles increases, indicating that polymer

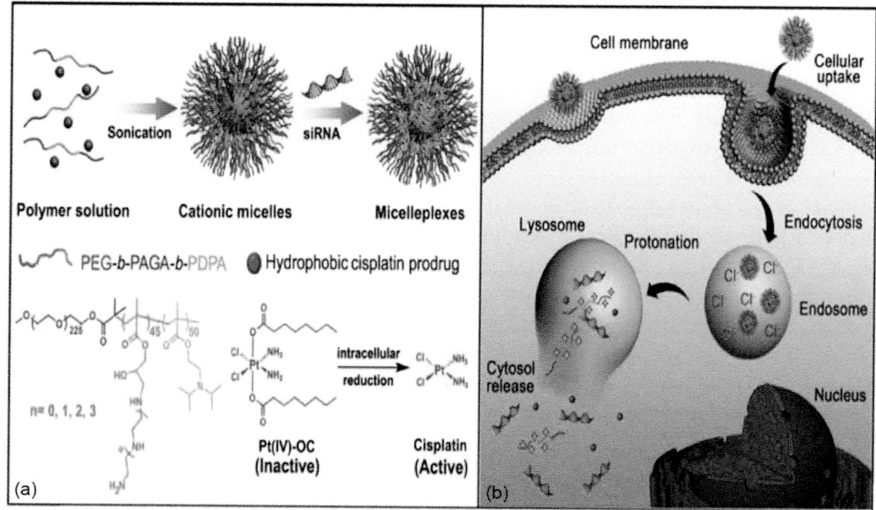

FIGURE 9.5 Schematic illustration for the preparation and intracellular activation of siRNA and cisplatin prodrug co-loaded pH-responsive micelleplexes: (a) Pt(IV)-OC was encapsulated in the hydrophobic core and siRNA was loaded on the intermediate layer of the micelleplexes; (b) Intracellular acidic pH-induced activation of micelleplexes. The micelles were dissociated inside acidic late endosome/lysosome due to protonation of the PDPA core. Pt(IV)-OC prodrug and siRNA were then simultaneously released into cytosol due to the strong hemolytic activity of the cationic triblock copolymers. Reused under creative common licenses. (From Yu, H. et al., *Theranostics*, 6, 14–27, 2016.)

is acid-sensitive and chlorin e6 at acid environment had redox status and fluorescence emitted used in diagnostic purpose (Lee and Jeong, 2018). The various studies showed that the acid-sensitive polymers are using for the preparation of different types of targeted delivery carriers for the treatment and diagnosis of cancer (Li et al., 2018; Sim et al., 2018; Yang et al., 2018).

9.4.3 Enzyme-Activated Prodrugs in Cancer Therapy

There are two approaches to develop the enzyme-activating prodrug therapy, i.e., drug activating enzymes either being present/expressed in that particular tumor or another is nontoxic prodrug along with the exogenous enzyme that is now expressed in the tumor, is systematically administered (Xu and McLeod, 2001).

The β-galactosidase is a lysosomal enzyme which is overstimulated in various cancers, i.e., liver (Kim et al., 2017), ovarian cancer (Bhattacharya and Barlow, 1979), colon cancer (Sharma et al., 2018), etc. The β-galactosidase enzymes stimulated prodrug is developed for the colon cancer treatment (Sharma et al., 2018). β-galactosidase enzymes also used to activate the prodrug which prove the chemiluminescence for the diagnostic purpose (Gnaim et al., 2018). Another enzyme, β-glucuronidase, is found in higher concentration in the microenvironment of solid tumors. β-glucuronidase provide a activation tool for the prodrug activation in cancer chemotherapy. Renoux et al., prepared the β-glucuronidase enzyme responsive prodrug of with monomethylauristatin E linked with glucuronide triggers and self-immolative linkers, to deliver in the TME of solid tumors. After administration by intravenous route, the prodrug binds covalently to plasmatic albumin enabling it to accumulate inside tumor cells and drug release is initiated by extracellular β-glucuronidase. The two-step activation are involved in this delivery system (Figure 9.6) (Renoux et al., 2017). Similarly, Compain et al. (2018) synthesized β-glucuronidase responsive albumin binding prodrug for inhibition of protein kinase and prodrug first bind with human serum albumin and subsequently released upon the enzymatic activation, provided the cytotoxic effect in solid tumor.

Cathepsins (Cat), best studied lysosomal hydrolases, found to increase activity and localized inside TME which lead to cancer progression, proliferation, invasion and metastasis (Haider et al., 2019; Olson and Joyce, 2015). Cat-B is a cysteine protease found abundantly in normal cells/tissues. It is one of the important intracellular proteases which is highly upregulated in malignant tumors and premalignant lesions at the level of mRNA and protein (Olson and Joyce, 2015; Podgorski and Sloane, 2003). Cat-B cleaves prodrug which is attached with linkers with Leu, Arg-Arg, Ala-Leu, Phe-Arg, Phe-Lys, Ala-Phe-Lys, Gly-Leu-Phe-Gly, Gly-Phe-Leu-Gly, and Ala-Leu-Ala-Leu between drugs and substrats (Zhong et al., 2013). Shim et al., prepared the doxorubicin Cat-B activating prodrug by using Phe-Lys, a Cat-B specific dipeptide, and para aminobenzyloxycarbonyl for cancer treatment and found that lower dose–dependent inhibitory effect on SGC-7901 cells (Shim et al., 2019). Similarly, Karnthaler-Benbakka et al. (2019) fabricated the Cat-B-Cleavable prodrug of the VEGFR Inhibitor Sunitinib by using Cat-B using peptide and self-immolative linker. The nanochemotherapeutic carriers are also using for the specific and selective targeting for the cancer treatment on Cat-B mechanistic approaches. Gotov et al. (2018) prepared docetaxel bearing hyaluronic acid-Cat-B-cleavable-peptide–AuNPs for the treatment of cancer. The system delivers the docetaxel in selective manner into the tumor to enhance its therapeutic effect.

The glutathione transferases (GSTs) overexpression in cancer cells provides unique opportunities for prodrug therapy. The GSTs overexpression on cancer cells provides distinctive possibilities for prodrug therapy. In reality, activation of prodrugs in GST overexpressing cells can lead to higher concentrations of an activated drug as compare to normal cells which having moderate enzyme intensity (Allocati et al., 2018). Johansson et al. (2011) synthesized the 2,4-dinitrobenzenesulfonyl DOX (DNS-DOX) and 4-mononitrobenzenesulfonyl DOX (MNS-DOX), are the GST-pi and microsomal GST1 sensitive prodrug which overcome the drug in doxorubicin resistant MCF7 cell line and improved the cytotoxicity. Ling et al., prepared the glutathione responsive Pt (IV) prodrug and constructed self-assembled nanoparticles (Ling et al., 2018).

The different endogenous enzymes are also utilized for prodrug activation. Various endogenous enzyme-activated prodrug had prepared during the decade, i.e., NADPH cytochrome P450 oxidoreductase activate the prodrug (Zhang et al., 2018a), FAPα-mediated activation (Sun et al., 2019),

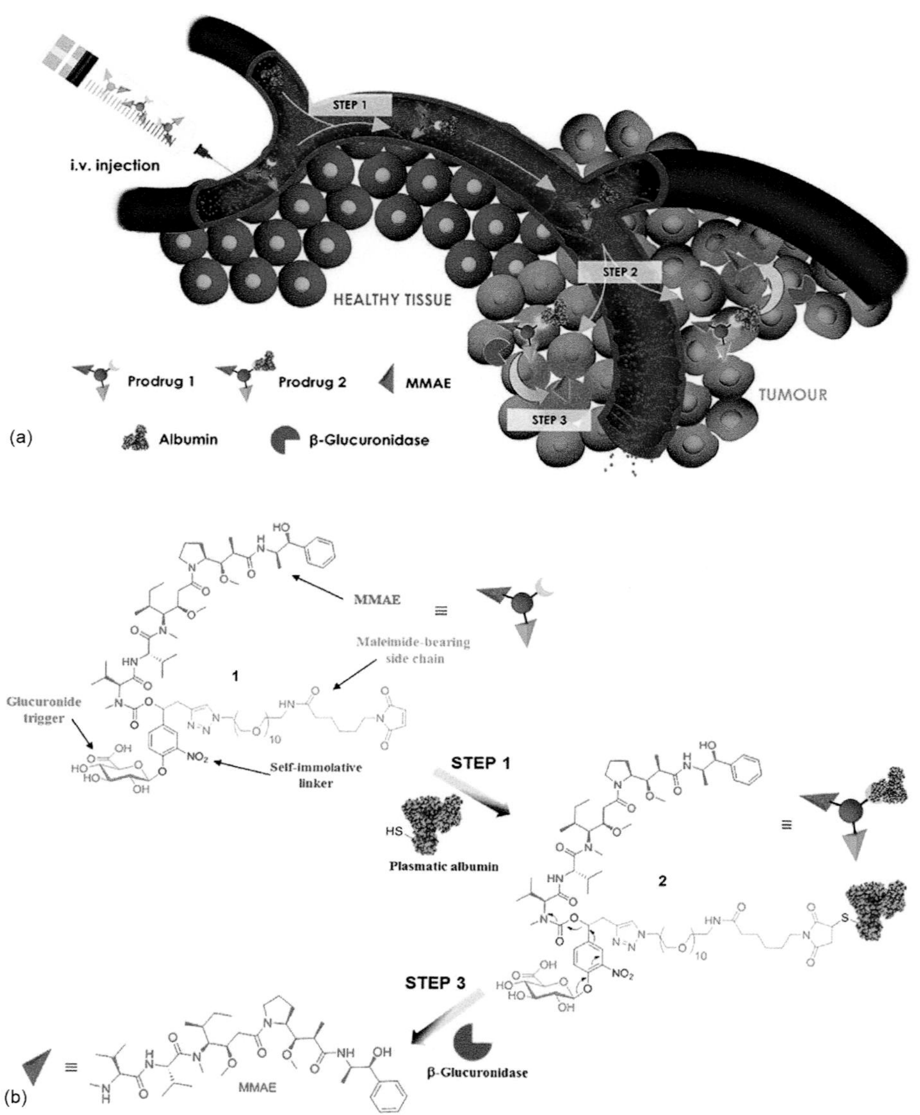

FIGURE 9.6 The principle of tumor targeting with the β-glucuronidase-responsive drug delivery system 1. (a) In the blood, prodrug 1 binds to circulating albumin (step 1). The resulting macromolecule 2 accumulates passively in malignant tissues (step 2) where the cleavage of the glucuronide by extracellular β-glucuronidase triggers the release of MMAE (step 3). (b) The maleimide-bearing side chain of prodrug 1 reacts with the thiol at the cysteine 34 position of albumin through Michael addition (step 1). Hydrolysis of the glucuronide trigger by β-glucuronidase induces the release of MMAE via a 1,6-elimination mechanism followed by a spontaneous decarboxylation (step 3). Reused under creative common licenses. (From Renoux, B. et al., *Chem. Sci.*, 8, 3427–3433, 2017.)

Aminopeptidase-N-activated theranostic prodrug (Xiao et al., 2018), DT-diaphorase activated prodrug (Chen et al., 2018), glucarpidases activated prodrug (Rashidi et al., 2018).

Chemotherapy exoenzymes used for the prodrug exogenously during the administration. The exoenzymes cytosine deaminase used for the activation of PEG-click nucleic acids-PLGA nanoparticles (Harguindey et al., 2019). A prodrug of 5-flourouracil bioactivated as active drug by tumor-associated enzyme thymidine phosphorylase and uridine phosphorylase administered exogenously (Singh et al., 2008).

9.5 Macromolecular Prodrugs

Prodrugs are the agents which get transformed on the administration, either through the metabolism or spontaneous chemical conversion, to convert into a pharmacologically active moiety (Denny, 2001).

9.5.1 Antibody-Drug Conjugates (ADCs)

ADCs are novel, an emerging, an effective and immunotherapy-based approach for cancer or solid tumor targeting which deliver cytotoxic drug in antigen positive tumor or cancer cells, receptor mediated endocytosis are required to activate ADCs. Monoclonal antibodies, linker and cytotoxin are three essential elements for the development of antibody-drug conjugates (Dong et al., 2019; Diamantis and Banerji 2016). Monoclonal antibodies specific for cell-surface antigens are conjugated to highly potent cytotoxins through various linkers (Perez et al., 2014). This complex reduces the toxicity as well as expanded the therapeutic window of anticancer drugs (Nittoli et al., 2018).

In the late 1950s, ADCs were evaluated for the first time by using antibodies which addressed to leukemia cells and conjugated with the anticancer drug methotrexate (Johnston and Scott, 2018). Currently, the US FDA (Food and Drug Administration) has approved four types of ADCs are available in market. Pfizer developed Mylotarg (Gemtuzumabozogamicin), was approved as first ADCs by the US FDA in 2000, to treat acute myeloid leukemia (AML), it targeted $CD33^+$ carried a calicheamicin (DNA fragmenting payload) (Leal et al., 2014), due to fatal hepatotoxicity caused by Mylotarg, it was withdrawn from the marketin 2010 and reapproved after seven years in 2017 when result of new clinical trials was in favor to Mylotarg and prove that Mylotarg has a favorable clinical risk-benefit ratio in the treatment for newly-diagnosed patients with $CD33^+$ AML, in 2010 FDA-approved Brentuximabvedotin (Adcetris), developed by Seattle Genetics for the treatment of anaplastic large-cell lymphoma (ALCL) and Hodgkin's lymphoma (HL), Brentuximabvedotin target to $CD30^+$ (Goli et al., 2018). Two more ADCs was approved by FDA in 2013 and 2017, respectively, ado-trastuzumabentansine (Kadcyla) target to HER-2^+ in the patients of breast cancer (Richardson et al., 2018) and Inotuzumabozogamicin (Besponsa), developed by Pfizer target to $CD22^+$ adult B-cell precursor in the patient with acute lymphoblastic leukemia (ALL) (Figure 9.7).

9.5.1.1 ADC Linkers

Drug molecules (toxophores) attached to antibody through linker in the ADCs. It provides stability to ADCs in systemic circulation. Linkers mainly divided into cleavable or noncleavable category. Noncleavable linkers release drug into lysosomes by the proteolytic catabolism of antibody. While cleavable linkers are stable in systemic circulation and are able to release drug within the antigen expressing

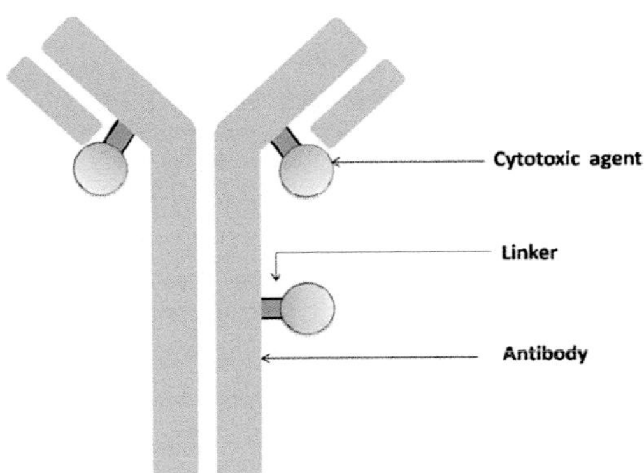

FIGURE 9.7 Antibody drug conjugate.

TABLE 9.1
Classification of Linkers Employed in ADCs

Linker Type		Examples	ADCs
Cleavable linker	Acid-sensitive	Hydrozone linker	Gematuzumab ozogamicin ADC
	Glutathione sensitive	Disulfide linker	Lorvotuzumab mertansine ADC
	Protease sensitive	Valine-citrulline (vc) dipeptide linker	Brentuximabvedotin ADC
Noncleavable linker		Maleimidocaproyl (mc) or thioether linkers	Ado trastuzumab emtansine (T-DM1) ADC

cells, due to the breakdown of connection between cytotoxins and antibodies, by the action of acidic intercellular pH of cancerous cell, proteolytic enzymes of lysosomes or reduction of disulfide bond by glutathione (Richardson et al., 2018). The various types of linkers employed in ADCs are listed in Table 9.1.

Mechanism for drug release from ADCs: ADCs binds with specific overexpressed antigen at the cell surface according to mAb when it comes into contact to the cancer cell, then internalized via endocytosis. Endocytic vesicle fussed with lysosomes and released drug in cytoplasm due to degradation of ADCs by lysosomal enzymes, free intercellular drug moiety bind with targets and causes cell death (Figure 9.8) (Shefet-Carasso and Benhar, 2015).

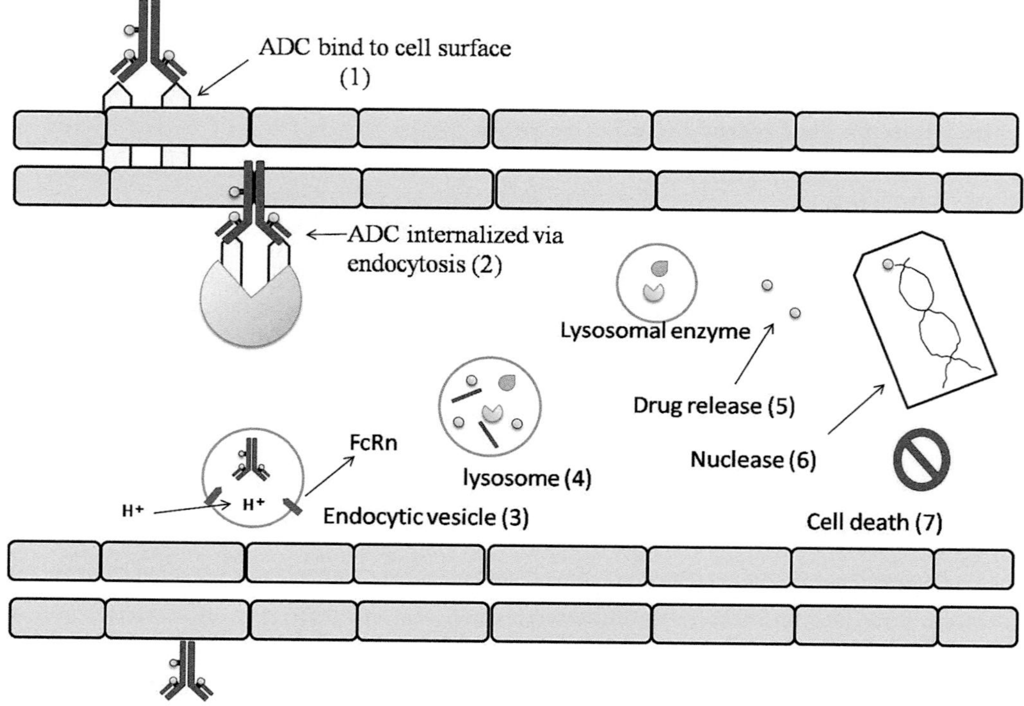

FIGURE 9.8 Mechanism of release by ADCs in cancer cell, (1) ADC binds to the cell surface using mAb on cell surface, (2) ADC internalized via endocytosis and cross the cell surface, (3) ADC Formed Endocytic vesicles, (4) Endocytic vesicles fused with lysosomes, (5) Drug release from lysosomes, (6) Drug enters into nuclease and bind with DNA, (7) Interaction between drug and DNA cause cell death.

9.5.1.2 *Appropriate Target for ADCs*

Various types of antigen receptors or specific proteins are overexpressed in the different types of cancer or solid tumors, which is the unique targets for the cancer treatment, identification and validation of these antigen targets are essential for the development of efficient ADCs. These antigens overexpressed in different part of the tumor i.e., ECM, stroma, vasculature, glycoproteins and gangliosides on the surface of tumor cell (Teicher, 2009; Vater and Goldmacher, 2010). Antigen which overexpressed in different part of cancer and solid tumor are show in Figures 9.9 and 9.10.

9.5.2 Polymer Drug Conjugate

A polymer-drug conjugate (PDCs) was designed by Helmut Ringsdorf (1975) to overcome the curbs of traditional chemotherapy, in which drug molecules covalently connected to the hydrophilic polymer (back bone as vehicle) via linkages that are specially designed to liberate drug at specific site or within structures.

Advantages of PDCs:

1. Enhance the drug targeting; reduce the dose size, provided sustained release profile and helpful in developing a multifunctional drug delivery system (Chandna et al., 2010).
2. To enhance drug solubility, in-vivo stability, pharmacokinetics properties and protect drug molecules against deactivation (Luo et al., 2012; Pang et al., 2014).

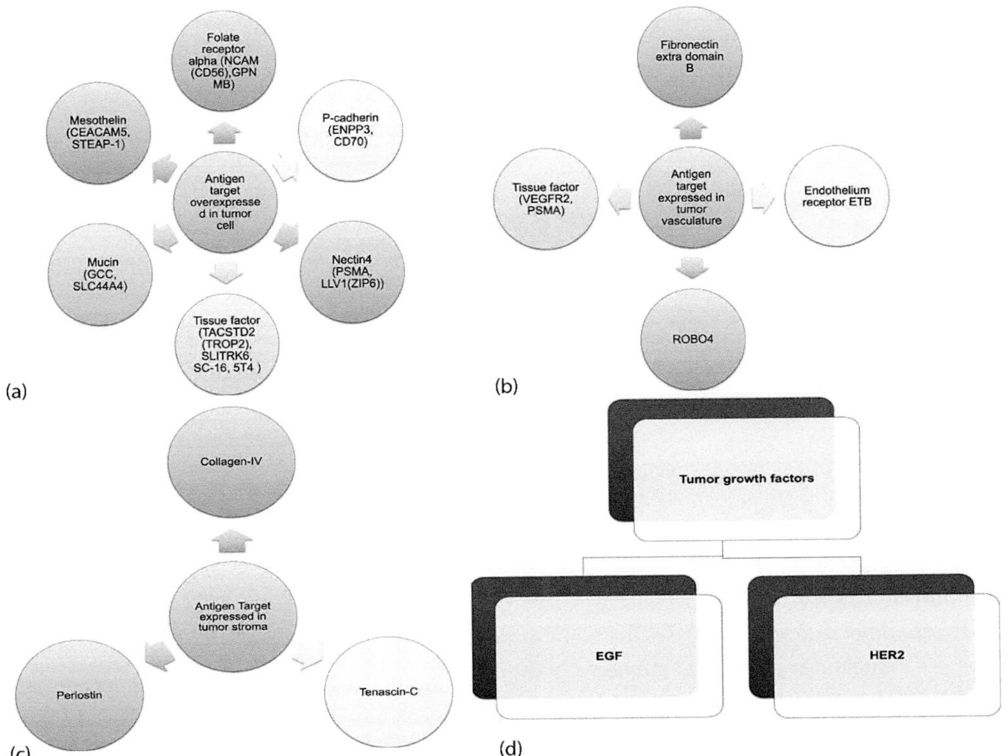

FIGURE 9.9 Antigen target in solid tumor: (a) Antigen target over-expressed in tumor cell, (b) antigen target expressed in tumor vasculature, (c) antigen target expressed in tumor stroma, and (d) tumor growth factors.

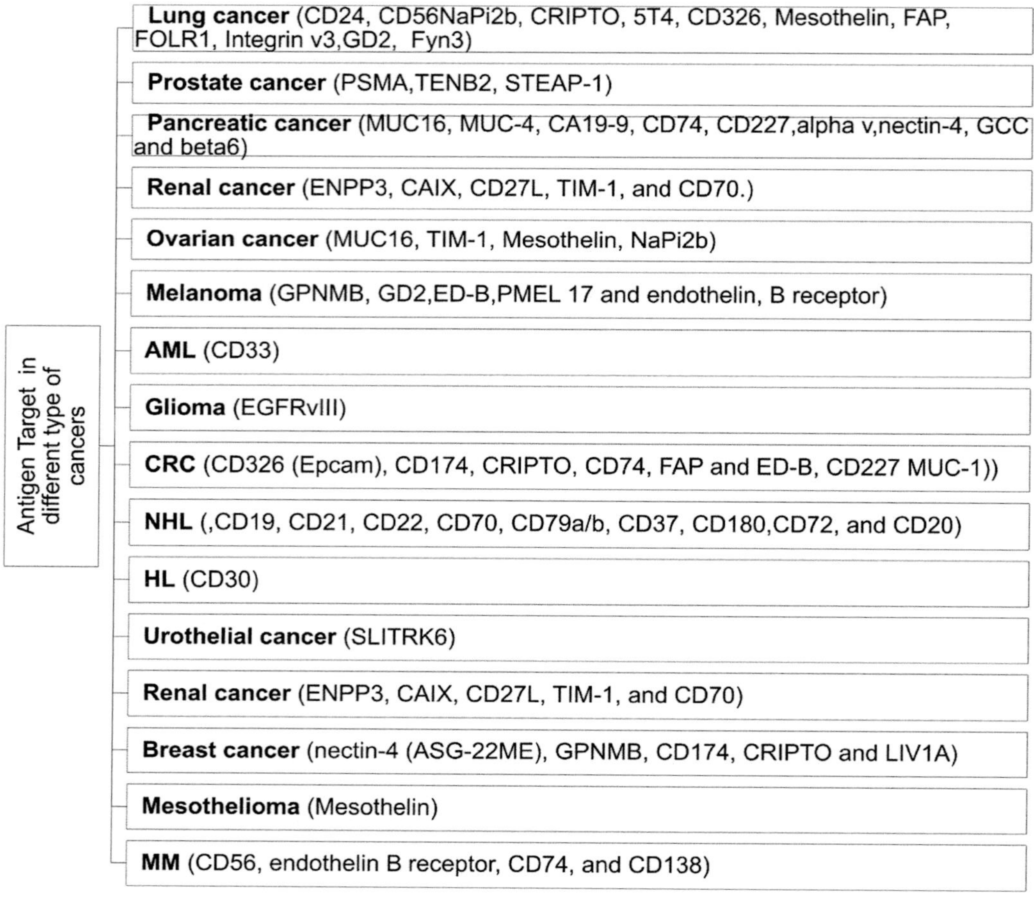

FIGURE 9.10 Antigen target in different types of cancer.

3. Enhanced therapeutic efficacy by prolonged plasma exposure, plasma half-life of drug and accumulation of drug in tumor tissue using EPR effect as well as reduced local and systemic side effects (Chau et al., 2006).
4. Polymer drug conjugation enhances the drug loading capacity of nanocarrier systems (Park et al., 2006).
5. PDCs is also a unique system which overcome the "burst drug release" problems associated with other polymeric drug carriers (Liposomes, ethosomes, niosomes etc.) in which drug is physically encapsulated by polymer (Allocati et al., 2018).

Methods for PDCs:

There are three different strategies for synthesize polymer-drug conjugates

1. Conjugating synthesized polymers to the drug molecules,
2. Conjugating a monomer to the drug, then polymerized by using reversible addition–fragmentation transfer polymerization, ring-opening metathesis polymerization or ring-opening polymerization.
3. Drug molecule having multifunctional groups Use as a monomer for poly-drug polymerization (Feng and Tong, 2016; Zelikin et al., 2016).

9.5.2.1 Polymers in PDCs

9.5.2.1.1 N-(2-hydroxypropyl) Methacrylamide (HPMA) Copolymer-Drug Conjugates

HPMA is a synthetic nonimmunogenic, nontoxic, and water-soluble copolymer, widely used in PDCs because it is not absorbed by plasma proteins and having extended period of systemic circulation. 20 to 30 kDa HPMA copolymers with the acid labile linkers or peptide cleavable linker are suitable for the synthesis of PDCs because HPMA-drug conjugates less than 40 kDA molecular weight can easily accumulates in tumor tissues by passive targeting and eliminated from the body through renal filtration. HPMA has also immune-modulatory effects, it inhibits the expression of Fas ligand by modifying cell-surface properties of tumor cells, Fas actively involved in the neutralization of leukocytes which attack or kill cancer cells (Říhová and Kovář, 2010). Immuno-modulatory effect of HPMA encourage its use in the development of PDCs, because immune system plays a essential role in the complete eradication of cancer cells. Thus, HPMA conjugation with anticancer drug provides dual advantage (Kovář et al., 2003).

Various HPMA-anticancer drug conjugates have been developed and entered in the clinical trial which have provided the promising results and safety of HPMC copolymer as a new platform for chemotherapy (Duncan and Vicent, 2010). After that PDCs of other cytotoxic drugs like camptothecin (CPT), paclitaxel (PTX), and dichloro(1,2-diaminocyclohexane)platinum(II) (DACHP) was also developed. All conjugates showed tolerable toxicity and bioavailability in phase I and II clinical trials (Vogus et al., 2017).

9.5.2.1.2 Polysaccharide-Drug Conjugates

Polysaccharides are water-soluble, biodegradable and biocompatible polymer; it is one the best choice for the synthesis of PDCs due to having many functional groups for conjugation with drug molecule. Pharmacokinetic property of polysaccharides depends upon its molecular weight, electrical charge, branching and polydispersity (Mehvar, 2003).

There are various types of naturally or synthetic water-soluble polysaccharides such as hyaluronic acid, dextran, chitosan, and heparin which have been exploited for development of PDCs of water-insoluble or hydrophobic drugs.

Hyaluronic acid: It is an anionic high molecular weight (>250 kDa) polysaccharides composed by (1→4) interglycosidic linkages between alternating units of N-acetyl-d-glucosamine and d-glucuronic acid, having anti-inflammatory effects. Carboxylic and hydroxyl groups at hyaluronate backbone provides specific sites for drug conjugation, it has also property to form hydrogel by absorbing water and great affinity to bind with CD44 and RHAMM (cell-surface markers). CD44 is specific cell-surface protein manly overexpressed at various types of cancer cells.

Dextran: Dextran is macromolecule, composed by repeating unit of glucose, was used as plasma expander. Primary and secondary hydroxyl groups of glucose are functional and active site for conjugation with drug molecules.

Chitosan: It is de-acetylated derivative of natural polysaccharide chitin; amino moieties of chitosan backbone is responsible for polymer drug conjugations. Chitosan has mucoadhesive properties which enhance the retention time of conjugated drugs in the respiratory mucosa (Yamamoto et al., 2005).

Heparin: It is an anticoagulant having the property to deactivates heparanase, prevents cancer cell adhesion, activates NK cells to attack and prevents tumor angiogenesis and metastasis by inhibits the activity of growth factors such as bFGF and VEGF. Heparin is the combination of 2-amino-deoxyglucopyranose (glucosamine) and pyranosyluronic acid residues (Li et al., 2012). Heparin–drug conjugate offers better solubility, long systemic circulation as well as more cytotoxicity than free drug. Heparin–drug conjugation reduces the anticoagulant property of heparin which decreases the risk of hemorrhagic complications in patients. Heparin-PTX conjugates was developed which exhibited more water solubility and cytotoxicity than free drug (Goodarzi et al., 2013).

9.5.2.1.3 PEG-Drug Conjugates

PEG is a hydrophilic FDA-approved polysaccharide for biological applications (Knop et al., 2010). "PEGylation" is an approach proposed by Davis and Abuchowski in the 1970s (Pelegri-O'Day et al., 2014), in which drug or protein molecules are conjugated with PEG to enhance the therapeutic index or plasma half-life by reducing reticulo-endothelial system (RES) uptake.

"PEGylation" also promote passive-targeting, EPR (enhanced permeability and retention) effects (Ajazuddin et al., 2013) and reduce the host cytotoxicity. "PEG-amino-acid-oligopeptide-irinotecan" drug conjugate was developed and patented by Xu et al. (2011), during the *in-vivo* study using human ovarian cancer cells (SKOV-3) in nude mice they founded that this conjugate inhibits the growth of tumor cells at low dose (45 mg/kg) and reduce host cytotoxicity of irrinotecan (Khan et al., 2018; Xu et al., 2014). "PEGylation" or PEG-drug conjugation improves the solubility of drug as well as drug carrying capacity of nanocarrier or nanomaterial. Solubility of anticancer drug SN38 was enhanced by preparation of micelle of PEG-drug conjugation (Kurzrock et al., 2012).

9.5.2.1.4 PG-Drug Conjugates

Poly(α,L-glutamic acid) (PG) is negatively charged, highly water-soluble naturally polypeptide of L-glutamic acid. PG degrades by lysosomal enzyme and mostly cleared by renal pathway so PG is an excellent polymeric carrier of therapeutic agents. Carboxylic group at glutamate backbone is responsible for PG-drug Conjugation. PG- drug conjugation enhances the solubility of poorly water-soluble drug.

Solubility of PTX (Taxol®, TXL) was improved by developing PG-TXL conjugate, in which PTX was conjugated via an ester linkage through 2′-hydroxyl group of poly(L-glutamic acid) (PG). Result of preclinical studies revealed that this conjugate has high water solubility, better antitumor efficacy and was more safe as compared PTX (Li et al., 1998).

Camptothecins is a broad-spectrum anticancer drug inhibit topoisomerase-1 activity, but having limited therapeutic efficacy due to poor aqueous solubility and in instability of the lactone ring in systemic fluid, for the enhance stability and solubility PG-Camptothecins conjugate was developed by coupling the carboxylic acid of PG with hydroxy group of camptothecins. This conjugation also stuck the growth of growth of established H322 human lung tumors grown subcutaneously in nude mice when administered intravenously (Zou et al., 2001).

9.5.3 Others

9.5.3.1 Prodrug with Other Therapeutic Agents

The prodrug nanoparticles can also be used for delivery of anticancer drugs along with the other therapeutic agents for enhancing chemotherapeutic effects, this combination also shows synergists effects of both (Luo et al., 2014). For example, micelles of doxorubicin and disulfiram was developed by employing polymer poly (styrene-co-maleic anhydride) and acid cleavable linker adipic-di-hydrazide for the treatment of MDR cancer, disulfiram was encapsulated in the DOX conjugate polymer (SMA-ADH-DOX), this combination enhances the accumulation of doxorubicin in tumor by fast release of DSF, DSF reduce the pumping out of the chemotherapeutic drug from the tumor (Duan et al., 2013).

9.5.3.2 Prodrug with Nucleic Acid

Combination of anticancer drug with nucleic acid like plasmid DNA, small interfering RNA, and antisense oligonucleotides reduce the side effects as well as drug resistance of anticancer drug and enhance the efficacy of cytotoxic drug by targeting the different pathways of disease (Li et al., 2013). Cationic polymer or lipids are used for the development of polymer-drug-nucleic acid complex. Cationic polymer-based drug delivery system cyclodextrin-polyethylenimine-doxorubicin conjugates were developed for the gene delivery, which deliver both p53 plasmid and doxorubicin. Tumor suppressor gene p53 induces apoptosis as well as increases the sensitivity of tumor cells to anticancer drugs through the P-gp inhibition (Lu et al., 2011).

9.5.3.3 Combination Chemotherapy and Phototherapy

Photodynamic therapy is a new approach for the cancer treatment where light-sensitive agents are used. They produce a reactive oxygen species in the presence of specific wavelength cause cell necrosis or apoptosis. The combination therapy of chemotherapy and phototherapy has synergistic therapeutic effects (Ge et al., 2016). For example, a prodrug nanoparticle-based chemo-photodynamic combination TCAD@Ce6 nanoparticles was developed in which photosensitizer chlorine e6 (Ce6) and acid-sensitive cis-aconitic anhydride modified doxorubicin was encapsulated in prodrug nanoparticle. System was easily hydrolyzed in TME and enhance cellular uptake of Ce6 and doxorubicin, Ce6 induce inhibition of A549 tumor in the presence near-infrared light irradiation (Hou et al., 2016).

9.6 Summary and Future Perspectives

Prodrug-based nano-drug delivery systems are the most prospective area of research for treatment of cancer, having the advantages of both nanotechnology and prodrug, in which nontoxic/cytotoxic drugs, enzymes and genes or nucleic acid can be efficiently delivered to the cancer cells or specific target sites. Prodrug served as the effective drug delivery platforms which have distinct advantages like innate tumor targeting, good biocompatibility, facilitated cellular uptake and improved drug availability; it is also provide stability, multiple payloads of therapeutic agents, triggered drug release in the target cells and codelivery of hydrophobic and hydrophilic drug. Prodrug based delivery systems utilize different types of tissue microenvironments like pH, enzymes, temperature, hypoxic condition and over expressed various receptors or antigen in cancer cells for the effective and site-specific delivery of active moiety, which enhance the efficacy and reduce the site effects of toxophores on normal cells.

Recent knowledge in multiple disciplines such as genomics, proteomics, cancer molecular biology, tumor chemistry and immunology, novel disease markers and associated mechanisms suggesting the combinatorial approaches to be the most hopeful strategies to control and treat multifactorial pathologies like cancer (Scomparin et al., 2017). Various types of prodrug like polymer drug conjugates, polymer antibody conjugates are in clinical trials, more than 25 prodrug-based polymer-drug conjugation have been successfully approved for human use (Luque-Michel et al., 2017), in which some of the conjugation like Brentuximabvedotin and ado-trastuzumabemtansine was successful reached in market after the approval by FDA for the treatment of certain cancers (Zhang et al., 2017). FDA was approved Doxil® in 1995 as first nanomedicine for the clinical use, drug (doxorubicin) was encapsulated in PEGylated liposomes which enhanced delivery of a drug to the cancer cell by passive targeting (extravasation-dependent) (Gabizon et al., 2003). Recently a new combination of cytarabin and daunorubicin encapsulated in liposome was approved by FDA for the treatment of acute myeloid leukemia (Gondi and Rao, 2013). Looking at the various examples presented in this chapter it can be claimed that prodrug-based drug delivery strategies possess great potential for improved therapy of cancer.

ACKNOWLEDGMENTS

The VP and TH would like to thanks the Indian Council of Medical Research (ICMR), New Delhi for providing the SRF funding for research work (grant no. for VP is 45/01/2018-NAN/BMS and for TH is 45/07/2018-NAN/BMS).

REFERENCES

Ajazuddin, A. Alexander, B. Amarji, and P. Kanaujia. 2013. Synthesis, characterization and in vitro studies of pegylated melphalan conjugates. *Drug Dev Ind Pharm* 39 (7):1053–1062.

Alimoradi, H., S. S. Matikonda, A. B. Gamble, G. I. Giles, and K. Greish. 2016. Hypoxia responsive drug delivery systems in tumor therapy. *Curr Pharm Des* 22 (19):2808–2820.

Allocati, N., M. Masulli, C. D. Ilio, and L. Federici. 2018. Glutathione transferases: Substrates, inhibitors and pro-drugs in cancer and neurodegenerative diseases. *Oncogenesis* 7 (1):8.

Bao, Y. Y., E. Guegain, J. Mougin, and J. Nicolas. 2018. Self-stabilized, hydrophobic or PEGylated paclitaxel polymer prodrug nanoparticles for cancer therapy. *Polym Chem* 9 (6):687–698.

Bhattacharya, M., and J. J. Barlow. 1979. Tumor markers for ovarian cancer. *Int Adv Surg Oncol* 2:155–176.

Brizel, D. M., T. Schroeder, R. L. Scher, et al. 2001. Elevated tumor lactate concentrations predict for an increased risk of metastases in head-and-neck cancer. *Int J Radiat Oncol Biol Phys* 51 (2):349–353.

Chandna, P., J. J. Khandare, E. Ber, L. Rodriguez-Rodriguez, and T. Minko. 2010. Multifunctional tumor-targeted polymer-peptide-drug delivery system for treatment of primary and metastatic cancers. *Pharm Res* 27 (11):2296–2306.

Chau, Y., R. F. Padera, N. M. Dang, and R. Langer. 2006. Antitumor efficacy of a novel polymer–peptide–drug conjugate in human tumor xenograft models. *Int J Cancer* 118 (6):1519–1526.

Chen, Z., B. Li, X. Xie, F. Zeng, and S. Wu. 2018. A sequential enzyme-activated and light-triggered pro-prodrug nanosystem for cancer detection and therapy. *J Mater Chem B* 6 (17):2547–2556.

Cheng, W., L. Gu, W. Ren, and Y. Liu. 2014. Stimuli-responsive polymers for anti-cancer drug delivery. *Mater Sci Eng C Mater Biol Appl* 45:600–608.

Choi, K. Y., M. Swierczewska, S. Lee, and X. Chen. 2012. Protease-activated drug development. *Theranostics* 2 (2):156.

Chouaib, S., Y. Messai, S. Couve, B. Escudier, M. Hasmim, and M. Z. Noman. 2012. Hypoxia promotes tumor growth in linking angiogenesis to immune escape. *Front Immunol* 3:21.

Compain, G., N. Oumata, J. Clarhaut, et al. 2018. A β-glucuronidase-responsive albumin-binding prodrug for potential selective kinase inhibitor-based cancer chemotherapy. *Eur J Med Chem* 158:1–6.

Connors, T. A., and R. J. Knox. 1995. Prodrugs in cancer chemotherapy. *Stem Cells* 13:501–511.

Connors, T. A., and M. E. Whisson. 1966. Cure of mice bearing advanced plasma cell tumours with aniline mustard: The relationship between glucuronidase activity and tumour sensitivity. *Nature* 210 (5038):866.

Delplace, V., P. Couvreur, and J. Nicolas. 2014. Recent trends in the design of anticancer polymer prodrug nanocarriers. *Polym Chem* 5 (5):1529–1544.

Denny, W. A. 2001. Prodrug strategies in cancer therapy. *Eur J Med Chem* 36 (7–8):577–595.

Denny, W. A. 2004. Tumor-activated prodrugs—a new approach to cancer therapy. *Cancer Invest* 22 (4):604–619.

Diamantis, N., and U. Banerji. 2016. Antibody-drug conjugates—an emerging class of cancer treatment. *Br J Cancer* 114 (4):362.

Dickson, B. D., W. W. Wong, W. R. Wilson, and M. P. Hay. 2019. Studies towards hypoxia-activated prodrugs of PARP inhibitors. *Molecules* 24 (8):1559.

Dong, W., J. Shi, T. Yuan, et al. 2019. Antibody-drug conjugates of 7-ethyl-10-hydroxycamptothecin: Sacituzumab govitecan and labetuzumab govitecan. *Eur J Med Chem* 167:583–593.

Duan, X., J. Xiao, Q. Yin, et al. 2013. Smart pH-sensitive and temporal-controlled polymeric micelles for effective combination therapy of doxorubicin and disulfiram. *ACS Nano* 7 (7):5858–5869.

Duncan, R., and M. J. Vicent. 2010. Do HPMA copolymer conjugates have a future as clinically useful nanomedicines? A critical overview of current status and future opportunities. *Adv Drug Deliv Rev* 62 (2):272–282.

Etrych, T., V. Subr, R. Laga, B. Rihova, and K. Ulbrich. 2014. Polymer conjugates of doxorubicin bound through an amide and hydrazone bond: Impact of the carrier structure onto synergistic action in the treatment of solid tumours. *Eur J Pharm Sci* 58:1–12.

Feng, Q., and R. Tong. 2016. Anticancer nanoparticulate polymer-drug conjugate. *Bioeng Transl Med* 1 (3):277–296.

Fu, L. H., C. Qi, Y. R. Hu, J. Lin, and P. Huang. 2019. Glucose oxidase-instructed multimodal synergistic cancer therapy. *Adv Mater* e1808325.

Gabizon, A., H. Shmeeda, and Y. Barenholz. 2003. Pharmacokinetics of pegylated liposomal Doxorubicin: Review of animal and human studies. *Clin Pharmacokinet* 42 (5):419–436.

Gao, D., and P. C. Lo. 2018. Polymeric micelles encapsulating pH-responsive doxorubicin prodrug and glutathione-activated zinc(II) phthalocyanine for combined chemotherapy and photodynamic therapy. *J Control Release* 282:46–61.

Ge, Y., Y. Ma, and L. Li. 2016. The application of prodrug-based nano-drug delivery strategy in cancer combination therapy. *Colloids Surface B* 146:482–489.

Giang, I., E. L. Boland, and G. M. Poon. 2014a. Prodrug applications for targeted cancer therapy. *AAPS J* 16 (5):899–913.

Giang, I., E. L. Boland, and G. M. K. Poon. 2014b. Prodrug applications for targeted cancer therapy. *AAPS J* 16 (5):899–913.

Gnaim, S., A. Scomparin, S. Das, R. Blau, R. Satchi-Fainaro, and D. Shabat. 2018. Direct real-time monitoring of prodrug activation by chemiluminescence. *Angew Chem Int Ed Engl* 57 (29):9033–9037.

Goli, N., P. K. Bolla, and V. Talla. 2018. Antibody-drug conjugates (ADCs): Potent biopharmaceuticals to target solid and hematological cancers-an overview. *J Drug Deliv Sci Technol* 48:106–117.

Gondi, C. S., and J. S. Rao. 2013. Cathepsin B as a cancer target. *Expert Opin Ther Tar* 17 (3):281–291.

Goodarzi, N., R. Varshochian, G. Kamalinia, F. Atyabi, and R. Dinarvand. 2013. A review of polysaccharide cytotoxic drug conjugates for cancer therapy. *Carbohydr Polym* 92 (2):1280–1293.

Gotov, O., G. Battogtokh, and Y. T. Ko. 2018. Docetaxel-loaded hyaluronic acid–cathepsin b-cleavable-peptide–gold nanoparticles for the treatment of cancer. *Mol Pharm* 15 (10):4668–4676.

Gurav, D. D., A. S. Kulkarni, A. Khan, and V. S. Shinde. 2016. pH-responsive targeted and controlled doxorubicin delivery using hyaluronic acid nanocarriers. *Colloids Surf B* 143:352–358.

Haider, T., R. Tiwari, S. P. Vyas, and V. Soni. 2019. Molecular determinants as therapeutic targets in cancer chemotherapy: An update. *Pharmacol Ther* 200:85–109.

Hargreaves, R. H., J. A. Hartley, and J. Butler. 2000. Mechanisms of action of quinone-containing alkylating agents: DNA alkylation by aziridinylquinones. *Front Biosci* 5:E172–E180.

Harguindey, A., S. Roy, A. W. Harris, et al. 2019. Click nucleic acid mediated loading of prodrug activating enzymes in PEG-PLGA nanoparticles for combination chemotherapy. *Biomacromolecules* 20 (4):1683–1690.

Harris, A. L. 2002. Hypoxia—a key regulatory factor in tumour growth. *Nat Rev Cancer* 2 (1):38.

Hong, Y., S. Che, B. Hui, et al. 2019. Lung cancer therapy using doxorubicin and curcumin combination: Targeted prodrug based, pH sensitive nanomedicine. *Biomed Pharmacother* 112:108614.

Hou, W., X. Zhao, X. Qian, et al. 2016. pH-Sensitive self-assembling nanoparticles for tumor near-infrared fluorescence imaging and chemo–photodynamic combination therapy. *Nanoscale* 8 (1):104–116.

Hufziger, K. T., F. S. Thowfeik, D. J. Charboneau, et al. 2014. Ruthenium dihydroxybipyridine complexes are tumor activated prodrugs due to low pH and blue light induced ligand release. *J Inorg Biochem* 130:103–111.

Iizasa, T., T. Fujisawa, M. Suzuki, et al. 1999. Elevated levels of circulating plasma matrix metalloproteinase 9 in nonsmall cell lung cancer patients. *Clin Cancer Res* 5 (1):149–153.

Ikeda, Y., H. Hisano, Y. Nishikawa, and Y. Nagasaki. 2016. Targeting and treatment of tumor hypoxia by newly designed prodrug possessing high permeability in solid tumors. *Mol Pharm* 13 (7):2283–2289.

Johansson, K., M. Ito, C. M. Schophuizen, et al. 2011. Characterization of new potential anticancer drugs designed to overcome glutathione transferase mediated resistance. *Mol Pharm* 8 (5):1698–1708.

Johnston, M. C., and C. J. Scott. 2018. Antibody conjugated nanoparticles as a novel form of antibody drug conjugate chemotherapy. *Drug Discov Today*.

Karnthaler-Benbakka, C., B. Koblmuller, M. Mathuber, et al. 2019. Synthesis, characterization and in vitro studies of a cathepsin b-cleavable prodrug of the VEGFR inhibitor sunitinib. *Chem Biodivers* 16 (1):e1800520.

Kato, Y., S. Ozawa, C. Miyamoto, et al. 2013. Acidic extracellular microenvironment and cancer. *Cancer Cell Int* 13 (1):89.

Khan, J., A. Alexander, S. Saraf, and S. Saraf. 2018. Exploring the role of polymeric conjugates toward anti-cancer drug delivery: Current trends and future projections. *Int J Pharm* 548 (1):500–514.

Kim, D., Z. G. Gao, E. S. Lee, and Y. H. Bae. 2009. In vivo evaluation of doxorubicin-loaded polymeric micelles targeting folate receptors and early endosomal pH in drug-resistant ovarian cancer. *Mol Pharm* 6 (5):1353–1362.

Kim, D., E. S. Lee, K. T. Oh, Z. G. Gao, and Y. H. Bae. 2008. Doxorubicin-loaded polymeric micelle overcomes multidrug resistance of cancer by double-targeting folate receptor and early endosomal pH. *Small* 4 (11):2043–2050.

Kim, E.-J., R. Kumar, A. Sharma, et al. 2017. In vivo imaging of β-galactosidase stimulated activity in hepatocellular carcinoma using ligand-targeted fluorescent probe. *Biomaterials* 122:83–90.

Kim, H. S., A. Sharma, W. X. Ren, J. Han, and J. S. Kim. 2018. COX-2 Inhibition mediated anti-angiogenic activatable prodrug potentiates cancer therapy in preclinical models. *Biomaterials* 185:63–72.

Knop, K., R. Hoogenboom, D. Fischer, and U. S. Schubert. 2010. Poly (ethylene glycol) in drug delivery: Pros and cons as well as potential alternatives. *Angew Chem* 49 (36):6288–6308.

Knox, R. J., and S. Chen. 2004. Quinone reductase–mediated nitro-reduction: Clinical applications. In *Methods in Enzymology*. Amsterdam, the Netherlands: Elsevier.

Korman, A. J., J. Engelhardt, J. Loffredo, et al. 2017. Abstract SY09-01: Next-generation anti-CTLA-4 antibodies. Paper read at Proceedings of the American Association for Cancer Research Annual Meeting 2017, at USA.

Kovář, M., T. Mrkvan, J. Strohalm, et al. 2003. HPMA copolymer-bound doxorubicin targeted to tumor-specific antigen of BCL1 mouse B cell leukemia. *J Control Release* 92 (3):315–330.

Kumar-Sinha, C., R. B. Shah, B. Laxman, et al. 2004. Elevated alpha-methylacyl-CoA racemase enzymatic activity in prostate cancer. *Am J Pathol* 164 (3):787–793.

Kumar, S., L. J. K. Henry, S. Natesan, and R. Kandasamy. 2017. Atrial natriuretic peptide-conjugated chitosan-hydrazone-mPEG copolymer nanoparticles as pH-responsive carriers for intracellular delivery of prednisone. *Carbohydr Polym* 157:1677–1686.

Kurzrock, R., S. Goel, J. Wheler, et al. 2012. Safety, pharmacokinetics, and activity of EZN-2208, a novel conjugate of polyethylene glycol and SN38, in patients with advanced malignancies. *Cancer* 118 (24):6144–6151.

Leal, M., P. Sapra, S. A. Hurvitz, et al. 2014. Antibody–drug conjugates: An emerging modality for the treatment of cancer. *Ann N Y Acad Sci* 1321 (1):41–54.

Lee, E. S., K. T. Oh, D. Kim, Y. S. Youn, and Y. H. Bae. 2007. Tumor pH-responsive flower-like micelles of poly(L-lactic acid)-b-poly(ethylene glycol)-b-poly(L-histidine). *J Control Release* 123 (1):19–26.

Lee, E. S., H. J. Shin, K. Na, and Y. H. Bae. 2003. Poly (l-histidine)–PEG block copolymer micelles and pH-induced destabilization. *J Control Release* 90 (3):363–374.

Lee, S. J., and Y. I. Jeong. 2018. Hybrid nanoparticles based on chlorin e6-conjugated hyaluronic acid/poly (L-histidine) copolymer for theranostic application to tumors. *J Mater Chem B* 6 (18):2851–2859.

Li, C., D.-F. Yu, R. A. Newman, et al. 1998. Complete regression of well-established tumors using a novel water-soluble poly (L-glutamic acid)-paclitaxel conjugate. *Cancer Res* 58 (11):2404–2409.

Li, J., Y. Wang, Y. Zhu, and D. Oupický. 2013. Recent advances in delivery of drug–nucleic acid combinations for cancer treatment. *J Control Release* 172 (2):589–600.

Li, L., J. K. Kim, K. M. Huh, Y.-K. Lee, and S. Y. Kim. 2012. Targeted delivery of paclitaxel using folate-conjugated heparin-poly (β-benzyl-l-aspartate) self-assembled nanoparticles. *Carbohydr Polym* 87 (3):2120–2128.

Li, Z., Q. Chen, Y. Qi, et al. 2018. Rational design of multifunctional polymeric nanoparticles based on poly (L-histidine) and d-α-Vitamin E Succinate for reversing tumor multidrug resistance. *Biomacromolecules* 19 (7):2595–2609.

Ling, X., J. Tu, J. Wang, et al. 2018. Glutathione-responsive prodrug nanoparticles for effective drug delivery and cancer therapy. *ACS Nano* 13 (1):357–370.

Liu, H. M., R. L. Zhang, Y. W. Niu, et al. 2015. Development of hypoxia-triggered prodrug micelles as doxorubicin carriers for tumor therapy. *RSC Adv* 5 (27):20848–20857.

Low, P. S., and A. Antony. 2004. Folate receptor-targeted drugs for cancer and inflammatory diseases. *Adv Drug Deliv Rev* 56 (8):1055–1058.

Lu, D., X. Wen, J. Liang, Z. Gu, X. Zhang, and Y. Fan. 2009. A pH-sensitive nano drug delivery system derived from pullulan/doxorubicin conjugate. *J Biomed Mater Res B Appl Biomater* 89 (1):177–183.

Lu, X., Q.-Q. Wang, F.-J. Xu, G.-P. Tang, and W.-T. Yang. 2011. A cationic prodrug/therapeutic gene nanocomplex for the synergistic treatment of tumors. *Biomaterials* 32 (21):4849–4856.

Luo, C., J. Sun, B. Sun, and Z. He. 2014. Prodrug-based nanoparticulate drug delivery strategies for cancer therapy. *Trends Pharmacol Sci* 35 (11):556–566.

Luo, Q., P. Wang, Y. Miao, H. He, and X. Tang. 2012. A novel 5-fluorouracil prodrug using hydroxyethyl starch as a macromolecular carrier for sustained release. *Carbohydr Polym* 87 (4):2642–2647.

Luque-Michel, E., E. Imbuluzqueta, V. Sebastian, and M. J. Blanco-Prieto. 2017. Clinical advances of nanocarrier-based cancer therapy and diagnostics. *Expert Opin Drug Deliv* 14 (1):75–92.

Mehvar, R. 2003. Recent trends in the use of polysaccharides for improved delivery of therapeutic agents: Pharmacokinetic and pharmacodynamic perspectives. *Curr Pharm Biotechnol* 4 (5):283–302.

Montcourrier, P., I. Silver, R. Farnoud, I. Bird, and H. Rochefort. 1997. Breast cancer cells have a high capacity to acidify extracellular milieu by a dual mechanism. *Clin Exp Metastasis* 15 (4):382–392.

Mukerabigwi, J. F., W. Yin, Z. Zha, et al. 2019. Polymersome nanoreactors with tumor pH-triggered selective membrane permeability for prodrug delivery, activation, and combined oxidation-chemotherapy. *J Control Release* 303:209–222.

Mura, S., D. T. Bui, P. Couvreur, and J. Nicolas. 2015. Lipid prodrug nanocarriers in cancer therapy. *J Control Release* 208:25–41.

Mürdter, T. E, G. Friedel, J. T. Backman, et al. 2002. Dose optimization of a doxorubicin prodrug (HMR 1826) in isolated perfused human lungs: Low tumor pH promotes prodrug activation by β-glucuronidase. *J Pharmacol Exp Ther* 301 (1):223–228.

Mürdter, T. E., B. Sperker, K. T. Kivistö, et al. 1997. Enhanced uptake of doxorubicin into bronchial carcinoma: β-glucuronidase mediates release of doxorubicin from a glucuronide prodrug (HMR 1826) at the tumor site. *Cancer Res* 57 (12):2440–2445.

Muz, B., P. de la Puente, F. Azab, and A. K. Azab. 2015. The role of hypoxia in cancer progression, angiogenesis, metastasis, and resistance to therapy. *Hypoxia (Auckl)* 3:83–92.

Nakamura, H., T. Etrych, P. Chytil, et al. 2014. Two step mechanisms of tumor selective delivery of N-(2-hydroxypropyl)methacrylamide copolymer conjugated with pirarubicin via an acid-cleavable linkage. *J Control Release* 174:81–87.

Nittoli, T., M. P. Kelly, F. Delfino, et al. 2018. Antibody drug conjugates of cleavable amino-alkyl and aryl maytansinoids. *Bioorg Med Chem* 26 (9):2271–2279.

Ohwada, J., S. Ozawa, M. Kohchi, et al. 2009. Synthesis and biological activities of a pH-dependently activated water-soluble prodrug of a novel hexacyclic camptothecin analog. *Bioorg Med Chem Lett* 19 (10):2772–2776.

Olson, O. C., and J. A. Joyce. 2015. Cysteine cathepsin proteases: Regulators of cancer progression and therapeutic response. *Nat Rev Cancer* 15 (12):712–729.

Pang, X., X. Yang, and G. Zhai. 2014. Polymer-drug conjugates: Recent progress on administration routes. *Expert Opin Drug Deliv* 11 (7):1075–1086.

Park, K., G. Y. Lee, Y.-S. Kim, et al. 2006. Heparin–deoxycholic acid chemical conjugate as an anticancer drug carrier and its antitumor activity. *J Control Release* 114 (3):300–306.

Patel, A., and S. Sant. 2016. Hypoxic tumor microenvironment: Opportunities to develop targeted therapies. *Biotechnol Adv* 34 (5):803–812.

Pelegri-O'Day, E. M, E.-W. Lin, and H. D. Maynard. 2014. Therapeutic protein–polymer conjugates: Advancing beyond PEGylation. *J Am Chem Soc* 136 (41):14323–14332.

Perche, F., S. Biswas, T. Wang, L. Zhu, and V. P. Torchilin. 2014. Hypoxia-targeted siRNA delivery. *Angew Chem Int Ed Engl* 53 (13):3362–3366.

Perez, H. L., P. M. Cardarelli, S. Deshpande, et al. 2014. Antibody–drug conjugates: Current status and future directions. *Drug Discov Today* 19 (7):869–881.

Podgorski, I., and B. F. Sloane. 2003. Cathepsin B and its role (s) in cancer progression. Paper read at Biochemical Society Symposia.

Polvani, S., M. Calamante, V. Foresta, et al. 2011. Acycloguanosyl 5′-thymidyltriphosphate, a thymidine analogue prodrug activated by telomerase, reduces pancreatic tumor growth in mice. *Gastroenterology* 140 (2):709–720.

Qu, F., S. Park, K. Martinez, et al. 2017. Ruthenium complexes are pH-Activated Metallo Prodrugs (pHAMPs) with light-triggered selective toxicity toward cancer cells. *Inorg Chem* 56 (13):7519–7532.

Rashidi, F. B., A. D. AlQhatani, S. S. Bashraheel, et al. 2018. Isolation and molecular characterization of novel glucarpidases: Enzymes to improve the antibody directed enzyme pro-drug therapy for cancer treatment. *PLoS One* 13 (4):e0196254.

Renoux, B., F. Raes, T. Legigan, et al. 2017. Targeting the tumour microenvironment with an enzyme-responsive drug delivery system for the efficient therapy of breast and pancreatic cancers. *Chem Sci* 8 (5):3427–3433.

Richardson, D. L., S. M. Seward, and K. N. Moore. 2018. Antibody drug conjugates in the treatment of epithelial ovarian cancer. *Hematol Oncol Clin North Am* 32 (6):1057–1071.

Říhová, B., and M. Kovář. 2010. Immunogenicity and immunomodulatory properties of HPMA-based polymers. *Adv Drug Deliv Rev* 62 (2):184–191.

Ringsdorf, H. 1975. Structure and properties of pharmacologically active polymers. Paper read at Journal of Polymer Science: Polymer Symposia.

Ross, D., and D. Siegel. 2018. Quinone Reductases. In *Comprehensive Toxicology (Third Edition)*, edited by C. A. McQueen. Oxford: Elsevier, Vol. 4, pp. 207–218.

Ross, D., and D. Siegel. 2004. NAD (P) H: Quinone oxidoreductase 1 (NQO1, DT-diaphorase), functions and pharmacogenetics. In *Methods in Enzymology*. Amsterdam, the Netherlands: Elsevier.

Ruan, Q., X. Zhang, X. Lin, X. Duan, and J. Zhang. 2018. Novel 99m Tc labelled complexes with 2-nitroimidazole isocyanide: Design, synthesis and evaluation as potential tumor hypoxia imaging agents. *MedChemComm* 9 (6):988–994.

Scomparin, A., H. F. Florindo, G. Tiram, E. L. Ferguson, and R. Satchi-Fainaro. 2017. Two-step polymer-and liposome-enzyme prodrug therapies for cancer: PDEPT and PELT concepts and future perspectives. *Adv Drug Deliv Rev* 118:52–64.

Sharma, A., E. J. Kim, H. Shi, J. Y. Lee, B. G. Chung, and J. S. Kim. 2018. Development of a theranostic prodrug for colon cancer therapy by combining ligand-targeted delivery and enzyme-stimulated activation. *Biomaterials* 155:145–151.

Shefet-Carasso, L. R., and I. Benhar. 2015. Antibody-targeted drugs and drug resistance—challenges and solutions. *Drug Resist Updat* 18:36–46.

Shim, M. K., J. Park, H. Y. Yoon, et al. 2019. Carrier-free nanoparticles of cathepsin B-cleavable peptide-conjugated doxorubicin prodrug for cancer targeting therapy. *J Control Release* 294:376–389.

Shinde, S. S, M. P. Hay, A. V. Patterson, W. A. Denny, and R. F. Anderson. 2009. Spin trapping of radicals other than the •OH radical upon reduction of the anticancer agent tirapazamine by cytochrome P450 reductase. *J Am Chem Soc* 131 (40):14220–14221.

Sim, T., C. Lim, Y. H. Cho, E. S. Lee, Y. S. Youn, and K. T. Oh. 2018. Development of pH-sensitive nanogels for cancer treatment using crosslinked poly (aspartic acid-graft-imidazole)-block-poly(ethylene glycol). *J Appl Polym Sci* 135 (20):46268.

Singh, N. K. M., S. R. White, S. Kalagara, and S. Kadavakollu. 2017. Inhibition of CYP2S1 mediated metabolism of anticancer prodrug AQ4N by liarozole. *J Appl Pharm Sci* 7 (02):001–007.

Singh, Y., M. Palombo, and P. J. Sinko. 2008. Recent trends in targeted anticancer prodrug and conjugate design. *Curr Med Chem* 15 (18):1802–1826.

Sonawane, S. J., R. S. Kalhapure, and T. Govender. 2017. Hydrazone linkages in pH responsive drug delivery systems. *Eur J Pharm Sci* 99:45–65.

Stornetta, A., K.-C. K. Deng, S. Danielli, et al. 2018. Drug-DNA adducts as biomarkers for metabolic activation of the nitro-aromatic nitrogen mustard prodrug PR-104A. *Biochem Pharm* 154:64–74.

Sun, J., D. Yang, S.-H. Cui, et al. 2019. Enhanced anti-tumor efficiency of gemcitabine prodrug by FAPα-mediated activation. *Int J Pharm* 559:48–57.

Teicher, B. A. 2009. Antibody-drug conjugate targets. *Curr Cancer Drug Tar* 9 (8):982–1004.

Thambi, T., V. G. Deepagan, H. Y. Yoon, et al. 2014. Hypoxia-responsive polymeric nanoparticles for tumor-targeted drug delivery. *Biomaterials* 35 (5):1735–43.

Vater, C. A., and V. S. Goldmacher. 2010. Antibody–cytotoxic compound conjugates for oncology. In *Macromolecular Anticancer Therapeutics*. New York: Springer.

Vogus, D. R, V. Krishnan, and S. Mitragotri. 2017. A review on engineering polymer drug conjugates to improve combination chemotherapy. *Curr Opin Colloid Interface Sci* 31:75–85.

Wallace, D. C. 2012. Mitochondria and cancer. *Nat Rev Cancer* 12 (10):685–98.

Wan, Y., Y. Bu, J. Liu, et al. 2018. pH and reduction-activated polymeric prodrug nanoparticles based on a 6-thioguanine-dialdehyde sodium alginate conjugate for enhanced intracellular drug release in leukemia. *Polym Chem* 9 (24):3415–3424.

Wang, H. B., F. M. Xu, D. D. Li, X. S. Liu, Q. Jin, and J. Ji. 2013. Bioinspired phospholipid polymer prodrug as a pH-responsive drug delivery system for cancer therapy. *Polym Chem* 4 (6):2004–2010.

Wang, J. L., A. Foehrenbacher, J. C. Su, et al. 2012. The 2-nitroimidazole EF5 is a biomarker for oxidoreductases that activate the bioreductive prodrug CEN-209 under hypoxia. *Clin Cancer Res* 18 (6):1684–1695.

Wang, S., T. Xu,Y. Yang, and Z. Shao. 2015. Colloidal stability of silk fibroin nanoparticles coated with cationic polymer for effective drug delivery. *ACS applied materials & interfaces* 7:21254–21262.

Wang, Y., Y. Xie, J. Li, et al. 2017. Tumor-penetrating nanoparticles for enhanced anticancer activity of combined photodynamic and hypoxia-activated therapy. *ACS Nano* 11 (2):2227–2238.

Weiss, G. J., J. R. Infante, E. G. Chiorean, et al. 2011. Phase 1 study of the safety, tolerability, and pharmacokinetics of TH-302, a hypoxia-activated prodrug, in patients with advanced solid malignancies. *Clin Cancer Res* 17 (9):2997–3004.

Wilson, W. R., and M. P. Hay. 2011. Targeting hypoxia in cancer therapy. *Nat Rev Cancer* 11 (6):393–410.

Xiao, M., W. Sun, J. Fan, et al. 2018. Aminopeptidase-N-activated theranostic prodrug for NIR tracking of local tumor chemotherapy. *Adv Funct Mater* 28 (47):1805128.

Xu, G., and H. L. McLeod. 2001. Strategies for enzyme/prodrug cancer therapy. *Clin Cancer Res* 7 (11):3314–3324.

Xu, L., W. Huang, and X. Zhao, X. 2011. Polyethylene glycol-amino acid oligopeptide-irinotecan drug conjugate and drug composition thereof. Google patents.

Xu, L., W. Huang, and X. Zhao. 2014. Peg-amino acid-oligopeptide-irinotecan drug conjugates and the pharmaceutical compositions. Google Patents.

Yamamoto, H., Y. Kuno, S. Sugimoto, H. Takeuchi, and Y. Kawashima. 2005. Surface-modified PLGA nanosphere with chitosan improved pulmonary delivery of calcitonin by mucoadhesion and opening of the intercellular tight junctions. *J Control Release* 102 (2):373–381.

Yang, Y., L. Xu, W. Zhu, et al. 2018. One-pot synthesis of pH-responsive charge-switchable PEGylated nanoscale coordination polymers for improved cancer therapy. *Biomaterials* 156:121–133.

Yu, H., C. Guo, B. Feng, et al. 2016. Triple-layered pH-responsive micelleplexes loaded with siRNA and cisplatin prodrug for NF-Kappa B targeted treatment of metastatic breast cancer. *Theranostics* 6 (1):14–27.

Zelikin, A. N, C. Ehrhardt, and A. M. Healy. 2016. Materials and methods for delivery of biological drugs. *Nat Chem* 8 (11):997.

Zhang, M. K., C. X. Li, S. B. Wang, et al. 2018a. Tumor starvation induced spatiotemporal control over chemotherapy for synergistic therapy. *Small* 14 (50):e1803602.

Zhang, R., L. Feng, Z. Dong, et al. 2018b. Glucose & oxygen exhausting liposomes for combined cancer starvation and hypoxia-activated therapy. *Biomaterials* 162:123–131.

Zhang, X., X. Li, Q. You, and X. Zhang. 2017. Prodrug strategy for cancer cell-specific targeting: A recent overview. *Eur J Med Chem* 139:542–563.

Zheng, X., H. Tang, C. Xie, J. Zhang, W. Wu, and X. Jiang. 2015. Tracking cancer metastasis in vivo by using an iridium-based hypoxia-activated optical oxygen nanosensor. *Angew Chem* 54 (28):8094–8099.

Zhong, Y. J., L. H. Shao, and Y. Li. 2013. Cathepsin B-cleavable doxorubicin prodrugs for targeted cancer therapy (Review). *Int J Oncol* 42 (2):373–383.

Zou, Y., Q. P. Wu, W. Tansey, et al. 2001. Effectiveness of water soluble poly(L-glutamic acid)-camptothecin conjugate against resistant human lung cancer xenografted in nude mice. *Int J Oncol* 18 (2):331–336.

10

Nitroreductases in Prodrug Therapies

Tuğba Güngör, Ferah Cömert Önder, and Mehmet Ay

CONTENTS

10.1 Introduction

There has been a significantly increase in the prodrug development to improve the efficiency of a drug due to its desired physicochemical, pharmacokinetic and/or pharmacological properties. Prodrugs that are chemically modified bioreversible derivatives of drug molecules, are important tools for developing these properties. The aim of prodrug design is depended on optimizing the conditions to generate rational drug. The nitroreductases have been developed for prodrug strategies and nitro aromatic compounds are mostly used in this enzyme/prodrug combinations. The reduction of nitro groups to hydroxylamine and amine active metabolites are successfully performed by the activation of NTRs. Additionally, the most important usage of nitroreductases can be considered to become an alternative to drug treatment in Gene-Directed Enzyme Prodrug Therapy (GDEPT). One of the novel approaches in gene therapy, GDEPT is developed in place of conventional chemotherapy and other therapies. Many of nitroreductases from different sources such as *Escherichia coli*, *Pseudomonas pseudoalcaligenes* and *Staphylococcus saphrophyticus* etc. have been developed to be used in various applications such as medicine, fluorescent probe, analytic and biotechnology. Among these applications, cancer applications are widely used by NTRs. For this purpose, nitro-containing aromatic amide substrate, CB1954 (5-(aziridin-1-yl)-2,4-dinitrobenzamide) which are currently in clinical cancer trials (phase I/II), is the best-known example with the combination of NfsB from *E. coli*. Nitro containing effective, selective, and potential prodrugs have been designed to eliminate some of the limitations in prodrug therapies with NTRs compared to CB1954 and SN23862. The investigation of these enzyme/prodrug combinations on NTR-based GDEPT strategies has attracted attention by multidisciplinary researchers in recent years. The current chapter focuses on these new approaches in prodrug therapies developed by nitroreductases (Anlezark et al. 1992; Hajnal et al. 2016; Rautio et al. 2008; Prosser et al. 2013; Vass et al. 2009; Christofferson and Wilkie 2009).

10.2 Gene-Directed Enzyme Prodrug Therapy (GDEPT)

To overcome the limitations of conventional chemotherapy and other cancer therapies, researchers have focused on novel approaches such as gene therapy. Gene-directed enzyme prodrug therapy (GDEPT) is a type of gene therapy methods that has three essential steps targeting the cancer therapy, specific gene delivery, controlled conversion of prodrugs to cytotoxic drugs in target cancer cells and spread out the toxicity to the neighbors of cancer cells via bystander effect (Zhang et al. 2015a; Denny 2003; Springer and Niculescu-Duvaz 2002; Williams et al. 2015). Although several GDEPT products are at the different stages of clinical cancer trials over 20 years, there is no GDEPT drug in the market. GDEPT is consisted of three components including the prodrug to be activated, the activating enzyme and the carrier system (Denny 2002; Zhang et al. 2015a). At the first step of this concept, the coding gene is cloned into a vector and delivered to a tumor cell with or without carriers (Denny 2002; Dachs et al. 2009). Then, the gene is transcribed into mRNA and translated into the enzyme inside the tumor cell. In the third step, a prodrug is administered systemically and taken by the same cell. Finally, the conversion of nontoxic prodrug is occurred to a cytotoxic drug by the enzyme inside the cell (Denny 2002; Dachs et al. 2009). At the following step, formed toxic drug diffuses to the other neighbor cells with bystander effect (Zhang et al. 2015a). To obtain an effective bystander effect, metabolites should have good diffusion properties or transport effectively by neighbor cells through the intersection of gaps (Duarte et al. 2012). The main advantage of GDEPT-based cancer therapy is the higher concentration of toxic drug in tumor cells and the very low concentration in normal tissues (Denny 2002; Hamstra et al. 2004; Dachs et al. 2009; Russell and Khatri 2006). Enzymes for GDEPT should have some properties such as the high ratio expression in tumor cells compared to healthy cells and high catalytic activity to able to convert prodrugs even at low concentration. An ideal prodrug for GDEPT should be nontoxic or minimally toxic before the activation with enzyme but highly toxic after the enzymatic process (Zhang et al. 2015a).

In summary, GDEPT studies that may be considered an alternative to drug treatment or complementary treatment are very important in cancer therapy.

10.3 Nitroreductases for GDEPT

The use of nitroreductases (NTR) in Gene-Directed Enzyme Prodrug Therapy (GDEPT) studies is remarkable in recent days. NTRs are from oxidoreductase family of homodimeric flavoenzymes of 24 kDa subunits with tightly bound FMN (Flavin mononucleotide) or FAD (flavin adenine dinucleotide) (Denny 2002). They use nicotinamide cofactors such as NADH and NADPH to catalyze the reduction reaction of nitro-containing compounds (prodrugs) via ping-pong mechanism. While nitroreductase enzymes are absent in humans, they are commonly found in some bacterial species and rarely in eukaryotes (Denny 2002).

Nitroreductases can be classified into oxygen-insensitive (Type I) and oxygen-sensitive (Type II) reductases (Patterson and Wyllie 2014; Oliveira et al. 2010). So the reduction of the nitro groups by NTR enzymes can occur in two ways, as seen in Figure 10.1. Type I nitroreductase reduce the nitro group with two-electron transfer system to produce nitroso, hydroxylamine and amine, respectively (Roldán et al. 2008; Miller et al. 2018). Nitro compounds, hydroxylamine and amine metabolites are stable, while nitroso intermediate is unstable due to the second two-electron transfer is faster than the first two-electron reduction that can react with biomolecules to form toxic and mutagenic products (Roldán et al. 2008; Valiauga et al. 2018). On the other hand, Type II nitroreductase is carried out the reduction reactions with single electron transfer system to give unstable nitro radical anion which reoxidize to starting nitro compound and produce superoxide anion with futile cycle at aerobic conditions. The two nitro radical anions form starting nitro compound and nitroso derivative in the absence of oxygen. It is thought that nitroso compounds observed in biological systems can be formed in this way (Oliveira et al. 2010; Qin et al. 2018). Oxygen-sensitive nitroreductases are found in *Escherichia coli* and some *Clostridium strains* (Roldán et al. 2008).

The most important reason why nitroreductases are preferred in the treatment of GDEPT is that the electronic range of reduction of aromatic nitro compounds is very wide (Denny 2002, 2003). While the

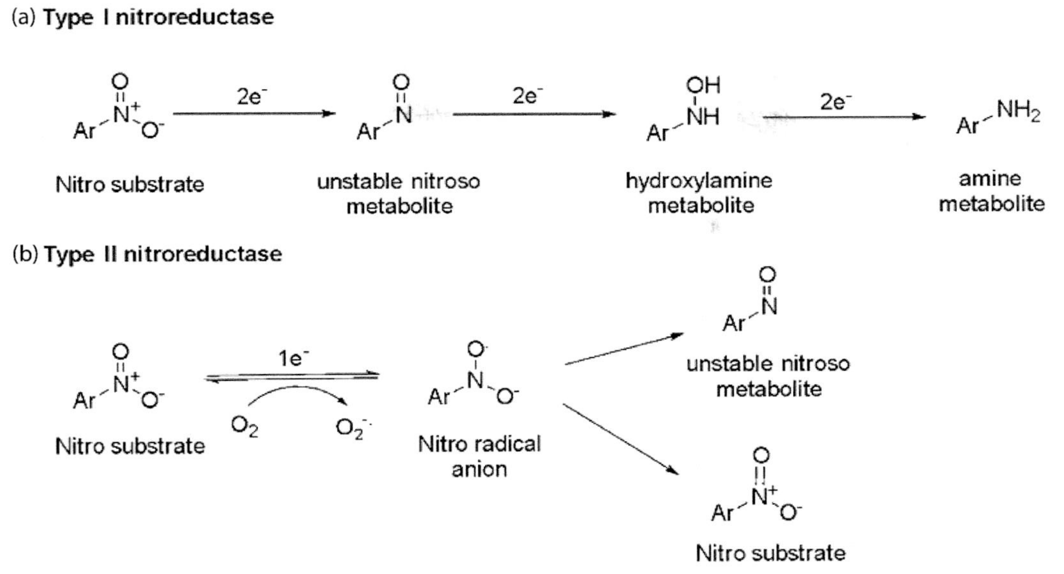

(a) **Type I nitroreductase**

Nitro substrate → (2e⁻) → unstable nitroso metabolite → (2e⁻) → hydroxylamine metabolite → (2e⁻) → amine metabolite

(b) **Type II nitroreductase**

Nitro substrate ⇌ (1e⁻; O_2 → $O_2^{-\cdot}$) Nitro radical anion → unstable nitroso metabolite / Nitro substrate

FIGURE 10.1 Reduction mechanism of nitroaromatic compounds. (Modified from Patterson and Wylie. *Trends in Parasitology*, 30, 289–298, 2014.)

Hammed σp electronic parameter of the nitro group is 0.78, the σp value of the hydroxylamine derivative formed by the reduction of four electrons is 0.34 and the σp value of the amino derivative formed by the reduction of six electrons is −0.66. Therefore, the electronic exchange (Δσp) values are 1.12 and 1.44 for hydroxylamine and amino derivatives, respectively (Denny 2002, 2003). This wide range allows the electrophilic metabolites to be potentially cytotoxic and act as a robust "electronic switch" at the activation of aromatic nitro compounds. So, it can be concluded that nitroreductases have broad substrate scope.

Bacterial nitroreductases, which are not oxygen-sensitive, have many applications such as synthesis of commercially important industrial chemicals, biological cleaning of soil pollutants, application of biosensors and detection of explosives (Race et al. 2007; Oliveira et al. 2010; Nadeau et al. 2000; Roldán et al. 2008; Zenno et al. 1996).

10.3.1 Types of Bacterial Nitroreductases

Nitroreductase genes found most commonly in bacterial genomes and nitroreductase-like proteins also found rarely in archaea and some eukaryotic species (Roldán et al. 2008). Discovery of nitroreductases in bacteria has been realized to reduce chloramphenicol and p-nitrobenzoic acid to their corresponding metabolites (Saz and Marina 1959; Cartwright and Cain 2015; Villanueva 1964). In recent years, a lot of different nitroreductases from various sources such as gram negative-positive bacteria, aerobic-anaerobic bacteria, pathogens, heterotrophs-phototrophs, mesophilic-thermophilic genus are identified and characterized (Roldán et al. 2008; Denny 2002; Çelik and Yetiş 2012).

Oxygen-insensitive nitroreductases (Type I) can be divided into two groups including *E. coli* nitroreductases NfsA identified as group A and NfsB identified as group B (Roldán et al. 2008). Almost all nitroreductases occur as homodimers with 24–30 kDa size, applicability on wide substrate scope, use FMN as cofactor and catalyze the nitro reduction through ping-pong kinetic mechanism. While group A generally uses NADPH, group B uses both NADH and NADPH as electron donors (Roldán et al. 2008).

Three oxygen-insensitive nitroreductases, the NfsA, NfsB and NfsC have been isolated and purified from *E. coli* (Zenno et al. 1996). While the major nitroreductase is NfsA, that is a homodimeric globular protein of 26.8 kDa, the minor nitroreductase is the NfsB of *E. coli*, that is most common, well characterized and has been studied extensively on nitroreductase concept of GDEPT. There is not much study on NfsC enzyme. While NfsA family members typically use NADPH, the NfsB members can use both cofactors, NADPH and NADH. NfsB reduces both the 2- and 4-nitro positions of the CB1954 prodrug, whereas NfsA preferentially reduces the 2-nitro group with 8-fold higher sensitivity (Dachs et al. 2009). The expression of the NfsB in human tumor cells increases the sensitivity to CB1954 about 2500-fold. *E. coli* nitroreductase NfsB is a Type 1 oxygen-insensitive flavoenzyme (MW: 23.9 kDa) that can interact various type of prodrugs such as nitroaromatic, nitroheterocyclic and quinone based chemicals (Denny 2002; Patterson et al. 2003). It has a close sequence homology to the nitroreductase of *S. typhimurium*, Frase I major flavin reductase in *V. fischeri* and the nitroreductase of Enterobacter cloacae (Denny 2002, 2003; Roldán et al. 2008). The NfsB enzyme (NTR) is a dimer protein that has 217 amino acids and a FMN cofactor per subunit. There are two active sites of NTR at dimer interface occurred by both monomers. The protein is a relatively rigid structure that has a role at the reaction between FMN and the prodrug. The oxidized and reduced forms of the enzyme are a little different due to the butterfly angle of the isoalloxazine ring across the N5–N10 axis of FMN from 16 to 25 (Christofferson and Wilkie 2009; Race et al. 2005). According to the kinetic studies, the enzyme reduces the nitro-containing prodrugs efficiently using NADH or NADPH via bi-bi (ping-pong) mechanism (Christofferson and Wilkie 2009; Race et al. 2005). At the first step, NAD(P)H interacts with the enzyme and transfer two electrons to FMN cofactor and the formation of $NAD(P)^+$ occurs. At the following step, the nitro prodrug binds to the active site of reduced enzyme and nitro groups reduce with two electrons for producing to the corresponding metabolites (Christofferson and Wilkie 2009; Race et al. 2005).

Various studies including random or site directed mutagenesis have been conducted by many researchers to increase the catalytic activity of NfsB, selective reduction of the 4-nitro group of CB1954 and cell sensitization (Grove et al. 2003; Guise et al. 2007; Jaberipour et al. 2010; Race et al. 2007; Bai et al. 2015). For example, it has been determined that the substitution of residue 124 with tryptophan increased substantially the selectivity for the nitro group at 4-position of CB1954 (Bai et al. 2015). To obtain better catalytic activity, two triple mutants (T41L/N71S/F124W and F123A/N71S/F124W) are prepared randomly and kinetic data are recorded. According to these results, mutant enzymes showed a 9.2- to 17.2-fold increase at k_{cat}/K_m compared to the wild-type enzyme and regioselectively reduced the 4-NO_2 group of CB1954. While F124W mutation make powerful hydrophobic interaction with the aziridinyl group of CB1954 to be helpful the reduction of the 4-NO_2 group, the F123A and T41L mutations provides better kinetic data on enzyme activity (Bai et al. 2015).

Homologs of NfsB are classical nitroreductase (Cnr) (MW: 24 kDa) in the *Salmonella enterica* (*typhimuriu*) (Watanabe et al. 1990; Roldán et al. 2008) and Frase I, a flavin oxidoreductase in the *V. Fischeri* (Zenno and Saigo 1994; Koike et al. 1998). Mutant Cnr nitroreductase is cloned from *S. typhimurium* strain TA1538NR and the enzymatic properties have been compared with the wild type. It is concluded that the mutant protein shows weak enzymatic activity at excess amount of FMN conditions and the substrate specificity is close to the wild type enzyme (Watanabe et al. 1990; Roldán et al. 2008). Frase I of *V. fischeri* can use both NADH and NADPH as electron donors and FMN as the most effective, FAD as the medium effective and riboflavin as the weak effective electron acceptors. Frase I is showed >90% of the total flavin reductase activity in crude extracts of *V. fischeri* cells at the conditions using FMN and NADH as an electron acceptor and donor whereas the NfsB shows very low-level activity (Zenno and Saigo 1994).

An alternative novel nitroreductase, Ssap-NtrB has an unusual cold active property, which has not been described too much previously (Çelik and Yetiş 2012). This nitroreductase have been identified and characterized from uropathogenic staphylococcus, *Staphylococcus saprophyticus* as a FMN-containing flavoenzyme. The gene encoding the Ssap-NtrB has been cloned and the recombinant protein overexpressed in *E. coli*. There are 20% identity and 25% similarity Ssap-between NtrB and NfsB. While both enzymes have common features such as using either NADH or NADPH as cofactors, pH activity-profiles and molecular sizes, the optimum catalytic activity of NfsB is at 30–40°C and Ssap-NtrB is at 15°C (Çelik and Yetiş 2012). At very low temperature, it reduces basic known substrates including nitrofurazone, CB1954, SN23862, $K_3[Fe(CN)_6]$. The reduction reaction of Ssap-NtrB with CB1954 gives the main metabolites of 4-hydroxylamine and 2-hydroxylamine isomer mixture with around 1:1 ratio and small amounts of two corresponding amine metabolites with full conversion in 30 min reaction time (Çelik and Yetiş 2012). In addition, Ssap-NtrB catalyzes the reduction reaction of a wide scope of nitro-containing

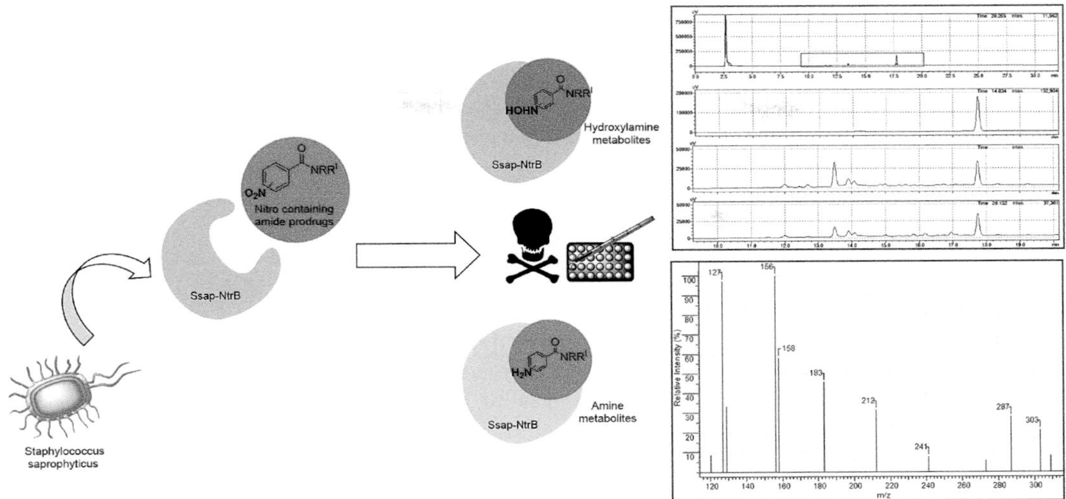

FIGURE 10.2 The schematic representation for the activation of nitro-containing amide prodrugs by Ssap-NtrB.

aliphatic, aromatic, heterocyclic amide prodrugs efficiently and better Ntr/prodrug combinations are presented to the Ntr literature, as seen in Figure 10.2 (Güngör et al. 2018, 2019). It is reported that cold adopted Ssap-NtrB can be an ideal nitroreductase candidate such conditions including cryotheraphy which the tissues are exposed to 4–10°C or temperature sensitive prodrugs, intermediates or other pharmacological compounds. Also, this unique property can reduce the risk of damaging healthy tissues surrounding tumor cells (Çelik and Yetiş 2012).

Another novel oxygen-insensitive nitroreductase, NbzA has been purified from extracts of bacteria *Pseudomonas pseudoalcaligenes* JS45 by using precipitation, anion-exchange and gel filtration chromatography techniques (Somerville et al. 1995). This enzyme which reduces different types of nitroaromatic compounds shows relatively stability below 40°C, active at a wide pH range and against to inhibitors. The enzyme is a flavoprotein which bound flavin mononucleotide (2 mole/per mole protein) as cofactor and its estimated size is 30 kDa. The reduction reaction of nitrobenzene with enzyme gives hydroxylaminobenzene as only metabolite and the corresponding amine derivative does not occur (Somerville et al. 1995).

Enzyme immobilization provides the reusability of enzymes, facilitates the recovery of substrate and/or product, prevents byproduct formation and increase the efficiency of biocatalyst (Grazú et al. 2005; Cao 2005). There are some applications of immobilized enzymes such as biocatalytic applications, microfluidic applications and analytical devices using at peptide mapping (Berne et al. 2006). The immobilization of NbzA has been done polyethylenimine-mediated silica formation with high yields and high loading capacity. It is observed that approximately 80% NbzA is immobilized and enzyme activity of immobilized version is more stable than NbzA in solution. The resulting immobilized NbzA is packed into a microreactor and investigated enzymatic activity on nitro-containing prodrug candidates such as nitrobenzene, CB1954 using a continuous-flow technique. This system is reported as a fast and reusable screening method to the catalytic activation of prodrugs by NbzA (Berne et al. 2006).

Salmonella typhimurium nitroreductase A (*SnrA*), is the second and major nitroreductase from *S. enterica serovar typhimurium* TA1535 that has been cloned and characterized (Nokhbeh et al. 2002). SnrA which is a homolog of *E. coli* nitroreductase-NfsA (87% sequence identity) has 240 amino acids and catalyze the reduction reaction with NADPH like NfsA and flavin mononucleotide (FMN). Kinetic results of SnrA are similar to the kinetics of nitroreductase activity of NfsA and FMN reductase activity of Frp (Nokhbeh et al. 2002).

A nitroreductase, Frm2 (YCL026c-A) has been identified from the source of the yeast *Saccharomyces cerevisiae* (Bang et al. 2012). It is reported that Frm2 can be associated with the lipid signaling pathway and cellular homeostasis. To understanding the role of the Frm2, the nitroreductase activity of enzyme with 4-nitroquinoline-*N*-oxide (4-NQO) using NADH and the effect of oxidative stress defense system is investigated in detail. It is concluded according to MS results of metabolites, Frm2 can catalyze the reduction of NQO to 4-aminoquinoline-*N*-oxide through 4-hydroxylaminoquinoline applying two-electron reduction and increase the resistance to oxidative stress (Bang et al. 2012). Also, the crystal structure of Frm2 is expressed as having minimal architecture with simple configuration in the FMN binding site like as CinD and YdjA bacterial nitroreductases (Song et al. 2015).

PnrA is another nitroreductase isolated from the TNT-degrading strain *Pseudomonas putida* with 28 kDa weight (Caballero et al. 2005). Enzyme can reduce 1 mole 2,4,6-trinitrotoluene (TNT) using 2 mole of NADPH to give to 4-hydroxylamine-2,6-dinitrotoluene. PnrA is applicable a large number of nitro-containing prodrug including nitrotoluene, nitrobenzoate or nitroaniline derivatives. According to the V_{max}/K_m kinetic parameters of prodrugs, TNT found to be the most efficient substrate catalyzed by PnrA (Caballero et al. 2005).

Gox0834, FMN-containing and NADPH-dependent nitroreductase in dimer form with 31.4 molecular mass kDa is cloned from Gluconobacter oxydans and heterogeneously overexpressed in *E. coli* (Yang et al. 2016). While the enzyme reduces various mononitro, polynitro, and polycyclic nitroaromatic compounds, the best activity is showed at the reduction of CB1954 with the k_{cat}/K_m value (0.020 $s^{-1}/\mu M$). The optimum activity of Gox0834 is at 40–60°C and pH 7.5–8.5 for the reduction of CB1954. Broad substrate scope, good catalytic activity and high stability produce potential applications including prodrug activation, bioremediation or other biocatalytic processes to this enzyme (Yang et al. 2016).

A new nitroreductase (SNR) is cloned and identified from bacterium *Streptomyces mirabilis* DUT001 (Yang et al. 2019). Enzyme is a homodimer with 24 kDa molecular weight and uses NADH as a cofactor at reduction reaction. SNR catalyze the reduction of nitro compounds such as nitrofurazone, pollutants such as 2,4-dinitrotoluene and 2,4,6-trinitrotoluene or 3-nitrophthalimide, 4-nitrophthalimide, and 4-nitro-1,8-naphthalic anhydride. Kinetic data of best catalytic activity is observed the reduction of FMN with 0.234 $\mu M^{-1}sec^{-1}$ k_{cat}/K_m values. The reasons of high activity to FMN can be the absence of the helix entrance to the substrate pocket and this property supports to generate larger substrate-binding pocket, remove steric effect and facilitate the interaction of big substrates with the active center of enzymes (Yang et al. 2019).

Other known nitroreductases and their bacterial sources are given below (Roldán et al. 2008). FRP nitroreductase-*Vibrio harveyi* (Lei et al. 1994), NprA and NprB-*Rhodobacter capsulatus* (Roldán et al. 2008), NfrA1 and YwrO-*Bacillus subtilis* (Zenno et al. 2005), *B. Licheniformis* nitroreductase-*Bacillus licheniformis* (Emptage et al. 2009), NitA and NitB-*Clostridium acetobutylicum* (Kutty and Bennett 2005), YdgI-Bc and YfkO-Bc-*Bacillus cereus* (Gwenin et al. 2015), nitroreductase-heterotrophic bacterias (*Thermus thermophilus, Comamonas acidovorans, Selenomonas ruminantium* (Park et al. 1992; Anderson et al. 2002).

In recent years, studies on artificial enzymes known as "nanozymes" that have many advantages like high efficiency, high stability, ease of recycling, difficulty of denaturation and low cost etc. compared to natural enzymes have increased. For instance, magnetic platinum(II)-based artificial nitroreductase is designed, obtained at four steps and investigated the catalytic reduction behaviour of 4-nitrophenol by sodium borohydride in aqueous solution. While highly stable platinum(II) complexes with special ligand scaffold provide a catalytic active site and halogenated ligand acting as a protein mimic provides the catalysis through binding substrates and efficiency and long-term resistance. Also, it is reported that this artificial nitroreductase is stable even after twenty catalytic cycles with excellent activity and good turnover frequency according to kinetic parameters (Aghahosseini et al. 2019).

10.3.2 Applications of Nitroreductases

The NAD(P)H-dependent bacterial nitroreductases have different types of applications such as bioremediation and biodegradation of nitroaromatic compounds, medical therapies especially cancer with GDEPT concept and fluorescent probe, antibiotics, some genotoxicity tests, listeria, the synthesis of commercially high-value chemicals, some analytical applications, and biotechnological products (Somerville et al. 1995; Rieger et al. 2002; Lewis et al. 2004; Ramos et al. 2005; Knox et al. 1993; Knox and Connors 1995; Chen et al. 2004; Searle et al. 2004; Nadeau et al. 2000; Roldán et al. 2008).

Cancer therapies based on the reduction of an inactive nitro-containing prodrug and the conversion to highly cytotoxic hydroxylamino and/or amino metabolites in the tumor cells is the most studied medical application of the bacterial nitroreductases (Roldán et al. 2008; Oliveira et al. 2010; Russell and Khatri 2006). The development of new nitroreductase from different sources and novel nitro functional group bearing prodrug candidates lead to improve the nitroreductase/prodrug combinations for the different types of cancer therapy. Detailed information about cancer therapy of various nitroreductases with prodrugs are given at Section 10.4.

Another promising application of nitroreductases is clinical diagnosis and drug screening. It is well known that overexpression of NTR occurs at hypoxic conditions that means reduced or insufficient oxygen medium (Li et al. 2013; Kumari et al. 2019). The development of new specific fluorescent probe for NTR detection at hypoxia conditions in the biological systems is very important for researchers due to hypoxia is primarily associated with solid tumors (Harvey et al. 2011; Kumari et al. 2019). In fact, the basic approach of this system is same as known about NTRs. Firstly, nitro group of small nonfluorescent molecules reduces to hydroxylamine or amine metabolites by nitroreductase enzyme and various electron rearrangement and/or cleavage of some bonds occurs. At the end of this process, the molecule is reconstructed and a fluorophore releases showing strong fluorescence property (Qin et al. 2018). Three basic mechanisms are reported on the fluorescent probes for NTR, photo-induced electron transfer (PET), domino reaction and intramolecular charge transfer (ICT) (Qin et al. 2018). The method of fluorescence detection and imaging of NTR is more selective, sensitive and advantageous compared to conventional method to determine of NTR activity such as Clark electrode, nuclear magnetic resonance (NMR) and electron paramagnetic resonance (EPR) methods (Qin et al. 2018). For instance; a novel conjugated with 4-nitrobenzylcarbamate moiety and 2,5-bis(methylsulfinyl)-1,4-diaminobenzene based probe (BBP) is developed as nitroreductase (NTR) responsive fluorescent probe to identify hypoxic tumor. It is reported that fluorescent probe shows the excellent specificity to NTRs with a rapid response time, distinguish the hypoxic tumor cells from healthy cells and can be used as an effective agent to detect the early malignant tumor formation (Zhang et al. 2019b). Another example is 1*H*-Benzo[*de*]isoquinoline-1,3-(2*H*)-diones or naphthalimides which has a lot of biological applications such as antitumor, antitrypanosomal, antiviral, local anesthetic, analgesic etc. Perfect fluorophore properties of naphthalimides is due to their good chemical stability, high fluorescence quantum yield, and multiple region for chemical changes (Kumari et al. 2019). Therefore, nitro naphthalimide based NTR-targeted fluorescent sensors can image the hypoxia selectively in cancer (Kumari et al. 2019). Moreover, many research teams are working to develop new, more effective NTR-based fluorescent probe including a fluorescent moiety and an aromatic nitro group (Yang et al. 2017; Luo et al. 2017; Guo et al. 2013; Yuan et al. 2014; Cui et al. 2011; Zhang et al. 2015b; Li et al. 2015; Jiang et al. 2013; Fan et al. 2019; Li et al. 2013).

A new, different approach on using nitroreductases to reduce nitro group of pre-antibiotics that is prodrug form of existing antibiotics in the market had been developed (Çelik et al. 2016). To prove this idea, sulfamethoxazole which is a known sulfonamide bacteriostatic antibiotic has been redesigned and the pre-antibiotic that have nitro group instead of amino has been synthesized as a prodrug. It is determined that after interaction with Ssap-NtrB nitroreductase, the conversion of nitro group to amine occurs and antimicrobial capacity of the drug increases. This strategy can be applied commercially available antibiotics to increase the antibiotic specificity across the target pathogen, reduce the resistance, use of low-dose drug and obtain more efficient-more selective antibiotics (Çelik et al. 2016).

Another example using this concept is fluoroquinolones which are widely used at many bacterial infections. Due to their toxicity and side effects, the usage of these fluoroquinolone antibiotics are restricted (Pardeshi et al. 2019). To overcome this problem, a strategy that uses a bacterial enzyme for activation of a prodrug to generate the active antibiotic is proposed (Pardeshi et al. 2019). For this purpose, a Ciprofloxacin-latent fluorophore conjugate molecule is designed and synthesized as a prodrug and specifically activated by bacterial nitroreductase to deliver an antibiotic, Ciprofloxacin. After an effective reduction reaction and the following rearrangements at the molecule by nitroreductase enzyme, a fluorescence signal caused by the fluorophore moiety is recorded by confocal microscopy. It is reported that conjugate molecules have potent bactericidal activity nearly identical to Ciprofloxacin (Pardeshi et al. 2019).

Soil and groundwater have been polluted by the polynitroaromatic compounds that produce and dispose from the industrial sources such as explosives, pesticides-herbicides, dyes, solvents, textiles, paper, pharmaceuticals and plastic-polymer (Oliveira et al. 2010; Maksimova et al. 2018; Roldán et al. 2008). In addition, public health has been affected from these contaminations of the ecosystem in the short or long term. So, the degradation of nitroaromatic compounds and removing pollutants from the environment has been an important issue for researchers. The bacterial nitroreductases is one of the important solutions of this problem. They catalyze the reduction of nitro-containing pollutants using a flavin mononucleotide as cofactor at aerobic and anaerobic conditions and form the nitroso, hydroxylamino and amino derivatives (Roldán et al. 2008). Biodegradation of TNT which is a strong explosive using since 1902 and the most widely nitroaromatic pollutant has been investigated using methanogens, clostridia, denitrifiers, sulfate reducing bacteria and enzymes including nitroreductases, oxidases, hydrogenases and peroxidases (Lewis et al. 2004; Ramos et al. 2005; Roldán et al. 2008; Maksimova et al. 2018). Also, different aerobic-anaerobic conditions, phytoremediation, transgenic plants bearing bacterial nitroreductases, genetically modified bacterial strains have been applied to remediation of TNT and other polynitroaromatic compounds (Roldán et al. 2008; Oliveira et al. 2010).

Air pollution is one of the important reasons of lung cancer which cause to death worldwide (Murray et al. 2018). 3-Nitrobenzanthrone (3-NBA) is the most mutagenic and lung carcinogen in rodents identified in diesel exhaust particulate and polluted air. It is known that NAD(P)H:quinone oxidoreductase 1 (NQO1) can activate the 3-NBA in liver *in vitro* as the responsible major nitroreductase. According to the recent literature acknowledgements, human aldo-keto reductase (AKR1C3) shows nitroreductase activity toward the chemotherapeutic agent PR-104A although this type of enzymes typically reduce the two-electron on carbonyl groups (Guise et al. 2010; Murray et al. 2018). So, AKR1C isoforms is used for nitroreductive activation of 3-nitrobenzanthrone in this study (Murray et al. 2018). According to HPLC and kinetic results, it is reported that AKR1C1-1C3 catalyze the reduction of 3-NBA successfully and form the reduced metabolite, 3-aminobenzanthrone and AKR1C1, AKR1C3 and NQO1 enzymes have very close catalytic efficiencies (Murray et al. 2018).

In another study on bioremediation, three disperse dyes (Disperse Red 73, Disperse Red 78, Disperse Red 167) that include nitro and azo groups at structures are selected because of their chemical similarity with minor functional group changes such as cyano, chlorine or amide (Franco et al. 2018). These dyes are synthetic organic dyes using commonly in textile industries and expected to be harmful to both humans and the ecology due to the mutagenicity of nitro compounds. The catalytic activation of these dyes with nitroreductase immobilized on Fe_2O_3 magnetic particles modified with the tosyl group is investigated in detail. The metabolite of dyes are analyzed by liquid chromatography coupled to diode array detectors (HPLC-DAD) and mass spectrometry (LC-MS/MS). The conversion rate of main amine metabolites are found as 50, 98 and 99% for DR 73, DR 78 and DR 167, respectively (Franco et al. 2018).

Using Umu test to the detection of DNA damaging carcinogenic chemicals in environmental genotoxicity field has increased rapidly in the last decades due to their superior properties such as simplicity, sensitivity, rapidity and reproducibility (Oda 2016). Ames test (Salmonella/microsome assay), Biochemical prophage induction assay, SOS Chromotest and Umu test based on different concepts have been used over times (Oda 2016; Oliveira et al. 2010). Genetically engineered umu tester strains are overexpressed in bacterial nitroreductase or *O*-acetyltransferase enzymes for the detection of nitroarenes and arylamines and human phase I drug metabolic enzyme (cytochrome P450) and glutathione *S*-transferase, *N*-acetyltransferases and sulfotransferases as rat/human phase II drug enzymes are developed successfully for genotoxicity assays later (Oda 2016). It is known that carcinogenic nitroarenes activated to genotoxins through the reduction reaction to arylhydroxylamine metabolites by bacterial nitroreductase and at the following step, these arylhydroxylamines are further activated by *O*-acetyltransferase (O-AT) to obtain the final reactive electrophiles in bacterial or mammalian cell systems. To increase the sensitivity across to carcinogenic nitroarenes and arylamines, drug-metabolizing enzymes overproduced in the subcloned nitroreductase (NR) and/or *O*-acetyltransferase (O-AT) genes and developed new tester strains NM2009 (from O-AT) and NM3009 (from NR/O-AT) (Oda 2016). NM3009 showed the highest sensitivity to nitroarene carcinogens including 1-nitronaphthalene, 2-nitrofluorene, 3,7-dinitrofluoranthene, 5-nitroacenaphthene, 1-nitropyrene, 1,6-dinitropyrene, 3,9-dinitrofluoranthene, 4,4-dinitrophenyl-1,8-dinitropyrene, m-dinitrobenzene and 2,4-dinitrotoluene etc. Highly sensitive tester strain, NM3009, have many advantages to detect the genotoxic nitroarenes in environmental samples (Oda 2016).

Listeria monocytogenes (*Listeria*) which can cause important bacterial infections resulting in death is a gram-positive facultative bacterium and discovery of selective and sensitive diagnosis method is essential (Zhang et al. 2019a). Applying methods to detect *Listeria* have some disadvantages such as long-term analysis, tentative, not identifying bacterial viability and antibiotic efficacy. To detect *Listeria*, a new approach is developed *in vitro* and *in vivo* using nitroreductase responsive fluorescent probe with a fluorescence off-on property. It is concluded that this probe shows a selective detection (12.5 ng/mL) to real-time monitoring NTRs rapidly in 10 min with high specificity and sensitivity. Also, it is determined that this probe does not only detect *Listeria* infections but distinguish *Listeria* from *E. coli* according to the results of animal studies. Fluorescent nitroreductase probe has a potential in diagnostic and therapeutic applications of severe bacterial infections (Zhang et al. 2019a).

Nitroreductases can be used for various types of analytical methods at different applications. For example, mesotrione (2-(4-methylsulphonyl-2-nitrobenzyl)cyclohexane-1,3-dione) is a selective herbicide to use the pre- and postformed weed control in maize crops (Hdiouech et al. 2019). Surface and ground waters have generally some mesotrione and it is a requirement the selective detection of this herbicide which can be harmful to human health and environment. There are some methods such as fluorometric immunosensors, liquid chromatography coupled with UV–Vis or fluorescence via 2-amino-4-methylsulfonylbenzoic acide (AMBA), colorimetric cell bioassay with 4-hydroxyphenylpyruvate dioxygenase (HPPD) that are highly sensitive, quite expensive and not applicable to an external large plantation area. A new mesotrione monitoring biosensor is developed using a nitroreductase (NfrA2) immobilized on a glassy carbon electrode using the adsorption technique on nanoparticles of $Mg_2Al\text{-}NO_3$ layered double hydroxides (MgAl-NPs) (Hdiouech et al. 2019). A competitive enzymatic reduction between $Fe(CN)_6^{-3}$ and mesotrione is occurred by the activation with NfrA2. While $Fe(CN)_6^{-3}$ is enzymatically reduced to the $Fe(CN)_6^{-4}$ at NADH-containing medium initially, a competitive enzymatic reaction occurs to cause the decreases of the anodic current at the nitroreductase immobilized modified electrode when mesotrione is added. Mesotrione can be define by chronoamperometry at 0.375 V/SCE with a sensitivity of 18.4 mA $M^{-1}cm^{-2}$ over a linear concentration range between 5–60 μM and a detection limit (3σ) of 3 μM (Hdiouech et al. 2019).

10.4 Prodrug Concept and Nitro-Containing Prodrugs

Prodrugs are described as chemically modified forms of the compounds that have low biological activity. Prodrugs are especially designed as pharmacologically active agents when they are transformed into their active derivatives by using chemical or enzymatic processes (Hajnal et al. 2016). The prodrugs can be easily converted to its cytotoxic agent and exhibit desired pharmacological effects. However, prodrugs can be defined as inactive forms of active drugs with optimized their important properties. They have specific protective groups in order to prevent undesirable properties of the parent molecule. Hereby, a new active molecule is generated following the biotransformation of the prodrug (Hajnal et al. 2016).

In drug design, prodrugs are generally designed to solve the problems for the development of the most active drug molecule and it depends on various common physical properties such as insolubility in water, chemical instability, low lipophilicity and pharmacokinetic properties such as poor or no bioavailability, changing of absorption/elimination and target specificity (Hajnal et al. 2016; Rautio et al. 2008; Yang et al. 2014. Furthermore, pharmacodynamic properties like toxicity and therapeutic index should be decreased and improved in prodrug research, respectively (Zawilska et al. 2013). Designing of valuable prodrugs are needed in terms of eliminating these problems.

As is known, there are approximately 5–8% approved drugs that are used as prodrugs in the worldwide therapy (Yang, n.d.; Hajnal et al. 2016). They may occur from natural compounds such as phytochemical/botanical components or endogenous substances and may be produced synthetic or semi-synthetic in drug design. For instance, aspirin, psilocybin, parathion, codeine, heroin, L-DOPA and various antiviral nucleosides are the most important known naturally occurring and synthetic prodrugs (Wu 2009).

Additionally, the most known valuable prodrugs are indicated to be used in different types of therapeutic areas including anti-influenza, antihypertensive, antibiotics, antiulcer, anticoagulant, anti-inflammatory, antifungal, anesthetic and anticancer. For instance, enalapril is a prodrug that is used for improving the availability of its active metabolite and Angiotensin Converting Enzyme (ACE) inhibitor, enalaprilat. The oral bioavailability of enalapril is similar with enalaprilat about 40% and the structural difference of enalapril stems from the presence of ethyl ester group on its structure (Yang et al. 2014; Todd and Heel 1986; Rautio et al. 2008). Miproxifene phosphate (TAT-59) that is a phosphate ester derivative of tamoxifen, is designed for breast cancer patients in antiestrogen therapy and the growth inhibitory activity of this prodrug is indicated as stronger in *in vivo* xenograft tumor models like tamoxifen (Yang et al. 2014; Rautio et al. 2008; Shibata et al. 2000). For another example, omeprazole is a proton pump inhibitor (PPI) and have been used in the treatment of *Helicobacter pylori* infection in children as antiulcer prodrug (Litalien et al. 2005; Yang et al. 2014). Table 10.1 shows some of the known prodrugs and their therapeutic areas.

TABLE 10.1

Known Prodrugs and Therapeutic Areas

A. ESTER PRODRUGS			
Prodrug Name	**Chemical Structure**	**Therapeutic Area**	**References**
Ximelagatran	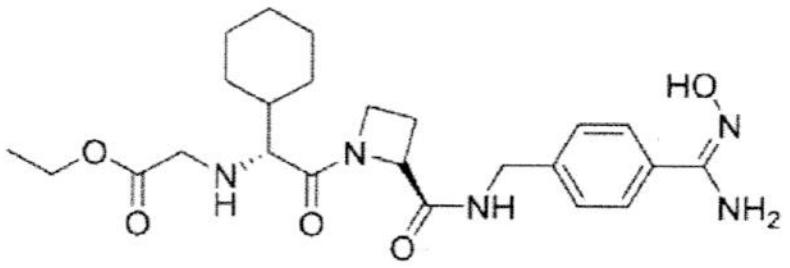	Anticoagulant	Clement et al. (2003), Rautio et al. (2008)
Famciclovir	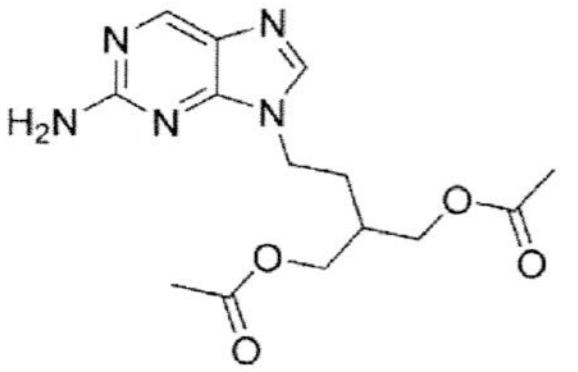	Antiviral	Hodge et al. (1989), Rautio et al. (2008)
Oseltamivir	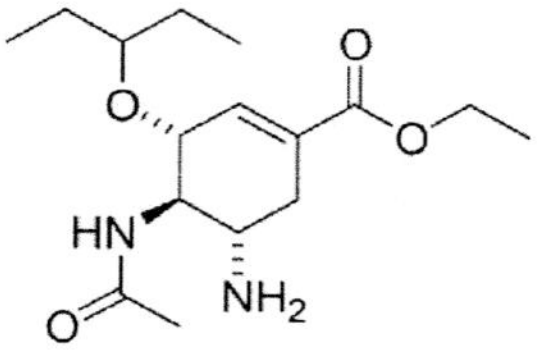	Anti-influenza	He et al. (1999), Rautio et al. (2008), Yang et al. (2014)
B. AMIDE PRODRUGS			
Prodrug Name	**Chemical Structure**	**Therapeutic Area**	**References**
Enalapril	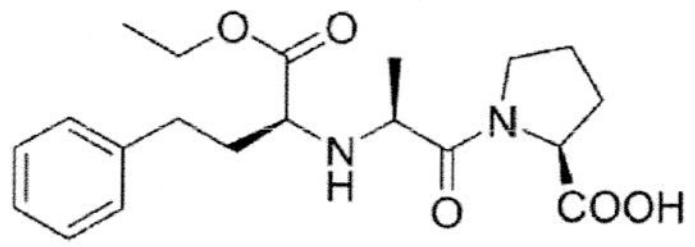	ACE inhibitor	Rautio et al. (2008), Todd et al. (1986), Yang et al. (2014)
Pivampicillin	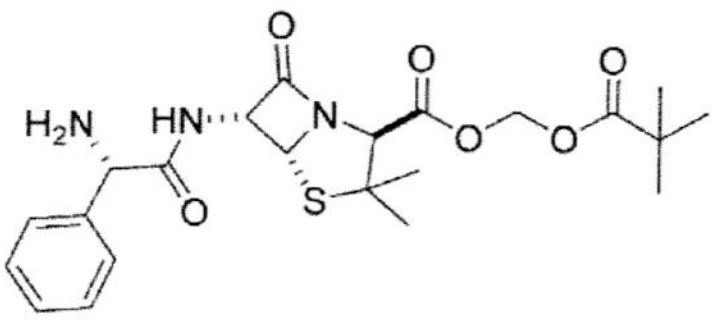	β-Lactam antibiotic	Ehrnebo et al. (1979), Rautio et al. (2008), Yang et al. (2014)
Midodrine	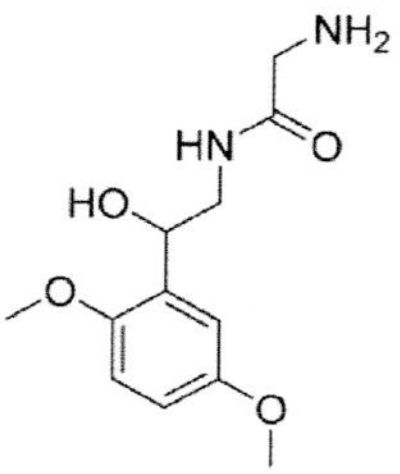	Vasoconstrictor	Rautio et al. (2008), Tsuda et al. (2006)

(*Continued*)

TABLE 10.1 (*Continued*)

Known Prodrugs and Therapeutic Areas

C. PHOSPHATE ESTER PRODRUGS

Prodrug Name	Chemical Structure	Therapeutic Area	References
Fosamprenavir	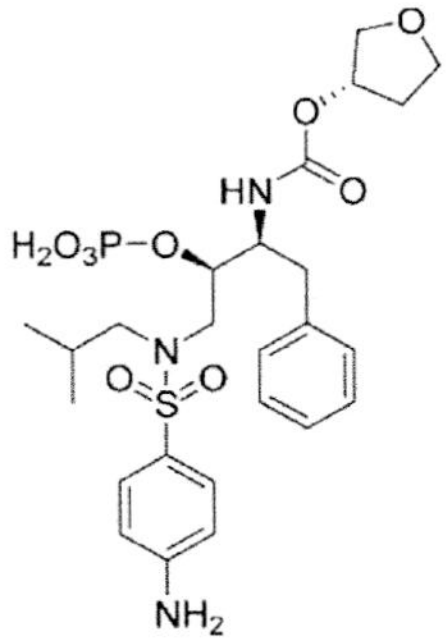	Protease inhibitor	Zawilska (2013), Rautio et al. (2008)
Estramustine phosphate	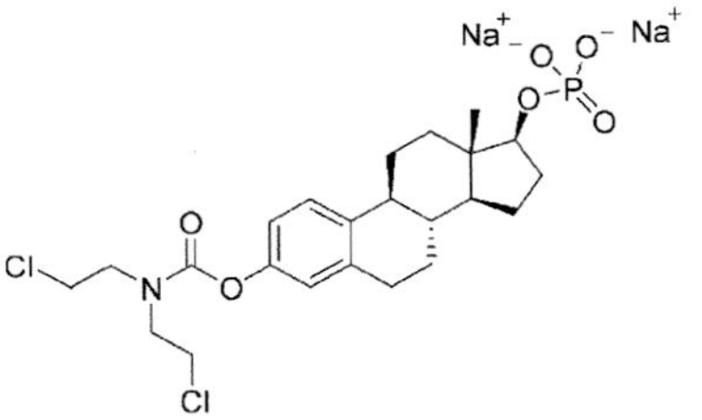	Anticancer	Zawilska et al. (2013)
Phosphonooxymethyl propofol	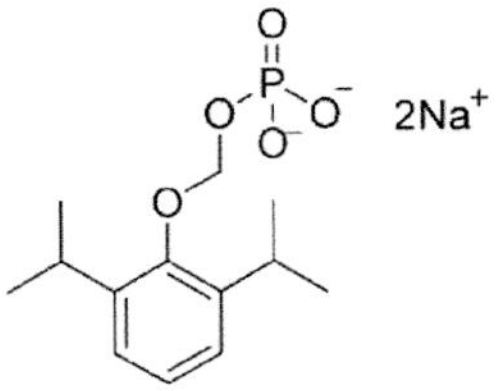	Anesthetic	Rautio et al. (2008), Schywalsky et al. (2003), Stella et al. (2005), Yang et al. (2014),

D. CARBAMATE PRODRUGS

Prodrug Name	Chemical Structure	Therapeutic Area	References
Irinotecan	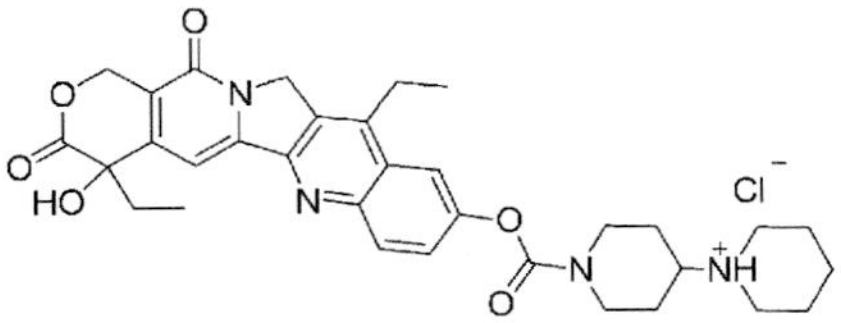	Anticancer	Rautio et al. (2008), Rothenberg et al. (2011)
Capecitabine	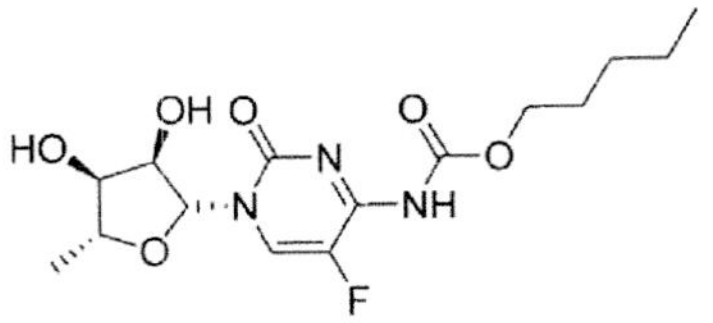	Anticancer	Desmoulin et al. (2002), Yang et al. (2014)
Bambuterol	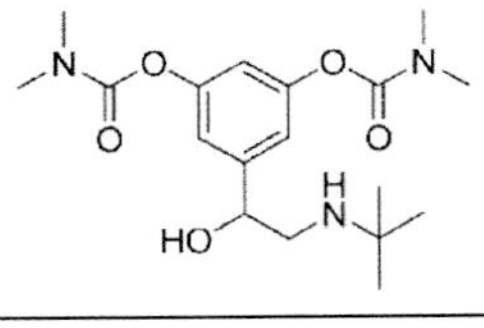	Asthma	Rautio et al. (2008), Tunek et al. (1988)

Nitroaromatic compounds are most important groups of industrial chemicals and widely used. These compounds contain at least one -NO_2 group which has strong electronegativity, in their core structures. There are many compounds that have been identified for biologically active agents in medicine (Ju and Parales 2010). Chloramphenicol, azomycin, orinocin, Thaxtomin A and B can be given as nitroaromatic antibiotics that are produced by bacteria of the genus of Streptomyces (Ju and Parales 2010; Roldán et al. 2008). Many nitro aromatic compounds are identified with their toxic and mutagenic properties. The toxicity of these compounds is related with the various reduction products like hydroxylamines that can interact with biomolecules (Roldán et al. 2008).

Nitroaromatic compounds are acutely toxic and mutagenic, DNA directly can be damaged through oxidation and reduction products of nitroaromatic compounds or these compounds can lead to induce of mutagenesis during DNA synthesis. Mutagenicity, hepatotoxicity and carcinogenicity can be depended on the position of nitro groups on the aromatic ring (Ju and Parales 2010; Patterson and Wyllie 2014). Bioactivation can be affected for many nitroaromatic compounds and when nitro groups are activated by human enzymes, the toxicity can be reduced. Nitroaromatics selectively activated by the bacterial-like NTRs to obtain the metabolites (Patterson and Wyllie 2014; Atwell et al. 2007) and nitro-containing prodrugs are generally used for nitroreduction by NTRs. This enzyme/prodrug combination is depended on the nitroreductase (NTR) enzymes due to their potential applications in biocatalysis, biomedicine and prodrug activation for cancer treatments (Roldán et al. 2008). NTR enzyme classes as Type I and Type II and four types of prodrugs have been used in these enzyme/prodrug systems. Nitroaromatic prodrugs are divided into four classes for NTR applications. These are:

- Dinitroaziridinylbenzamides
- Dinitrobenzamide mustards
- 4-Nitrobenzylcarbamates
- Nitroindoles (Denny 2002; Malekshah et al. 2016)

Using the functional groups is an important tool for designing of prodrugs. The majority of these active groups such as carboxylic, amine, phosphate/phosphonate and carbonyl have been successfully used in the designing of prodrugs. Especially, prodrugs are generated by the modification of these groups including amides, carbamates, esters etc. Amide groups which are the most valuable groups in prodrug/drug design, are generally more stable and they can have some of the limited usage in enzymatic studies. However, commonly known clinically approved prodrugs are determined in this category. The carbamate derivatives can be stable via enzymatic reactions and its structure tends to hydrolysis (Rautio et al. 2008).

10.4.1 Dinitroaziridinylbenzamides in Nitroreductase Based Therapies

The clinically advanced nitroaromatic prodrugs in GDEPT studies that can be activated by nitroreductase enzymes, identified as 2,4-dinitrobenzamide CB1954 (**1**) (5-(aziridin-1-yl)-2,4-dinitrobenzamide) (Prosser et al. 2013). The most studied prodrug with NfsB from *Escherichia coli* is CB1954 (**1**) that is developed in Chester Beatty Research Institute in the 1960s, and it is currently in clinical trials (phase I/II) (Çelik et al. 2016; Prosser et al. 2013). CB1954 (**1**) has a lipophilic character (log P: +1.54) (Denny 2003). This prodrug is successfully transplanted into Walker 256 rat carcinoma and reported as the first therapy. It is converted to potent form by nitroreduction (Knox et al. 1993; Prosser et al. 2013). In this enzyme/prodrug combination systems, the initial conversion is identified from a strongly electron-withdrawing nitro group to an electron-donating hydroxylamine to transform the prodrug into a cytotoxic compound (Christofferson and Wilkie 2009). When NfsA reduces 2-NO_2 group of CB1954 (**1**) to the hydroxylamine (-NHOH), NfsB can be generated the same amounts of the 2- and 4-NHOH products (Vass et al. 2009; Ball et al. 2019). The 2-NO_2 metabolites of CB1954 (**1**) can be found as highly effective at bystander cell killing and 4-NO_2 metabolite is found more cytotoxic (Vass et al. 2009). 2-Amino metabolite that is thought to inhibit growth and development of cancer cells, showed more diffusion and higher stability (Helsby et al. 2004a, 2004b). The results of NfsB nitroreductase with CB1954 (**1**) have been reported for published K_m and peak serum concentration in humans as 862 mM and 5–10 mM, respectively. NTR/CB1954 combination has been determined as inefficient in *in vivo* studies (Vass et al. 2009; Anlezark et al. 1992, 1995; Chung-Faye et al. 2001).

CB1954 (**1**) derived prodrugs are synthesized by forming the amide side chain on the corresponding chloro(di)nitroacid starting materials and then followed by the aziridine ring. These prodrugs have higher selectivity against the cells that is expressed with NfsB. However, the decrease in lipophilicity cause *in vitro* weakening of bystander effect. The decrease in lipophilicity caused *in vivo* studies to reduce the effect against NfsB expressing xenografts (Helsby et al. 2004a, 2004b). Figure 10.3 shows the structures of the known nitro prodrugs used in NTR applications.

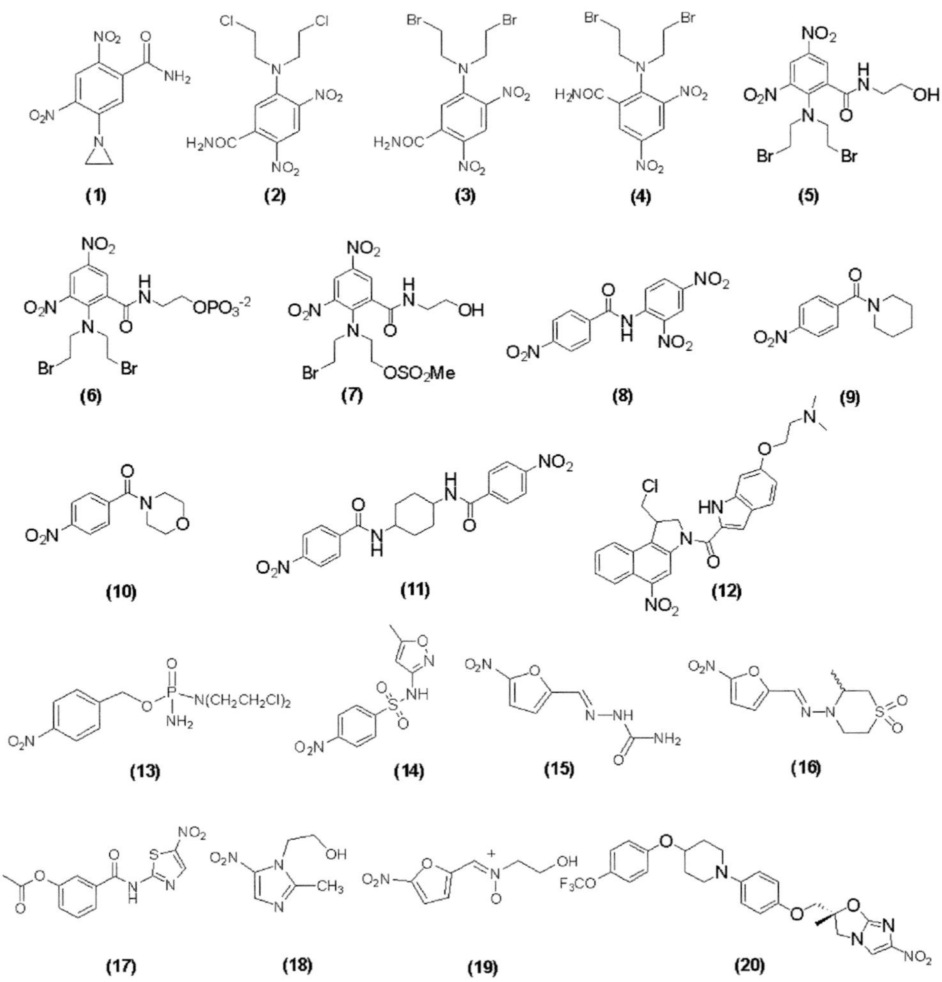

FIGURE 10.3 Structures of the prodrugs used NTR/Prodrug combinations. CB1954 **(1)**, SN23862 **(2)**, SN24927 **(3)**, SN26209 **(4)**, SN27686 **(5)**, SN28343 **(6)**, PR-104A **(7)**, *N*-(2,4-Dinitrophenyl)-4-nitrobenzamide **(8)**, 4-(4-Nitrobenzoyl) piperidine **(9)**, 4-(4-Nitrobenzoyl) morpholine **(10)**, *N*,*N*′-(1,4-Cyclohexyl)bis(4-nitrobenzamide) **(11)**, Nitro-CBI-DEI **(12)**, LH-7 **(13)**, *N*-(5-Methylisoxazol-3-yl)-4-nitrobenzenesulfonamide **(14)**, Nitrofurazone **(15)**, Nifurtimox **(16)**, Nitazoxanide **(17)**, Metronidazole **(18)**, ADC111 **(19)**, Delamanid **(20)**.

10.4.2 Dinitrobenzamide Mustards and Dinitrobenzamides in Nitroreductase Based Therapies

Designing of alternative prodrug compounds for nitroreductases is one of the active drug development field. Low solubility in water and moderate kinetic data from the reduction with NfsB have encouraged the development of analogs of CB1954 (**1**) (Helsby et al. 2004a, 2004b).

The dinitrobenzamide mustard, SN23862 (**2**) (5-[*N,N*-bis(2-chloroethyl)amino]2-hydroxyamino-4-nitrobenzamide) is reduced by nitroreductase at 2-position to generate hydroxylamine product. SN23862 (**2**) has the highest K_m and k_{cat} values than CB1954 (**1**). There are some differences between the activation mechanism of CB1954 (**1**) and SN23862 (**2**) prodrugs although they are reduced by nitroreductase enzymes. The difference depends on the alkylating moiety of SN23862 (**2**) (Friedlos et al. 1997). The combination with NfsB nitroreductase of prodrug exhibits more higher cytotoxic properties (Searle et al. 2004; Anlezark et al. 1992, 1995). SN23862 (**2**) that is chosen as a model prodrug to synthesize many dinitrobenzamide prodrugs, has some potential advantages compared to CB1954 (**1**). These prodrugs that are cytotoxic to tumor cells at high concentrations, are reduced by NfsB to 2-hydroxylamine metabolites (Anlezark et al. 1995). Also, cobalt (III) complexes of nitro containing aromatic and aliphatic mustards have the highest cytotoxicity compared to spheroid.

Another prodrug from the class of DNBMs is activated by NfsB, SN24927 (**3**) that is developed to use in combination with NfsB nitroreductase (Helsby et al. 2004a) is the bromo group containing derivative of SN23862 (**2**). Using NfsB expression to activate the 2,4-dinitrobenzamide-5-mustard, SN24927 (**3**), has exhibited a superior bystander effect compared to CB1954 (**1**). It is supported with *in vivo* analysis for its antitumor activity (Atwell et al. 2007). SN26209 (**4**) is a 3,5-dinitrobenzamide-2-mustard has been shown as a potent NfsB prodrug (Singleton et al. 2007). The activation of DNBMs depends on the reduction of the nitro group at the para position to the mustard for the bioactivation. When the reduction amounts are equal at ortho and para positions of CB1954 (**1**), it can be reduced by NfsA. Expressing of NfsA or NfsB in SKOV3 cancer cells displays more sensitivity to CB1954. DNBMs exhibited higher effect in SKOV3-NfsA cells (Vass et al. 2009; Helsby et al. 2004a).

SN27686 (**5**) is a 3,5-dinitrobenzamide mustard derivative of CB1954 (**1**) and shows more selectivity and higher potential compared to CB1954 (**1**). Also, the prodrug can exhibit higher effect in NfsB-modified xenograft tumor models (Singleton et al. 2007). The phosphate ester derived SN28343 (**6**) that is developed as a most effective prodrug for NTR applications, can show low toxicity within *in vivo* studies (Singleton et al. 2007; Chung-Faye et al. 2001).

New dinitrobenzamide mustards as phosphate esters have been designed due to low solubility of first generation dinitrobenzamides in water. PR-104 (phosphate ester preprodrug form of PR-104A (2-((2-bromoethyl)-2-{[(2-hydroxyethyl)amino]carbonyl}-4,6-dinitro anilino)ethylmethanesulfonate)) that has been firstly developed as a tumor hypoxia-activated (GDEPT independent) therapeutic, is tolerated by humans at up to twenty times the maximum dose of CB1954 (**1**) on a molar basis. NfsB as bacterial nitroreductase exhibits strong reducing for PR-104A (**7**) prodrug (Prosser et al. 2013).

It is well known that various benzamide derivatives are used as drug at a great number of therapeutic agents including some types of cancer. Early studies show that newly designed and synthesized nitro substituted aromatic amide prodrugs using CB1954 (**1**) as a model prodrug have been studied to produce their metabolites that is activated by Ssap-NtrB. Hereby, nitro group(s) containing prodrugs at the different positions on the aromatic ring are examined with their interaction with Ssap-NtrB compared to model prodrugs CB1954 (**1**) and SN23862 (**2**). It is followed by determining of prodrugs for their antiproliferative effects in Hep3B, HT-29 and PC3 cells. *In vitro* findings showed that prodrug (**8**) that is a 2,4-dinitroaniline derivative, is found more active than CB1954 (**1**) (137 fold) and SN23862 (**2**) (31 fold) with Ssap NtrB (Güngör et al. 2018).

To determine the new prodrug/NTR combinations, four different amide substituted core structures such as phenylacetamide, 4-(4-nitrobenzoyl)morpholine/piperidine, *N*-substituted phenyl benzamide, bis(4-nitrobenzamide) have been investigated to be used for nitroreductase based cancer therapy. Designing of these prodrugs that have different pharmacophore groups, are actualized to detect the most effective structures that are activated with Ssap-NtrB. 4-(4-nitrobenzoyl)piperidine (**9**), 4-(4-nitrobenzoyl)morpholine (**10**) and *N,N'*-(1,4-cyclohexyl)bis(4-nitrobenzamide) (**11**) are indicated potential prodrugs which exhibited low IC_{50} values in cancer cells. Both of the studies may show new areas to develop novel prodrugs for NTR based cancer therapy (Güngör et al. 2018, 2019).

To date, the nitroreductase GDEPT studies have been mainly depended on the prodrug CB1954 (**1**). The nitro-chloromethylbenzindoline prodrug (**12**) (nitro-CBI-DEI) can be an excellent prodrug to be used for GDEPT therapy. This is depended on the formation product which is highly cytotoxic cell-permeable by activated *E. coli* NfsB. The nitro-chloromethylbenzindolines (nitro-CBIs) are designed as hypoxia-activated prodrugs of amino analogs of the cyclopropylindoline antitumor antibiotics. Nitro-CBI-DEI substrates have showed high cytotoxicity and bystander effect by the NfsB-PA/nitro-CBI-DEI combination that can be potential for GDEPT (Green et al. 2013). Figure 10.3 shows the structures of the known nitro prodrugs used in NTR applications.

10.4.3 4-Nitrobenzylcarbamates in Nitroreductase Based Therapies

4-Nitrobenzylcarbamates that have low reduction potential (490 mV) (Denny 2003) are the main class of identified prodrugs by NTR (Hay et al. 2005). 4-Nitrobenzylcarbamates are reduced to hydroxylamine via enzymatic reaction and starting material turns to amine group (Denny 2002; Asche et al. 2006). The fragmentation step of the carbamates are highly faster in the presence of electron-donating substituents on the benzyl ring. Different types of cytotoxic amines have been examined as potential GDEPT prodrugs (Denny 2003).

To discovery of carbamates/enzyme therapies, designed and synthesized of nitrobenzylcarbamate and nitroimidazole derivatives of doxorubicin (DOX) that is a well-known anticancer drug, have been performed in multiple reaction steps and examined with the combination of NTR in GDEPT systems. These new nitro-containing carbamate derivatives of anticancer drug exhibit strong cytotoxic effect than DOX in SKOV3 and WiDr cancer cells. Although carbamate prodrugs containing nitrobenzyl units directly bound to DOX can show average selectivity for NTR, the selectivity is increased 10–370 more fold by 4-aminobenzyl-containing carbamate-DOX analogs. However, 2-nitroimidazoylmethyl carbamate prodrug shows moderate selectivity for NTR. The selective cytotoxicity can be observed in NfsB-modified cells within *in vitro* studies. This circumstance is not acceptable for *in vivo* analysis compared to CB1954 (**1**) (Hay et al. 2003; Hay et al. 2005).

Cyclophosphamide nitroaryl and nitrobenzyl phosphoramide mustards are synthesized in multiple steps and the combination with *E. coli* NTR has been examined in V79 cancer cells (Jiang et al. 2006; Hu et al. 2003). Selective cytotoxicity and the efficiency of substrate may be needed for the combination of 4-nitrophenylcyclophosfamide analogs with *E. coli* nitroreductase. LH-7 (**13**) (acyclic 4-nitrobenzylphosphoramide) prodrug has 170.000 fold more selective cytotoxicity in modified cells with NfsB compared to CB1954 (**1**) and therapeutic effect is found higher than CB1954 (**1**) (Vass et al. 2009; Prosser et al. 2013; Malekshah et al. 2016). In early studies, various phosphoramides have been designed to transfer phosphoramide mustard alkylating agent to the cells. When electron-withdrawing nitro group exposed to reduction by Type 1 NTRs, it can introduce into the cells. This event results from the cleavage of benzylic C-O bond. Hence, 4-nitrobenzyl phosphoramide mustards (LH-7 derivatives) are effective substrates for NfsB (Hu et al. 2011). Figure 10.3 shows the structures of the known nitro prodrugs used in NTR applications.

10.4.4 Other Prodrugs in Nitroreductase Based Therapies

Antibiotic activation is performed to determine the potential usage of NTR enzymes. Synthesized prodrug *N*-(5-methylisoxazol-3-yl)-4-nitrobenzenesulfonamide (**14**) that is a prodrug of sulfamethoxazole, has been reported for its directly reductive activation with Ssap-NtrB obtained from the pathogen *S. saprophyticus* by using *in vitro* available conditions. The findings indicate that nitroso (NO), hydroxylamine (NHOH) and amine (NH_2) derivatives of metabolites have been occurred and amine derivative could be used as essential antibiotic (Çelik et al. 2016).

A kind of a topical anti-infective agent, nitrofurazone (**15**) exhibits higher activity with the combination of nitroreductases and different reaction products such as amines and open chain nitriles from nitrofurans have been determined by using Type 1 nitroreductase when compared with nitro-containing prodrugs such as misonidazole and nitracrine (Hall et al. 2011). Nitrofuran furaltadone and nitroimidazole metronidazole compounds have anti-trypanosomal activity. The interests for these compounds can not continue due to their low efficacy and neurotoxicity. Hence, 5-nitrofuran nifurtimox has been discovered for therapeutic applications (Patterson and Wyllie 2014). One of the reported study suggests that the reduction of nitrofurans by Type 1 nitroreductase generates nitroso and hydroxylamine metabolites. It is the evidence that trypanosomal Type 1 nitroreductase can successfully catalyst the reduction of this prodrug. Hereby, the reduction of nifurtimox, is a nitroheterocyclic compound has been identified to understand its trypanocidal effects for the ability of nitroreductase. Nifurtimox (**16**) metabolism has been found more effective in anaerobic conditions (Hall et al. 2011).

NfsB is originally identified due to its bacterial sensitivity for nitrofuran antibiotics (Sastry and Jayaraman 1984). Nitrofurans, nitrothiazoles and nitroimidazoles are known for their *in vitro* antimicrobial activity. Among this action, the reduction of 5-nitro group is performed to produce hydroxylamine metabolites that is a redox active nitrothiazolyl-salicylamide prodrug, nitazoxanide (**17**) (NTZ). This prodrug is identified with its *in vitro* activity against *H. pylori* that is used to determine antimicrobial action of NTZ (**17**) and nitrofuran due to its sensitivity to these prodrugs. Three enzymes pyruvate oxidoreductase (PorGDAB), RdxA, an oxygen-insensitive NADPH nitroreductase, and FrxA, oxygen-insensitive flavin NADPH nitroreductase can be reduced NTZ (**17**) to form of biologically active metabolites. NTZ (**17**) can be used to kill anaerobic bacterial, protozoan, and helminthic species. Prodrugs are therapeutic agents for many microorganisms and the best-known example is Metronidazole (**18**) (MTZ) that is a nitroimidazole prodrug and is often used in combination therapies against *H. pylori* (Sisson et al. 2002). (MTZ) (**18**) has been used for antiprotozoal and antibacterial therapies since 1960s. The combination of *NfsB*/MTZ is thought to be an encouraging prodrug. ADC111 (**19**) is a nitrofurazone analog of a type of nitroaromatic compound nitrofurantoin which is a prodrug and is used to treat urinary tract infections (UTIs). Bactericidal activity of ADC111 (**19**) depended on the presence of nitroreductases in *E. coli*. Prodrugs may be the most important class to be designed in antibiotic therapies using NTRs (Fleck et al. 2014).

A series of synthesized nitro-containing derivatives, 2-nitroaryl-1,2,3,4-tetrahydroisoquinoline, nitro substituted 5,6-dihydrobenzimidazo[2,1-a]isoquinoline-*N*-oxide, nitropyrido and nitropyrido-imidazole (isoquinoline derivative) demonstrate moderate interaction with *E. coli* nitroreductase. Trifluoromethylated derivative and *N*-nitroaryl-1,2,3,4-tetrahydroisoquinoline derivatives have been found more effective and it can be reduced more faster with NQO1 enzyme than NTR compared to CB1954 (Burke et al. 2011). Additionally, quinone derivatives show low cytotoxic activity as well as higher substrate effect (Bailey and Hart 1997; Denny 2002). The most efficient strategy that uses bio-reductive enzyme for activation of prodrug to produce the active form as antibiotic, is developed in early studies. For instance, designing of nitroreductase-activated conjugate of Ciprofloxacin to delivery of fluoroquinolones is successfully activated by nitroreductases as mentioned in detail Section 10.3.2 (Pardeshi et al. 2019).

Diazeniumdiolates that are often designed using proper functional groups into stable compounds, are generally used in biological studies as reliable sources of NO. These compounds are known with their therapeutic applications. One of the important diazeniumdiolate compound, O^2-(4-Nitrobenzyl)-1-(2-methylpiperidine-1-yl)diazene-1-yum-1,2-diolate which has stability in aqueous buffer and can metabolize with NTR to produce nitric oxide (NO), is designed to be substrate for *E. coli* nitroreductase. Due to the strong effect of the substrate in cancer cells, it is used NO based therapies to transport of therapeutic NO to the target site and decrease the high concentrations of NO that trigger apoptosis and cause death of the cells (Sharma et al. 2013).

The known antitubercular prodrugs are isoniazid, ethionamide, prothionamide, thiacetazone, isoxyl, para-aminosalicylic acid, pyrazinamide and delamanid. All of them have not a transport moiety in their molecular structures and are not intentionally designed as prodrugs initially, but the prodrugs properties of compounds are discovered in later times. The activation mechanism of prodrugs in tuberculosis treatment including oxidation, hydrolysis, condensation and reduction by various enzyme sources are different from each other (Laborde et al. 2017). Delamanid (**20**) (OPC-67683) is a nitroimidazole derivative and licensed by two global approvals by EMA and MHWL in 2014. Prodrug can be used in combination with an optimized background antituberculosis regimen in multidrug resistant tuberculosis when the resistance or tolerability problem occur. The antibacterial activity of delamanid is specific for mycobacteria and MIC values are typically in the range of 0.011–0.045 mm against tuberculosis and MDR-TB. Pretomanid (PA-824) which is a bicyclic 4-nitroimidazole derivative similarly delamanid (**20**) is currently in clinical development for MDR-TB (Laborde et al. 2017). Delamanid (**20**) is reduced by the mycobacterial deazaflavin (F420)-dependent nitroreductase (Ddn) encoded by the Rv3547 gene (Laborde et al. 2017; Liu et al. 2018). According to proposed mechanism, the reduction of the nitroimidazole by nitroreductase gives the nitronate (or nitronic acid) intermediate, firstly. Some following rearrangements occur in the structure and nitrous acid and the inactive des-nitro delamanid derivative as major stable metabolite release. Different type rearrangements of intermediates gives a minor inactive metabolite and further oxidations produces another inactive metabolite. The risk of resistance results from mutations in the activating enzyme leading to the decrease of prodrug activation. Mutations in one of the five genes, ddn, fgd1, fbiA, fbiB, and fbiC in bioactivation pathway can result in resistance to delamanid (Laborde et al. 2017; Liu et al. 2018).

Finally, the cytotoxic effect of Tallimustin analogs as anticancer drug with the interaction of nitroreductases are examined in ovarian cancer cells and mouse models, many prodrugs show similar or higher effect with CB1954 (**1**) (Güngör 2016; Denny 2002; Hay et al. 2003). Figure 10.3 shows the structures of the known nitro prodrugs used in NTR applications.

In summary, nitroreductases and their effective enzyme/prodrug combinations have been examined from *E. coli* including NfsA since years (Prosser et al. 2010). NfsA-modified cells exhibit 10-fold more selectivity to prodrug CB1954 (**1**) compared to NfsA (Vass et al. 2009). Although, NfsB is able to reduce the nitro groups at 2- and 4- positions on the structure, NfsA generally reduced the 2-nitro group (Dachs et al. 2009). The efficiency of prodrugs with NfsB is limited due to its poor catalytic activity in the applications of clinically approved drug doses. Development of new enzymes are needed to interact with the prodrugs (Gwenin et al. 2015). Some structural changes and mutagenesis on nitroreductases have been performed to increase the activity and site selectivity of NfsB against CB 1954 (**1**). The research can increase the selectivity at 4-position to generate hydroxylamine metabolite (Bai et al. 2015). NTR based *in vitro* studies indicate that nitro aromatic amide prodrugs including CB1954, SN23862 and their mustards, phosphoramides and prodrugs **8–11**, are more effective and potent in the therapeutic areas, especially for cancer therapies. The metabolites of mono- and dinitro-benzamides can be highly cytotoxic for the cancer cells in GDEPT based nitroreductase/prodrug systems. Different functional groups at prodrug structures can cause the changing of activity in the therapy.

Some of the researchers mentioned that nitroreductases have been studied to develop nitroreductase based biosensors in recent years (Chaignon et al. 2006). For this purpose, prodrug, 4,7-bis(4-dodecylthiophen-2-yl)-5,6-dinitrobenzo[c][1,2,5]thiadiazole (BTTD-NO_2) that is a benzothiazole derivative, can be reduced by using P450 nitroreductase to obtain BTTD-NH_2 metabolite to be observed solid tumor cells that is hypoxic. According to flow cytometry analysis, MG63 tumor cells can be labeled approximately 65% with strong red fluorescence in hypoxic conditions. BTTD-NH_2 can be used as a potent fluorescence probe to determine and label of the tumors (Jiang et al. 2013). The derivatives of 2-nitroarylbenzothiazole and 2-nitroarylbenzoxazole are designed as potential nitroreductase substrates to determine the various microorganisms and yeasts through fluorescence effect of the substrates (Cellier et al. 2011).

As a result, nitroreductase based prodrugs are widely used in the different therapeutic areas and other applications that is mentioned in Section 10.3.2. The new prodrugs are still needed for nitroreductase based enzyme/prodrug combinations and discovering of selective and potential drugs to compare with the best known prodrugs CB1954, SN23862 and PR104. This is achieved with some of the early studies determined by nitroreductases for prodrug therapies.

10.5 Conclusion and Future Perspectives

It is well known, designing and development of novel and effective prodrugs are the most important strategy to solve the various types of problems such as insolubility, physicochemical and pharmaceutical properties, specify to target, bioavailability, pharmacokinetic and pharmacodynamic profiles in drug discovery and drug permeation mainly depends on increasing lipophilicity. Several prodrugs which can be generally activated valuable enzymes, are used in various clinical applications. Especially, anticancer prodrugs focus on target specific molecules like enzymes due to overexpressed in tumor cells. This circumstance can lead to promising chemotherapeutic prodrugs including enzyme activated prodrugs. Enzymes are generally used for the activation of prodrugs. Nitroreductases that can reduce the nitro groups to active metabolites such as nitroso, hydroxylamine and amines come from fungi or bacteria such as *E. coli* NTR. It is the most important class of oxidoreductases and different types of NTRs and newly strategies have been developed in the recent years.

Consequently, prodrugs are a valuable part of the drug design and delivery process, dramatically increasing the number of approved prodrugs is the evidence of this approach. The most important enzyme class for the combination of enzyme and nitro-containing prodrug systems is indicated as nitroreductases. This chapter contains the determining of nitroreductases and their therapeutic applications in enzyme/prodrug therapies. In addition, the different and valuable applications of NTRs are successfully given. Thereby, it will lead to new areas to discover more effective nitroreductase based enzyme prodrug concepts.

REFERENCES

Aghahosseini, Hamideh, Seyed Jamal Tabatabaei Rezaei, Mahshid Maleki, Davod Abdolahnjadian, Ali Ramazani, and Hashem Shahroosvand. 2019. "Pt(II)-Based Artificial Nitroreductase: An Efficient and Highly Stable Nanozyme." *ChemistrySelect* 4 (4): 1387–93. doi:10.1002/slct.201803776.

Anderson, Peter J., Lindsay J. Cole, David B. McKay, and Barrie Entsch. 2002. "A Flavoprotein Encoded in *Selenomonas ruminantium* Is Characterized after Expression in *Escherichia coli*." *Protein Expression and Purification* 24 (3): 429–38. doi:10.1006/prep.2001.1581.

Anlezark, Gillian M., Roger G. Melton, Roger F. Sherwood, Brian Coles, Frank Friedlos, and Richard J. Knox. 1992. "The Bioactivation of 5-(Aziridin-1-Yl)-2,4-Dinitrobenzamide (CB1954)—I: Purification and Properties of a Nitroreductase Enzyme from *Escherichia coli*—A Potential Enzyme for Antibody-Directed Enzyme Prodrug Therapy (ADEPT)." *Biochemical Pharmacology* 44 (12): 2289–95. doi:10.1016/0006-2952(92)90671-5.

Anlezark, Gillian M., Roger G. Melton, Roger F. Sherwood, William Robert Milson, William Denny, Brian D. Palmer, Richard J. Knox, Frank Friedlos, and A. Williams. 1995. "Bioactivation of Dinitrobenzamide Mustards by an *E. coli* B Nitroreductase." *Biochemical Pharmacology* 50 (5): 609–18. doi:10.1016/0006-2952(95)00187-5.

Asche, Christian, Pascal Dumy, Danièle Carrez, Alain Croisy, and Martine Demeunynck. 2006. "Nitrobenzylcarbamate Prodrugs of Cytotoxic Acridines for Potential Use with Nitroreductase Gene-Directed Enzyme Prodrug Therapy." *Bioorganic and Medicinal Chemistry Letters* 16 (7): 1990–94. doi:10.1016/j.bmcl.2005.12.089.

Atwell, Graham J., Shangjin Yang, Frederik B. Pruijn, Susan M. Pullen, Alison Hogg, Adam V. Patterson, William R. Wilson, and William A. Denny. 2007. "Synthesis and Structure-Activity Relationships for 2,4-Dinitrobenzamide-5- Mustards as Prodrugs for the *Escherichia coli* NfsB Nitroreductase in Gene Therapy." *Journal of Medicinal Chemistry* 50 (6): 1197–212. doi:10.1021/jm061062o.

Bai, Jing, Jun Yang, Yong Zhou, and Qing Yang. 2015. "Structural Basis of *Escherichia coli* Nitroreductase NfsB Triple Mutants Engineered for Improved Activity and Regioselectivity toward the Prodrug CB1954." *Process Biochemistry* 50 (11): 1760–66. doi:10.1016/j.procbio.2015.08.012.

Bailey, Susan M., and Ian R. Hart. 1997. "Nitroreductase Activation of CB1954–An Alternative 'suicide' Gene System." *Gene Therapy* 4 (2): 80–81. doi:10.1038/sj.gt.3300400.

Ball, Patrick, Emma Thompson, Simon Anderson, Vanessa Gwenin, and Chris Gwenin. 2019. "Time Dependent HPLC Analysis of the Product Ratio of Enzymatically Reduced Prodrug CB1954 by a Modified and Immobilised Nitroreductase." *European Journal of Pharmaceutical Sciences* 127: 217–24. doi:10.1016/j.ejps.2018.11.001.

Bang, Seo Young, Jeong Hoon Kim, Phil Young Lee, Kwang Hee Bae, Jong Suk Lee, Pan Soo Kim, Do Hee Lee, Pyung Keun Myung, Byoung Chul Park, and Sung Goo Park. 2012. "Confirmation of Frm2 as a Novel Nitroreductase in *Saccharomyces cerevisiae*." *Biochemical and Biophysical Research Communications* 423 (4): 638–41. doi:10.1016/j.bbrc.2012.05.156.

Berne, Cécile, Lorena Betancor, Heather R. Luckarift, and Jim C. Spain. 2006. "Application of a Microfluidic Reactor for Screening Cancer Prodrug Activation Using Silica-Immobilized Nitrobenzene Nitroreductase." *Biomacromolecules* 7 (9): 2631–36. doi:10.1021/bm060166d.

Burke, Philip J., Lai Chun Wong, Terence C. Jenkins, Richard J. Knox, and Stephen P. Stanforth. 2011. "The Synthesis of 2-Nitroaryl-1,2,3,4-Tetrahydroisoquinolines, Nitro-Substituted 5,6-Dihydrobenzimidazo[2,1-a] Isoquinoline N-Oxides and Related Heterocycles as Potential Bioreducible Substrates for the Enzymes NAD(P)H: Quinone Oxidoreductase 1 and *E. coli*." *Bioorganic and Medicinal Chemistry Letters* 21 (24): 7447–50. doi:10.1016/j.bmcl.2011.10.044.

Caballero, Antonio, Juan J. Lázaro, Juan L. Ramos, and Abraham Esteve-Núñez. 2005. "PnrA, a New Nitroreductase-Family Enzyme in the TNT-Degrading Strain *Pseudomonas putida* JLR11." *Environmental Microbiology* 7 (8): 1211–19. doi:10.1111/j.1462-2920.2005.00801.x.

Cao, Linqiu. 2005. "Immobilised Enzymes: Science or Art?" *Current Opinion in Chemical Biology* 9 (2): 217–26. doi:10.1016/j.cbpa.2005.02.014.

Cartwright, Nancy J., and Ronald B. Cain. 2015. "Bacterial Degradation of the Nitrobenzoic Acids." *Biochemical Journal* 71 (2): 248–61. doi:10.1042/bj0710248.

Çelik, AAyhan, GGülden Yetiş, Mehmet Ay, and Tuğba Güngör, 2016. "Modification of Existing Antibiotics in the Form of Precursor Prodrugs That Can Be Subsequently Activated by Nitroreductases of the Target Pathogen." *Bioorganic and Medicinal Chemistry Letters* 26 (16): 4057–60. doi:10.1016/j.bmcl.2016.06.081.

Çelik, Ayhan, and Gülden Yetiş. 2012. "An Unusually Cold Active Nitroreductase for Prodrug Activations." *Bioorganic and Medicinal Chemistry* 20 (11): 3540–50. doi:10.1016/j.bmc.2012.04.004.

Cellier, Marie, Olivier J. Fabrega, Elizabeth Fazackerley, Arthur L. James, Sylvain Orenga, John D. Perry, Vindhya L. Salwatura, and Stephen P. Stanforth. 2011. "2-Arylbenzothiazole, Benzoxazole and Benzimidazole Derivatives as Fluorogenic Substrates for the Detection of Nitroreductase and Aminopeptidase Activity in Clinically Important Bacteria." *Bioorganic and Medicinal Chemistry* 19 (9): 2903–10. doi:10.1016/j.bmc.2011.03.043.

Chaignon, Philippe, Sylvie Cortial, Aviva P. Ventura, Philippe Lopes, Frédéric Halgand, Olivier Laprevote, and Jamal Ouazzani. 2006. "Purification and Identification of a Bacillus Nitroreductase: Potential Use in 3,5-DNBTF Biosensoring System." *Enzyme and Microbial Technology* 39 (7): 1499–506. doi:10.1016/j.enzmictec.2006.04.023.

Chen, Ming-Jer, Nicola K. Green, Gary M. Reynolds, Joanne R. Flavell, Vivien Mautner, David James Kerr, Lawrence S. Young, and Peter F. Searle. 2004. "Enhanced Efficacy of *Escherichia coli* Nitroreductase/CB1954 Prodrug Activation Gene Therapy Using an E1B-55K-Deleted Oncolytic Adenovirus Vector." *Gene Therapy* 11 (14): 1126–36. doi:10.1038/sj.gt.3302271.

Christofferson, Andrew, and John Wilkie. 2009. "Mechanism of CB1954 Reduction by *Escherichia coli* Nitroreductase." *Biochemical Society Transactions* 413–18. doi:10.1042/bst0370413.

Clement, Bernd and Katrin Lopian. 2003. "Characterization of in vitro biotransformation of new, orally active, direct thrombin inhibitor ximelagatran, an amidoxime and ester prodrug." *Drug Metabolism and Disposition* 31 (5): 645–51.

Cui, Lei, Ye Zhong, Weiping Zhu, Yufang Xu, Qingshan Du, Xin Wang, Xuhong Qian, and Yi Xiao. 2011. "A New Prodrug-Derived Ratiometric Fluorescent Probe for Hypoxia: High Selectivity of Nitroreductase and Imaging in Tumor Cell." *Organic Letters* 13 (5): 928–31. doi:10.1021/ol102975t.

Dachs, Gabi U., Michelle A. Hunt, Sophie Syddall, Dean C. Singleton, and Adam V. Patterson. 2009. "Bystander or No Bystander for Gene-Directed Enzyme Prodrug Therapy." *Molecules* 14 (11): 4517–45. doi:10.3390/molecules14114517.

Denny, William A. 2003. "Prodrugs for Gene-Directed Enzyme-Prodrug Therapy (Suicide Gene Therapy)." *Journal of Biomedicine and Biotechnology* 2003 (1): 48–70. doi:10.1155/s1110724303209098.

Denny, William A. 2002. "Nitroreductase-Based GDEPT." *Current Pharmaceutical Design* 8: 1349–61.

Desmoulin, Franck, Véronique Gilard, Myriam Malet-Martino, and Robert Martino. 2002. "Metabolism of capecitabine, an oral fluorouracil prodrug: (19)F NMR studies in animal models and human urine." *Drug Metabolism and Disposition* 30 (11): 1221–29. doi:10.1124/dmd.30.11.1221.

Duarte, Sónia, Georges Carle, Henrique Faneca, Maria C. Pedroso de Lima, and Valérie Pierrefite-Carle. 2012. "Suicide Gene Therapy in Cancer: Where Do We Stand Now?" *Cancer Letters* 324 (2): 160–70. doi:10.1016/j.canlet.2012.05.023.

Ehrnebo, Maths, Sten-Ove Nilsson, and Lars O. Boreus. 1979. "Pharmacokinetics of ampicillin and its prodrugs bacampicillin and pivampicillin in man." *Journal of Pharmacokinetics and Biopharmaceutics* 7 (5): 429–51. doi:10.1007/BF01062386.

Emptage, Caroline D., Richard J. Knox, Michael J. Danson, and David W. Hough. 2009. "Nitroreductase from *Bacillus licheniformis*: A Stable Enzyme for Prodrug Activation." *Biochemical Pharmacology* 77 (1): 21–29. doi:10.1016/j.bcp.2008.09.010.

Fan, Yunshi, Mi Lu, Xie An Yu, Miaoling He, Yu Zhang, Xiao Nan Ma, Junping Kou, Bo Yang Yu, and Jiangwei Tian. 2019. "Targeted Myocardial Hypoxia Imaging Using a Nitroreductase-Activatable Near-Infrared Fluorescent Nanoprobe." *Analytical Chemistry* 91: 6585–592. doi:10.1021/acs.analchem.9b00298.

Fleck, Laura E., E. Jeffrey North, Richard E. Lee, Lawrence R. Mulcahy, Gabriele Casadei, and Kim Lewis. 2014. "A Screen for and Validation of Prodrug Antimicrobials." *Antimicrobial Agents and Chemotherapy* 58 (3): 1410–19. doi:10.1128/aac.02136-13.

Franco, Jefferson Honorio, Bianca F. da Silva, Alexandre A. de Castro, Teodorico C. Ramalho, María Isabel Pividori, and Maria Valnice Boldrin Zanoni. 2018. "Biotransformation of Disperse Dyes Using Nitroreductase Immobilized on Magnetic Particles Modified with Tosyl Group: Identification of Products by LC-MS-MS and Theoretical Studies Conducted with DNA." *Environmental Pollution* 242: 863–71. doi:10.1016/j.envpol.2018.07.054.

Friedlos, Frank, William A. Denny, Brian D. Palmer, and Caroline J. Springer. 1997. "Mustard Prodrugs for Activation by *Escherichia coli* Nitroreductase in Gene-Directed Enzyme Prodrug Therapy." *Journal of Medicinal Chemistry* 40 (8): 1270–75. doi:10.1021/jm960794l.

Chung-Faye, G., D. Palmer, D. Anderson, J. Clark, M. Dowries, J. Baddeley, S. Hussain, et al. 2001. "Virus-Directed, Enzyme Prodrug Therapy with Nitroimidazole Reductase: A Phase I and Pharmacokinetic Study of Its Prodrug, CB1954." *Clinical Cancer Research* 7 (9): 2662–68.

Grazú, Valeria, Olga Abian, Cesar Mateo, Francisco Batista-Viera, Roberto Fernández-Lafuente, and José Manuel Guisán. 2005. "Stabilization of Enzymes by Multipoint Immobilization of Thiolated Proteins on New Epoxy-Thiol Supports." *Biotechnology and Bioengineering* 90 (5): 597–605. doi:10.1002/bit.20452.

Green, Laura K., Sophie P. Syddall, Kendall M. Carlin, Glenn D. Bell, Christopher P. Guise, Alexandra M. Mowday, Michael P. Hay, Jeffrey B. Smaill, Adam V. Patterson, and David F. Ackerley. 2013. "*Pseudomonas aeruginosa* NfsB and Nitro-CBI-DEI - a Promising Enzyme/Prodrug Combination for Gene-Directed Enzyme Prodrug Therapy." *Molecular Cancer* 12 (1): 4–9. doi:10.1186/1476-4598-12-58.

Grove, Jane I, Andrew L Lovering, Christopher Guise, Paul R Race, Christopher J Wrighton, Scott A White, Eva I Hyde, and Peter F Searle. 2003. "Generation of *Escherichia coli* Nitroreductase Mutants Conferring Improved Cell Sensitization to the Prodrug CB1954." *Cancer Research* 63 (17): 5532–37.

Guise, Christopher P., Jane I. Grove, Eva Ilona Hyde, and Peter F. Searle. 2007. "Direct Positive Selection for Improved Nitroreductase Variants Using SOS Triggering of Bacteriophage Lambda Lytic Cycle." *Gene Therapy* 14 (8): 690–98. doi:10.1038/sj.gt.3302919.

Guise, Christopher P., Maria R. Abbattista, Rachelle S. Singleton, Samuel D. Holford, Joanna Connolly, Gabi U. Dachs, Stephen B. Fox, et al. 2010. "The Bioreductive Prodrug PR-104A Is Activated under Aerobic Conditions by Human Aldo-Keto Reductase 1C3." *Cancer Research* 70 (4): 1573–84. doi:10.1158/0008-5472.CAN-09-3237.

Güngör, Tugba. 2016."Development of Aromatic and Heterocyclic Prodrug Compounds for Nitroreductase Based Cancer Therapy," PhD Thesis, Çanakkale Onsekiz Mart University, Turkey.

Güngör, Tuğba, Ferah Cömert Önder, Esra Tokay, Ünzile Güven Gülhan, Nelin Hacıoğlu, Tuğba Taşkın Tok, Ayhan Çelik, Feray Köçkar, and Mehmet Ay. 2019. "PRODRUGS FOR NITROREDUCTASE BASED CANCER THERAPY- 2: Novel Amide/NTR Combinations Targeting PC3 Cancer Cells." *European Journal of Medicinal Chemistry* 383–400. doi:10.1016/j.ejmech.2019.03.035.

Güngör, Tugba, Gulden Yetis, Ferah C. Onder, Esra Tokay, Tugba T. Tok, Ayhan Celik, Mehmet Ay, and Feray Kockar. 2018. "Prodrugs for Nitroreductase Based Cancer Therapy-1: Metabolite Profile, Cell Cytotoxicity and Molecular Modeling Interactions of Nitro Benzamides with Ssap-NtrB." *Medicinal Chemistry* 14 (5): 495–507. doi:10.2174/1573406413666171129224424.

Guo, Ting, Lei Cui, Jiaoning Shen, Weiping Zhu, Yufang Xu, and Xuhong Qian. 2013. "A Highly Sensitive Long-Wavelength Fluorescence Probe for Nitroreductase and Hypoxia: Selective Detection and Quantification." *Chemical Communications* 49 (92): 10820–22. doi:10.1039/c3cc45367g.

Gwenin, Vanessa V., Paramasivan Poornima, Jennifer Halliwell, Patrick Ball, George Robinson, and Chris D. Gwenin. 2015. "Identification of Novel Nitroreductases from *Bacillus cereus* and Their Interaction with the CB1954 Prodrug." *Biochemical Pharmacology* 98 (3): 392–402. doi:10.1016/j.bcp.2015.09.013.

Hajnal, Kelemen, Hancu Gabriel, Rusu Aura, Varga Erzsébet, and Székely Szentmiklósi Blanka. 2016. "Prodrug Strategy in Drug Development." *Acta Medica Marisiensis* 62 (3): 356–62. doi:10.1515/amma-2016-0032.

Hall, Belinda S., Christopher Bot, and Shane R. Wilkinson. 2011. "Nifurtimox Activation by Trypanosomal Type I Nitroreductases Generates Cytotoxic Nitrile Metabolites." *Journal of Biological Chemistry* 286 (15): 13088–95. doi:10.1074/jbc. M111.230847.

Hamstra, Daniel A., Kuei C. Lee, Joseph M. Tychewicz, Victor D. Schepkin, Bradford A. Moffat, Mark Chen, Kenneth J. Dornfeld, et al. 2004. "The Use of 19F Spectroscopy and Diffusion-Weighted MRI to Evaluate Differences in Gene-Dependent Enzyme Prodrug Therapies." *Molecular Therapy* 10 (5): 916–28. doi:10.1016/j.ymthe.2004.07.022.

Harvey, Tracey J., Ivo M. Hennig, Steven D. Shnyder, Patricia A. Cooper, N. Ingram, G. D. Hall, P. J. Selby, and John D. Chester. 2011. "Adenovirus-Mediated Hypoxia-Targeted Gene Therapy Using HSV Thymidine Kinase and Bacterial Nitroreductase Prodrug-Activating Genes In Vitro and In Vivo." *Cancer Gene Therapy* 18 (11): 773–84. doi:10.1038/cgt.2011.43.

Hay, Michael P., Graham J. Atwell, William R. Wilson, Susan M. Pullen, and William A. Denny. 2003. "Structure-Activity Relationships for 4-Nitrobenzyl Carbamates of 5-Aminobenz[e]Indoline Minor Groove Alkylating Agents as Prodrugs for GDEPT in Conjunction with *E. coli* Nitroreductase." *Journal of Medicinal Chemistry* 46 (12): 2456–66. doi:10.1021/jm0205191.

Hay, Michael P., William R. Wilson, and William A. Denny. 2005. "Nitroarylmethylcarbamate Prodrugs of Doxorubicin for Use with Nitroreductase Gene-Directed Enzyme Prodrug Therapy." *Bioorganic and Medicinal Chemistry* 13 (12): 4043–55. doi:10.1016/j.bmc.2005.03.055.

Hdiouech, Slim, Felipe Bruna, Isabelle Batisson, Pascale Besse-Hoggan, Vanessa Prevot, and Christine Mousty. 2019. "Amperometric Detection of the Herbicide Mesotrione Based on Competitive Reactions at Nitroreductase@layered Double Hydroxide Bioelectrode." *Journal of Electroanalytical Chemistry* 835: 324–28. doi:10.1016/j.jelechem.2019.01.054.

He, George, Joseph Massarella, and Penelope Ward. 1999. "Clinical Pharmacokinetics of the Prodrug Oseltamivir and its Active Metabolite Ro 64-0802." *Clinical Pharmacokinetics* 37 (6): 471–84. doi:10.2165/00003088-199937060-00003.

Helsby, Nuala Ann, Dianne M. Ferry, Adam V. Patterson, Susan M. Pullen, and William R. Wilson. 2004a. "2-Amino Metabolites Are Key Mediators of CB 1954 and SN 23862 Bystander Effects in Nitroreductase GDEPT." *British Journal of Cancer* 90 (5): 1084–92. doi:10.1038/sj.bjc.6601612.

Helsby, Nuala A., Graham J. Atwell, Shangjin Yang, Brian D. Palmer, Robert F. Anderson, Susan M. Pullen, Dianne M. Ferry, Alison Hogg, William R. Wilson, and William A. Denny. 2004b. "Aziridinyldinitrobenzamides: Synthesis and Structure-Activity Relationships for Activation by *E. coli* Nitroreductase." *Journal of Medicinal Chemistry* 47 (12): 3295–307. doi:10.1021/jm0498699.

Hodge, R. Anthony, Vere Hodge, David Sutton, Malcolm R. Boyd, Michael R. Harnden, and Richard L. Jarvest. 1989. "Selection of an oral prodrug (BRL 42810; famciclovir) for the antiherpesvirus agent BRL 39123 [9-(4-hydroxy-3-hydroxymethylbut-1- yl)guanine; penciclovir]." *Antimicrobial Agents and Chemotherapy* 33 (10): 1765–73. doi:10.1128/AAC.33.10.1765.

Hu, Longqin, Xinghua Wu, Jiye Han, Lin Chen, Simon O. Vass, Patrick Browne, Belinda S. Hall, et al. 2011. "Synthesis and Structure-Activity Relationships of Nitrobenzyl Phosphoramide Mustards as Nitroreductase-Activated Prodrugs." *Bioorganic and Medicinal Chemistry Letters* 21 (13): 3986–91. doi:10.1016/j.bmcl.2011.05.009.

Hu, Longqin, Chengzhi Yu, Yongying Jiang, Jiye Han, Zhuorong Li, Patrick Browne, Paul R. Race, Richard J. Knox, Peter F. Searle, and Eva I. Hyde. 2003. "Nitroaryl Phosphoramides as Novel Prodrugs for *E. coli* Nitroreductase Activation in Enzyme Prodrug Therapy." *Journal of Medicinal Chemistry* 46 (23): 4818–21. doi:10.1021/jm034133h.

Jaberipour, Mansooreh, Simon O. Vass, Christopher P. Guise, Jane I. Grove, Richard J. Knox, Longqin Hu, Eva I. Hyde, and Peter F. Searle. 2010. "Testing Double Mutants of the Enzyme Nitroreductase for Enhanced Cell Sensitisation to Prodrugs: Effects of Combining Beneficial Single Mutations." *Biochemical Pharmacology* 79 (2): 102–11. doi:10.1016/j.bcp.2009.07.025.

Jiang, Qian, Zhanyuan Zhang, Jiao Lu, Yan Huang, Zhiyun Lu, Yanfei Tan, and Qing Jiang. 2013. "A Novel Nitro-Substituted Benzothiadiazole as Fluorescent Probe for Tumor Cells under Hypoxic Condition." *Bioorganic and Medicinal Chemistry* 21 (24): 7735–41. doi:10.1016/j.bmc.2013.10.019.

Jiang, Yongying, Jiye Han, Chengzhi Yu, Simon O. Vass, Peter F. Searle, Patrick Browne, Richard J. Knox, and Longqin Hu. 2006. "Design, Synthesis, and Biological Evaluation of Cyclic and Acyclic Nitrobenzylphosphoramide Mustards for *E. coli* Nitroreductase Activation." *Journal of Medicinal Chemistry* 49 (14): 4333–43. doi:10.1021/jm051246n.

Ju, Kou-San, and Rebecca E. Parales. 2010. "Nitroaromatic Compounds, from Synthesis to Biodegradation." *Microbiology and Molecular Biology Reviews* 74 (2): 250–72. doi:10.1128/mmbr.00006-10.

Knox, Richard J., and Torn A. Connors. 1995. "Antibody-Directed Enzyme Prodrug Therapy: Potential in Cancer." *Clinical Immunotherapeutics* 3 (2): 136–53. doi:10.1007/BF03259275.

Knox, Richard J., Frank Friedlos, and Marion P. Boland. 1993. "The Bioactivation of CB 1954 and Its Use as a Prodrug in Antibody-Directed Enzyme Prodrug Therapy (ADEPT)." *Cancer and Metastasis Reviews* 12 (2): 195–212. doi:10.1007/BF00689810.

Koike, Hideaki, Hiroshi Sasaki, Toshiro Kobori, Shuhei Zenno, Kaoru Saigo, Michael E.P. Murphy, Elinor T. Adman, and Masaru Tanokura. 1998. "1.8 Å Crystal Structure of the Major NAD(P)H:FMN Oxidoreductase of a Bioluminescent Bacterium, *Vibrio fischeri*: Overall Structure, Cofactor and Substrate-Analog Binding, and Comparison with Related Flavoproteins." *Journal of Molecular Biology* 280 (2): 259–73. doi:10.1006/jmbi.1998.1871.

Kumari, Rashmi, Dhanya Sunil, and Raghumani S. Ningthoujam. 2019. "Naphthalimides in Fluorescent Imaging of Tumor Hypoxia – An Up-to-Date Review." *Bioorganic Chemistry* 88: 102979. doi:10.1016/j.bioorg.2019.102979.

Kutty, Razia, and George N. Bennett. 2005. "Biochemical Characterization of Trinitrotoluene Transforming Oxygen-Insensitive Nitroreductases from *Clostridium acetobutylicum* ATCC 824." *Archives of Microbiology* 184 (3): 158–67. doi:10.1007/s00203-005-0036-x.

Laborde, Julie, Céline Deraeve, and Vania Bernardes-Génisson. 2017. "Update of Antitubercular Prodrugs from a Molecular Perspective: Mechanisms of Action, Bioactivation Pathways, and Associated Resistance." *ChemMedChem* 12 (20): 1657–76. doi:10.1002/cmdc.201700424.

Lei, Benfang, Mengyao Liu, Shouqin Huang, and Shiao-Chun Tu. 1994. "Vibrio Harveyi NADPH-Flavin Oxidoreductase: Cloning, Sequencing and Overexpression of the Gene and Purification and Characterization of the Cloned Enzyme." *Journal of Bacteriology* 176 (12): 3552–58. doi:10.1128/jb.176.12.3552-3558.1994.

Lewis, Thomas A., David A. Newcombe, and Ronald L. Crawford. 2004. "Bioremediation of Soils Contaminated with Explosives." *Journal of Environmental Management* 70 (4): 291–307. doi:10.1016/j.jenvman.2003.12.005.

Li, Yuhao, Yun Sun, Jiachang Li, Qianqian Su, Wei Yuan, Yu Dai, Chunmiao Han, Qiuhong Wang, Wei Feng, and Fuyou Li. 2015. "Ultrasensitive Near-Infrared Fluorescence-Enhanced Probe for In Vivo Nitroreductase Imaging." *Journal of the American Chemical Society* 137 (19): 6407–16. doi:10.1021/jacs.5b04097.

Li, Zhao, Xiaohua Li, Xinghui Gao, Yangyang Zhang, Wen Shi, and Huimin Ma. 2013. "Nitroreductase Detection and Hypoxic Tumor Cell Imaging by a Designed Sensitive and Selective Fluorescent Probe, 7-[(5-Nitrofuran-2-Yl)Methoxy]-3 H -Phenoxazin-3-One." *Analytical Chemistry* 85 (8): 3926–32. doi:10.1021/ac400750r.

Litalien, Catherine, Yves Théorêt, and Christophe Faure. 2005. "Pharmacokinetics of Proton Pump Inhibitors in Children." *Clinical Pharmacokinetics* 44 (5): 441–66. doi:10.2165/00003088-200544050-00001.

Liu, Yongge, Makoto Matsumoto, Hidekaza Ishida, Kinue Ohguro, Masuhiro Yoshitake, Rajesh Gupta, Lawrence Geiter, and Jeffrey Hafkin. 2018. "Delamanid: From Discovery to Its Use for Pulmonary Multidrug-Resistant Tuberculosis (MDR-TB)." *Tuberculosis* 111: 20–30. doi:10.1016/j.tube.2018.04.008.

Luo, Shenzheng, Rongfeng Zou, Junchen Wu, and Markita P. Landry. 2017. "A Probe for the Detection of Hypoxic Cancer Cells." *ACS Sensors* 2 (8): 1139–45. doi:10.1021/acssensors.7b00171.

Maksimova, Yu. G., A. Yu. Maksimov, and V. A. Demakov. 2018. "Biotechnological Approaches to the Bioremediation of an Environment Polluted with Trinitrotoluene." *Applied Biochemistry and Microbiology* 54 (8): 767–79. doi:10.1134/S0003683818080045.

Malekshah, Obeid M., Xuguang Chen, Alireza Nomani, Siddik Sarkar, and Arash Hatefi. 2016. "Enzyme/Prodrug Systems for Cancer Gene Therapy." *Current Pharmacology Reports* 2 (6): 299–308. doi:10.1007/s40495-016-0073-y.

Miller, Anne Frances, Jonathan T. Park, Kyle L. Ferguson, Warintra Pitsawong, and Andreas S. Bommarius. 2018. "Informing Efforts to Develop Nitroreductase for Amine Production." *Molecules* 23 (2): 211. doi:10.3390/molecules23020211.

Murray, Jessica R., Clementina A. Mesaros, Volker M. Arlt, Albrecht Seidel, Ian A. Blair, and Trevor M. Penning. 2018. "Role of Human Aldo-Keto Reductases in the Metabolic Activation of the Carcinogenic Air Pollutant 3-Nitrobenzanthrone." Research-article. *Chemical Research in Toxicology* 31 (11): 1277–88. doi:10.1021/acs.chemrestox.8b00250.

Nadeau, Lloyd J., Zhongqi He, and Jim C. Spain. 2000. "Production of 2-Amino-5-Phenoxyphenol from 4-Nitrobiphenyl Ether Using Nitrobenzene Nitroreductase and Hydroxylaminobenzene Mutase from *Pseudomonas pseudoalcaligenes* JS45." *Journal of Industrial Microbiology and Biotechnology* 24 (4): 301–5. doi:10.1038/sj.jim.2900821.

Nokhbeh, M. R., Shahram Boroumandi, Nicholas Pokorny, Peter Koziarz, E. Suzanne Paterson, and Iain B. Lambert. 2002. "Identification and Characterization of SnrA, an Inducible Oxygen-Insensitive Nitroreductase in *Salmonella enterica* Serovar Typhimurium TA1535." *Mutation Research - Fundamental and Molecular Mechanisms of Mutagenesis* 508 (1–2): 59–70. doi:10.1016/S0027-5107(02)00174-4.

Oda, Yoshimitsu. 2016. "Development and Progress for Three Decades in Umu Test Systems." *Genes and Environment* 38 (1): 1–14. doi:10.1186/s41021-016-0054-8.

Oliveira, Iuri, Diego Bonatto, João Antonio, and Pêgas Henriques. 2010. "Nitroreductases: Enzymes with Environmental, Biotechnological and Clinical Importance." *Current Research, Technology and Education Topics in Applied Microbiology and Microbial Biotechnology*, 1008–1019.

Pardeshi, Kundansingh A., T. Anand Kumar, Govindan Ravikumar, Manjulika Shukla, Grace Kaul, Sidharth Chopra, and Harinath Chakrapani. 2019. "Targeted Antibacterial Activity Guided by Bacteria-Specific Nitroreductase Catalytic Activation to Produce Ciprofloxacin." Research-article. *Bioconjugate Chemistry* 30 (3): 751–59. doi:10.1021/acs.bioconjchem.8b00887.

Park, Ho-Jin -J, Christian O.A. Reıser, Simone Kondruweıt, Helmut Erdmann, Rolf D. Schmıd, and Mathias Sprınzl. 1992. "Purification and Characterization of a NADH Oxidase from the Thermophile *Thermus thermophilus* HB8." *European Journal of Biochemistry* 205 (3): 881–85. doi:10.1111/j.1432-1033.1992.tb16853.x.

Patterson, Adam, Mark Saunders, and Olga Greco. 2003. "Prodrugs in Genetic Chemoradiotherapy." *Current Pharmaceutical Design* 9 (26): 2131–54. doi:10.2174/1381612033454117.

Patterson, Stephen, and Susan Wyllie. 2014. "Nitro Drugs for the Treatment of Trypanosomatid Diseases: Past, Present, and Future Prospects." *Trends in Parasitology* 30 (6): 289–98. doi:10.1016/j.pt.2014.04.003.

Prosser, Gareth A., Janine N. Copp, Alexandra M. Mowday, Christopher P. Guise, Sophie P. Syddall, Elsie M. Williams, Claire N. Horvat, et al. 2013. "Creation and Screening of a Multi-Family Bacterial Oxidoreductase Library to Discover Novel Nitroreductases That Efficiently Activate the Bioreductive Prodrugs CB1954 and PR-104A." *Biochemical Pharmacology* 85 (8): 1091–1103. doi:10.1016/j.bcp.2013.01.029.

Prosser, Gareth A., Janine N. Copp, Sophie P. Syddall, Elsie M. Williams, Jeff B. Smaill, W. R. Wilson, A. V. Patterson, and David F. Ackerley. 2010. "Discovery and Evaluation of *Escherichia coli* Nitroreductases That Activate the Anti-Cancer Prodrug CB1954." *Biochemical Pharmacology* 79 (5): 678–87. doi:10.1016/j.bcp.2009.10.008.

Qin, Wenjing, Chenchen Xu, Yanfei Zhao, Changmin Yu, Sheng Shen, Lin Li, and Wei Huang. 2018. "Recent Progress in Small Molecule Fluorescent Probes for Nitroreductase." *Chinese Chemical Letters* 29 (10): 1451–55. doi:10.1016/j.cclet.2018.04.007.

Race, Paul R., Andrew L. Lovering, Richard M. Green, Abdelmijd Ossor, Scott A. White, Peter F. Searle, Christopher J. Wrighton, and Eva I. Hyde. 2005. "Structural and Mechanistic Studies of *Escherichia coli* Nitroreductase with the Antibiotic Nitrofurazone." *Journal of Biological Chemistry* 280 (14): 13256–64. doi:10.1074/jbc.m409652200.

Race, Paul R., Andrew L. Lovering, Scott A. White, Jane I. Grove, Peter F. Searle, Christopher W. Wrighton, and EvaI Hyde. 2007. "Kinetic and Structural Characterisation of *Escherichia coli* Nitroreductase Mutants Showing Improved Efficacy for the Prodrug Substrate CB1954." *Journal of Molecular Biology* 368 (2): 481–92. doi:10.1016/j.jmb.2007.02.012.

Ramos, Juan L., M. Mar González-Pérez, Antonio Caballero, and Pieter Van Dillewijn. 2005. "Bioremediation of Polynitrated Aromatic Compounds: Plants and Microbes Put up a Fight." *Current Opinion in Biotechnology* 16 (3 SPEC. ISS.): 275–81. doi:10.1016/j.copbio.2005.03.010.

Rautio, Jarkko, Hanna Kumpulainen, Tycho Heimbach, Reza Oliyai, Dooman Oh, Tomi Järvinen, and Jouko Savolainen. 2008. "Prodrugs: Design and Clinical Applications." *Nature Reviews Drug Discovery* 7 (3): 255–70. doi:10.1038/nrd2468.

Rieger, Paul Gerhard, Helmut Martin Meier, Michael Gerle, Uwe Vogt, Torsten Groth, and Hans Joachim Knackmuss. 2002. "Xenobiotics in the Environment: Present and Future Strategies to Obviate the Problem of Biological Persistence." *Journal of Biotechnology* 94 (1): 101–23. doi:10.1016/S0168-1656(01)00422-9.

Roldán, María Dolores, Eva Pérez-Reinado, Francisco Castillo, and Conrado Moreno-Vivián. 2008. "Reduction of Polynitroaromatic Compounds: The Bacterial Nitroreductases." *FEMS Microbiology Reviews* 32 (3): 474–500. doi:10.1111/j.1574-6976.2008.00107.x.

Rothenberg, Mace L. 2011. "Irinotecan (CPT-11): recent developments and future directions — colorectal cancer and beyond." *Oncologist* 6: 66–80. doi:10.1634/theoncologist.6-1-66.

Russell, Pamela J, and Aparajita Khatri. 2006. "Novel Gene-Directed Enzyme Prodrug Therapies against Prostate Cancer." *Expert Opinion on Investigational Drugs* 15 (8): 947–61. doi:10.1517/13543784.15.8.947.

Sastry, Srinivas S., and R. Jayaraman. 1984. "Nitrofuratoin-Resistant Mutants of *Escherichia coli*: Isolation and Mapping." *MGG Molecular & General Genetics* 196 (2): 379–80. doi:10.1007/BF00328076.

Saz, Arthur K. and Martinez L. Marina. 1959. "Enzymatic Basis of Resistance to Aureotmycin Iii. Inhibition By Aureomycin of Protein-Stimulated Electron Transport in Escherichia coli. *Journal of Bacteriology* 79: 527–31.

Schywalsky, Michael, Harald Ihmsen, Alexander Tzabazis, Fechner J., Eric Burak, James Vornov, Helmut Schwilden, and Jürgen Schüttler. 2003. "Pharmacokinetics and pharmacodynamics of the new propofol prodrug GPI 15715 in rats." *European Journal of Anaesthesiology* 20 (3): 182–90. doi:10.1017/S0265021503000322.

Searle, Peter F., Ming Jen Chen, Longqin Hu, Paul R. Race, Andrew L. Lovering, Jane I. Grove, Chris Guise, et al. 2004. "Nitroreductase: A Prodrug-Activating Enzyme for Cancer Gene Therapy." In *Clinical and Experimental Pharmacology and Physiology*. doi:10.1111/j.1440-1681.2004.04085.x.

Sharma, Kavita, Kundan Sengupta, and Harinath Chakrapani. 2013. "Nitroreductase-Activated Nitric Oxide (NO) Prodrugs." *Bioorganic and Medicinal Chemistry Letters* 23 (21): 5964–67. doi:10.1016/j.bmcl.2013.08.066.

Shibata, Jiro, Toshiyuki Toko, Hitoshi Saito, Akio Fujioka, Kouji Sato, Akihiro Hashimoto, Konstanty Wierzba, and Yuji Yamada. 2000. "Estrogen Agonistic/Antagonistic Effects of Miproxifene Phosphate (TAT-59)." *Cancer Chemotherapy and Pharmacology* 45 (2): 133–41. doi:10.1007/s002800050021.

Singleton, Dean C., Da-Qiang Li, Sally Yan Bai, Sophie P. Syddall, Joshua Ballantyne Smaill, Yuqiao Shen, William A. Denny, William Renay. Wilson, and Adam V. Patterson. 2007. "The Nitroreductase Prodrug SN 28343 Enhances the Potency of Systemically Administered Armed Oncolytic Adenovirus ONYX-411 NTR." *Cancer Gene Therapy* 14 (12): 953–67. doi:10.1038/sj.cgt.7701088.

Sisson, Gary, Avery Goodwin, Ausra Raudonikiene, Nicky J. Hughes, Asish K. Mukhopadhyay, Douglas E. Berg, and Paul S. Hoffman. 2002. "Enzymes Associated with Reductive Activation and Action of Nitazoxanide, Nitrofurans, and Metronidazole in *Helicobacter pylori*." *Antimicrobial Agents and Chemotherapy* 46 (7): 2116–23. doi:10.1128/AAC.46.7.2116-2123.2002.

Somerville, Christopher C., Sherley F. Nishino, and Jim C. Spain. 1995. "Purification and Characterization of Nitrobenzene Nitroreductase from *Pseudomonas pseudoalcaligenes* JS45." *Journal of Bacteriology* 177 (13): 3837–42. doi:10.1128/jb.177.13.3837-3842.1995.

Song, Hyung Nam, Dae Gwin Jeong, Seo Young Bang, Se Hwan Paek, Byoung Chul Park, Sung Goo Park, and Eui Jeon Woo. 2015. "Crystal Structure of the Fungal Nitroreductase Frm2 from *Saccharomyces cerevisiae*." *Protein Science* 24 (7): 1158–63. doi:10.1002/pro.2686.

Springer, Caroline J, and Ion Niculescu-Duvaz. 2002. "Chapter 8 – Gene-Directed Enzyme Prodrug Therapy." *Anticancer Drug Development* 137–55. doi:10.1016/B978-012072651-6/50009-7.

Stella, Valentino J., Jan J. Zygmunt, Ingrid Gundo Georg, and Muhammed S. Safadi. 2005. "Water soluble prodrugs of hindered alcohols or phenols."US19980131385.

Todd, Peter A., and Rennie C. Heel. 1986. "Enalapril. A Review of Its Pharmacodynamic and Pharmacokinetic Properties, and Therapeutic Use in Hypertension and Congestive Heart Failure." *Drugs* 31 (3): 198–248.

Tsuda, Masahiro, Tomohiro Terada, Megumi Irie, Toshiya Katsura, Ayumu Niida, Kenji Tomita, Nobutaka Fujii, and Ken-ichi Inui. 2006. "Transport characteristics of a novel peptide transporter 1 substrate, antihypotensive drug midodrine, and its amino acid derivatives." *Journal of Pharmacology and Experimental Therapeutics* 318 (1): 455–60. doi:10.1124/jpet.106.102830.

Tunek, Anders, Eva Levin, and Leif-A. Svensson. 1988. "Hydrolysis of 3H-bambuterol, a carbamate prodrug of terbutaline, in blood from humans and laboratory animals in vitro." *Biochemical Pharmacology* 20 (37): 3867–76. doi:10.1016/0006-2952(88)90068-8.

Valiauga, Benjaminas, Lina Misevičiene, Michelle H. Rich, David F. Ackerley, Jonas Šarlauskas, and Narimantas Čenas. 2018. "Mechanism of Two-/Four-Electron Reduction of Nitroaromatics by Oxygen-Insensitive Nitroreductases: The Role of a Non-Enzymatic Reduction Step." *Molecules* 23 (7): 1–9. doi:10.3390/molecules23071672.

Vass, Simon O., David Jarrom, William R. Wilson, Eva I. Hyde, and Peter F. Searle. 2009. "*E. coli* NfsA: An Alternative Nitroreductase for Prodrug Activation Gene Therapy in Combination with CB1954." *British Journal of Cancer* 100 (12): 1903. doi:10.1038/sj.bjc.6605094.

Villanueva, Julio Rodríguez. 1964. "The Purification of a Nitro-Reductase of Nocardia V." *The Journal of Biological Chemistry* 239 (3): 773–76.

Watanabe, Masahiko, Motoi Ishidate, and Takehiko Nohmi. 1990. "Nucleotide Sequence of *Salmonella typhimurium* Nitroreductase Gene." *Nucleic Acids Research* 18 (4): 1059. doi:10.1093/nar/18.4.1059.

Williams, Elsie M., Rory F. Little, Alexandra M. Mowday, Michelle H. Rich, Jasmine V. E. Chan-Hyams, Janine N. Copp, Jeff B. Smaill, Adam V. Patterson, and David F. Ackerley. 2015. "Nitroreductase Gene-Directed Enzyme Prodrug Therapy: Insights and Advances toward Clinical Utility." *Biochemical Journal* 471 (2): 131–53. doi:10.1042/bj20150650.

Wu, Kuei Meng. 2009. "A New Classification of Prodrugs: Regulatory Perspectives." *Pharmaceuticals* 2 (3): 77–81. doi:10.3390/ph2030077.

Yang, Dan, Hang Yu Tian, Tie Nan Zang, Ming Li, Ying Zhou, and Jun Feng Zhang. 2017. "Hypoxia Imaging in Cells and Tumor Tissues Using a Highly Selective Fluorescent Nitroreductase Probe." *Scientific Reports* 7 (1): 2–9. doi:10.1038/s41598-017-09525-2.

Yang, Yanghui, Yu Chen, Herve Aloysius, Daigo Inoyama, and Hu Longqin. 2014. "Enzymes and Targeted Activation of Prodrugs." In *Enzyme Technologies : Pluripotent Players in Discovering Therapeutic Agents*. Hsiu-Chiung Yang, Wu-Kuang Yeh, and James R. McCarthy (eds.). John Wiley & Sons, Inc. pp. 165–235. doi:10.1002/9781118739907.ch5.

Yang, Jun, Jing Bai, Mingbo Qu, Bo Xie, and Qing Yang. 2019. "Biochemical Characteristics of a Nitroreductase with Diverse Substrate Specificity from *Streptomyces mirabilis* DUT001." *Biotechnology and Applied Biochemistry* 66 (1): 33–42. doi:10.1002/bab.1692.

Yang, Yuanyuan, Jinping Lin, and Dongzhi Wei. 2016. "Heterologous Overexpression and Biochemical Characterization of a Nitroreductase from *Gluconobacter oxydans* 621H." *Molecular Biotechnology* 58 (6): 428–40. doi:10.1007/s12033-016-9942-1.

Yuan, Jun, Yu Qiong Xu, Nan Nan Zhou, Rui Wang, Xu Hong Qian, and Yu Fang Xu. 2014. "A Highly Selective Turn-on Fluorescent Probe Based on Semi-Cyanine for the Detection of Nitroreductase and Hypoxic Tumor Cell Imaging." *RSC Advances* 4 (99): 56207–10. doi:10.1039/c4ra10044a.

Zawilska, Jolanta B., Jakub Wojcieszak, and Agnieszka B. Olejniczak. 2013. "Prodrugs: A Challenge for the Drug Development." *Pharmacological Reports* 65 (1): 1–14. doi:10.1016/S1734-1140(13)70959-9.

Zenno, Shuhei, and Kaoru Saigo. 1994. "NAD (P) H-Flavin Oxidoreductases That Are Similar in Sequence to *Escherichia coli* Fre in Four Species of Luminous Bacteria: Identification of the Genes Encoding NAD (P) H-Flavin Oxidoreductases that are Similar in Sequence to *Escherichia coli* Fre In." *Journal of Bacteriology* 176 (12): 3544–51.

Zenno, Shuhei, Toshiro Kobori, Masaru Tanokura, and Kaoru Saigo. 2005. "Purification and Characterization of NfrA1, a *Bacillus subtilis* Nitro/Flavin Reductase Capable of Interacting with the Bacterial Luciferase." *Bioscience, Biotechnology, and Biochemistry* 62 (10): 1978–87. doi:10.1271/bbb.62.1978.

Zenno, Shuhei, Hideaki Koike, Ajit N. Kumar, Ramamirth Jayaraman, Masaru Tanokura, and Kaoru Saigo. 1996. "Biochemical Characterization of NfsA, the *Escherichia coli* Major Nitroreductase Exhibiting a High Amino Acid Sequence Homology to Frp, a *Vibrio harveyi* Flavin Oxidoreductase." *Journal of Bacteriology* 178 (15): 4508–14. doi:10.1128/jb.178.15.4508-4514.1996.

Zhang, Jin, Vijay Kale, and Mingnan Chen. 2015a. "Gene-Directed Enzyme Prodrug Therapy." *The AAPS Journal* 17 (1): 102–10. doi:10.1208/s12248-014-9675-7.

Zhang, Jing, Hong Wen Liu, Xiao Xiao Hu, Jin Li, Li Hui Liang, Xiao Bing Zhang, and Weihong Tan. 2015b. "Efficient Two-Photon Fluorescent Probe for Nitroreductase Detection and Hypoxia Imaging in Tumor Cells and Tissues." *Analytical Chemistry* 87 (23): 11832–39. doi:10.1021/acs.analchem.5b03336.

Zhang, Lei, Leilei Guo, Xue Shan, Xiaohong Lin, Tingting Gu, Jikang Zhang, Junliang Ge, et al. 2019a. "An Elegant Nitroreductase Responsive Fluorescent Probe for Selective Detection of Pathogenic *Listeria* In Vitro and In Vivo." *Talanta* 198: 472–79. doi:10.1016/j.talanta.2019.02.026.

Zhang, Lei, Xue Shan, Leilei Guo, Jikang Zhang, Junliang Ge, Qing Jiang, and Xinghai Ning. 2019b. "A Sensitive and Fast Responsive Fluorescent Probe for Imaging Hypoxic Tumors." *Analyst* 144 (1): 284–89. doi:10.1039/c8an01472h.

11

Prodrug Approach Using Carboxylesterase: Different Substrate Specificities of Human Carboxylesterase Isozymes

Masato Takahashi and Masakiyo Hosokawa

CONTENTS

11.1 Introduction

Generally, in order for an orally administrated drug to exert its medicinal effects, it needs to be rapidly absorbed from the digestive tract and transferred to circulation blood without undergoing metabolism in the digestive tract. Although some drugs have excellent pharmacological activity, they are poorly absorbed due to their water solubility or high molecular weight, and many are rapidly metabolized in the digestive tract and liver. Therefore, various techniques have been developed to overcome the drawbacks of such drugs, and they have been used to improve the pharmacokinetics of drugs. The development of prodrugs has evolved with the development of drug delivery systems. Drug-metabolizing enzymes involved in xenobiotics *in vivo* have been used for the development of prodrugs. Therefore, it is considered very important to examine the correlation between the activity of an enzyme that metabolically activates a prodrug and the structure of the prodrug. The present chapter focuses on important issues in the design and development of prodrugs such as a different substrate specificities of human carboxylesterase (CES) isozymes and other hydrolases.

11.2 Purpose of Prodrugs

Prodrugs were proposed by Albert in 1958 as medicinal products that have no activity *per se* and only convert to active forms *in vivo* to show their efficacy. Table 11.1 shows some prodrugs that are currently being developed for various purposes. As shown in this table, among the currently marketed drugs, there are many drugs that have been made into prodrugs (Based on Stella 2010). In fact, 5%–7% of the drugs that are currently marketed are considered to be prodrugs. Generally, when considering absorption of medicine, substances with high lipid solubility are more likely to permeate biological membranes (having properties close to oil) by mechanisms such as passive diffusion. Prodrugs that

TABLE 11.1

Metabolic Activation and Purpose of Representative Prodrugs

Prodrug	Active Form	Purpose
Hydrocortisone succinate	Hydrocortisone	Improvement in solubility
Methylprednisolone hemisuccinate	Methylprednisolone	
Dexamethasone phosphate ester	Dexamethasone	
Fosfluconazole	Fluconazole	
Tarampicillin	Ampicillin	Improved digestive tract absorption (including stabilization)
Bacampicillin		
Vibampicillin		
Lenampicillin		
Oseltamivir	Oseltamivir carboxylate	
Candesartan cilexetil	Candesartan	
Fursultiamine	Tiamine	
Imidapril	Imidaprilat	
Temocapril	Temocaprilat	
Cefpodoxime proxetyl	Cefpodoxime	
Cefteram pivoxil	Cefteram	
Simvastatin	Active form	
Dabigatran etexilate	Dabigatran	
Clopidogrel	Active form	
Prasugrel	Active form	
Olmesartan medoxomil	Olmesartan	
Flutamide	Active form	
Laninamivir octanoate	Laninamivir	Transpulmonary absorption improvement
Cytarabine ocfosphate	Cytarabine	Improvement in sustainability of pharmacological action
Aracepril	Captopril	
Testosterone enanthate	Testosterone	
Fluphenazine decanoate	Fluphenazine	
Valaciclovir	Aciclovir	
Aciclovir	Acyclovir triphosphate	Reduction of side effects/ Reaching the target organ
Irinotecan (CPT-11)	SN-38	
Indometacin farnesil	Indometacin	
Acemetacin		
Proglumetacin		
Doxifluridine	5-fluorouracil	
Tegafur		
Futraful		
Chloramphenicol palmitate	Chloramphenicol	Bitter alleviation
Quinine ethylcarbonate	Quinine	

have been chemically modified with substances having high lipid solubility by utilizing such properties are intended to improve absorption. If a drug has excellent pharmacological activity but its polarity is high and it cannot pass through a biological membrane, it can be combined with a highly lipid-soluble modifying chemical group to make it into a prodrug. Such a prodrug has improved lipid-solubility of all the molecules of the drug and it can pass through a biological membrane. A prodrug taken through the biological membrane into the living body converted to the original drug and exhibits pharmacological activity by degrading the modified compound. In the case of prodrugs, the bioavailability (BA) is higher in efficacy when the BA of the prodrug itself is lower, but the efficacy is lower when BA is higher. The main purposes of considering prodrugs are shown in Table 11.2 (Based on Imai and Hosokawa 2010; Imai and Ohura 2010).

Here, we will briefly describe the situation of the development of prodrugs. Prodrugs mask functional groups involved in the efficacy of drugs with ester bonds and the like to reduce the drug efficacy and, at the same time, alter the properties of the whole molecule. A prodrug is designed to be metabolically converted to a pharmacologically active parent drug in target organs after being taken *in vivo*. However, some medicines were not intended to be prodrugs from the beginning, but were actually found to be prodrugs as a result of examining the activity of metabolites. Such prodrugs are often activated by oxidation or reduction. For example, the anticancer drugs Futraful and Tegafur are hydroxylated by cytochrome P450 and then chemically hydrolyzed to form 5-fluorouracil as shown in Table 11.1. In addition, Loxoprofen (an anti-inflammatory and analgesic drug) is pharmacologically active in the trans-alcohol formed by reduction of the ketone group by keto-reductases. In many cases, an ester bond or an amide bond is used as a modifying group of the prodrug. For example, methylprednisolone hemi-succinate is a drug that was developed as an anti-shock agent at the time of surgery, and is a prodrug that has increased water-solubility because it needs to be an injection. Although valacyclovir is a prodrug of acyclovir, acyclovir needs to be taken 5 times a day because of the rapid excretion of acyclovir, but valacyclovir having valine ester-linked has been developed. Valacyclovir has a long-lasting effect and is effective if taken in three times a day. Prodrugs aimed at reaching the target organ and reducing side effects include capecitabine, irinotecan, indomethacin farnesyl. Although capecitabine is finally activated to 5-fluorouracil (5-FU), it is an orally administrable anticancer agent, and can be treated at any time and place. This prodrug is activated in three steps: it is first hydrolyzed by CES and then metabolically activated by cytidine deaminase followed by thymidine phosphorylase. Among these, cytidine deaminase is expressed in both cancer cells and normal cells, and thymidine phosphorylase is mainly expressed in cancer cells, and this prodrug is therefore prodrug aimed at reaching the target organ. The anti-influenza drugs Oseltamivir and Laninamivir octanoate are examples of fully acting and locally acting prodrugs. Oseltamivir is a systemic prodrug intended to increase oral absorption. It is not hydrolyzed in the small intestine, and the parent drug produced by the first hepatic metabolism shifts to systemic circulation and

TABLE 11.2

Main Purposes of Considering Prodrugs

1. Improving bioavailability
   Reasons for low bioavailability
   - Low solubility
   - Large first pass effects in the intestine and liver
   - Excretion into the intestinal tract by transporters of intestinal epithelial cells
2. Targeting specific organs
   Conditions that allow targeting
   - Easily reaching the target organ and uptake speed being very fast
   - Metabolic activation to the parent compound occurs selectively at the target site
   - The parent compound that is activated at the target site remaining at that site
3. Reduction of side effects
   Improvement of taste (bitter), protection of the gastric mucosa, prevention of cytotoxicity of anticancer drugs

exhibits pharmacological effects (Ose et al. 2009; Hosokawa 2008). On the other hand, Laninamivir octanoate is a hydrophobic octanoate that is directly administered as a powder inhalant to the lung where the anti-influenza drug is acting. Laninamivir octanoate translocates to lung cells, and it is hydrolyzed by *S*-formylglutathione hydrolase to exert its pharmacological effects. This prodrug maintains its full efficacy after a single dose. It is expected to stay in the lungs for a long time due to various factors such as increased lung uptake and retention due to hydrophobicity, mild activation with esterase D, and the intracellular lock-in phenomenon.

11.3 Enzymes Responsible for *in vivo* Hydrolysis of Prodrugs

There are many types of hydrolytic enzymes *in vivo*. Enzymes involved in metabolic activation of prodrugs include carboxylesterases (CESs), allylacetamide deacetylase (AADAC), butyrylcholinesterase (BuChE), and paraoxonase (PON). Among these enzymes, CESs plays the most important role as metabolic activation enzymes *in vivo*. CESs are composed of multiple gene family groups with different substrate specificities, and there are large animal species differences in addition to organ differences (Hosokawa 2008; Satoh and Hosokawa 2006; Satoh and Hosokawa 1998). On the other hand, since there are many cases in which the active form of a prodrug, such as the ACE inhibitor temocapril, is converted to an organic anion when it is metabolized by a CES, the metabolic activation by a CES sometimes has high specificity for the organic anion. It also leads to the production of transporter substrates. Furthermore, metabolites of CES are likely to be substrates for conjugate enzymes and drug transporters. For example, SN-38, a hydrolyzed metabolite of irinotecan, is a substrate for uridine diphosphate glucuronosyltransferase (UGT), and conjugated metabolites (SN-38 glucuronide) are more likely to be excreted. On the other hand, the organic anion metabolite of CES becomes a substrate for the organic anion transporter and is easily excreted.

The CES is assigned an enzyme number of EC 3.1.1.1 as a representative enzyme for hydrolyzing an ester of carboxylic acid and has long been known as an enzyme catalyzing a hydrolysis reaction of a carboxyl ester. However, this enzyme is involved in the detoxification and metabolic activation of xenobiotics including many pharmaceuticals, pesticides, and environmental chemicals as well as prodrugs because the substrate specificity is diverse such as hydrolysis of amide and thioester bonds. In addition, since this enzyme is involved in the metabolism of long chain acyl compounds *in vivo*, it is considered to play an important role in lipid metabolism *in vivo* (Ose et al. 2009; Hosokawa 2008; Satoh and Hosokawa 2006). This enzyme is expressed in the liver and as well as in small intestinal epithelial cells, kidney, lung, brain, adipose tissue, muscle, nasal mucosa, and plasma. The highest activity of this enzyme is observed in the liver, but relatively high levels of activity are also observed in small intestine epithelial cells, the kidney and plasma. Therefore, as described later, differences in substrate specificity and enzyme activity of this enzyme are ester and amide types. The enzyme has a great influence on the pharmacokinetics of prodrugs. In addition, there is a remarkable species difference in the enzyme, particularly for CES in plasma, and characteristics of the enzyme are different in experimental small animals such as rats and mice and in humans. Furthermore, in beagle dogs used in nonclinical studies, little expression of CES was found in the small intestine, and caution should therefore be taken when using newly developed ester and amide prodrugs in clinical trials.

11.4 Tissue Distribution of CESs

CESs are classified into five types, CES1 to CES5, according to the difference in amino acid sequences. CES1 and CES2 are mainly expressed in mammals. It is known that there is a large species difference in the tissue distributions of CES1 and CES2. Table 11.3 shows the tissue distributions of CES1 and CES2 in each species (Based on Hosokawa 2008). In humans, CES1 is mainly expressed in the liver and lung but is not expressed in the small intestine. On the other hand, CES1 is also expressed in the small intestine in guinea pigs and monkeys. In humans, CES2 is highly expressed in the small intestine and

TABLE 11.3

Tissue-Specific Expression Profiles of CES1 and CES2 Isozymes

Species	Isozyme	Liver	Small Intestine	Kidney	Lung
Guinea Pig	CES1	+++	+++	++	NT
	CES2	–	+	–	NT
Beagle Dog	CES1	+++	–	NT	+++
	CES2	++	–	NT	+
Monkey	CES1	+++	++	–	NT
	CES2	+	+++	+	NT
Human	CES1	+++	–	+	+++
	CES2	+	+++	+++	–

–, undetectable; +, weakly expressed; ++, moderately expressed; +++, strongly expressed; NT, not tested.

kidney, while there is little expression of CES2 in the small intestine in guinea pigs and beagle dogs. Due to such species differences, the results obtained in nonclinical studies using animals are often different in clinical trials in humans. In order to avoid such problems, it is necessary to conduct prodrug development with consideration of species differences.

11.5 Substrate Specificities of CESs

In the structural design of a prodrug, it is very important to investigate the correlation between the structure of the prodrug and the substrate recognition ability of the metabolic activation enzyme. In the case of a prodrug intended to improve absorption, it is required that the prodrug be metabolically activated rapidly after absorption from the small intestine and show its effect immediately. On the other hand, in the case of a prodrug for sustained release, the prodrug is required to be metabolically activated slowly, and it is desirable that a metabolite be gradually formed. In the case of a prodrug aimed at reducing side effects, it is necessary for the structure to be maintained as it is without being metabolized at the site where the side effects occur. Thus, the metabolic activation rate of the prodrug needs to be controlled according to each purpose. In order to control the rate of metabolic activation, it is necessary to investigate the relationship between the enzyme and substrate.

As described above, CES1 and CES2 are mainly expressed in humans, but there are significant differences in their substrate specificities. In the case of the ester structure of the substrate being divided into an acyl site and an alkoxy site, CES1 recognizes a substrate having a large acyl group and a small alkoxy group, and it can efficiently catalyze hydrolysis reactions. In contrast, CES2 is likely to recognize substrates with large alkoxy groups and small acyl groups and can catalyze hydrolysis reactions. Table 11.4 shows the structures of drugs and natural products that are hydrolyzed by CESs (Based on Hosokawa 2008). The prodrugs will be described later. First, cocaine is a compound having a two-ester structure, but the enzymes that hydrolyze the ester sites are different (Brzezinski et al. 1994, 1997). The hydrolysis proceeds by CES1 since the methyl ester site has a smaller alkoxy group (methoxy group) than the acyl group. On the other hand, since the benzoyl ester moiety has a smaller acyl group (benzoyl group) than the alkoxy group, hydrolysis proceeds by CES2. Since pethidine and methylphenidate also have smaller alkoxy groups such as an ethoxy group and a methoxy group than acyl groups, they are hydrolyzed by CES1 (Burnell et al. 1999; Zhang et al. 1999; Sun et al. 2004; Zejin et al. 2004). As for heroin, CES2 is selectively hydrolyzed since the acyl group (acetyl group) is overwhelmingly smaller than the alkoxy group (Kamendulis et al. 1996). Thus, it is possible to make a prodrug specifically activated in each tissue by using information on substrate specificity and tissue distribution of CESs. For example, it is sufficient to structurally design the prodrug so that it is hydrolyzed by CES1 in the case of exerting an effect in the liver or lung and hydrolyzed by CES2 in the case of exerting an effect in the small intestine or kidney.

TABLE 11.4

Substrate Specificities of CES1 and CES2 Families

Substrate	Product (Carboxylic Acid)	Product (Alcohol)	Substrate Specificity
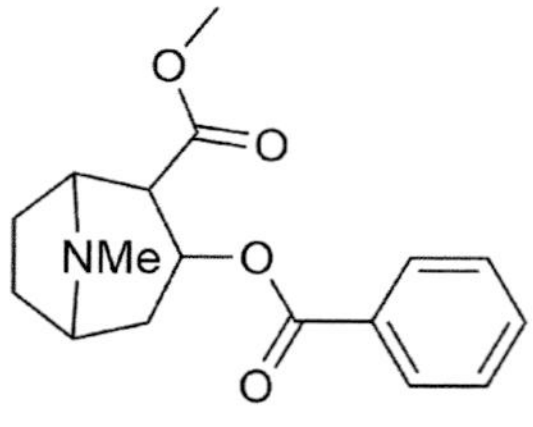 cocaine (methyl ester)	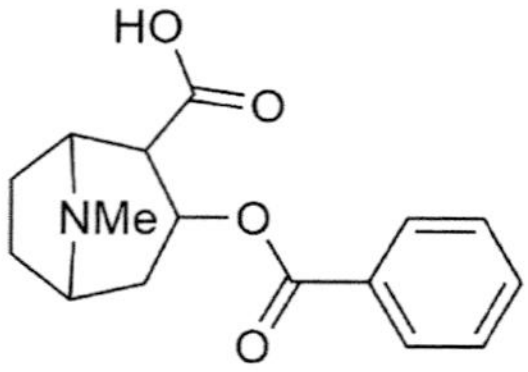	H_3C-OH	CES1>>CES2
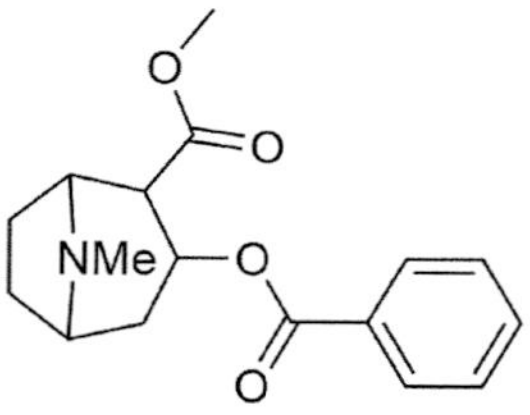 cocaine (benzoyl ester)	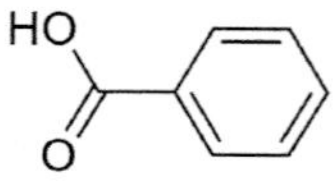	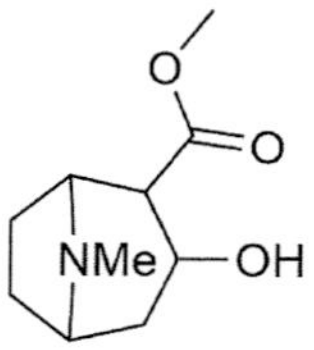	CES2>>CES1
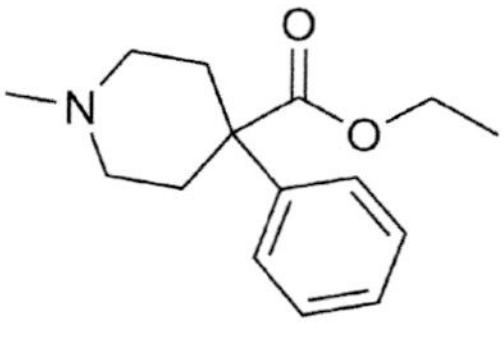 pethidine	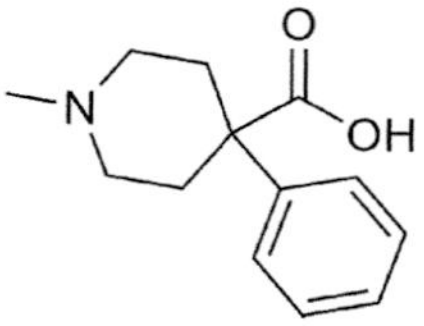	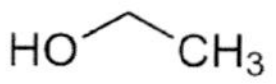 $HO\frown CH_3$	CES1>>CES2
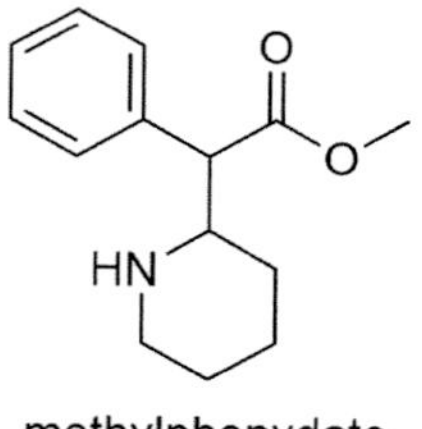 methylphenydate	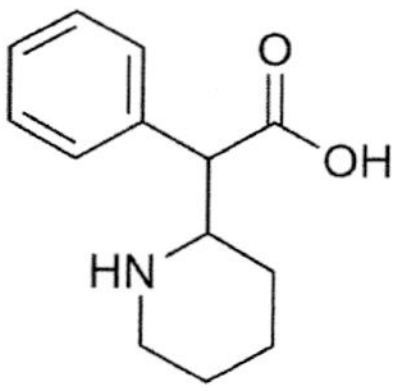	H_3C-OH	CES1>>CES2
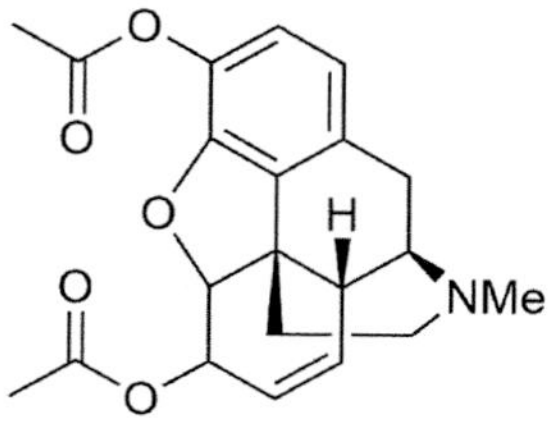 heroin	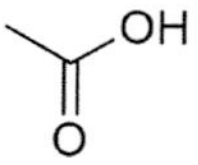	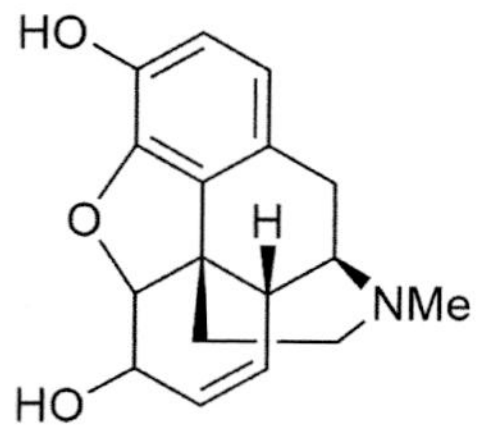	CES2>>CES1

11.6 Prodrug Examples

11.6.1 Anti-Influenza Virus Agent

Oseltamivir (Figure 11.1) is an influenza therapeutic drug that suppresses the growth of influenza virus by inhibiting neuraminidase. The active form of oseltamivir (oseltamivir carboxylate) is ionized in the small intestine, and it has low absorbability with only 5% or less being absorbed by oral administration (Shi et al. 2007; Yang et al. 2007). The bioavailability of oseltamivir has been increased by enhancing its lipid solubility by ethyl esterification of the carboxyl group of oseltamivir carboxylate, which is the active substance, and increasing its absorbability (He et al. 1999). The acyl moiety of oseltamivir is a highly functionalized cyclohexene ring and the alkoxy moiety is an ethyl group. It is known that they are metabolically activated mainly by CES1 because the alkoxy group side is relatively small. Therefore, oseltamivir can be metabolically activated efficiently in the liver, in which CES1 is mainly expressed.

Raninamivir octanoate is also an influenza drug with the same neuraminidase inhibitory activity as that of oseltamivir. After inhaled administration, raninamivir octanoate has been shown to exhibit pharmacological activity by being metabolically activated to raninamivir in the lung and to be effective for both treatment and prevention of influenza (Yamashita et al. 2009; Watanabe et al. 2010). A prodrug in which an octanoyl group is introduced to the hydroxyl group at the 2- or 3-position of laninamivir is effective after inhaling only once at the time of onset and is used as a long-acting drug that is gradually metabolically activated. Raninamivir octanoate is known as an enzyme in which esterase D and acyl protein thioesterase, but not carboxylesterase 1, are involved in metabolic activation (Koyama et al. 2014). Raninamivir octanoate is a mixture of a 3-acylated compound and a 2-acylated compound because they are interconverted by an acyl transfer reaction, both of which are converted to raninamivir by hydrolysis.

11.6.2 ACE Inhibitors

Angiotensin-converting enzyme (ACE) inhibitors have an antihypertensive effect by inhibiting the enzyme that converts angiotensin I to angiotensin II. Temocapril (Sierakowski et al. 1997) (Figure 11.2), imidapril (Geshi et al. 2005; Yun et al. 2005), cilazapril (Williams et al. 1989), and quinapril (Breslin et al. 1996) are prodrugs in which a carboxyl group is converted to an ethyl ester, and the lipophilicity of each prodrug is higher than that of the parent compound. Each active form contains an amino group and a carboxyl group in the structure, and it therefore exists as a zwitterion in the small intestine. Converting to prodrugs has

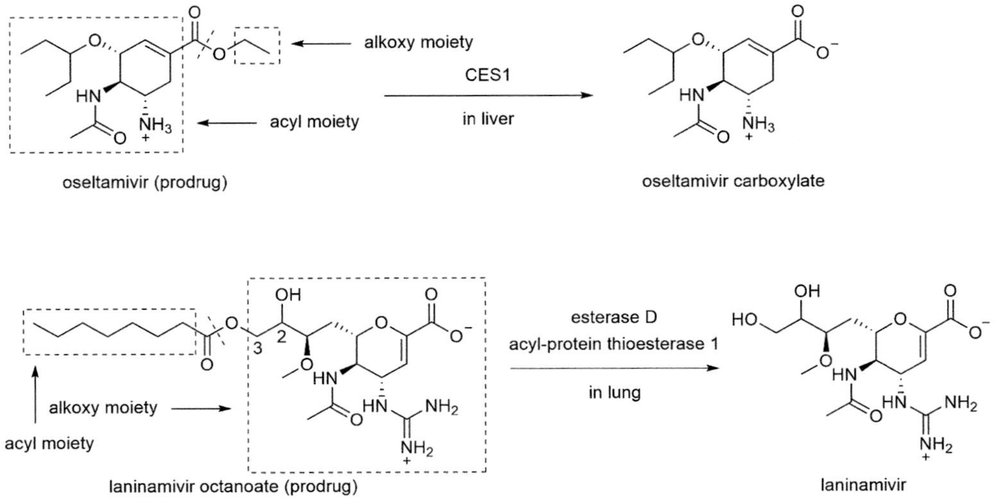

FIGURE 11.1 Examples of prodrugs of anti-influenza virus agents.

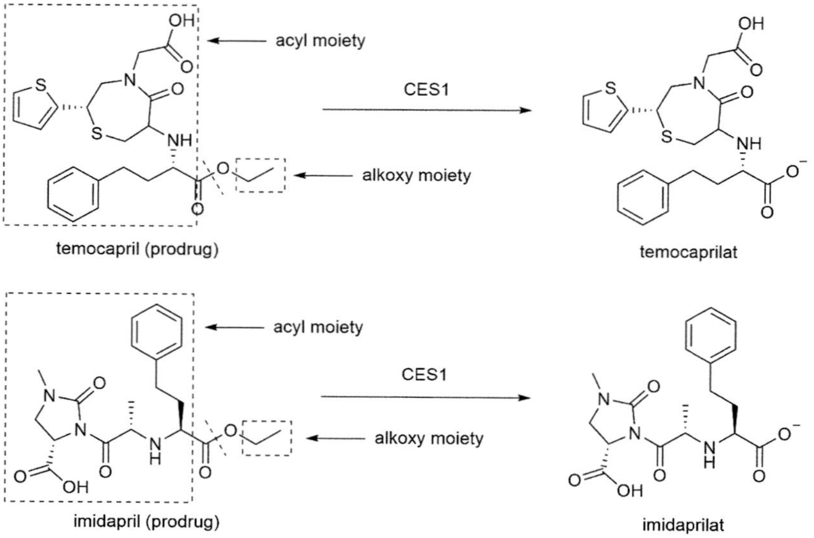

FIGURE 11.2 Examples of prodrugs of ACE inhibitors.

succeeded in enhancing lipid solubility and improving oral absorption. These prodrugs are metabolically activated rapidly by CES1 in the liver (Takai et al. 1997). Temocapril has the fastest rate of metabolic activation by CES1 among ACE inhibitors. Ethyl esters are relatively stable under neutral conditions but are highly susceptible to hydrolysis by CES1 due to the small size of the alkoxy group. Furthermore, since ethanol by-produced after hydrolysis has low toxicity, the ethyl group is often used as a modifying group for prodrugs.

11.6.3 NSAIDs

Non-steroidal anti-inflammatory drugs (NSAIDs) such as aspirin, diclofenac, ibuprofen and indomethacin (Figure 11.3) are among the most widely prescribed drugs. However, long-term use of NSAIDs can cause gastric mucosal injury (Graham et al. 2005). Prodrugs such as acemetacin and indomethacin farnesyl have been developed to reduce the side effects of indomethacin. The metabolic activation rates of newly synthesized indomethacin ester or thioester prodrugs in which various alkoxy groups were introduced into the carboxy group of indomethacin have been determined (Takahashi et al. 2018). Since the synthesized indomethacin prodrug has a relatively large indole ring at the acyl group side and a relatively small alcohol introduced at the alkoxy group side, it is metabolized by CES1 rather than CES2. In addition, the rate of metabolic activation of a prodrug having a linear alkoxy group with less steric hindrance near the ester is faster. This is in accordance with the fact that CES1 is likely to recognize esters with small alkoxy groups, indicating that the steric hindrance of the alkoxy groups can be used to modulate the rate of metabolic activation. Thioester prodrugs have also been studied, and it has been shown that they have a faster metabolic activation rate than that of ester prodrugs in a low concentration solution.

11.6.4 Statins

Statins inhibit the synthesis of mevalonic acid by inhibiting HMG-CoA reductase in the process of converting HMG-CoA to mevalonic acid in the rate-limiting step of cholesterol synthesis. Commonly, statin drugs have a 3,5-dihydroxycarboxylic acid structure similar to that of HMG-CoA. While this structure is essential for pharmacological action, it is a factor that reduces absorption from the small intestine. Simvastatin (Figure 11.4) and lovastatin are prodrugs in which the hydroxyl group at the

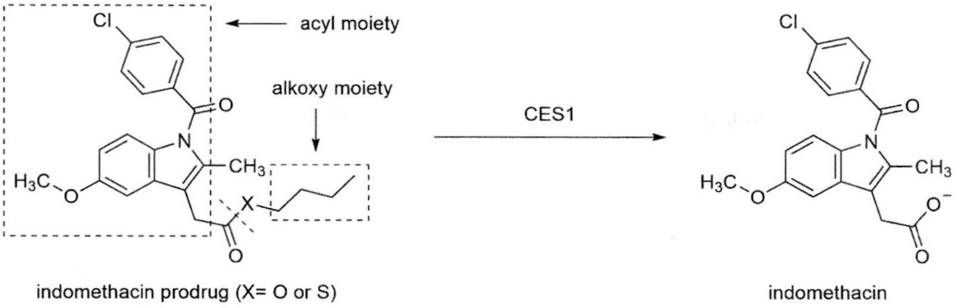

FIGURE 11.3 Example of a prodrug of indomethacin.

5-position and the carboxyl group are cyclically esterified by intramolecular dehydration condensation, and they have enhanced lipid solubility. These cyclic ester (lactone) prodrugs are metabolically activated by paraoxonases (PONs) that are present in the human liver and plasma (Gonzalvo et al. 1998; Draganov et al. 2005). PONs play important roles of cyclic ester and carbonate compounds such as pilocarpine (Hioki et al. 2011), olmesartan medoxomil (Ishizuka et al. 2012), spironolactone and prulifloxacin (Tougou et al. 1998).

Prodrugs in which a hydroxyl group is esterified by intermolecular condensation at the carboxyl group of atorvastatin have been studied (Mizoi et al. 2016). The atorvastatin prodrugs are metabolically activated by CESs, but not PONs. The metabolic activation rates of atorvastatin ester, amide and thioester prodrugs were examined and they were found to be metabolically activated by CES1 or CES2. It was confirmed that there is little metabolic activation of the amide but that the thioester is metabolically activated, though not to the extent of the ester. Since CES increased the steric hindrance of the alkoxy group in the vicinity of the ester, the metabolic activation rate decreased dramatically, but hCES2 had less influence than that of CES1. Instead, hCES2 is susceptible to the electron density of the ester, and the rate of metabolic activation by CES2 was greatly increased when the electron density of the ester decreased.

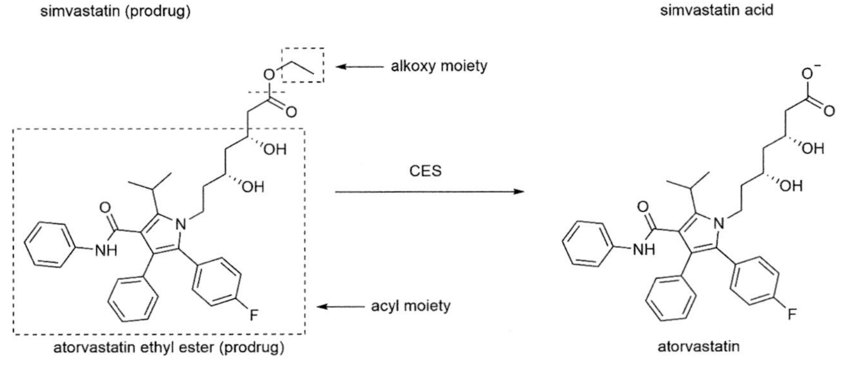

FIGURE 11.4 Examples of prodrugs of statins.

11.6.5 Haloperidol

Haloperidol (Figure 11.5) ameliorates the positive symptoms of schizophrenia by blocking the dopamine D_2 receptor. Haloperidol has been used for a long time, but forgetfulness to take the medication has often been a problem. Haloperidol decanoate is a decanoated prodrug of the alcohol moiety of haloperidol, and it takes a long time for metabolic activation. Since it is given intramuscularly once every four weeks, compliance has been increased by decreasing the number of medications (Beresford and Ward 1987). In addition to decanoic acid, it is known that 2-methyl butanoic acid takes an equal time for metabolic activation. The haloperidol ester is easily recognized by CES2 because the acyl group is smaller than the alkoxy group having a piperidine ring. Since CES2 can reduce the metabolic activation rate by making the acyl group bulky, the introduction of a long chain carboxylic acid or a branched carboxylic acid is effective for sustained action of the prodrug (Takahashi et al. 2019).

With haloperidol as the parent compound, there are prodrugs in which the carbonyl group is converted. The prodrug enol-esterified with pivalic acid has an enol-containing alkoxy group and an acyl group having pivalic acid. Overall, the alkoxy group is larger than the acyl group, but the metabolic rate by CES has been found to be susceptible to steric hindrance near the ester. Therefore, unlike haloperidol decanoate, the acyl group having pivalic acid is recognized to be larger than the alkoxy group and thus is more susceptible to metabolic activation by CES1 than by CES2. Thus, even on the same substrate, the metabolic activation enzyme may differ depending on the site of introduction of the acyl group. Furthermore, a prodrug in which 4-phenylbutyric acid is introduced into carbonyl-reduced haloperidol metabolite II has been developed to reduce the survival rate of human retinal endothelial cells (Sozio et al. 2015).

11.6.6 Antitumor Drugs

Capecitabine (Figure 11.6) is a fluoropyrimidine-type antimetabolite and has antitumor activity. Capecitabine is a prodrug developed to reduce the toxicity in bone marrow and the gastrointestinal tract, and it has a unique metabolic activation pathway. After capecitabine is absorbed, the carbamate

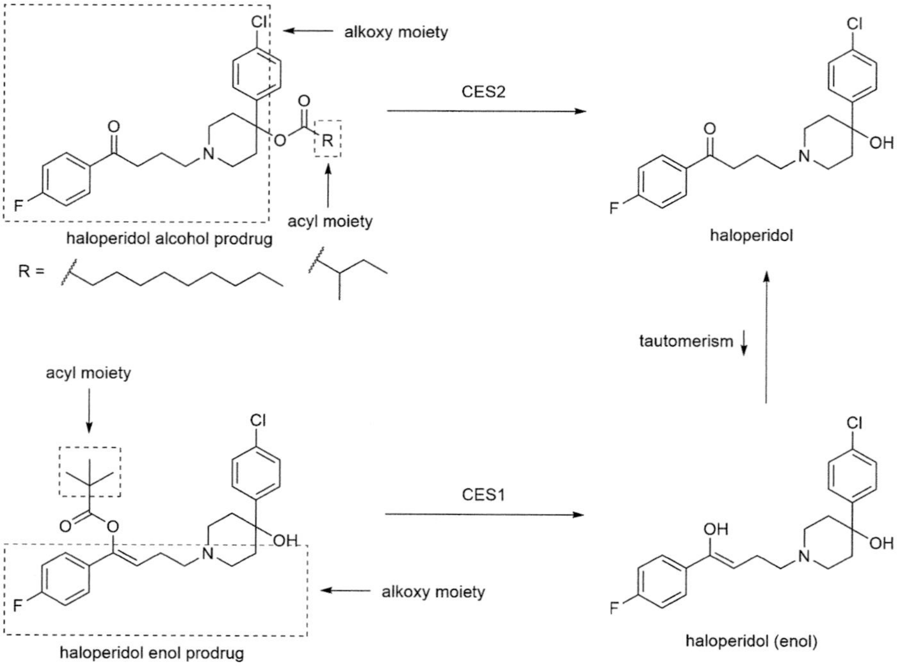

FIGURE 11.5 Examples of prodrugs of haloperidol.

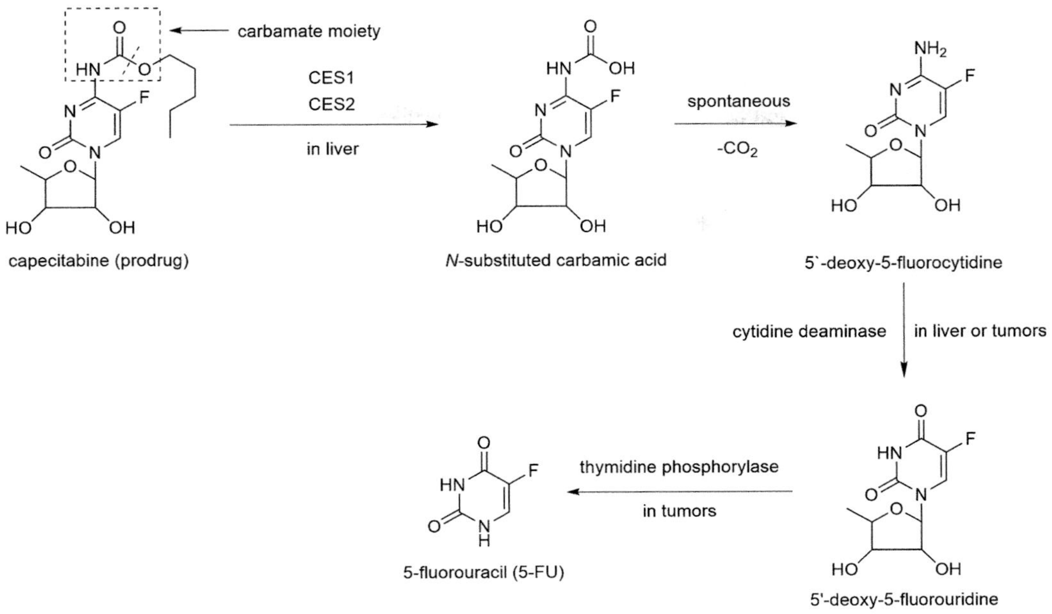

FIGURE 11.6 Example of a prodrug of 5-FU.

moiety of capecitabine is first biotransformationed to *N*-substituted carbamic acid by hydrolysis of the carbamate site by CES1 or CES2 in the liver (Quinney et al. 2005). The carbamic acid site is decomposed automatically to form 5′-deoxy-5-fluorocytidine. Next, deamination proceeds by cytidine deaminase that is present in the liver and tumor tissue to form 5′-deoxy-5-fluorouridine. Furthermore, thymidine phosphorylase, which is expressed at high levels in tumor tissue, becomes an activator, 5-FU, and exerts an antitumor effect (Miwa et al. 1998). Thus, targeting the drug to the tumor is achieved by gradual activation of capecitabine to 5-FU.

CPT-11 (Figure 11.7) is an antitumor agent synthesized using camptothecin as a lead compound and has topoisomerase I inhibitory activity. CPT-11, which has a carbamate structure, is also a prodrug of SN-38 that exerts its efficacy by being metabolized by a CES (Satoh et al. 1994). Since there is no classification of an acyl group or an alkoxy group like an ester in a carbamate structure, there seems to be no selectivity between CES1 and CES2. However, CPT-11 has been found to be preferentially catalyzed by CES2 (Mathijssen et al. 2001; Hatfield et al. 2011). In the case of carbamate, after the C-O bond is hydrolyzed by a CES, decomposition of the *N*-substituted carbamic acid site produces SN-38. Viewed as a carbamate, the alkoxy group is smaller than the acyl group and is therefore catalyzed by CES2. Hydrolysis reactions of carbamates and carbonates as well as esters and amides may also be catalyzed by a CES.

11.6.7 Antiplatelet Agents

Prasugrel (Figure 11.8) is a thienopyridine-type prodrug that has inhibitory action on platelet aggregation. Prasugrel has a single metabolic activation pathway (Rehmel et al. 2006; Farid et al. 2007; Williams et al. 2008), with CES2 in the small intestine first leading to hydrolysis of the ester. Although this reaction is slightly hydrolyzed by CES1, the ester moiety of plasugrel consists of an acetyl group (small acyl group) and a thienopyridine group (a large alkoxy group), and most of them undergo metabolic activation by CES2. It is known that CES2 is at least 25-times faster than CES1 in the first-step hydrolysis of prasugrel. The product from the hydrolysis is isomerized to R-95913. R-95913 is converted to the active form R-138727 in the liver mainly through cleavage by

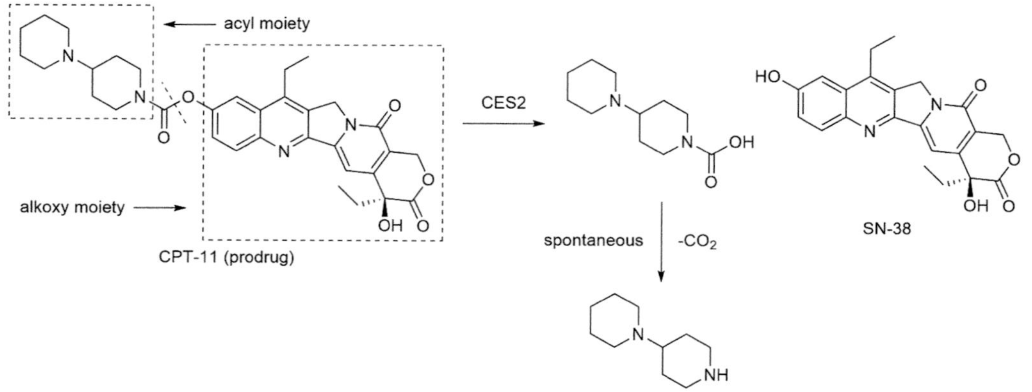

FIGURE 11.7 Example of a prodrug of SN-38.

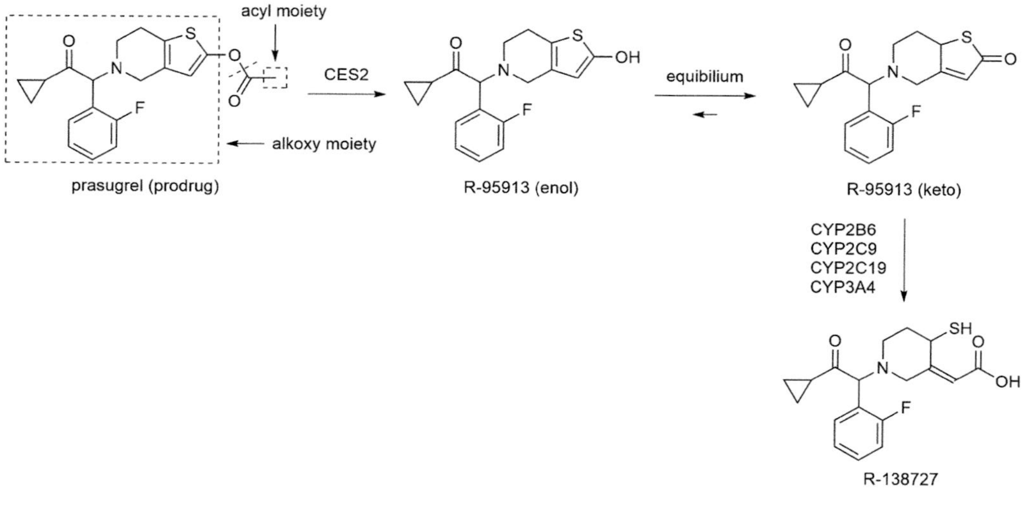

FIGURE 11.8 Example of a prodrug of R-138727.

CYP2B6, CYP2C9, CYP2C19 and CYP3A4. The maximum concentration of R-138727 was confirmed at 0.5 hours after oral administration, and the rapid formation of R-138727 is due to its high affinity to CES2.

In contrast to prasugrel, clopidogrel (Figure 11.9) has two major metabolic pathways. The main metabolic pathway is the formation of inactive metabolites by hydrolysis of the ester (Hagihara et al. 2009). The ester moiety of clopidogrel is composed of a bulky acyl group with a thienopyridine and a benzene ring and a small alkoxy group, which is a structure that can be easily recognized by CES1. The minor metabolic pathway is a two-step reaction pathway by CYP. First, hydroxylation at the 2-position of thiophene proceeds by the functions of CYP1A2, CYP2B6 and CYP2C19. The product is converted to the stable keto-type thiolactone by keto-enol tautomerism. Next, by the action of CYP2B6, CYP2C9, CYP2C19 and CYP3A4 from thiolactone, cleavage of the thiolactone site results in the formation of an active metabolite (Kazui et al. 2010). Thus, clopidogrel has two metabolic pathways in which metabolic inactivation and activation compete. It has also been found that interaction occurs with the same CES1 as a substrate ACE inhibitor (Kristensen et al. 2014).

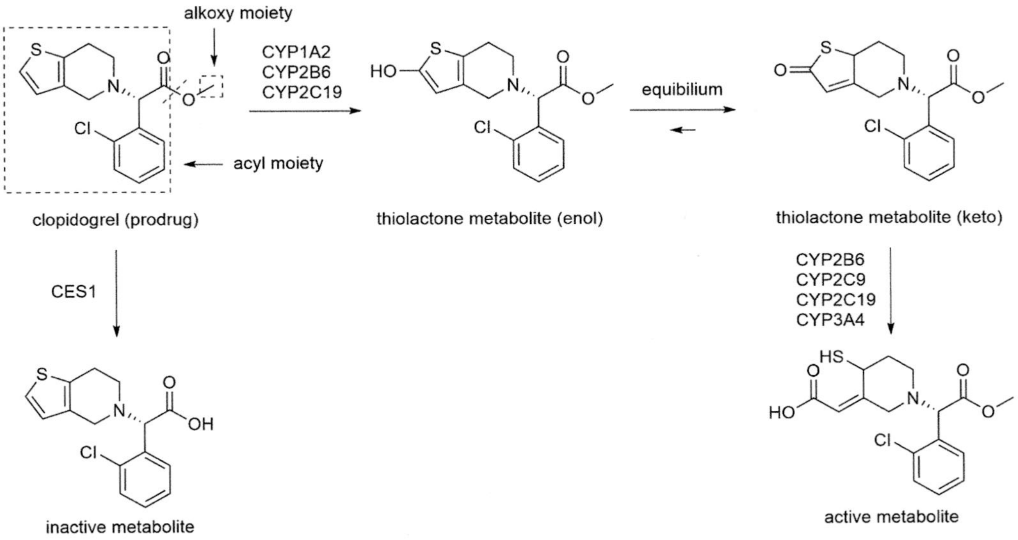

alkoxy moiety

CYP1A2
CYP2B6
CYP2C19

equibilium

acyl moiety

clopidogrel (prodrug)

thiolactone metabolite (enol)

thiolactone metabolite (keto)

CES1

CYP2B6
CYP2C9
CYP2C19
CYP3A4

inactive metabolite

active metabolite

FIGURE 11.9 Example of clopidogrel.

11.7 Conclusion

This chapter showed the different substrate specificities and tissue distributions of carboxylesterase isozymes and other hydrolases responsible for the metabolic activation of ester and carbamate prodrugs and examples of prodrugs that have been investigated. The relationship between prodrug structure and metabolic rate was also described. What ester the hydrolytic enzyme recognizes and can be hydrolyzed varies depending on the type and molecular species of the hydrolytic enzyme. Further experiments to understand the detailed relationships between CES isozymes and hydrolysable prodrugs will contribute to the design of ideal prodrugs. Future drug discovery will require a deep understanding of the properties of metabolic enzymes and theoretical prodrug design and synthesis based on those properties.

REFERENCES

Beresford, R., and Ward, A. 1987. Haloperidol decanoate: A preliminary review of its pharmacodynamic and pharmacokinetic properties and therapeutic use in psychosis. *Drugs* 33: 31–49.

Breslin, E., Posvar, E., Neub, M., et al. 1996. A pharmacodynamic and pharmacokinetic comparison of intravenous ouinaprilat and oral quinapril. *The Journal of Clinical Pharmacology* 36: 414–421.

Brzezinski, M. R., Abraham, T. L., Stone, C. L., et al. 1994. Purification and characterization of a human liver cocaine carboxylesterase that catalyzes the production of benzoylecgonine and the formation of cocaethylene from alcohol and cocaine. *Biochemical Pharmacology* 48: 1747–1755.

Brzezinski, M. R., Spink, B. J., Dean, R. A., et al. 1997. Human liver carboxylesterase hCE-1: binding specificity for cocaine, heroin, and their metabolites and analogs. *Drug Metabolism and Disposition* 25: 1089–1096.

Burnell, J., Bosron. W., Zhang, J., and Dumaual, M. 1999. Binding and hydrolysis of meperidine by human liver carboxylesterase HCE-1. *Journal of Pharmacology and Experimental Therapeutics* 290: 314–318.

Draganov, D., Teiber, J., Speelman, A., et al. 2005. Human paraoxonases (PON1, PON2, and PON3) are lactonases with overlapping and distinct substrate specificities. *Journal of Lipid Research* 46: 1239–1247.

Farid, N., Smith, R., Gillespie, T., et al. 2007. The disposition of prasugrel, a novel thienopyridine, in humans. *Drug Metabolism and Disposition* 35: 1096–1104.

Geshi, E., Kimura, T., Yoshimura, M., et al. 2005. A single nucleotide polymorphism in the carboxylesterase gene is associated with the responsiveness to imidapril medication. *Hypertension Research* 28: 719–725.

Gonzalvo, M.C., Gil, F., Hernandez, A.F., et al. 1998. Human liver paraoxonase (PON1): Subcellular distribution and characterization. *Journal of Biochemical and Molecular Toxicology* 12: 61–69.

Graham, D.Y., Opekun, A.R., Willingham, F.F., et al., 2005. Visible small-intestinal mucosal injury in chronic NSAID users. *Clinical Gastroenterology and Hepatology* 3: 55–59.

Hagihara, K., Kazui, M., Kurihara, A., et al. 2009. A possible mechanism for the differences in efficiency and variability of active metabolite formation from thienopyridine antiplatelet agents, prasugrel and clopidogrel. *Drug Metabolism and Disposition* 37: 2145–2152.

Hatfield, M.J., Tsurkan, L., Garrett, M., et al. 2011. Organ-specific carboxylesterase profiling identifies the small intestine and kidney as major contributors of activation of the anticancer prodrug CPT-11. *Biochemical Pharmacology* 81: 24–31.

He, G., Massarella, J., Ward, P. 1999. Clinical pharmacokinetics of the prodrug oseltamivir and its active metabolite ro 64-0802. *Clinical Pharmacokinetics* 37: 471–484.

Hioki, T., Fukami, T., Nakajima, M., et al., 2011. Human paraoxonase 1 is the enzyme responsible for pilocarpine hydrolysis. *Drug Metabolism and Disposition* 39: 1345–1352.

Hosokawa, M. 2008. Structure and catalytic properties of carboxylesterase isozymes involved in metabolic activation of prodrugs. *Molecules* 13: 412–431.

Imai, T., and Hosokawa, M. 2010. Prodrug approach using carboxylesterases activity: Catalytic properties and gene regulation of carboxylesterase in mammalian tissue. *Journal of Pesticide Science* 35: 229–239.

Imai, T., and Ohura, K. 2010. The role of intestinal carboxylesterase in the oral absorption of prodrugs. *Current Drug Metabolism* 11: 793–805.

Ishizuka, T., Fujimori, I., Nishida, A., et al. 2012. Paraoxonase 1 as a major bioactivating hydrolase for olmesartan medoxomil in human blood circulation: Molecular identification and contribution to plasma metabolism. *Drug Metabolism and Disposition* 40: 374–380.

Kamendulis, L.M., Brzezinski, M.R., Pindel, E.V., et al. 1996. Metabolism of cocaine and heroin is catalyzed by the same human liver carboxylesterases. *The Journal of Pharmacology and Experimental Therapeutics* 279: 713–717.

Kazui, M., Nishiya, Y., Ishizuka, T., et al. 2010. Identification of the human cytochrome P450 enzymes involved in the two oxidative steps in the bioactivation of clopidogrel to its pharmacologically active metabolite. *Metabolism Clinical and Experimental* 38: 92–99.

Koyama, K., Ogura, Y., Nakai, D., et al. 2014. Identification of bioactivating enzymes involved in the hydrolysis of laninamivir octanoate, a long-acting neuraminidase inhibitor, in human pulmonary tissue. *Drug Metabolism and Disposition* 42: 1031–1038.

Kristensen, K.E., Zhu, H.J., Wang, X., et al. 2014. Clopidogrel bioactivation and risk of bleeding in patients cotreated with angiotensin-converting enzyme inhibitors after myocardial infarction: A proof-of-concept study. *Clinical Pharmacology and Therapeutics* 96: 713–722.

Mathijssen, R.H.J., Alphen, R.J., Verweij, J., et al. 2001. Clinical pharmacokinetics and metabolism of irinotecan (CPT-11). *Clinical Cancer Research* 7: 2182–2194.

Miwa, M., Ura, M., Nishida, M., et al. 1998. Design of a novel oral fluoropyrimidine carbamate, capecitabine, which generates 5 fluorouracil selectively in tumours by enzymes concentrated in human liver and cancer tissue. *European Journal of Cancer* 34: 1274–1281.

Mizoi, K., Takahashi, M., Haba, M., and Hosokawa, M. 2016. Synthesis and evaluation of atorvastatin esters as prodrugs metabolically activated by human carboxylesterases. *Bioorganic and Medicinal Chemistry Letters* 26: 921–923.

Ose, A., Ito, M., Kusuhara, H., et al. 2009. Limited brain distribution of [3*R*,4*R*,5*S*]-4-acetamido-5-amino-3-(1-ethylpropoxy)-1-cyclohexene-1-carboxylate phosphate (ro 64-0802), a pharmacologically active form of oseltamivir, by active efflux across the blood-brain barrier mediated by organic anion transporter 3 (OAT3/SLC22AS) and multidrug resistance-associated protein 4 (MRP4/ABCC4). *Drug Metabolism and Disposition* 37: 315–321.

Quinney, S.K., Sanghani, S.P., Davis, W.I., et al. 2005. Hydrolysis of capecitabine to 5'-deoxy-5-fluorocytidine by human carboxylesterases and inhibition by loperamide. *Journal of Pharmacology and Experimental Therapeutics* 313: 1011–1016.

Rehmel, J.L.F., Eckstein, J.A., Farid, N.A., et al. 2006. Interactions of two major metabolites of prasugrel, a thienopyridine antiplatelet agent, with the cytochromes P450. *Drug Metabolism and Disposition* 34: 600–607.

Satoh, T., and Hosokawa, M. 1998. The mammalian carboxylesterases: From molecules to functions. *Annual Review of Pharmacology and Toxicology* 38: 257–288.

Satoh, T., and Hosokawa, M. 2006. Structure, function and regulation of carboxylesterases. *Chemico-Biological Interactions* 162: 195–211.

Satoh, T., Hosokawa, M., Atsumi, R., et al. 1994. Metabolic activation of CPT-11, 7-ethyl-10-[4-(1-piperidino)-1-piperidino]carbonyloxycamptothecin, a novel antitumor agent, by carboxylesterase. *Biological & Pharmaceutical Bulletin* 17: 662–664.

Shi, D., Yang, J., Yang, D., et al. 2007. Response to comments on anti-influenza prodrug oseltamivir is activated by carboxylesterase human carboxylesterase 1, and the activation is inhibited by antiplatelet agent clopidogrel. *Journal of Pharmacology and Experimental Therapeutics* 322: 424–425.

Sierakowski, B., Püchler, K., Witte, P. U., et al. 1997. Single-dose pharmacokinetics of temocapril and temocapril diacid in subjects with varying degrees of renal impairment. *European Journal of Clinical Pharmacology* 53: 215–220.

Stella, V.J. 2010. Prodrugs: Some thoughts and current issues. *Journal of Pharmaceutical Sciences* 99: 4755–4765.

Sun, Z., Murry, D. J., Sanghani, S. P., et al. 2004. Methylphenidate is stereoselectively hydrolyzed by human carboxylesterase CES1A1. *Journal of Pharmacology and Experimental Therapeutics* 310: 469–476.

Sozio, P., Fiorito, J., Di Giacomo, V., et al. 2015. Haloperidol metabolite II prodrug: Asymmetric synthesis and biological evaluation on rat C6 glioma cells. *European Journal of Medicinal Chemistry* 90: 1–9.

Takai, S., Matsuda, A., Usami, Y., et al. 1997. Hydrolytic profile for ester- or amide-linkage by carboxylesterases PI 5.3 and 4.5 from human liver. *Biological & Pharmaceutical Bulletin* 20: 869–873.

Takahashi, M., Ogaw, a T., Kashiwagi, H., et al. 2018. Chemical synthesis of an indomethacin ester prodrug and its metabolic activation by human carboxylesterase 1. *Bioorganic and Medicinal Chemistry Letters* 28: 997–1000.

Takahashi, M., Uehara, T., Nonaka, M., et al. 2019. Synthesis and evaluation of haloperidol ester prodrugs metabolically activated by human carboxylesterase. *European Journal of Pharmaceutical Sciences* 132: 125–131.

Tougou, K., Nakamura, A., Watanabe, S., et al. 1998. Paraoxonase has a major role in the hydrolysis of prulifloxacin (NM441), a prodrug of a new antibacterial agent. *Drug Metabolism and Disposition: The Biological Fate of Chemicals* 26: 355–359.

Watanabe, A., Chang, S-C., Kim, M.J., et al. 2010. Long-acting neuraminidase inhibitor laninamivir octanoate versus oseltamivir for treatment of influenza: A double-blind, randomized, noninferiority clinical trial. *Clinical Infectious Diseases* 51: 1167–1175.

Williams, E.T., Ponsler, G.D., Lowery, S.M., et al. 2008. The biotransformation of prasugrel, a new thienopyridine prodrug, by the human carboxylesterases 1 and 2. *Drug Metabolism and Disposition* 36: 1227–1232.

Williams, P.E., Brown, A.N., Rajaguru, S., et al. 1989. The pharmacokinetics and bioavailability of cilazapril in normal man. *British Journal of Clinical Pharmacology* 27: 181S–188S.

Yamashita, M., Tomozawa, T., Kakuta, M., et al. 2009. CS-8958, a prodrug of the new neuraminidase inhibitor R-125489, shows long-acting anti-influenza virus activity. *Antimicrobial Agents and Chemotherapy* 53: 186–192.

Yang, J., Shi, D., Yang, D., et al. 2007. Interleukin-6 alters the cellular responsiveness to clopidogrel, irinotecan, and oseltamivir by suppressing the expression of carboxylesterases hCE1 and hCE2. *Molecular Pharmacology* 72: 686–694.

Yun, J.H., Myung, J.H., Kim, H.J., et al. 2005. LC-MS determination and bioavailability study of imidapril hydrochloride after the oral administration of imidapril tablets in human volunteers. *Archives of Pharmacal Research* 28: 463–468.

Zhang, J., Burnell, J. C., Dumaual, N., and Bosron, W. F. 1999. Binding and hydrolysis of meperidine by human liver carboxylesterase hCE-1. *Journal of Pharmacology and Experimental Therapeutics* 290: 314–318.

Zejin, S., Murry, D.J., Sanghani, S.P., et al. 2004. Methylphenidate is stereoselectively hydrolyzed by human carboxylesterase CES1A1. *Journal of Pharmacology and Experimental Therapeutics* 310: 469–476.

12

Mathematical Models for ADME of Prodrugs

Vipin Bhati, Navyashree V., Shamsher Singh, Pooja Chawla, and Anoop Kumar

CONTENTS

12.1 Introduction

Prodrugs are pharmacologically inactive molecules which are converted to active molecule under *in vivo* biotransformation by enzymatic or chemical reactions. The active molecules act on particular target (receptor, enzymes, ion channels etc.) and exert the desired pharmacological effect (Satoskar et al. 2015; Wink 2008; Alagarsamy 2013). The molecules which have poor solubility and permeability are generally converted into prodrugs by introduction of chemically or enzymatically labile functional groups. The various drugs approved by the United State Food and Drug Administration (US FDA) as prodrugs are presented in Table 12.1.

The molecules that have poor physiochemical, biopharmaceutical or pharmacokinetic properties are generally converted into prodrug (Pérez-Urizar et al. 2000; Vig et al. 2013; Rautio et al. 2018). Currently, prodrug approach is widely used in the early stage of drug discovery. Approximately 5%–7% of approved drugs are in the form of prodrugs. Esters and amides groups are generally used to convert a drug into a prodrug. These groups mask the polar functional groups and hydrogen bonds of the active moiety. The inactive moiety of the ideal prodrug should be nontoxic and excreted out rapidly from the body (Rautio et al. 2008; Hamman et al. 2005). On the basis of different strategies, prodrugs are classified into two main classes i.e., the carrier linked prodrug and bio precursor prodrug (Lambert 2000). In carrier linked prodrug, carrier group is attached to the active drug which changes the physiochemical properties and the active drug moiety released by the enzymatic or chemical reactions (Khandare and Minko 2006). The bio precursor prodrug is transformed metabolically or chemically by hydration (e.g., lactones such as some statins), oxidation (e.g., dexpanthenol, nabumetone) or reduction (e.g., sulindac, platinum (IV) complexes) to the active agent.

TABLE 12.1

List of Prodrugs Approved by Food and Drug Administration (FDA) During 2008–2019

Year	Trade Name	Prodrug	Active Form	Activation Mechanism	Prodrug Strategy/Gain	Indication
2008	Toviaz	Fesoterodine fumarate	5-hydroxymethyl tolterodine	Hydrolysis by nonspecific esterase	Avoid variability in CYP2D6 activity.	Overactive bladder
	Lusedra	Fospropofol disodium	Propofol	Conversion by alkaline phosphatase	Increased aqueous solubility for IV injection by phosphorylation	Anaesthesia
	Emend	Fosaprepitant dimeglumine	Aprepitant	Dephosphorylation by phosphatase	Increased aqueous solubility by phosphorylation.	Prevention of chemotherapy-induced nausea and vomiting
2009	Effient	Prasugrel	R-138727	2 steps: Hydrolysis by esterase and CYP450 metabolism	Faster and more efficient conversion of parent molecule compared to Clopidogrel. Increased potency.	Prevention and reduction of thrombotic and cardiovascular events
	Istodax	Romidepsin	Metabolite with free thiol group	Activation by intracellular glutathione	—	Cutaneous T-cell lymphoma
2010	Gilenya	Fingolimod	Fingolimod-phosphate	Phosphorylation by sphingosine kinase	Structure activity relationship-based design to optimize activity.	Multiple sclerosis
	Pradaxa	Dabigatran etexilate	Dabigatran	Hydrolysis by esterase	Structure activity relationship based-design to optimize activity.	Thromboembolism
	Teflaro	Ceftaroline fosamil	Ceftaroline	Plasma phosphatase	N-phosphono-prodrug to improve aqueous solubility.	Acute bacterial skin and skin structure infections and community acquired pneumonia
2011	Edarbi	Azilsartan medoxomil	Azilsartan	Hydrolysis by esterase during absorption	—	Hypertension

(Continued)

TABLE 12.1 (*Continued*)

List of Prodrugs Approved by Food and Drug Administration (FDA) During 2008–2019

Year	Trade Name	Prodrug	Active Form	Activation Mechanism	Prodrug Strategy/Gain	Indication
	Horizant	Gabapentin enacarbil	Gabapentin	Hydrolysis by esterase	Transport by intestinal monocarboxylate transporter type 1 and sodium dependent multivitamin transporter. Hydrolysis in tissues.	Restless leg syndrome and postherpetic neuralgia
	Zytiga	Abiraterone acetate	Abiraterone sulfate and N-oxide abiraterone sulfate	2 steps: Hydrolysis by esterase Metabolism CYP3A4 and SULT2A1	—	Hormone refractory prostate cancer
2012	Zioptan	Tafluprost	Tafluprost acid	Hydrolysis by esterase	Good corneal penetration. Better activity Lesser pigmentation	Glaucoma
2013	Tecfidera	Dimethyl fumarate	Monomethylfumarate	Hydrolysis by esterase	—	Multiple sclerosis
	Sovaldi	Sofosbuvir	GS-461203	Intracellular metabolism by: Cathepsin-Acarboxylesterase 1Histidine triade nucleotide-binding protein 1Uridine monophosphate-cytidine monophosphate kinase.	Phosphoramidate prodrug to bypass first step phosphorylation required by nucleoside analogs.	Hepatitis C infection
	Aptiom	Eslicarbazepine acetate	Eslicarbazepine	Hydrolysis during first pass metabolism	Avoid the formation of epoxide following metabolism.	Epilepsy
2014	Northera	Droxidopa	Norepinephrine	Decarboxylation by L-aromatic-amino-acid decarboxylase	—	Neurogenic orthostatic hypotension and intradialytic hypotension Parkinsonism disease (off-label)

(*Continued*)

TABLE 12.1 (*Continued*)

List of Prodrugs Approved by Food and Drug Administration (FDA) During 2008–2019

Year	Trade Name	Prodrug	Active Form	Activation Mechanism	Prodrug Strategy/Gain	Indication
	Sivextro	Tedizolid phosphate	Tedizolid	Dephosphorylation by plasma phosphatase	Structure activity relationship design. Increased monoamineoxidase inhibitory profile. Improved aqueous solubility. Improved bioavailability	Acute bacterial skin and skin infections
2015	Aristada	Aripiprazole lauroxil	Aripiprazole	2 steps: Hydrolysis by esterase Nonenzymatic hydrolysis	Increase lipid solubility Prolonged action for IM injection.	Schizophrenia
	Xuriden	Uridine triacetate	Uridine	Deacetylation by esterase	Catabolism resistance Enhanced absorption	Hereditary oroticaciduria
	Entresto	Sacubitril	LBQ657	De-ethylation by liver carboxylesterase 1	—	Heart failure
	Ninlaro	Ixazomib citrate	Active boronic form	Rapid hydrolysis post administration	Improved affinity	Multiple myeloma
	Cresemba	Isavuconazonium sulfate	Isavuconazole	2 steps: Hydrolysis by esterase Intramolecular cyclization leading to N-dealkylation	Improved aqueous solubility	Invasive aspergillosis and invasive mucormycosis
	Genvoya Odefsey Descovy	Tenofovir alafenamide	Tenofovir	Hydrolysis by lysosomal protective protein or liver carboxylesterase 1	Phosphoramidate prodrug. Increased lipophilicity	HIV
	Uptravi	Selexipag	ACT-333679 (MRE-269)	Hydrolysis by carboxylesterase	Improved bioavailability	Pulmonary arterial hypertension
2016	Emflaza	Deflazacort	21-desacetyldeflazacort	Hydrolysis by esterase	—	Duchenne muscular dystrophy
	Xermelo	Telotristat ethyl	Lp-778902	Hydrolysis by carboxylesterase	Improved bioavailability	Carcinoid syndrome diarrhoea
2017	Austedo	Deutetrabenazine	Mainly Î±-dihydrotetrabenazine and Î²-dihydrotetrabenazine	CYP450 metabolism	Deuterated to retard hepatic metabolism	Huntington's disease

(*Continued*)

TABLE 12.1 (*Continued*)

List of Prodrugs Approved by Food and Drug Administration (FDA) During 2008–2019

Year	Trade Name	Prodrug	Active Form	Activation Mechanism	Prodrug Strategy/Gain	Indication
	Ingrezza	Valbenazine tosylate	Mainly Î±-dihydrotetrabenazine	CYP450 metabolism	Improved pharmacokinetics	Tardive dyskinesia
	Benznidazole	Benznidazole	Various electrophilic metabolites	Reduction by Trypanosoma cruzi nitroreductase	—	Chagas disease
	Solosec	Secnidazole	(Active metabolite)	Reduction by bacterial nitroreductase	—	Bacterial vaginosis
	Vyzulta	Latanoprostene bunod	Latanoprostacid Butanediol mononitrate	Hydrolysis by corneal esterase	Good corneal penetration Delivery of NO releasing species	Glaucoma
2018	Xofluza	Baloxavir marboxil	S-033447	Hydrolysis by esterase	—	Influenza A and B infection
	Krintafel	Tafenoquine	5,6-ortho-quinonetafenoquine	Bioconversion by CYP2D6	—	Malaria
	Akynzeo	Fosnetupitant	netupitant	Dephosphorylation by plasma phosphatase	Phosphorylation for IV injection	Chemotherapy induced nausea and vomiting
2019	Rocklatan	Netarsudil Latanoprost	AR13503 Active acid of latanoprost	Metabolized by esterases in eye 2 steps: Hydrolyzed by esterases in cornea Fatty acid β oxidation	To increase plasma concentration	To reduce elevated intra ocular pressure in patients with open angle glaucoma or ocular hypertension.

Further, based on cellular site of bioactivation, prodrugs can also be classified into two broad categories i.e., Type 1 and Type 2 (Zawilska et al. 2013; Jin and Penning 2007). Type 1 prodrugs are bioactivated intracellularly (e.g., antivirals and lipid lowering drugs, statins) (Lohar et al. 2012), whereas type 2 are bioactivated extracellularly particularly by the digestive fluids and systemic circulation (e.g., valganciclovir and fosamprenavir) (Dubey and Valecha 2014; Puris et al. 2019).

A good prodrug should contains number of characteristics such as partial P-glycoprotein efflux, ability to hydrolyze in the GI lumen and intestinal cells, biliary excretion, nonesterase metabolism in the liver and transformation of the parent moiety (Zawilska et al. 2013). Further, it should have weak or no activity against target, should be specific at physiological pH, easily soluble in aqueous environment, high passive permeability, resistant to hydrolysis during absorption, release active drug rapidly and qualitatively after absorption with no toxic or undesired pharmacological activity (Jornada et al. 2015; Balant 2003). Pro-moiety contained by the prodrug is generally released enzymatically or chemically while some prodrugs free their active drug after molecular modification such as an oxidation or reduction reactions (Stella and Nti-Addae 2007; Karaman 2013).

Absorption, Distribution, Metabolism and Excretion (ADME) is the pharmacokinetic study of the drugs (Kumar et al. 2016; Kant et al. 2019; Gupta et al. 2018). These studies are critical part of any drug development program, and essential for compliance with regulatory guidelines. The poor pharmacokinetic properties of the more potent compound make it far less active *in vivo* (Di and Kerns 2015). In 1991, approximately 40% of the molecules are failed due to poor pharmacokinetic profile. Thus, *in vitro* assay has been developed to decrease the attrition rate of the molecules due to poor pharmacokinetic profile (Penner et al. 2012). Currently, various economical and easy methods have been used by most of the pharmaceutical companies to screen the drug candidates for pharmacokinetic profile. The movement of prodrugs in the body is most important aspect. Thus, this chapter familiarizes the readers with the various mathematical models for ADME of prodrugs. Further, current challenges and future perspectives have been also discussed.

12.2 Mathematical Models for ADME of Prodrugs

Various types of pharmacokinetic models are used to describe drug absorption, distribution, metabolism, and elimination from the body (i.e., ADME) (Daina et al. 2017). Mathematical models have ability to predict the pharmacokinetics of drugs. These models can also use to check the pharmacokinetics of the prodrugs. Most commonly used mathematical models are compartment models, physiological models, pharmacodynamic models shown in Figure 12.1 which is explained below.

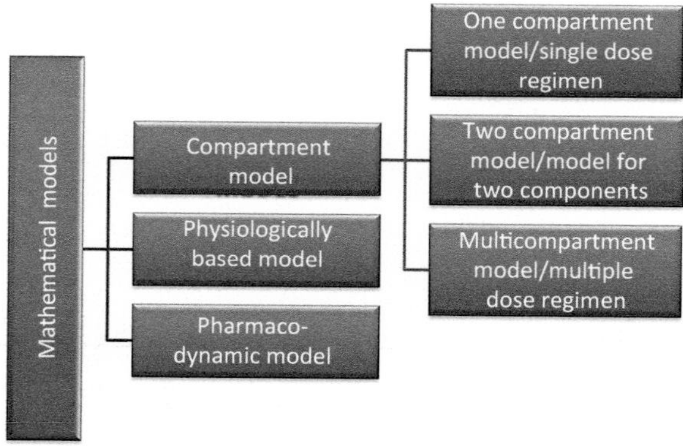

FIGURE 12.1 Types of mathematical models.

12.2.1 Compartmental Models

Compartmental investigation is the most commonly used strategy to predict the pharmacokinetic profile. In compartmental models, groups of tissues that have equal blood flow and drug affinity are considered as single compartments. The compartment does not represent any specific anatomic region within the body. Similar and uniform drug distribution is supposed within each compartment and simple first-order rate equations are used to explain the delivery of drug into and out of the compartment. Compartmental models are also called "open" models (Reisman et al. 2018). These models simply interpose the experimental data and allow an empirical formula to calculate the changes in the concentration of drug with time. Number of required compartments is empirically determined by the experimental data. The route of administration and experimental data defines the structure of model (Nestorov 2007). The various types of compartmental models are mentioned below.

12.2.1.1 Single Dose Regimen (One Compartment Model)

This model provides the basis for developing more complex models. Thus, it is crucial to understand this reaction kinetics thoroughly for further study (Cho and Yoon 2018).

The one-compartment open model is the simplest mammillary model that describes drug distribution and elimination. In the one-compartment model, the body is considered as a single, uniform compartment into which the drug is administered and from which it is eliminated. This model does not predict actual drug concentrations which will be proportional to the drug plasma concentrations (Miller et al. 2019).

In the one-compartment open model, after IV administration, the entire drug dose enters the blood circulation directly and is then distributed very rapidly in whole body. The drug equilibrates rapidly in the body, and it is assumed that the concentration throughout the compartment is equal to the plasma concentration (Brochot and Quindroit 2018). The simplest administration (pharmacokinetically) is a single rapid intravenous bolus and simplest compartment model is the one-compartment model, as shown in Figure 12.3. In one-compartment model, drug is added and eliminated through a central compartment (Brantuoh 2014). The central compartment is used to represent plasma and it is necessary that in highly perfused tissues the concentrations throughout the body are in rapid equilibrium according to the apparent volume of distribution (Riegelman et al. 2007).

$$V = \frac{\text{dose administered}}{\text{initial concentration}} \tag{12.1}$$

If we assume first-order elimination of the drug from the single compartment, the rate of change of drug concentration can be described by Equation (12.2), with k_{el} the elimination rate constant (Ma 2013).

$$\frac{dc}{dt} = -k_{el} \cdot C \tag{12.2}$$

This simple one-compartment model is represented in Figure 12.2 with the initial condition derived from Equation (12.1) (Ma 2013).

$$C_0 = \frac{dose}{V} \tag{12.3}$$

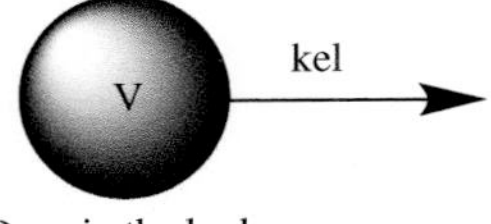

FIGURE 12.2 One-compartment model with first-order elimination after an IV bolus.

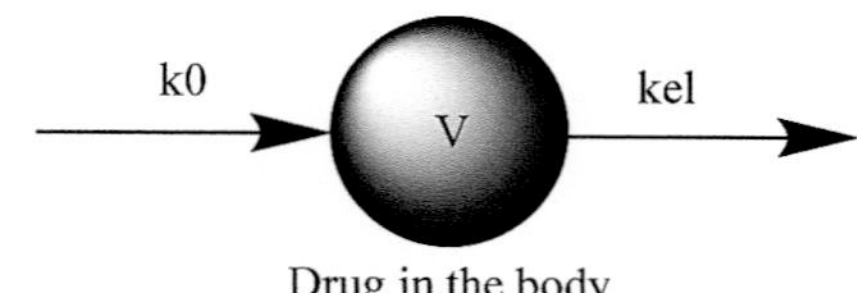

FIGURE 12.3 One-compartment model with first-order elimination during an IV infusion.

Equation (12.2) is integrated to give Equation (12.4) for drug concentration versus time.

$$C = C_0 e^{-Kel \cdot t} = \frac{dose}{V} \cdot e^{-Kel \cdot t} \tag{12.4}$$

When the drug is given as a zero-order, IV infusion (amount per time, mg/hr), an infusion rate constant *ko*, is added to the model as shown in Figure 12.3. The differential equation for drug concentration during the infusion can be derived and is shown in Equation (12.5) (Tallarida and Murray 2012).

$$\frac{dc}{dt} = ko - k_{el} \cdot t \tag{12.5}$$

Integration of this equation gives

$$C = \frac{K_0}{K_{el} \cdot V} \cdot \left[1 - e^{-Kel \cdot t}\right] \tag{12.6}$$

Concentrations up to time T can be calculated using Equation (12.6). Equation (12.7) can be used to calculate the drug concentrations after the infusion has stopped (Mager and Jusko 2001).

$$C = \frac{K_0}{K_{el} \cdot V} \cdot \left[1 - e^{-Kel \cdot T}\right] \cdot e^{-Kel(t-T)} \tag{12.7}$$

In its simplest form of absorption that might be described by a first-order rate constant Ka and an extent factor, the bioavailability F shown in Figure 12.4 (Mager and Jusko 2001).

12.2.1.2 Model for Two Components (Two Compartment Model)

In the two-compartment open model, the body is considered as two hypothetical compartments: (1) central compartment, which represent blood, extracellular fluid, and highly perfused tissues, and (2) tissue compartment, which contains tissues such as skin and fat tissues, in which the drug diffuses more slowly. The heart, liver, kidney, lungs and brain are highly perfused organs, which comes under central compartment. The drug enters the body and leaves it via the central compartment (Cho and Yoon 2018). In two-compartment model, drug can move between central and or plasma compartment and from the tissue compartment shown in Figure 12.5. In this model, the total quantity of drug is the sum of drug present in the central compartment and the drug present

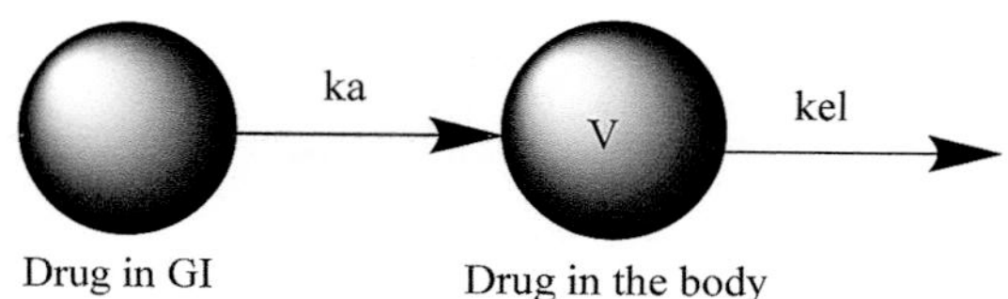

FIGURE 12.4 One-compartment model with first-order elimination and first-order absorption.

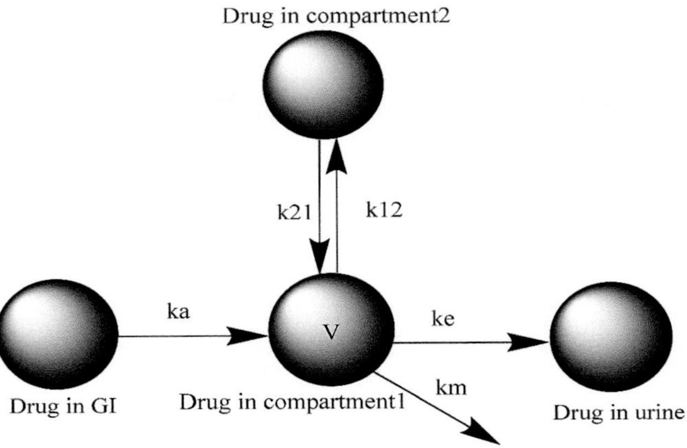

FIGURE 12.5 Two-compartment model with first-order absorption and two first-order elimination pathways.

in the tissue compartment. This model consists two components drug in the GI tract and drug in the body—thus two differential equations can be used to describe the scheme mathematically (Gerlowski and Jain 1983).

$$\frac{dXGI}{dt} = -ka \cdot K_a \cdot X_{GI} \tag{12.8}$$

and

$$\frac{dX_B}{dt} = K_a \cdot X_{GI} - K_{el} \cdot X_B \tag{12.9}$$

where X_{GI} and X_B are the amount of drug in the GI tract and the amount of drug in the body (single compartment), respectively. The initial conditions for these equations are X_{GI} (O) = F-dose and X_B (0) = 0. These differential equations can be integrated to give Equation (12.10) (Gerlowski and Jain 1983).

$$C = \frac{f \cdot dose \cdot ka}{V \cdot (ka - K_{el})} \cdot \left[e^{-Kel \cdot t - e^{-Ka \cdot t}} \right] \tag{12.10}$$

Note the development of the differential Equations (12.8) and (12.9). If we look at Figure 12.5 and focus on the drug in the GI component we can say that there is a negative transfer of drug because one arrow is leaving the circle. If we assume that the rate process is first-order, then we should multiply the rate constant *Ka* by the amount of drug in the component attached to the tail of the arrow (drug in GI). From this we obtain Equation (12.8). Similarly, we can derive the—$k_{el} \cdot X_B$ term of Equation (12.9) from the arrow leaving the drug in blood component. Because there is also an arrow pointing to this component, we have a positive (input) term. Since this input is first-order we multiply K_a by the amount in the component at the tail of the arrow.

12.2.1.3 Multiple Dose Regimen

The previous models explain drug concentrations after a single dose. Additional doses can be easily added when describing pharmacokinetic models using differential equations. A new dose is added at each dosing time to the dose compartment. Most drugs are administered with sufficient frequency that measurable and often pharmacologically significant levels of drug persist in the body when a subsequent dose is administered in a fixed dose at a constant dosing interval (e.g., every 6 h or once a day), the peak plasma level following the second level after the first dose, and therefore, the drug accumulates in the body relatives to initial dose. To calculate a multiple dose regimen for a patient or patients' pharmacokinetic parameters are first obtained from the plasma level-time curve generated by single-dose drug studies. With these pharmacokinetic parameters and knowledge of the size of the dose and dosage interval, the complete plasma level-time curve or the plasma level may be predicted at any time after the beginning of the dosage-regimen (Rudek et al. 2013). For example, multiple oral dosing can be described as new doses (as F* dose,) to the drug in the GI component at various dosing times. This method is quite easy and it can consist linear or nonlinear models and flexible dosing regimens (Levine et al. 2001).

Simple multiple dose regimens, typically uniform dose and dosing interval, can be added to the integrated equation dosing using a multiple dosing function $[(C1e^{-n\cdot kel\cdot T})/\ (1-e^{-Kel\cdot t})]$. Thus, for linear systems such as a one-compartment model with uniform multiple IV bolus doses, the drug concentration can be calculated as

$$C = \frac{dose}{V}\cdot\left[\frac{1-e^{-n\cdot kel\cdot T}}{1-e^{-n\cdot kel\cdot T}}\right]\cdot e^{-n\cdot kel\cdot t} \tag{12.11}$$

where T is the dosing interval, n is the number of doses, and t is the time since the last dose. Drug concentrations after uniform multiple oral dosing can be calculated as (Levine et al. 2001).

$$C = \frac{f\cdot dose\cdot ka}{V\cdot(ka-K_{el})}\cdot\left\{\left[\frac{1-e^{-n\cdot kel\cdot T}}{1-e^{-n\cdot kel\cdot T}}\right]\cdot e^{-n\cdot kel\cdot t}-\left[\frac{1-e^{-n\cdot ka\cdot T}}{1-e^{-ka\cdot T}}\right]\cdot e^{-ka\cdot t}\right\} \tag{12.12}$$

12.2.1.4 Nonlinear Elimination

At therapeutic or nontoxic plasma concentrations, the pharmacokinetics of most drugs can adequately explain by first-order or linear processes. However, there are a small number of well-established examples of drugs which have nonlinear absorption or distribution characteristics [e.g., ascorbic acid and naproxen] and several examples of drugs that are eliminated from the body in a nonlinear manner (Cho and Yoon 2018). It is not necessary that elimination and other processes are linear or first-order. Number of elimination processes are enzyme mediated and, can be saturate. Michaelis and Menton replaced the first-order eliminating process by saturable Michaelis-Menten (MM) type processes (Michaelis and Menten 1913). The one-compartment model, shown in Figure 12.6 has one first-order elimination process and one MM process Figure 12.7 (Levine et al. 2001). The differential equation for drug in body is given by Equation (12.13).

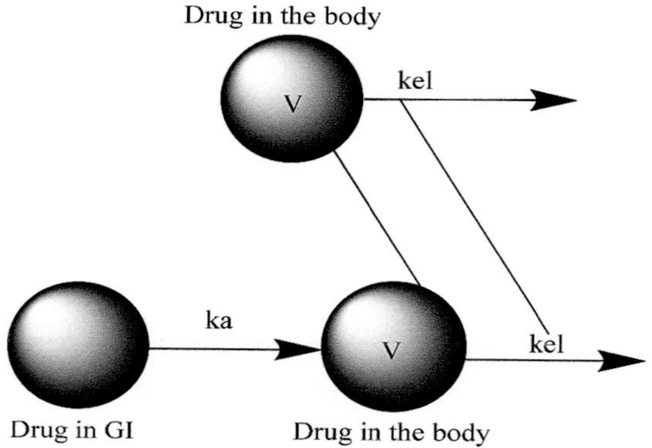

FIGURE 12.6 Linked one-compartment model with first-order elimination after IV and oral administration.

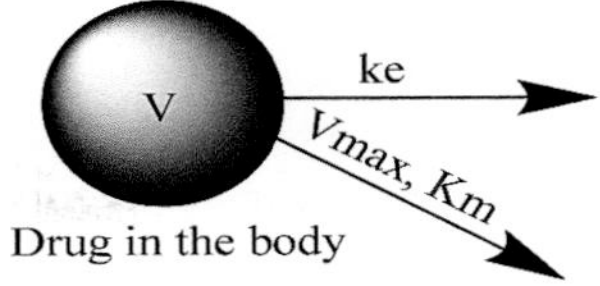

FIGURE 12.7 One-compartment model with first-order and Michaelis-Menten elimination.

$$\frac{dc}{dt} = -k_e \cdot C - \frac{Vmax \cdot C}{Km + c} \tag{12.13}$$

where *Vmax* and *Km* are the maximum rate and the Michaelis constant for the saturable process, respectively. Equation (12.13) includes one term to describe the linear elimination and another to describe the saturable process. Saturable elimination produces a drug concentration versus time curve that has a reduced slope at high concentrations and a faster slope at lower concentrations where the elimination appears linear (Levine et al. 2001).

12.2.1.5 Saturable Protein Binding

One more process which produces nonlinear elimination kinetics is saturable protein binding. If we assume that protein binding as a rapid equilibration process with an association constant *Ka*, and the total protein concentration is *Pt*, the equilibrium in Equation (12.14) can be used to derive the expression in Equation (12.15) for free drug concentration (Peters 2012).

$$[\text{Drug}] + [\text{Protein}] \overset{Ka}{\leftrightarrow} [\text{Drug} - \text{Protein}] \tag{12.14}$$

$$C_{free} = \frac{A \pm \sqrt{A^2 + 4 \cdot Ka \cdot C_{total}}}{2ka} \tag{12.15}$$

where

$$A = Ka \cdot C_{total} - 1 - Ka \cdot P_t \tag{12.16}$$

Finally, the concentration C_{free} can be included in the differential equation for drug amount versus time as shown in Equation (12.17).

$$\frac{dD_{total}}{dt} = -K_{el} \cdot C_{free} \cdot V \tag{12.17}$$

These equations estimate a drug concentration versus time curve that leads in a rapid early drop followed by a more gradual linear elimination as the drug concentration falls below protein saturation levels. The influence of saturable protein binding can be investigated by measuring free and total drug concentration and/or by measuring drug amounts excreted into urine (Peters 2012).

12.2.2 Physiologically Based Models

Other field of pharmacokinetic modeling that has achieved high acceptance is physiologically based pharmacokinetic (PBPK) models. PBPK models emulate the structure of the living organism being studied and represent the various organs and tissues as compartments in the model that are connected via a blood circulation loop which is subdivided into arterial and venous pools. In this model some organs or tissues are represented by single compartments, whereas others are combined into paired compartments. The blood flow to each organ, as well as the drug uptake to the organ, has to be known. Typically, these compartments include the main tissues of the body, namely, adipose, bone, brain, gut, heart, kidney, liver, lung, muscle, skin, and spleen (Glassman and Balthasar 2018). A physiological pharmacokinetic model is composed of a series of lumped compartments (body regions) representing organs or tissue spaces whose drug concentrations are assumed to be uniform (Bassingthwaighte et al. 2012). Based on this approach, the body is not just regarded as compartments, but is considered in better physiologically based terms. The tissues and organs of the body are depicted by their volumes and flow of blood. For example, distribution throughout the body is speculated by blood flow (i.e., flow-limited). Accordant with this approach it is imagined that the drug concentration in the blood leaving an organ is in equilibrium with the drug concentration in the organ (Peters 2012). Thus, the equilibrium constant or partition constant (R) can be depicted as

$$R = \frac{C_{tissue}}{C_{blood}} \tag{12.18}$$

These differential equations are based on the principle of mass balance and basic physiology. Thus, the rate of change of the drug amount in the tissue of interest is defined by a mass balance equation.

$$\text{rate of change} = \text{rate in} - \text{rate out} \tag{12.19}$$

The rate in term will consist drug transport into the tissue via blood flow, and for some tissues and drug administration. The rate out term may consists drug transport from the tissue through blood flow, and for certain tissues, drug elimination or clearance terms. With this background we can make a simplified PBPK model. For the blood component there is drug leaving to the tissues and drug returning from the tissues. For an IV infusion there would be an infusion rate constant, whereas for an IV bolus the initial condition for the blood component would be the dose. The differential equation for the blood component is given by Equation (12.20).

$$\frac{dA_b}{dt} = \frac{C_t}{R_t} \cdot Q_t + \frac{C_k}{R_k} \cdot Q_k + \frac{C_m}{R_m} \cdot Q_m + \frac{C_o}{R_o} \cdot Q_o - C_b \cdot Q_b \tag{12.20}$$

Dividing both sides of Equation (12.20) by the blood volume V_b gives the rate of change of drug concentration in the blood component C_b.

$$\frac{dA_b}{dt} = \left[\frac{C_t}{R_t} \cdot Q_t + \frac{C_k}{R_k} \cdot Q_k + \frac{C_m}{R_m} \cdot Q_m + \frac{C_o}{R_o} \cdot Q_o - C_b \cdot Q_b \right] / V_b \tag{12.21}$$

For the two noneliminating components, muscle and other, the differential equations are relatively simple.

$$\frac{dA_b}{dt} = \left[\left(C_b - \frac{C_m}{R_m} \right) \cdot Q_m \right] / V_m \tag{12.22}$$

$$\frac{dC_o}{dt} = \left[\left(C_b - \frac{C_o}{R_o} \right) \cdot Q_o \right] / V_o \tag{12.23}$$

A physiological model can be elaborated to consists diffusion limited transfer and protein or tissue binding. Depending on the aim of the modeling and the data available, additional tissues may be added in the model. In other cases, part of the model may be represented by simple exponential functions that act as input functions to certain tissues of interests, such as brain regions. These models have been called hybrid models because they include the complexity of the physiological approach for the region of interest and the simplicity of the compartmental approach for the remainder of the body (Clewell and Clewell 2008.)

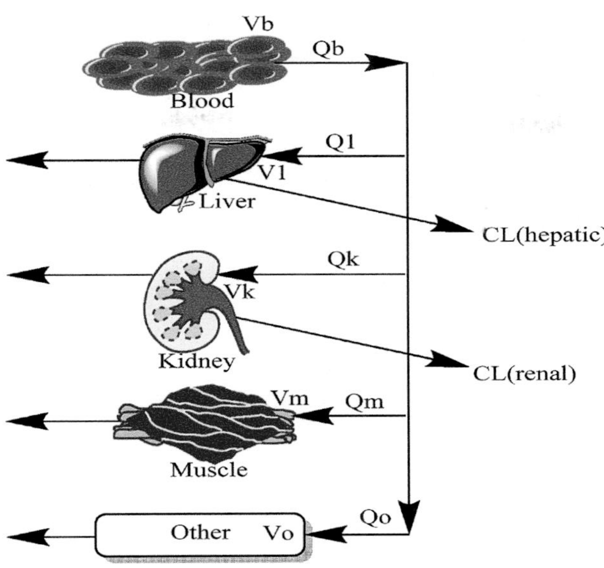

FIGURE 12.8 Simple physiologically based pharmacokinetic model.

It is possible to simulate drug disposition in various pathophysiologic conditions because PBPK models include physiological parameters. Compartmental blood flow can change, binding can be altered, or clearance terms may alter in various pathological conditions shown in Figure 12.8. The effect of these alterations on drug metabolism and excretion can be readily enhanced (Mangoni and Jackson 2004). Thus, much of the experimental work could be completed with laboratory animals, and the human experiments could consist of only blood and urine collection (for metabolism and other clearance information). From the combined data it should be possible to simulate drug concentrations in various human tissue regions and under various physiological conditions (Mangoni and Jackson 2004).

12.2.3 Pharmacodynamic Models

In many instances there may be quantifiable changes in different pharmacological parameters. These parameters include muscle relaxation, blood pressure, heart rate, EEG parameters, and pain relief. The response is observed direct drug concentration against effect relationship. If drug concentration increases the intensity of the pharmacological effect will increase. The mathematical model or equation for this model is Figure 12.9 and Equation (12.24).

$$\text{Effect} = \frac{E_{max} \cdot C^{\gamma}}{E_{C^{\gamma}_{50\%} + C^{\gamma}}} \tag{12.24}$$

where E_{max} is the maximum possible effect, $EC_{50\%}$ is the drug concentration that will produce an effect equal to $E_{max}/2$, and γ is a slope factor. Equation (12.24) shows the pharmacological response versus concentration or pharmacological model. Thus, it should be possible to combine these models to construct an overall pharmacological effect versus time or pharmacodynamic model shown in Figure 12.9. The calculated drug concentration in the effect compartment can be used with Equation (12.24) to calculate the drug effect. Thus, the drug concentration and the effect of the drug as it changes with time can be calculated and compared with observed values.

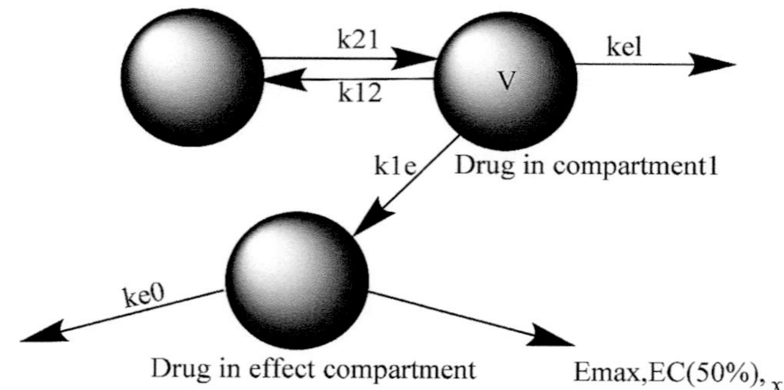

FIGURE 12.9 Pharmacodynamic model based on a two-compartment pharmacokinetic model.

12.3 Conclusion

In conclusion, the effect of any prodrug is depending upon plasma concentration of active moiety which depends upon dose, dose regimen, infusion rate etc. which can be predicted with the help of suitable mathematical model. Thus, these models can be used to program computers to deliver a variable rate infusion to achieve predetermined plasma level and hence a desired therapeutic effect.

ACKNOWLEDGMENTS

Authors are thankful to Dr. S.J.S. Flora, Director, NIPER, Raebareli and Mr. Parveen Garg, Chairman, ISFCP, Moga for their constant support and motivation. Author, Anoop Kumar is also thankful to Dr. Ruchika Sharma for editing of this chapter.

REFERENCES

Alagarsamy, V., 2013. *Textbook of Medicinal Chemistry Vol I-E-Book* (Vol. 1). Elsevier Health Sciences, Amsterdam, the Netherlands.

Balant, L.P., 2003. Metabolic considerations in prodrug design. *Burger's Medicinal Chemistry and Drug Discovery*, pp. 499–532.

Bassingthwaighte, J.B., Butterworth, E., Jardine, B. and Raymond, G.M., 2012. Compartmental modeling in the analysis of biological systems. In *Computational Toxicology* (pp. 391–438). Humana Press, Totowa, NJ.

Brantuoh, G.K., 2014. Linear pharmaco kinetic model of first order metabolism in the liver (Doctoral dissertation), Manakin.

Brochot, C. and Quindroit, P., 2018. Modelling the fate of chemicals in humans using a lifetime physiologically based pharmacokinetic (PBPK) model in MERLIN-expo. In *Modelling the Fate of Chemicals in the Environment and the Human Body* (pp. 215–257). Springer, Cham, Switzerland.

Cho, S. and Yoon, Y.R., 2018. Understanding the pharmacokinetics of prodrug and metabolite. *Translational and Clinical Pharmacology*, *26*(1), pp. 1–5.

Clewell, R.A. and Clewell III, H.J., 2008. Development and specification of physiologically based pharmacokinetic models for use in risk assessment. *Regulatory Toxicology and Pharmacology*, *50*(1), pp. 129–143.

Daina, A., Michielin, O. and Zoete, V., 2017. SwissADME: A free web tool to evaluate pharmacokinetics, drug-likeness and medicinal chemistry friendliness of small molecules. *Scientific Reports*, *7*, p. 42717.

Di, L. and Kerns, E.H., 2015. Drug-like properties: Concepts, structure design and methods from ADME to toxicity optimization. Elsevier.

Dubey, S. and Valecha, V., 2014. Prodrugs: A review. *World Journal of Pharmaceutical Research*, *3*(7), pp. 277–297.

Gerlowski, L.E. and Jain, R.K., 1983. Physiologically based pharmacokinetic modeling: Principles and applications. *Journal of Pharmaceutical Sciences*, *72*(10), pp. 1103–1127.

Glassman, P.M. and Balthasar, J.P., 2018. Physiologically-based modeling of monoclonal antibody pharmacokinetics in drug discovery and development. *Drug Metabolism and Pharmacokinetics*, *34*, pp. 3–113.

Gupta, M., Kant, K., Sharma, R. and Kumar, A., 2018. Evaluation of in silico anti-parkinson potential of β-asarone. *Central Nervous System Agents in Medicinal Chemistry (Formerly Current Medicinal Chemistry-Central Nervous System Agents)*, *18*(2), pp. 128–135.

Hamman, J.H., Enslin, G.M. and Kotzé, A.F., 2005. Oral delivery of peptide drugs. *BioDrugs*, *19*(3), pp. 165–177.

Jin, Y. and Penning, T.M., 2007. Aldo-ketoreductases and bioactivation/detoxication. *Annual Reviews Pharmacology and Toxicology*, *47*, pp. 263–292.

Jornada, D., dos Santos Fernandes, G., Chiba, D., de Melo, T., dos Santos, J. and Chung, M., 2015. The prodrug approach: A successful tool for improving drug solubility. *Molecules*, *21*(1), p. 42.

Kant, K., Lal, U.R., Kumar, A. and Ghosh, M., 2019. A merged molecular docking, ADME-T and dynamics approaches towards the genus of Arisaema as herpes simplex virus type 1 and type 2 inhibitors. *Computational Biology and Chemistry*, *78*, pp. 217–226.

Karaman, R., 2013. Prodrugs design by computation methods-a new era. *Journal of Drug Designing*, *1*, p. e113.

Khandare, J. and Minko, T., 2006. Polymer–drug conjugates: Progress in polymeric prodrugs. *Progress in Polymer Science*, *31*(4), pp. 359–397.

Kumar, A., Dubey, R.K., Kant, K., Sasmal, D., Ghosh, M. and Sharma, N., 2016. Determination of deltamethrin in mice plasma and immune organs by simple reversed-phase HPLC. *Acta Chromatographica*, *28*(2), pp. 193–206.

Lambert, D.M., 2000. Rationale and applications of lipids as prodrug carriers. *European Journal of Pharmaceutical Sciences*, *11*, pp. S15–S27.

Levine, H.A., Sleeman, B.D. and Nilsen-Hamilton, M., 2001. Mathematical modeling of the onset of capillary formation initiating angiogenesis. *Journal of Mathematical Biology*, *42*(3), pp. 195–238.

Lohar, V., Singhal, S. and Arora, V., 2012. Research article prodrug: Approach to better drug delivery. *International Journal of Pharmaceutical Research*, *4*(1), pp. 15–21.

Ma, L., 2013. Analysis of nonlinear pharmacokinetic systems and the nonlinear disposition of phenylbutazone in equine (horses), (Doctoral dissertation), Oregon State University, Corvallis, OR.

Mager, D.E. and Jusko, W.J., 2001. General pharmacokinetic model for drugs exhibiting target-mediated drug disposition. *Journal of Pharmacokinetics and Pharmacodynamics*, *28*(6), pp. 507–532.

Mangoni, A.A. and Jackson, S.H., 2004. Age-related changes in pharmacokinetics and pharmacodynamics: Basic principles and practical applications. *British Journal of Clinical Pharmacology*, *57*(1), pp. 6–14.

Michaelis, L. and Menten, M.M., 2013. The kinetics of invertin action. *FEBS Letters*, *587*, pp. 2712–2720.

Miller, G., Klumpp, J.A., Poudel, D., Weber, W., Guilmette, R.A., Swanson, J. and Melo, D.R., 2019. Americium systemic biokinetic model for rats. *Radiation Research*, *192*, pp. 75–91.

Nestorov, I., 2007. Whole-body physiologically based pharmacokinetic models. *Expert Opinion on Drug Metabolism & Roxicology*, *3*(2), pp. 235–249.

Penner, N., Xu, L. and Prakash, C., 2012. Radiolabeled absorption, distribution, metabolism, and excretion studies in drug development: Why, when, and how? *Chemical Research in Toxicology*, *25*(3), pp. 513–531.

Pérez-Urizar, J., Granados-Soto, V., Flores-Murrieta, F.J. and Castañeda-Hernández, G., 2000. Pharmacokinetic-pharmacodynamic modeling: Why? *Archives of Medical Research*, *31*(6), pp. 539–545.

Peters, S.A., 2012. *Physiologically-Based Pharmacokinetic (PBPK) Modeling and Simulations: Principles, Methods, and Applications in the Pharmaceutical Industry*. John Wiley & Sons, Hoboken, NJ.

Puris, E., Gynther, M., Huttunen, J., Auriola, S. and Huttunen, K.M., 2019. L-type amino acid transporter 1 utilizing prodrugs of ferulic acid revealed structural features supporting the design of prodrugs for brain delivery. *European Journal of Pharmaceutical Sciences*, *129*, pp. 99–109.

Rautio, J., Kumpulainen, H., Heimbach, T., Oliyai, R., Oh, D., Järvinen, T. and Savolainen, J., 2008. Prodrugs: Design and clinical applications. *Nature Reviews Drug Discovery*, *7*(3), p. 255.

Rautio, J., Meanwell, N.A., Di, L. and Hageman, M.J., 2018. The expanding role of prodrugs in contemporary drug design and development. *Nature Reviews Drug Discovery*, *17*(8), p. 559.

Reisman, S., Ritter, A.B., Hazelwood, V., Michniak, B.B., Valdevit, A. and Ascione, A.N., 2018. *Biomedical Engineering Principles*. CRC Press, Boca Raton, FL.

Riegelman, S., Loo, J.C.K. and Rowland, M., 1968. Shortcomings in pharmacokinetic analysis by conceiving the body to exhibit properties of a single compartment. *Journal of Pharmaceutical Sciences*, *57*(1), pp. 117–123.

Rudek, M.A., Connolly, R.M., Hoskins, J.M., Garrett-Mayer, E., Jeter, S.C., Armstrong, D.K., Fetting, J.H. et al. 2013. Fixed-dose capecitabine is feasible: Results from a pharmacokinetic and pharmacogenetic study in metastatic breast cancer. *Breast Cancer Research and Treatment*, *139*(1), pp. 135–143.

Satoskar, R.S., Rege, N. and Bhandarkar, S.D., 2015. *Pharmacology and Pharmacotherapeutics-E-Book*. Elsevier Health Sciences, Amsterdam, the Netherlands.

Stella, V.J. and Nti-Addae, K.W., 2007. Prodrug strategies to overcome poor water solubility. *Advanced Drug Delivery Reviews*, *59*(7), pp. 677–694.

Tallarida, R.J. and Murray, R.B., 2012. *Manual of Pharmacologic Calculations: With Computer Programs*. Springer Science & Business Media, New York.

Vig, B.S., Huttunen, K.M., Laine, K. and Rautio, J., 2013. Amino acids as promoieties in prodrug design and development. *Advanced Drug Delivery Reviews*, *65*(10), pp. 1370–1385.

Wink, M., 2008. Evolutionary advantage and molecular modes of action of multi-component mixtures used in phytomedicine. *Current Drug Metabolism*, *9*(10), pp. 996–1009.

Zawilska, J.B., Wojcieszak, J. and Olejniczak, A.B., 2013. Prodrugs: A challenge for the drug development. *Pharmacological Reports*, *65*(1), pp. 1–14.

13

Prodrug Approaches for Natural Products

Sudhir Kumar Thukral, Pooja Chawla, Alok Sharma, and Viney Chawla

CONTENTS

13.1 Introduction

There is an ever-increasing demand from drug scientists throughout the world to emphasize on basic strategies for searching natural products either as drugs or prodrugs (Cragg and Newman, 2013). Prodrug discovery from natural sources is characterized by identification, elucidation, derivatization and chemical modification of secondary metabolites obtained from natural resources, which undergo biotransformation prior to reaching binding site and cellular receptors so as to produce desired therapeutic effect compensating the limitation of physiological barrier. Mimicking the biosynthesizing of lead moiety at molecular-level remains the basic tool of research and development (Guo, 2017). More than 20 new drugs launched in the pharmaceutical market since 2000 onwards indicate that medicinal plants, vertebrates and invertebrates, microorganisms, marine organisms are the acceptable source of prodrugs and on the whole for novel drugs. Natural products claiming and expected as prodrugs are chemically altered and can be modified on basis of their biological and structural properties with hyphenated analytical and bioevaluation tools, before undergoing clinical trials. Natural prodrug-based discovery is characterized with a starting point of exploring and determination of active metabolites produced form secondary natural metabolites after metabolism within the body cells upon interaction with cellular enzymes and body fluids as significant bioconversion. Further, prodrug compounds necessitate "custom-made" or individualized elucidation of the structures of natural moieties so as to meet a drug criterion (Huttunen et al., 2011). Natural prodrugs must fulfill the criterion and confirm bioconversion into bioactive metabolites in order to cure numerous ailments along with optimized pharmacokinetic parameters and strengthening the idea of novel treatment strategies with fewer adverse effects. Existence of prodrugs in nature as phytopharmaceutical moieties as endogenous cellular inclusions, or semisynthetic compounds or intentionally synthesized prodrugs or unintentionally discovered compounds as a result of serendipity episodes during the novel drug development has uplifted the process of drug discovery. Examples of the discovery of the prodrugs obtained from nature include development of romidepsin, butyrin, psilocybin, salvestrols, spiruchostatin A, prontosil, melatonin, baicalin, matricin, sennosides, barbaloin, geniposide, lignans etc. demonstrating the importance of natural secondary metabolites serving as a base template for the discovery of a large number of prodrugs. Structural elucidation followed by *in vitro* and *in vivo* evaluation of active metabolites of natural products can give a new pathway of research. However, it seeks strict preclinical and clinical evaluation (Rautio et al., 2008). Numerous leads from nature possess therapeutic efficacy but can be put to use in clinical applications after preclinical and clinical evaluation. However, the molecular modification of natural products and conversion of prodrugs into drugs is very different from the strategy of conventional drug discovery. The strategy for prodrug development must include determination of potency and receptor selectivity, to check physicochemical, biochemical, and pharmacokinetic properties, to eliminate or reduce adverse effects, to simplify the structural complexity. Moreover, it also includes removal of redundant atoms and chirality while retaining bioactivities with a motive to generate molecules which can be patented and finally the development of novel therapeutic active metabolites along with pathway of biochemical conversion (Takagi et al., 2006). In order to make sure that these novel operations see the light of the day, sophisticated isolation, identification, purification, synthesis and skillful approved standard protocols as per standard ICH guidelines at preclinical and clinical levels are essential to follow, so as to confirm and assure the natural prodrugs for commercial use through drug research and development.

13.2 Fabrication of Prodrugs from Natural Resources

Drug researchers had investigated numerous prodrugs from natural resources, may be optimum wangling with the technical assistance of hyphenated analytical techniques and clinical protocols, providing feasible pathway of drug research without huge investment. Nature is only basic supplier of leads and moieties of so called prodrugs which get metabolized into therapeutic active metabolites upon catalysis by cellular enzyme within cells and body fluids (Vijayaraj et al., 2014). Occurrence of prodrugs as biosynthesized botanical inclusions as endogenous substances, reserve foods, defense molecules or secondary metabolites or phytochemicals flags the significance of natural drug discovery (Morgan, 2018). Moreover, mimicking the biosynthetic pathways and identifying enzyme system, prodrugs can be derived *in vitro* during semi synthetic reaction or synthetic-structured drug design and development process. However, unintentionally and blind efforts as serendipity had also demonstrated interesting desired outcome during the development of prodrugs. QSAR studies after identification of the active moiety generated during metabolism of natural prodrugs within the cell can be explored to reveal and assign new pharmacophores bonded together in natural prodrugs (Albert, 1958). It is a matter of active scientific discussion among researchers to bring about new strategies aimed at discovery and development of novel prodrug compounds obtained from nature with or without special emphasis on target receptors. Prodrugs can be designed by restricted modifications in the natural molecules, such as considering the direction of "North-South-East-West" within the metabolites obtained from natural resources. This can be seen during launching of two analogs of paclitaxel, namely docetaxel and cabazitaxel (Galletti et al., 2007). However, analog synthesis is always accepted as a matter of choice, usually performed for natural leads and prodrugs observing ideal properties such as appropriate molecular size, modulating pharmacokinetic parameters, improving solubility and partition property, bypassing or avoiding enterohepatic circulation, raising the activity strength, selectivity, increasing metabolic and chemical stability, checking chirality and removing unnecessary chiral centers which are not participating in the binding to targets and economical strategy of prodrug synthesis (Zawilska et al., 2013). Achieving prodrug novelty and making the way as intellectual property has been seen recently in drugs designed from natural prodrugs. Numerous illustrations have been seen in the last decade while exploring natural resources in isolating natural moieties which are well performing and explaining the role of natural prodrugs. A few are discussed in this chapter so as to impress upon the talent of nature in its ability towards guiding the mankind in the healing war.

13.2.1 Romidepsin as Prodrugs

Romidepsin, a standalone histone deacetylases (HDACs) inhibitor activated only after uptake into cells, was originally submitted as NSC 630176 in 1990 (Vander Molen et al., 2011). This stable disulfide prodrug owes its anti-proliferative activity to unique caged bicyclic peptide structure. This molecule is generated via fermentation presently, a process developed in technical collaboration with Fujisawa Pharmaceuticals, presently known as Astellas Pharma (Xiao et al., 2003). Soil of Yamagata-prefecture, Japan, was natural resource of *Chromobacterium violaceum*, a Gram-negative rod-shaped bacterium biosynthesizing the anticancer prodrug which was elucidated as romedepsin (Ueda et al., 1994). Romidepsin also undergoes extensive metabolism primarily by CYP3A4, with the minor contribution from CYP_3A_5, CYP_1A_1, CYP_2B_6 and CYP_2C_{19} (Vander Molen et al., 2011). Specifically, $HDAC_1$ and $HDAC_2$ enzymes (class I) are inhibited more robustly by the reduced form as compared to $HDAC_4$ and $HDAC_6$ enzymes (class II), however reduced form of romidepsin is inactivated rapidly in serum, probably due to sequestration by serum proteins (Furumai et al., 2002). Reduction of disulfide bridge in romidepsin by glutathione into a monocyclic dithiol allows the free thiol groups to interact with zinc ions in the active site of class I and II histone deacetylase enzymes. By inhibiting class HDAC1, romidepsin inhibits the removal of acetyl groups from the lysine residues of N-terminal histone tails thus maintaining a more open and transcriptionally active chromatin state (Figure 13.1). Acetylation of nuclear and cytoplasmic proteins by romedepsin to some extent can also define pathways, further affecting the cell cycle and further apoptosis was supported by the fact when romidepsin was developed initially by the U.S. National Cancer Institute as an anti-ras compound directly correlated with tumorigenic potential (Peart et al., 2005). Romidepsin is employed in the case of progression or refractory lymphoma in competition with other approved drugs viz. the proteasome inhibitor bortezomib, the retinoid receptor ligand bexarotene, the antifolate pralatrexate as well as the HDAC inhibitor vorinostat. Phase 1 clinical studies which began in 1997, claim that romidepsin is only bicyclic inhibitor to have undergone clinical assessment and is considered a promising prodrug, marketed as Istodax formulated by the Celgene

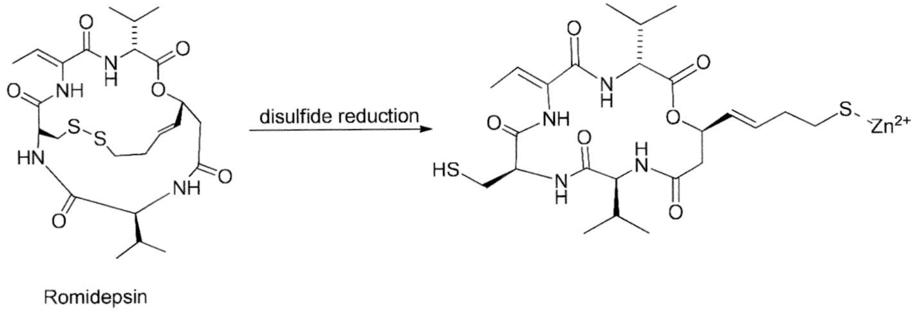

Romidepsin

disulfide reduction

FIGURE 13.1 Metabolism of romidepsin.

Corporation, USA (Marshall et al., 2002). After *in vitro* evaluation and *in vivo* safety studies and dosing schedule administered intravenously to patient at a dose of 14 mg/m^2 on days 1st, 8th and 15th of a 28-day cycle in systemic therapy for T-Cell lymphoma, injection containing romedepsin was approved as Istodax®, NSC 630176, FR901228, FK228 by the FDA in 2009 offering novel prodrug treatment.

13.2.2 Prontosil as Prodrug

Prontosil, introduced by Bayer in 1935, is chemically known as sulfamidochrysoidine and is a prodrug which led to the development of synthetic second-generation antibacterial sulfonamides (Enne et al., 2004). Discovery of prontosil fetched Domagk with a Nobel Prize in 1939. Its nonproprietary names include sulfamidochrysoidine, rubiazol, prontosil rubrum, prontosil flavum, aseptil rojo, streptocide, and sulfamidochrysoidine hydrochloride (Ghadage, 2013). Chemically, prontosil is an azo-dye with the sulfanilamide as its pharmacophore and in the human body, prontosil is metabolized into active moiety sulfanilamide under the action of cellular enzymes (Figure 13.2), the agent exhibiting broad-spectrum activity against systemic bacterial disease (Macielag et al., 2003). Limited success of treatment in patients with bacterial infections and paradox of *in vitro* failure in antibacterial activity was explained in 1935 by French scientists, who determined that active moiety of prontosil was sulfanilamide produced after bioconversion which possesses antibacterial effect (Macfarlane, 1984). Within a few years, sulfanilamide and its derivatives were available in market shelf. Before the first world war, Bayer had pursued a research strategy for development of prontosil as an anti-infective compound and the discovery of *in vivo* antimicrobial effect of prontosil in the early 1930s of the twentieth century mark the beginning of the chemotherapy development. Sulfanilamide that competes with p-aminobenzoic acid, the substrate of dihydropteroate synthetase in the bacterial synthetic pathway to folic acid (Dax, 2012). The discovery of novel antimicrobials presenting journey of prontosil (prodrug) and sulfonamides led to the Nobel Prize in Medicine, which added plethora of charm in exploring treasure of natural prodrugs.

13.2.3 Psilocybin as Prodrug

Psilocybin is a psychedelic prodrug that was isolated in 1957 and biosynthesized in *Psilocybe semilanceata Psilocybe azurescens* and *Psilcybe cyanescens*. With a structural backbone of "indole moiety," psilocybin is biosynthesized in various species of mushrooms with the most potent one, *Psilocybe mexicana*, yielding approximately 0.5% of psilocybin (Shirota et al., 2003). Psilocybin is a prodrug and is rapidly dephosphorylated to psilocin, is then further metabolized to psilocin-O-glucuronide being the main urinary metabolite (Figure 13.3). Dephosphorylation has been well observed in preclinical studies and conversion of Psilocybin into biologically active psilocin by cleavage of phosphoric ester group by alkaline phosphatase in the brain, i.e., dephosphorylation after administration as a key conclusion that complete conversion of psilocybin into the systemic circulation is yet to be reproved (Dahan et al., 2014). Chemically, psilocybin, an indole alkylamine and tryptamine moiety, the main active ingredient of the group of fungi (known as magic mushrooms), when given through oral route is almost entirely

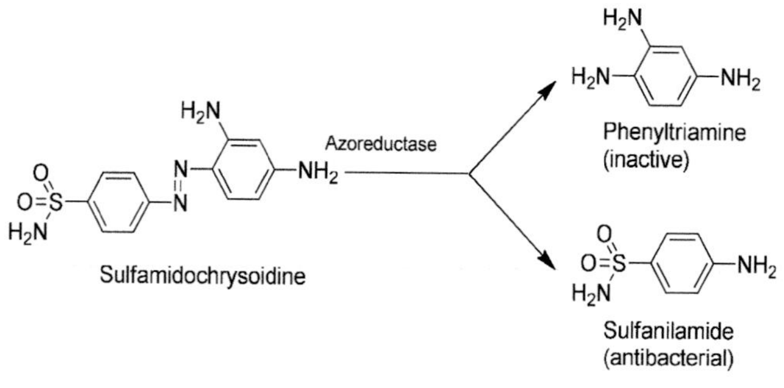

FIGURE 13.2 Metabolism of prontosil.

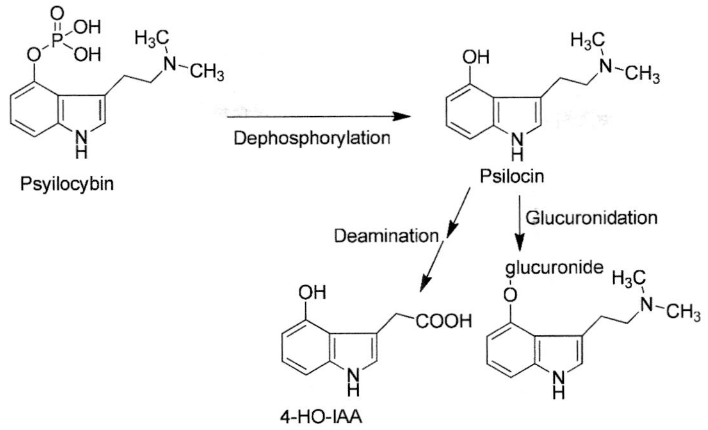

FIGURE 13.3 Metabolism of psilocybin.

transformed into psilocin during its first pass metabolism, the effect of which mimics to that of LSD and mescaline. As a consequence, psilocybin and psilocin are listed as Schedule I drugs under the United Nations 1971 Convention on psychotropic Substances. Neuropsychological effects of psilocin may be due to agonitism of serotonergic receptors, i.e., $5HT_2A$, $5HT_2C$, and $5HT_1C$ inside the brain and regulates the psychological motivation (Woods et al., 2013). Psilocin possess high-affinity as agonist at serotonin 5-HT_2A receptors, which are especially prominent in the prefrontal cortex increasing cortical activity secondary to down-stream postsynaptic glutamate but there is nil or negligible affinity of psilocin for dopamine D_2 receptors. However, dose-response and toxicological profile along with physiological variability between individuals are yet to be clinical evaluated. Moreover, studies are required to identify additional metabolites, and the influence of drug interactions of psilocybin.

13.2.4 Salvestrols as Prodrug

Discovered in 1998, salvestrols are a group of naturally occurring anticancer plant compounds that act as prodrugs of resveratrol. This nontoxic prodrug is activated to resveratrol in cytoplasm of human cancer cells by the enzyme, CYP_1B_1. The activated resveratrol inhibits growth of cancer cell, as it is cytotoxic to cancer cell with no harm to noncancerous body cells (Potter and Burke, 2006). Initial research indicates that anticancer effect of salvestrols cells is limited to only those body cells which are involved in progress of active disease. Salvestrols being a phytonutrient after getting metabolized by the tumor-specific CYP_1B_1 enzyme in cancer cells initiate a cascade of processes, including apoptosis, resulting in the arrest of growth of cancer cells (Tan et al., 2007). Hence, salvestrols are natural prodrugs compounds as the cytotoxic activity upon *in vivo* conversion through activation by CYP_1B_1 cancer markers (Figure 13.4). In their natural source, salvestrols

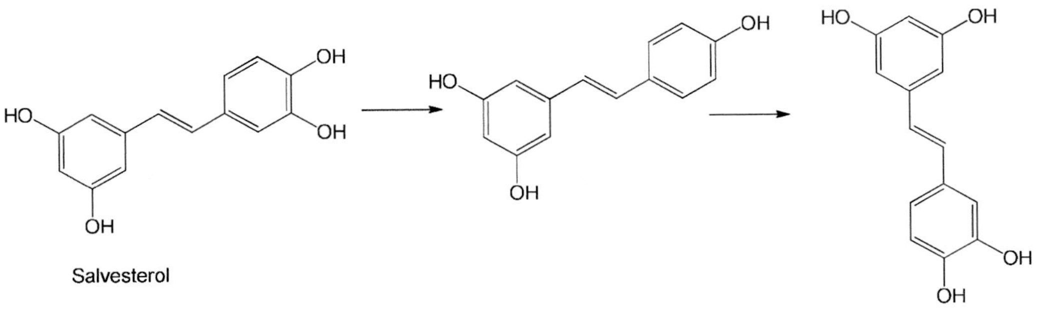

FIGURE 13.4 Metabolism of salvestrols.

form part of a plant's defense mechanism, some being hydrophilic while others are lipophilic but all are phytoalexins that are elicited by invading pathogens. Salvestrol on administration with other dietary supplements can provide promising positive outcomes for patients undergoing oncotherapy. Nutritional-based therapy approach through salvesterol is well advised in patients recovering from cancer treatment, as it resolves the deficiencies of nutrients caused by chemotherapy. The site-specificity of salvestrols can be attributed to the fact that the toxins releases during metabolism of salvestrols by CYP_1B_1 are confined to the cancer cells and are exhausted through the destruction of the cancer cell (Downie et al., 2005). Secondly, it is a food-based mechanism that depends mainly on enzymatic activation and certain co-factors which are often a part of one's daily nutrition. CYP_1B_1 is now widely acclaimed as a universal cancer marker due to its pervasiveness throughout the various cancers and stages of cancer. Overwhelming active moieties such as piceatannol produced during metabolism of resveratrol by cytochrome P_{450} possess antileukemic effect (Daniel et al., 1999). Resveratrol enjoys the reputation of one of the most widely studied polyphenols with a plethora of metabolic effects coupled with promising health benefits. Resveratrol is available over the counter with single capsules containing between 20 to 500 mg pure resveratrol. It is rapidly metabolized by sulfotransferases and UDP glucuronosyltransferases and gut microflora therefore therapeutic benefits of oral resveratrol could also be derived from its metabolites, which have exhibited better anticancer potential than resveratrol (Murray et al., 1997). The primary metabolite in humans is the sulfated metabolite resveratrol-3-sulfate. Other sulfated and glucuronidated metabolites include resveratrol-4′-O-sulfate, resveratrol-3O-4′O-disulfate, resveratrol-3-O-glucuronide and resveratrol-4′-O-glucuronide (Gibson et al., 2003).

13.2.5 Spiruchostatin A as Prodrug

Spiruchostatin A biosynthesized in *Pseudomonas* species is a gene expression enhancing substance and is known to exist in two main forms such as spiruchostatin A and spiruchostatin B. Spiruchostatin A is a depsipeptide natural product having close structural similarity to fk228, a HDAC inhibitor. Spiruchostatin A contains bicyclic depsipeptides composed of 4-amino-3-hydroxy-5-methylhexanoic acid and 4-amino-3-hydroxy-5-methylheptanoic acid residue (Figure 13.5) a potent inhibitor of the growth of various cancer cell lines and potent *in vitro* inhibitor of Class I HDAC enzyme activity (Salvador and Luesch, 2012).

13.2.6 Melatonin as Prodrug

Melatonin, as a hormone prodrug of N1-acetyl-5-methoxykynuramine moiety (AMK), is naturally occurring compound biosynthesized in animals, plants and microbes. In animals, circulating levels of melatonin regulates the entrainment of the circadian rhythms of many distinct physiological functions and therapeutic effects of melatonin are produced via activation of melatonin receptors, however being a powerful antioxidant, it is involved in protection of nuclear and mitochondrial DNA (Acuña et al., 2003). N1-acetyl-N2-formyl-5-methoxykynuramine (AFMK) and N1-acetyl-5-methoxykynuramine (AMK) are produced on oxidation by free radicals (Figure 13.6) and these bioactive metabolites of melatonin further lead for biochemical reaction (Hardeland and Pandi-Perumal, 2005). AMK interacts and protect the mitochondria, by reducing and repairing electrophilic radicals with inhibiting and down regulating cyclooxygenase-2enzyme. Chemically melatonin is N-acetyl-5-methoxytryptamine, secreted as a terminal antioxidant from the vertebrate pineal, is produced in multiple cells and organs act as paracoid, autocoid, tissue factor and antioxidant due to presence of electron-rich aromatic indole ring functions which play role as an electron donor donating an electron at a potential of 715 mV as indicated by cyclic voltametry (Jou et al., 2004). Moreover, with ability to escape from redox cycling, melatonin acts as free radical scavenger following irreversibly oxidization generating metabolites such as C3-OHM, AFMK and AMK throughout the organism, thereby crossing all physiological barriers and ensuring easily penetration in all cells (Naji et al., 2004).

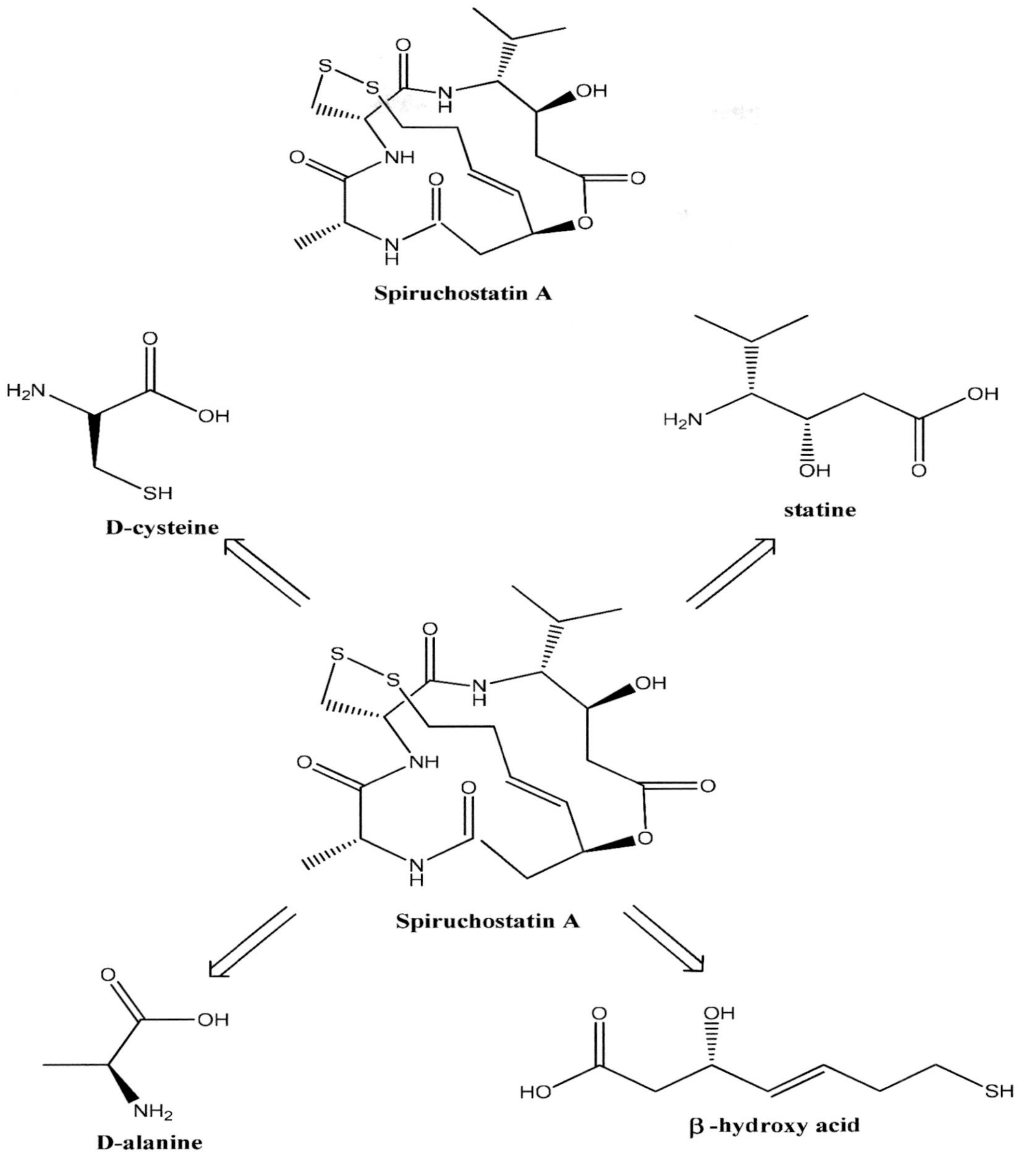

FIGURE 13.5 Metabolism of spiruchostatin A.

13.2.7 Baicalin as Prodrug

Baicalin happens to be a major constituent in species of *Scutellaria*, namely *S. lateriflora*, *S. galericulata*, and *S. rivularia*, as well as in *Oroxylum indicum* (Bignoniaceae). Different species of *Scutellaria* are primarily distributed in Asian countries, such as China, Russia, Mongolia, Japan, and Korea. Baicalin, aglycosidal prodrug is metabolized to baicalein (active moiety) through intestinal bacteria and entero-hepatic circulation (Figure 13.7). Baicalein, the aglycone moiety of baicalin, able to inhibit prolyloligo-peptidase, has a documented history of safe administration to humans and is a highly attractive base to develop novel therapies for neurodisorders such as schizophrenia, bipolar affective disorder and related

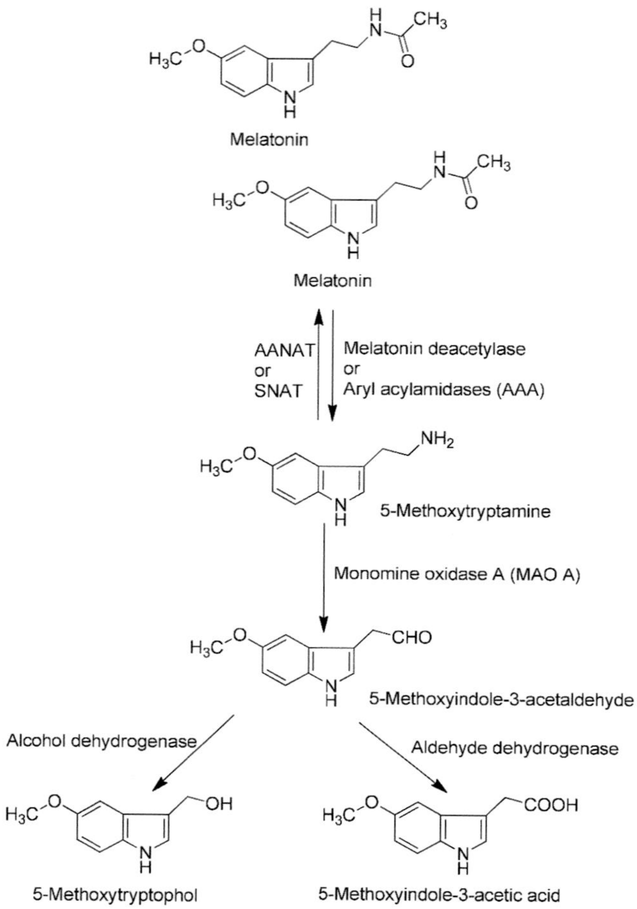

Melatonin

Melatonin

AANAT or SNAT

Melatonin deacetylase or Aryl acylamidases (AAA)

5-Methoxytryptamine

Monomine oxidase A (MAO A)

5-Methoxyindole-3-acetaldehyde

Alcohol dehydrogenase

Aldehyde dehydrogenase

5-Methoxytryptophol

5-Methoxyindole-3-acetic acid

FIGURE 13.6 Metabolism of melatonin.

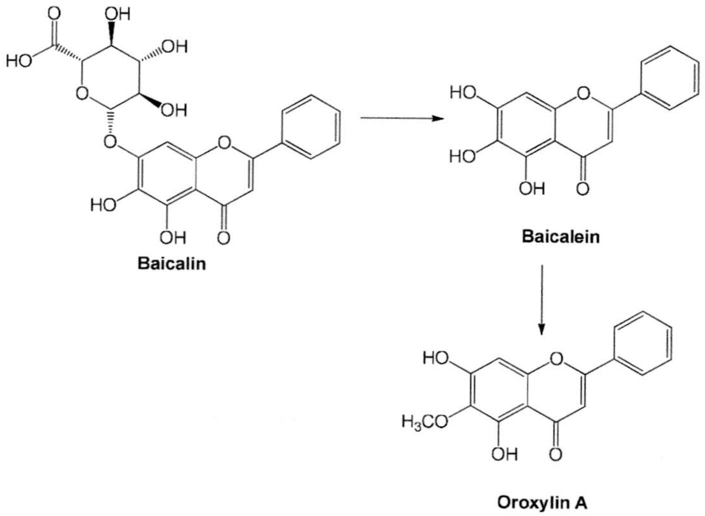

Baicalin

Baicalein

Oroxylin A

FIGURE 13.7 Metabolism of baicalin.

neuropsychiatric diseases (Tarrago et al., 2008). Baicalin chemically known as 5,6,7-trihydroxyflavone7-O-beta-d-glucuronide or baicalein 7-O-β-d-glucuronic acid or 7-d-glucuronic acid-5,6-dihydroxyflavone is one of the most important flavonoid component, biosynthesized in roots of *Scutellaria baicalensis* Georgi, an important Chinese medicinal herb as it is investigated for neuroprotective effect (Patel et al., 2012). In particular, baicalin effectively prevents neurodegenerative diseases via different pharmacological mechanisms, in addition to other effects such as anti-apoptotic, antioxidant, anti-inflammatory, neuro rejuvenator, etc. (Kalyani et al., 2013). Baicalin has been extensively used in pharmaceutical and food industries due to its outstanding bioactivities antitumor, hypoglycemic, antithrombotic, cardiac, hepatic and neuroprotective (Patel et al., 2012). Baicalin, a prodrug able to reach the CNS, is basically a prolyloligopeptidase inhibitor has been evaluated for schizophrenia, bipolar affective disorder, and related neuropsychiatric disorders and therefore may have important clinical implications (Peterson et al., 2010). Baicalin, along with its aglycone baicalein, is a positive allosteric modulator of the benzodiazepine and non-benzodiazepine-specific site of $GABA_A$ receptor (Kumar et al., 2014).

13.2.8 Matricin as Prodrug

Matricin, a prodrug of chamazulene exists as a colorless, crystalline substance biosynthesized in the chamomile flowers as a blue essential oil with anti-inflammatory action, in which polycyclic aromatic hydrocarbons and the terpene derivative can be converted to chamazulene in contact with gastric fluid (Franke and Schilcher, 2006). Matricin and chamazulene, both belong to sesquiterpenes, consisting of three isoprene units (Figure 13.8). Chamazulene carboxylic acid is a natural profen possessing anti-inflammatory activity and a degradation product of proazulenic sesquiterpene lactones e.g., matricin. Chamazulene carboxylic acid and proazulenes occur in chamomile (*Matricaria recutita*), yarrow (*Achillea millefolium*) and a few other *Asteraceae* species (Reininger and Bauer, 1998). Chamazulene

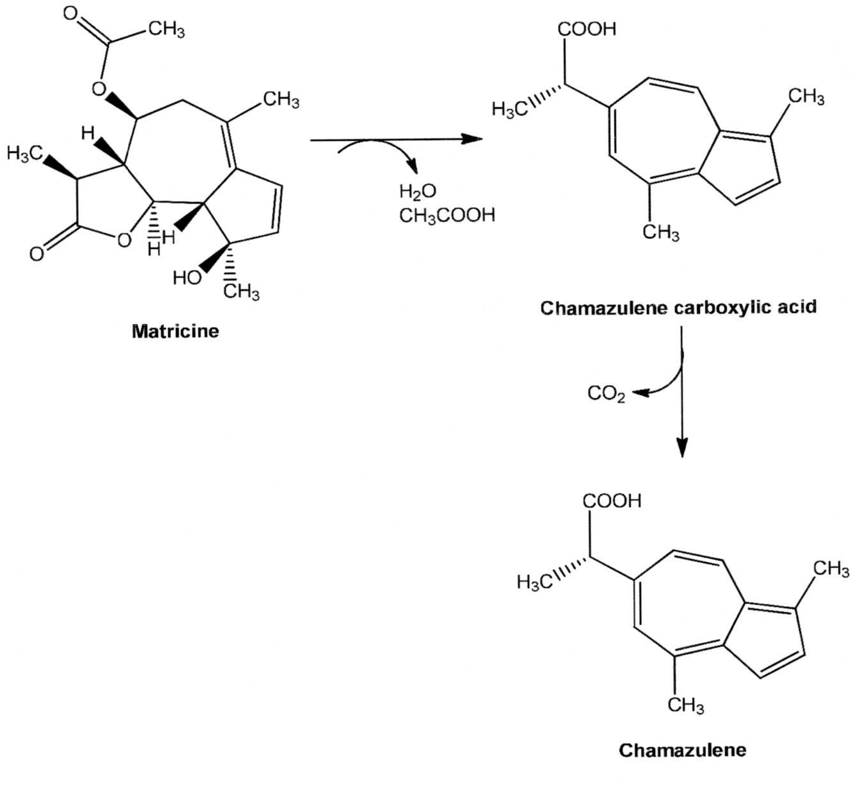

FIGURE 13.8 Metabolism of Matricin.

carboxylic acid was first isolated from yarrow in 1953, many years before the development of the fully synthetic anti-inflammatory, analgesic, and antipyretic profens, e.g., ibuprofen and naproxen. Its structural resemblance to these analgesic drug substances was overlooked, although yarrow and chamomile have been in use as antiphlogistic agents for ages (Ramadan et al., 2006).

13.2.9 Sennosides as Prodrug

Sennosides are dianthrones comprising of two anthrone units, each bearing only one carbonyl group, can be called as dimeric anthraquinone glycosides used as laxatives. Nonquinones β-linked sennosides biosythesized in *Cassia* species are O-glycosides, are inactive and remain unabsorbed in the upper part of the intestine but on conversion into rhein anthrone (active metabolite) in colon by *Bifidobacterium* species they are easily absorbed and act directly on colon wall region to produce the desired laxative effect. Sennosides are not traced in living plant, as they are formed during harvesting and drying from monomeric anthrone glycosides (Bone and Mils, 2013). Sennosides alter the motility of the large intestine through stimulation of peristaltic contractions and inhibition of local contractions leading to accelerated colonic transit, thereby reducing fluid absorption. Moreover, there is an influence on secretion processes with effect in absorption and secretion of water; retention of potassium ions, stimulation of active chlorine secretion resulting in enhanced fluid secretion along with an irritation of the intestinal mucosa and endothelial cells (Leng-Peschlow, 1988). Sennosides undergo metabolism into active compound on reaching the colon so as to induce defecation (Figure 13.9). After oral administration, anthraquinone aglycones are absorbed and metabolized as their corresponding glucuronide and sulfate derivatives in the blood but most of the sennosides are excreted in the feaces as unchanged sennosides in form of polyquinones (Leng-Peschlow, 1993).

13.2.10 Barbaloin as Prodrug

Anthraquinones are natural phytochemicals comprising three benzene rings fused together with a carbonyl group at apex and a quinone moiety. Barbaloin, C-glucoside of aloe emodin anthrone, found in *Aloe vera* (also called the healing plant) is a mixture of two diastereomers, termed barbaloin (aloin A) and isobarbaloin i.e., aloin B (Patel et al., 2012). Barbaloin as prodrug of anthrone is an

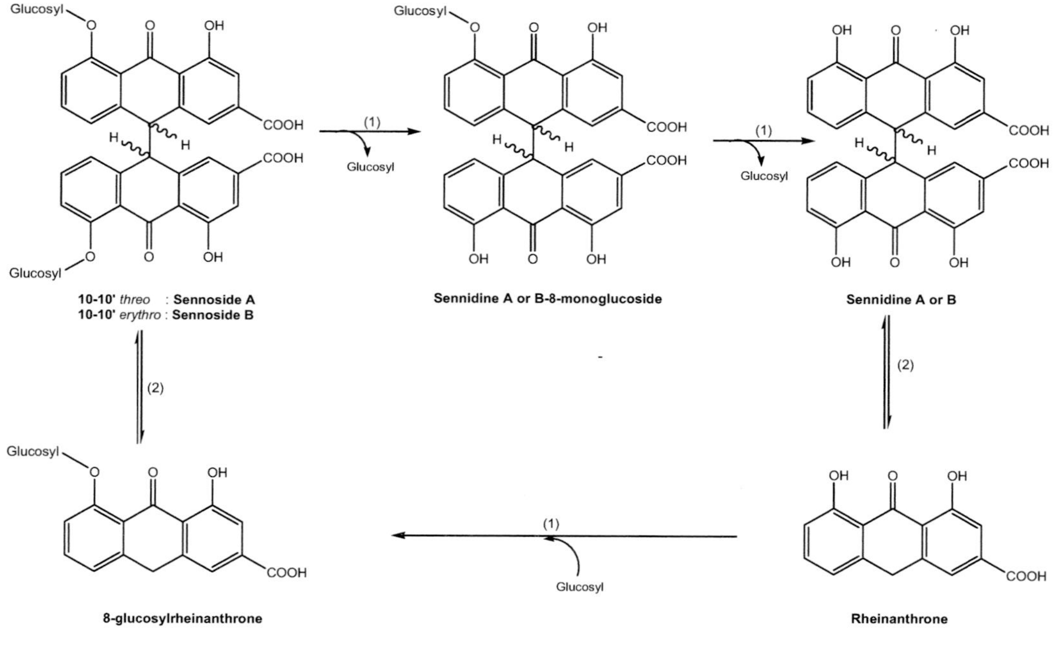

FIGURE 13.9 Metabolism of sennosides.

anthraquinone glycoside which is converted into active anthrone (aloe emodin) in the gastro intestinal tract by β-glucuronides *Eubacterium* sp., (Figure 13.10) and acts by increasing water content of large intestines, a causative factor to treat constipation. Barbaloin is known to possess a plethora of therapeutic activities such as strong inhibitory effect on histamine release, antiviral, antimicrobial, anticancer, anti-inflammatory, cathartic, antioxidant activity and as an alternative for pharmaceutical or cosmetic applications (Balunas and Kinghorn, 2005).

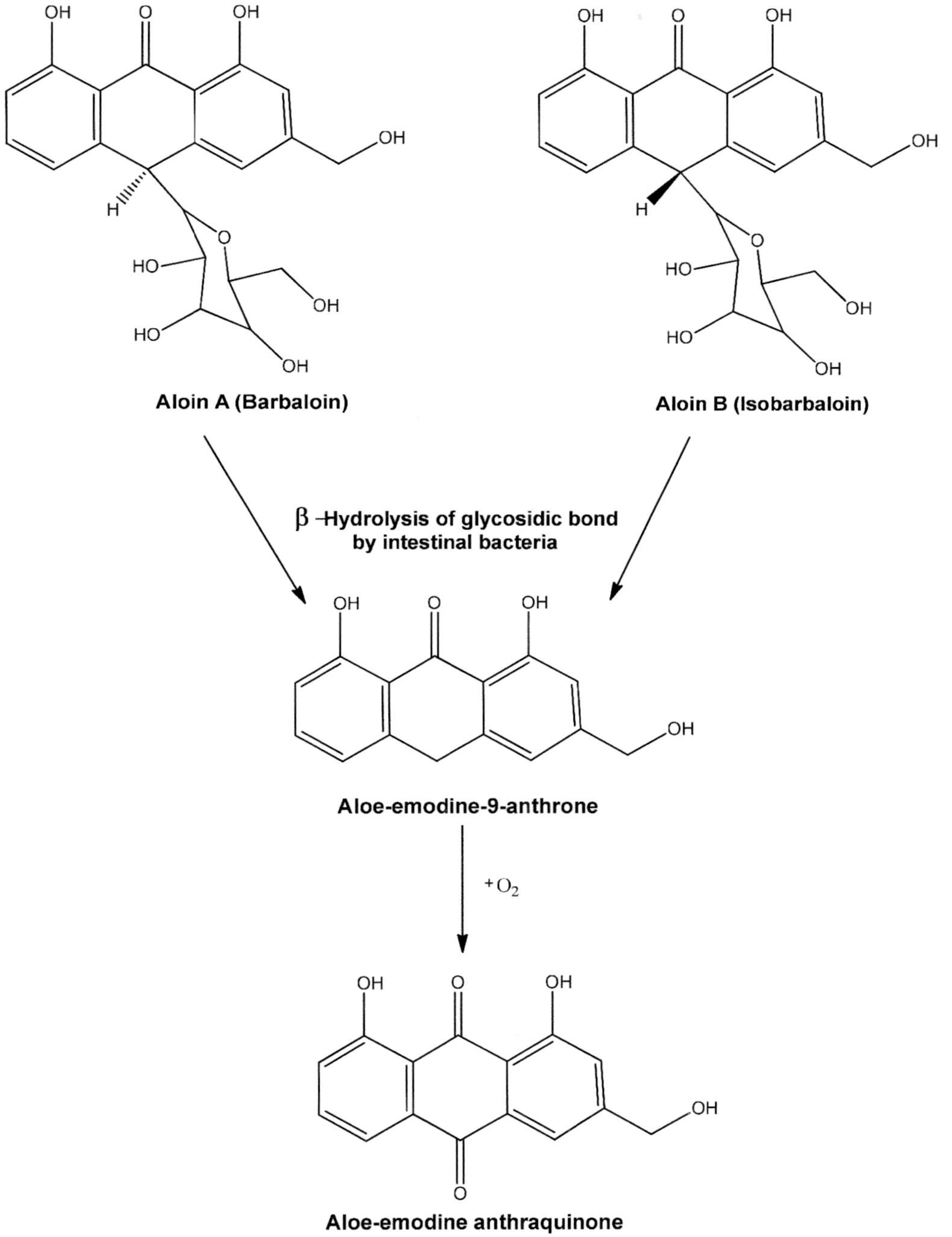

FIGURE 13.10 Metabolism of barbaloin.

13.2.11 Geniposide as Prodrug

Genipin, obtained from the iridoid glucoside, geniposide, present in the dried ripe fruit of *Gardenia jasminoides* Ellis (red, blue and purple fruit), is lipophilic and able to permeate into intestinal mucosa, facilitating absorption and thereby producing varied pharmacological actions, such as anti-inflammatory, antitumor, antithrombotic, antimicrobial, antidiabetic, neuroprotective and antidepressant (Yang et al., 2011). Geniposide converted by human intestinal microflora enzymes such as β-glucosidase to the active drug genipin in the gut (Figure 13.11), also used as an alternative crosslinking agent for pericardial

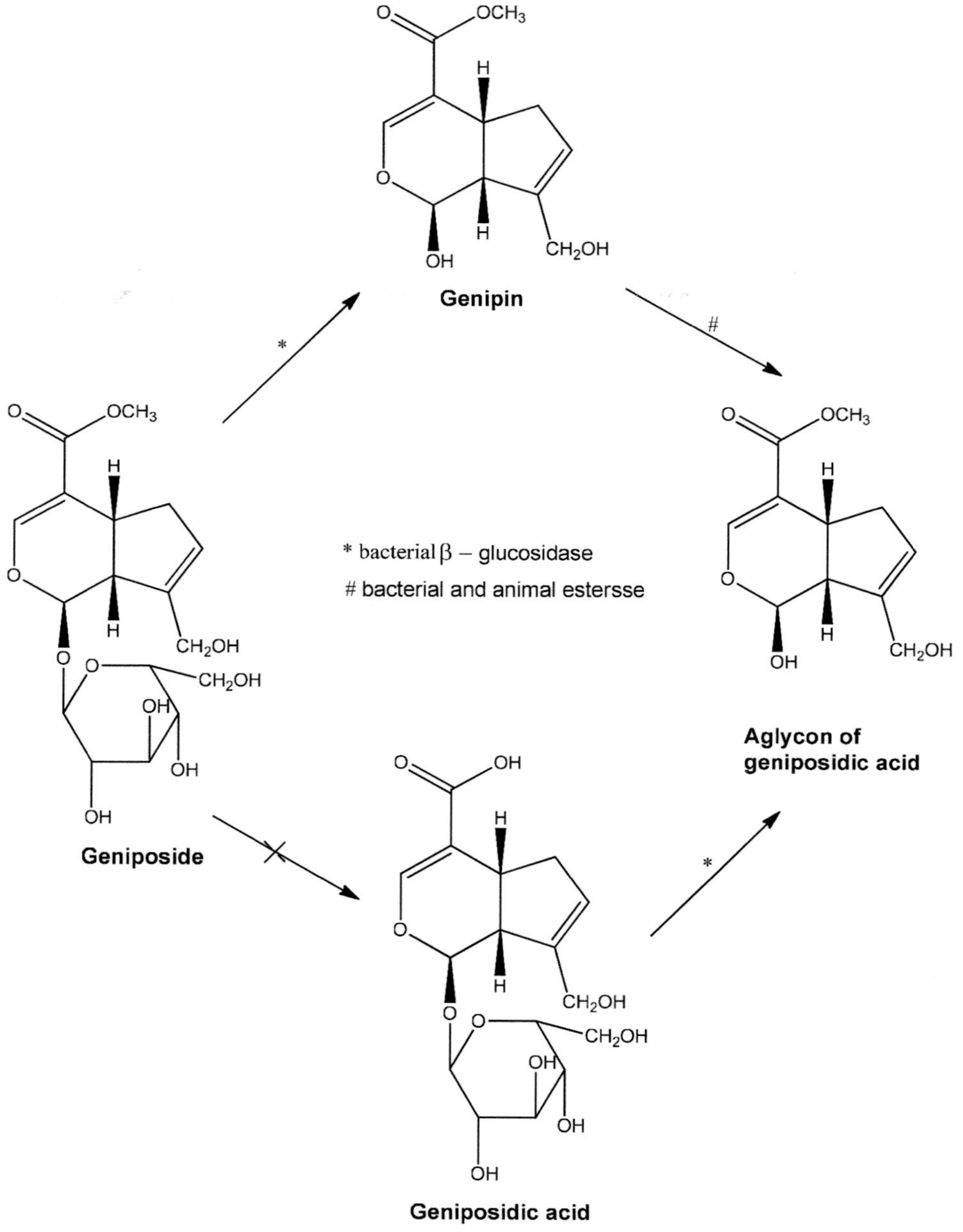

Genipin

Geniposide

Aglycon of geniposidic acid

Geniposidic acid

* bacterial β – glucosidase

bacterial and animal estersse

FIGURE 13.11 Metabolism of geniposide.

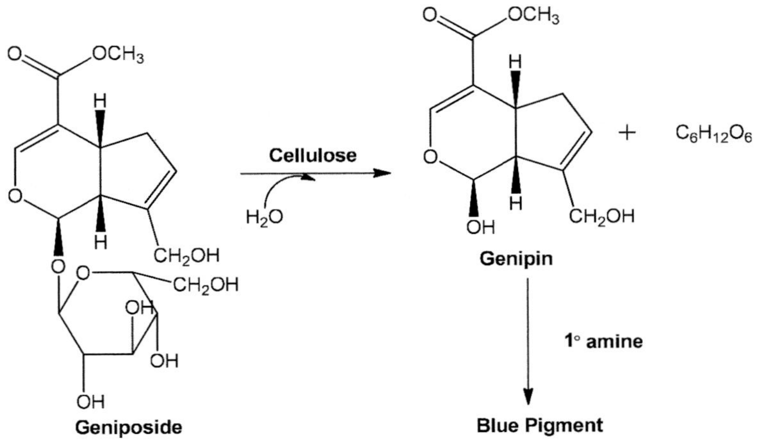

FIGURE 13.12 Biotransformation of genipin.

tissue, considering its physical, mechanical, biochemical characteristics and low cytotoxicity comparable to glutaraldehyde (Hung et al., 2008). Genipin itself is colorless but it reacts spontaneously with various peptides and amino acids and thus used to prepare a series of blue pigments (Figure 13.12), casein and whey protein films, also utilized in food industry (Seenivasan et al., 2008).

13.2.12 Lignans as Prodrug

The lignans belong to a group of natural prodrugs compounds found in certain plants of the family Asteraceae, Convolvulaceae including *Arctium lappa* (greater burdock), *Saussurea heteromalla* and *Ipomoea cairica* and are part of the human diet. Lignans are known to elicit antiviral, antioxidant and anticancer effects and have also been used in management of HIV infection. Lignans are metabolized to form mammalian lignans known as pinoresinol, lariciresinol, enterodiol secoisolariciresinol, matairesinol, hydroxymatairesinol, syringaresinol and sesamin and enterolactone by intestinal bacteria. Metabolites produced after metabolism of lignans possess various biological actions hence representing lignans as a vast group of prodrugs (Peterson et al., 2010).

13.2.13 Rohitukine as Prodrug

Rohitukine, a significant natural anticancer and anti-inflammatory chromone alkaloid, isolated from *Amoora rohituka* and *Dysoxylum binectariferum*, is highly hydrophilic in nature which hampers its oral bioavailability. Besides acting as an important precursor for the semisynthetic derivative flavopiridol, it is a very potent inhibitor of cyclin-dependent kinases i.e., CDK1, CDK2, CDK4, CDK7 and CDK9 (Kumar et al., 2016). Being nonmutagenic, nongenotoxic, noncardiotoxic, and nontoxic to normal cells, it has also been reported to possess several other pharmacological activities including antidyslipidemic, antiadipogenic, gastroprotective, antifertility, antileishmanial and immunomodulatory activities. Therefore, it is a promising prodrug strategy which may be displaced intact at gastric and intestinal pH in order to transform the parent moiety in plasma as desired for an ideal prodrug has explored is as a potential therapeutic candidate (Dąbrowska-Maś, 2017). Moreover, the polar properties imparted by hydroxyl groups further lead to phase II metabolism and high selective for cancer cells over normal fibroblast cells further support it as promising prodrug candidate (Figure 13.13). The inhibition of cell growth via caspase-dependent apoptosis and metabolically stable lead to down-regulation of the transcription of antiapoptotic proteins in cancer cells explore its anticancer potential. Hence design and synthesis of hexanoate ester prodrugs of rohitukine with improved drug-like properties such as chemical stability, enzymatic hydrolysis in plasma (esterase), optimum aqueous solubility and lipophilicity to achieve intestinal absorption support prodrug design strategy (Dhareshwar et al., 2007).

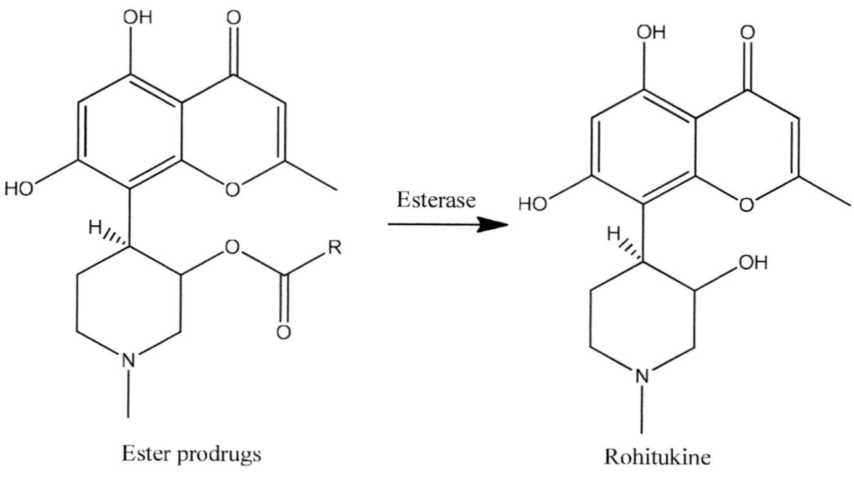

FIGURE 13.13 Metabolism of rohitukine.

13.2.14 Curcumin as Prodrug

Curcumin, a bright yellow colored polyphenolic principle biosynthesized in *Curcuma longa*, belongs to the group of curcuminoids. Chemically it is a diarylheptanoid natural phenol connected by two α, β-unsaturated carbonyl groups and is tautomeric compound (Nelson et al., 2017). In spite of vast clinical evaluation it has limited utility because of its instability and low bioavailability however it too reacts with several proteins including hERG, cytochrome P450s, and glutathione S-transferase but unfortunately may further increase the risk of adverse effects (Amalraj et al., 2017). Poor drug-like properties like nonselective target interaction, low bioavailability, undefined tissue distribution and extensive metabolism limit the bioactivity of this highly improbable lead (Oliveira et al., 2015). As a matter of fact, curcumin is a lead candidate with extreme therapeutic potential claiming investigation expense more than US$150 million in research as supported by National Center for Complementary and Integrative Health (Manolova et al., 2014), but its limited use in medical treatment, may be due to its poor drug properties. However, credited with antioxidant, anticancer, analgesic and anti-angiogenesis properties warrant its probability for clinical utility. Hence prodrug conversion as succinate ester of curcumin (Figure 13.14), emerged as better option to improve significant stability parameters in order to further explore their valuable moiety, as an inhibitor of tumor growth (Muangnoi et al., 2019). Evident from conclusions also support that ester derivative of curcumin was also found to be more stable in intestinal fluids (Muangnoi et al., 2018). Ability to induce apoptosis and alter the expression of Bax and Bcl-2 proteins to a higher extent than curcumin has also been reported for ester derivatives (Bhuket et al., 2017). Succinate ester

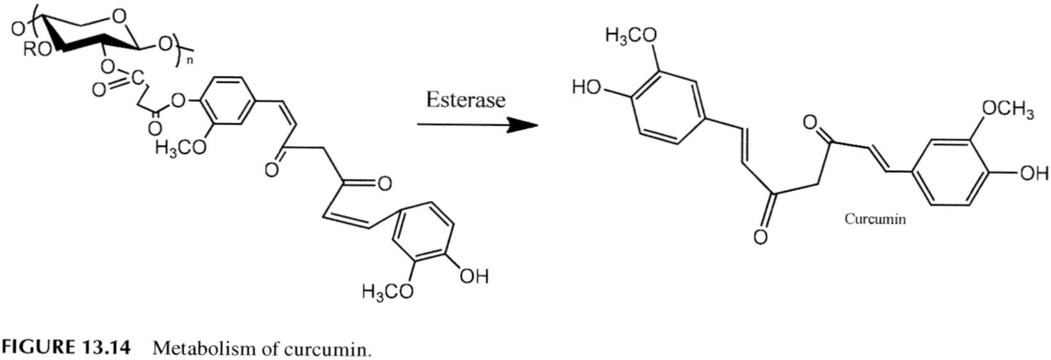

FIGURE 13.14 Metabolism of curcumin.

of curcumin as a promising prodrug, with a potential to act as an adjuvant in clinical studies, has been extensively investigated (Peng et al., 2005).

13.2.15 Carvacrol as Prodrug

Carvacrol also known as cymophenol, component of essential oil possessing antimicrobial, antitumor, antimutagenic, antigenotoxic, analgesic, antispasmodic, anti-inflammatory, angiogenic, antiparasitic, antiplatelet activities (Youssefi et al., 2019). However, poor physico-chemical properties such as low water solubility, high volatility, low chemical stability, hamper its potential therapeutic uses (Naghdi et al., 2017). Therefore, prodrug approach involving hydrophilic and lipophilic derivatization designed to promote membrane permeation and oral absorption. Studies reveals that succinic and glutaric acids linkers, improve the physical chemical properties of carvacrol with the aim of improving oral drug delivery. Moreover, prodrug conversion also enhances antibacterial profiles against gram-positive bacteria by interfering with the biofilm formation of *Staphylococcus aureus* and *Staphylococcus epidermidis* justify the prodrug strategy. Further esterification reaction (Figure 13.15) with a suitable acid chloride and prenylation enhances its lipophilicity thus ameliorating affinity, interaction and access with the bacterial membrane and could further enhance antibacterial activity (Muangnoi et al., 2018). Hence ester prodrugs design improves physico-chemical properties of carvacrol while retaining the biological activity (Marinelli et al., 2019).

13.2.16 Ursolic Acid and Other Triterpenoids as Prodrug

Ursolic acid, prodrug form of the ursonic acid and also referred to as urson, prunol, malol or 3-beta-3-hydroxy-urs-12-ene-28-oic-acid substituted by beta-hydroxy group at position 3, is biosynthesized in *Eriobotrya japonica, Rosmarinns officinalis, Melaleuca leucadendron, Ocimum sanctum* and *Glechoma hederaceae.* Chemically it is a hydroxy monocarboxylic acid, derived from a hydride of an ursane and is metabolized by *Aspergillus flavus.* Prodrug of Ursolic acid (UA), coded as US597 may suppress cancer cells adhesion by inhibition of STAT3 pathway and alteration of cancerous biomarkers resulting into cellular motility and as per reported trials no severe side effect were seen after metabolism of prodrugs (Yoon et al., 2016). Being a complementary ingredient of diet, it represents incredible interest in exploring therapeutic candidate with cytotoxic activities including derivitization of natural moiety into water-soluble derivatives with a motive to justify clinical use of these triterpenoids in various diseases including anticancer chemotherapies. Ursolic acid has been reported to produce antitumor and antioxidant activity, even too play significant role in regulating the apoptosis induced by high glucose presumably through scavenging of free radicals and treatment affects growth and apoptosis in cancer cells (Xiang et al., 2015). Studies reveals that ursolic acid inhibit DNA binding of NF-kappaB consisting of p50 and p65 further control IkappaB alpha degradation, IkappaB alpha phosphorylation, IkappaB alpha kinase activation and suppress tumor necrosis factor, phorbol ester, okadaic acid, H_2O_2 and even cigarette smoke. Ursolic acid also inhibited NF-kappaB-dependent reporter gene expression activated by TNF receptor, TNF receptor-associated death domain, TNF receptor-associated factor, NF-kappaB-inducing kinase, Ikappa Balpha kinase (Jia et al., 2015). Activation of NF-kappaB correlated with the

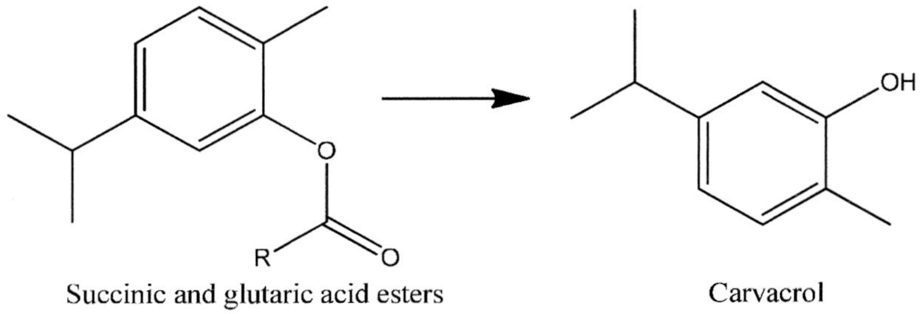

O
R
O
OH
Succinic and glutaric acid esters
Carvacrol

FIGURE 13.15 Metabolism of carvacrol.

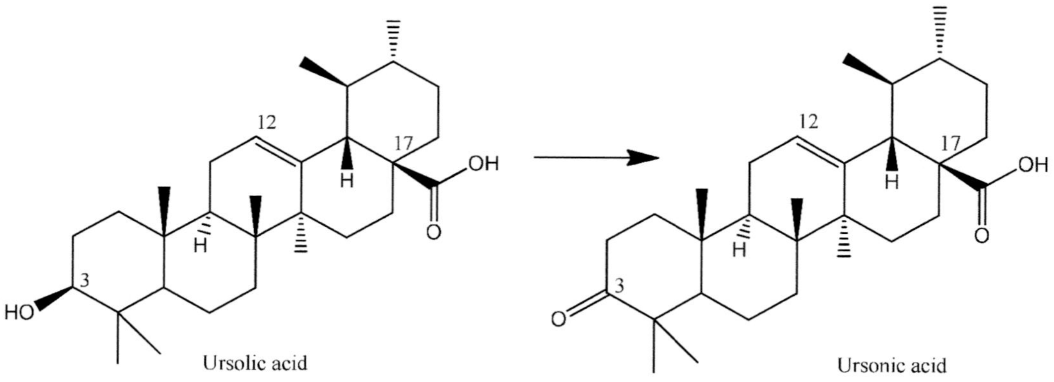

FIGURE 13.16 Metabolism of ursolic Acid.

suppression of NF-kappaB-dependent cyclin D1, cyclooxygenase and matrix metalloproteinase expression indicating that ursolic acid mediate its antitumorigenic and chemosensitizing effects after physiological metabolism (Figure 13.16) (Wang et al., 2011). Therefore, ursolic acid is a potential anticancer candidate and cytotoxic properties of this natural antitumor drug might be possible to increase with prodrugs designing strategy. Triterpenoids in particular, the lupane, oleanane and ursane type have been reported to exhibit significant pharmacological effects including hepatoprotective, hypoglycemic, immunomodulatory, anti-inflammatory, antioxidant and antitumor activities without prominent toxicity. However, poor drug properties like low solubility and selectivity, poor bioavailability, and short half-life severely limit their therapeutic applications. As a key resolution, prodrug strategy has been developed as an effective method to improve the selectivity of a wide range of triterpenoids.

13.2.17 Acteoside as Prodrug

Acteoside belongs to the class of organic compounds known as coumaric acids, an aromatic compound containing a cinnamic acid moiety, hydroxylated at the C2, C3 or C4 carbon atom of the benzene ring (Figure 13.17). It is a cinnamate ester, a disaccharide derivative, a member of catechols, a polyphenol

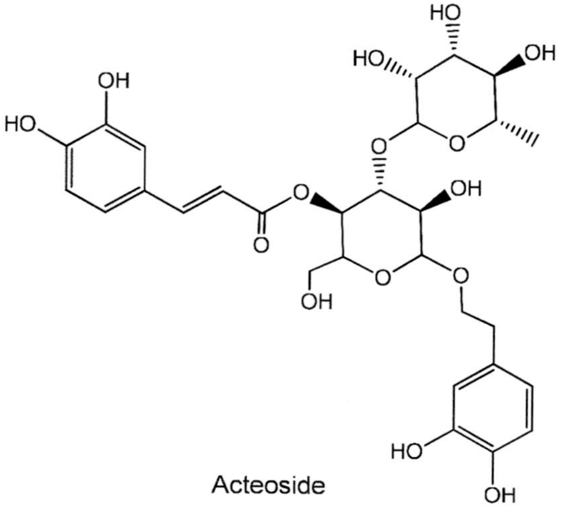

FIGURE 13.17 Structure of acteoside.

glycoside which is derived from a hydroxytyrosol and a trans-caffeic acid possessing neuroprotective, antileishmanial, anti-inflammatory and antibacterial agent (Shiao et al., 2017). Acteoside is under investigation in clinical trial NCT02662283 as a therapy for patients of IgA Nephropathy. This phenylethanoid glycosides isolated from *Herba cistanches* possesses has low oral bioavailability may serve as the potential natural drug which could be designed as prodrug. Acteoside (A) is an active compound in traditional herbal medicines and is structural isomer of isoacteoside which possess extensive biological activities including strong antioxidant, hepatoprotective and cell apoptosis regulation. Studies reveals that metabolites identified based on their retention times and fragmentation patterns during metabolism via intestinal enzymes by the intestinal bacteria gave an insight to clarify the metabolic pathway of acteoside (Cui et al., 2016). Fate of acteoside in the gut for both parent polyphenols and their degradation products, including small phenolic acid and aromatic catabolites is significantly important so as to formulate dosage form. Moreover, *in vivo* systematical investigation for metabolism characteristic profiles of acteoside by hyphenated chromatography technology had been utilized in various studies. In a study after oral administration of 200 mg/kg acteoside, a total of 44 metabolites were detected and identified, and through the comprehensive metabolites study in plasma, urine and feaces, acteoside systemical metabolites profiles and characteristics claim for its prodrug design strategy, proving that acteoside could exist stably and the process for biotransformation of acteoside in blood justifies strong drug-like properties (Qi et al., 2013). However, proposed metabolic pathway of acteoside after identification of the metabolites of acteoside produced by intestinal bacteria or intestinal enzyme gave an insight to clarify pharmacological mechanism of these traditional chinese medicines. Acteoside is the most widespread among the traditional chinese medicines including *Callicarpa kwangtungensis*, *Plantago asiatica* with remarkable biological properties such as hepatoprotective, vasorelaxant, hypoglycemic, hypolipidemic, neuroprotective and antioxidant effects. Thus studies on transformation by intestinal bacteria and intestinal enzyme, conducted to reveal metabolism mechanism of acteoside in intestinal tract screening confirm its design through prodrug strategy.

13.2.18 Harpagoside as Prodrugs

Harpagoside is a natural iridoid glycoside, biosynthesized through mevalonate pathway found in the plant *Harpagophytum procumbens*, also known as devil's claw. It is the active chemical constituent responsible for the medicinal properties of the plant, which have been used for centuries by the Khoisan people of southern Africa to treat diverse health disorders, including fever, diabetes, hypertension, and various blood related diseases. Harpagoside as prodrugs are active against pancreatic cancer cells *viz.* INS-1 and MIA PaCa-2, can be converted into potent anticancer candidate by selectively enhancing the activating enzyme's concentration in the tumor microenvironment (Figure 13.18) (Georgiev et al., 2013). Harpagoside is also widely used as a folk remedy to treat rheumatic complaints. Subsequent studies

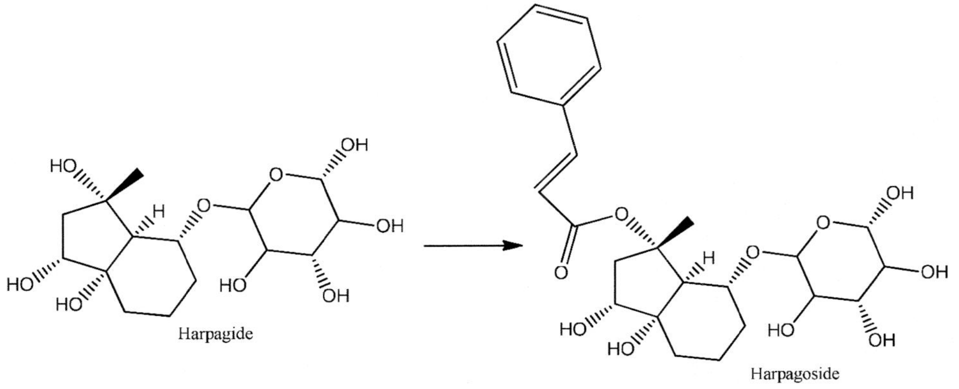

Harpagide

Harpagoside

FIGURE 13.18 Metabolism of herpagoside.

have shown that extracts have good anti-inflammatory and analgesic activities (Akhtar and Tariq, 2012). Clinical studies have confirmed an increase in pain relief for 60% of patients with an osteoarthritic hip or knee (Huang et al., 2006). Biochemical studies have elucidated the mechanism of action of pure harpagoside, showing that it moderately inhibited cyclooxygenases 1 and 2 of the arachidonic acid pathway and overall nitric oxide production in human blood and for treating rheumatoid arthritis. Several strategies suggest that to improve the therapeutic index of chemotherapy (i.e., conventional drugs) by their chemical transformation into prodrugs improving pharmacokinetic profiles and optimizing administration routes in comparison to the initial drug is need of hour (Park, 2016). The use of natural anticancer like hyperoside, quercetin, hypoxoside and rooperol in future studies will continue to explain their role as therapeutic agents against cancer and shed more light on their use as potential pancreatic anticancer drugs.

13.2.19 Hesperidin as Prodrug

Hesperidin, a prodrug of hesperitin was first isolated in 1828 by French chemist Lebreton is biosynthesized as flavonoid in *Citrus* species. It is metabolized into its aglycon hesperitin by 6-O-alpha-L-rhamnosyl-beta-D-glucosidase, an enzyme that uses hesperidin and H_2O to produce hesperetin and rutinose, is found in the *Ascomycetes* species (Figure 13.19) (Alam et al., 2014). As a fact, hesperidin fail to control the histamine release activated by immunoglobulins but the metabolite of hesperidin prevent histamine release from RBL2H3 cells (Lee et al., 2004) thus emphasizing the suitability of prodrug design research.

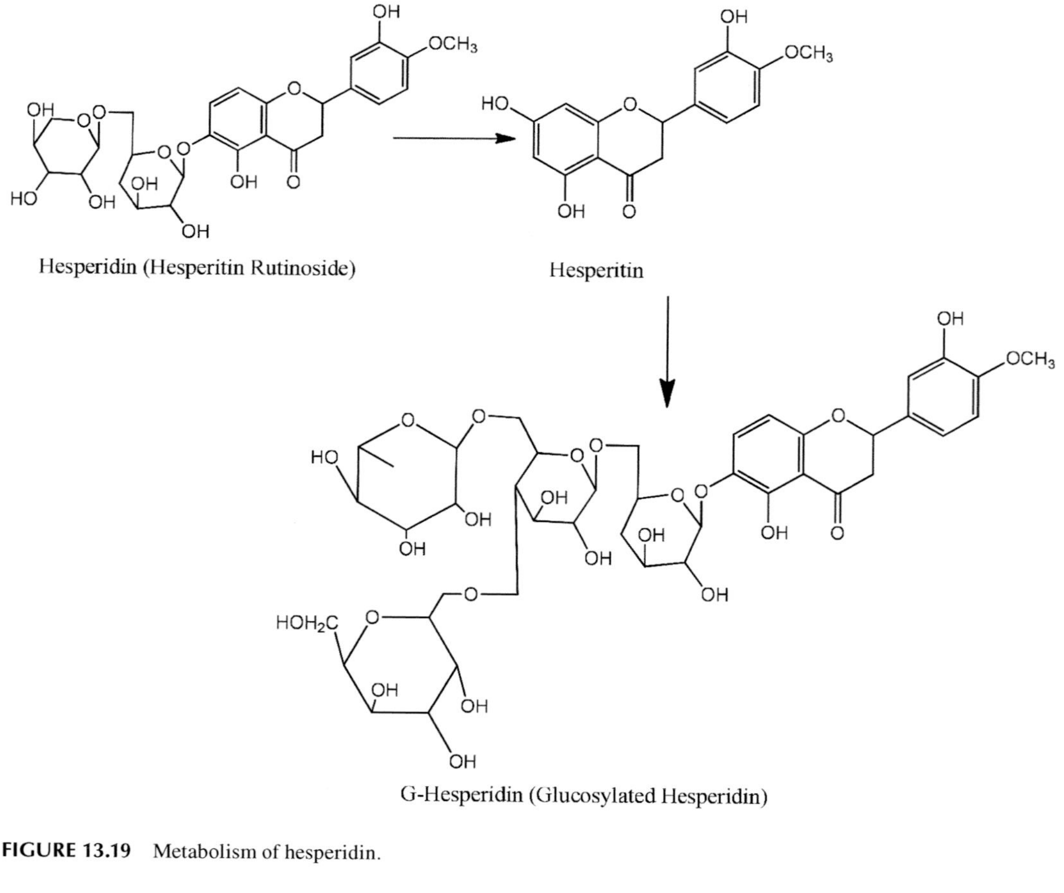

Hesperidin (Hesperitin Rutinoside)

Hesperitin

G-Hesperidin (Glucosylated Hesperidin)

FIGURE 13.19 Metabolism of hesperidin.

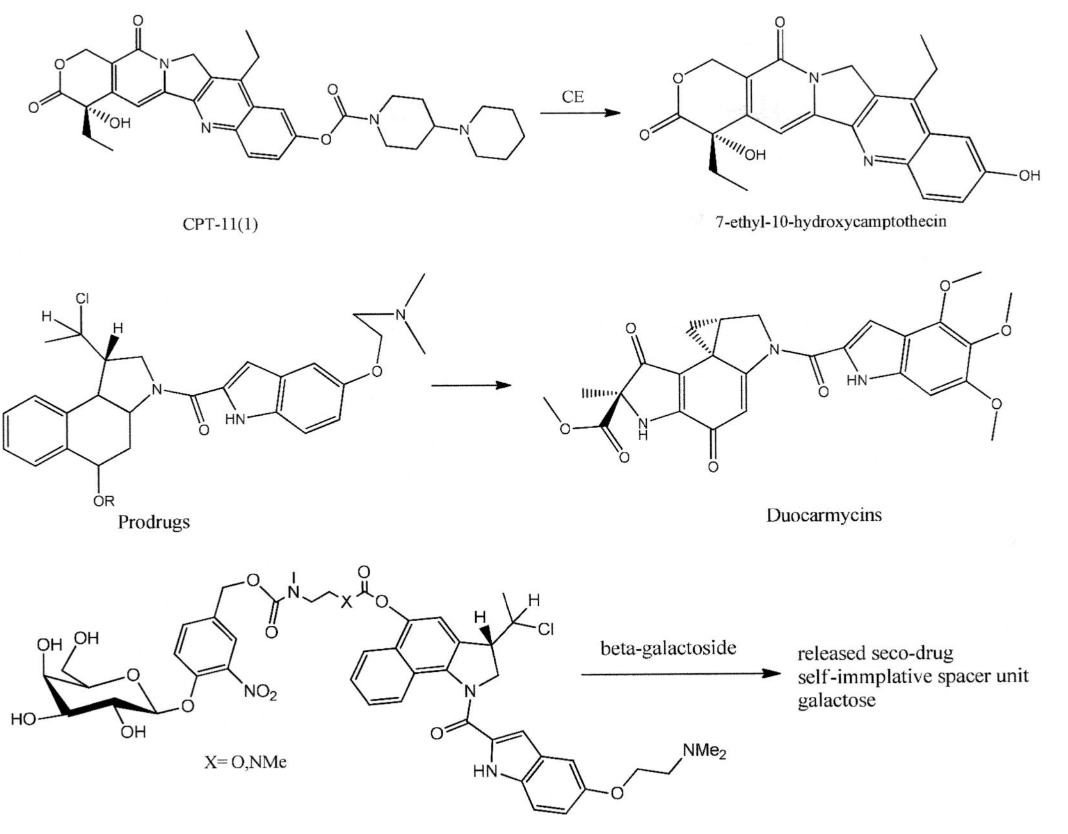

FIGURE 13.20 Metabolism of duocarmysin.

13.2.20 Duocarmycins as Prodrugs

Anticancer natural moieties such as duocarmycins were first isolated from bacterial source in 1978. They bind to the minor groove of DNA and exert their action by alkylating the nuclebase, mostly purine base at different phase of cellular division leading to DNA damage. Duocarmycins such as A, B1, B2, C1, C2, D are more efficacious in killing tumor cells as compared to tubulin binders (Searcey, 2002). Irreversible disruption of DNA and Galactose-modified duocarmycin (GMD) prodrugs as a new class of highly antineoplastic and senolytic agents. These GMD prodrugs are converted to their corresponding duocarmycin drugs in a manner dependent on processing by β-galactosidase (Figure 13.20). Since senescent cells display elevated levels of lysosomal b-galactosidase encoded by GLB1, GMD selectively affect senescent cells (Guerrero et al., 2019). mGMD prodrugs are also capable of eliminating by senescent cells caused by anticancer therapies and preneoplastic senescent cells in mouse models. The duocarmycins represent a new group of antitumor compounds produced by *Streptomyces* with significant activity in murine and human tumor models suggesting successful natural prodrug strategy.

13.2.21 *N*-Acetylcarnosine as Prodrug

N-Acetylcarnosine, a natural compound occurs in human muscular and other tissue which is metabolized into peptide carnosine (Figure 13.21). Its molecular structure is identical to carnosine with an additional acetyl group making it more resistant to degradation by carnosinase, an enzyme that breaks down carnosine into amino acids, beta-alanine and histidine (Babizhayev, 1996). *N*-acetylcarnosine as

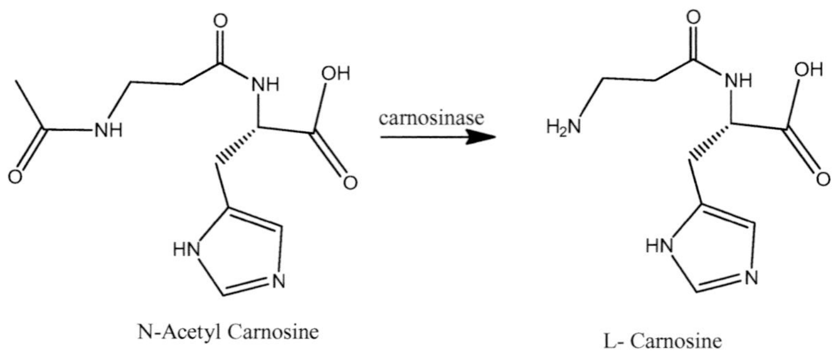

FIGURE 13.21 Metabolism of N-acetylcarmosine.

prodrug ophthalmic preparations, possess an antioxidant effect and are currently used for the treatment of cataract and other ophthalmic disorders associated with oxidative stress, including age-related and diabetic cataracts. Studies reveal that ophthalmic solutions containing 1% *N*-acetylcarnosine promote corneal absorption of carnosine and possess synergistic effect while controlling the extent of ocular and systemic absorption in prodrug-based designed ophthalmic formulations. *N*-Acetylcarnosine prodrug ocular systems including lubricating eye drops design marketed under numerous brand labels increase the intraocular uptake of the active principle *L*-carnosine from its ophthalmic carrier *N*-acetylcarnosine in the aqueous humor and further enhance the ocular bioavailability, bioactivating universal antioxidant and anti-cataract efficacy in humans represent success of this ophthalmic prodrug (Babizhayev, 2012).

13.2.22 Eriocitrin as Prodrug

Eriocitrin, a main flavonoid in lemon, citrus, grapefruit, vegetables, processed drinks and wine possesses strong antioxidant, lipid-lowering and anticancer activities. Eriocitrin also known as eriodictyol glycoside is a flavanone-7-*O*-glycoside is metabolized into eriodictyol (Li et al., 2019) which is converted into 3,4-dihydroxyhydrocinnamic and phloroglucinol in presence of intestinal bacteria (Figure 13.22) (*Bacteroides uniformis* or *Bacteroides distasonis*). Eriocitrin plays an important role in management of hyperlipidaemia, cardiovascular and cerebrovascular diseases. It is marketed as a

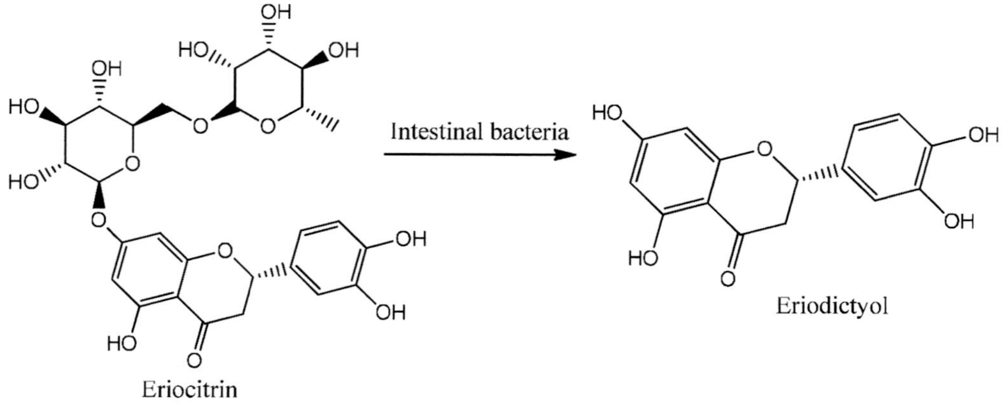

FIGURE 13.22 Metabolism of eriocitrin.

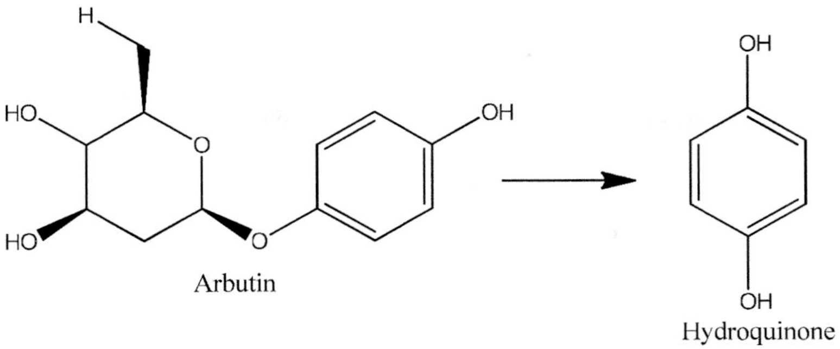

FIGURE 13.23 Metabolism of arbutin.

dietary supplement, usually in conjunction with vitamins B and vitamins C as antiobese compound. DNA microarray analysis revealed that eriocitrin increased mRNA of mitochondrial biogenesis genes, such as mitochondria transcription factor, nuclear respiratory factor, cytochrome C oxidase subunit and ATP synthase. Experimental studies reveal that eriocitrin increased mitochondrial size and mtDNA content, which resulted in ATP production in HepG2 cells and zebrafish with a conclusion that dietary eriocitrin ameliorates diet-induced hepatic steatosis with activation of mitochondrial biogenesis (Miyake et al., 2000).

13.2.23 Arbutin as Prodrug

Arbutin, a natural product found in foods, over-the-counter drugs, quince jam samples, herbal dietary supplements and cosmetic skin-lightening products, is available in both natural and synthetic forms as it inhibits tyrosinase and thus prevents the formation of melanin. Arbutin biosynthesized in *Bergenia crassifolia*, wheat, bearberry, bear grape and medicinal plants primarily belonging to the family Ericaceae. Arbutin is dermally hydrolyzed to hydroquinone in presence of bacteria like *Staphylococcus aureus* and *Staphylococcus epidermidis* and act as skin whitening agent (Bang, 2008). Arbutin converted to hydroquinone (Figure 13.23), which has antimicrobial, astringent, and disinfectant properties and is also used as a stabilizer for color photographic images and as an anti-infective and diuretic for the urinary system. It is also an inhibitor of melanin formation and a skin-lightening agent. Arbutin too had an inhibitory effect on tyrosinase activity and inhibited the production of melanin by both tyrosinase and autoxidation and may assist in the control of hyperpigmentary disorders (Pillaiyar, 2017).

13.2.24 Allin as Prodrug

Allin is a prodrug of allicin, an unstable compound which is converted into diallyl disulfide. This sulfoxide prodrug is a natural constituent of fresh garlic having both carbon- and sulfur-centered stereochemistry. Garlic has been used since antiquity as a therapeutic remedy for certain conditions and the enzyme alliinase converts alliin into allicin which is responsible for the aroma of fresh garlic and presumed to affect immune responses in blood (Mishra et al., 2001). Allicin is known to possess antiviral, antifungal, antibacterial, antioxidant action and has been used to alleviate a variety of health problems due to its high content of organosulfur compounds activity. Allin has been reported to reduce cytokine levels, serum glucose level and possess antioxidant, cardioprotective and neuroprotective activity. Metabolism of allin suggest that these garlic organosulfur compounds may behave like prodrugs and

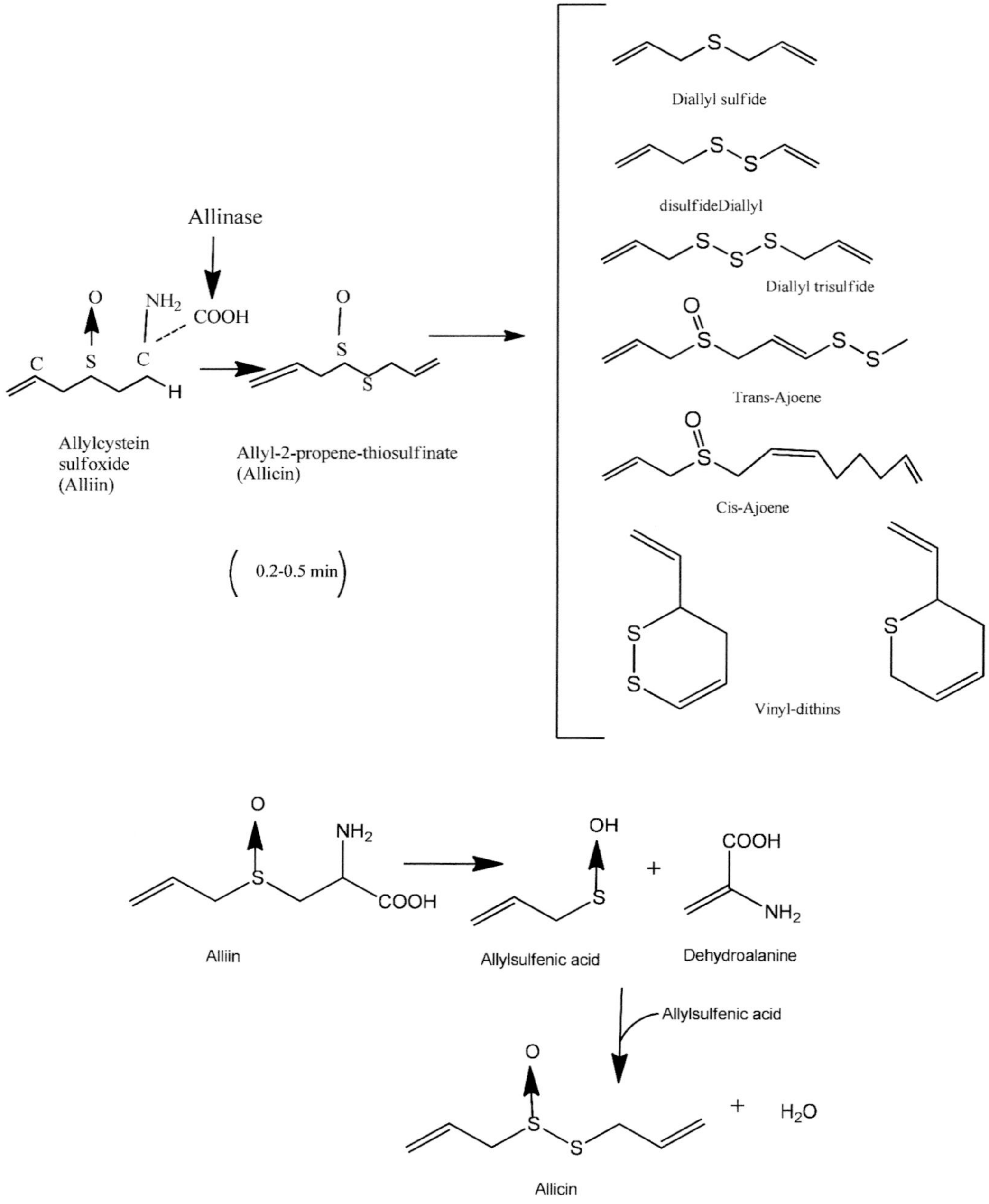

FIGURE 13.24 Metabolism of allicin.

their multiple active metabolites with different biological properties could provide mechanistic evidences for multiple targets in physiological systems (Figure 13.24). With potentially *in vivo* conversion the active metabolites of garlic organosulfur compounds could be considered as biomarker compounds to establish the relationship between pharmacokinetics and pharmacodynamics in models in clinical trial models (Gao et al., 2013).

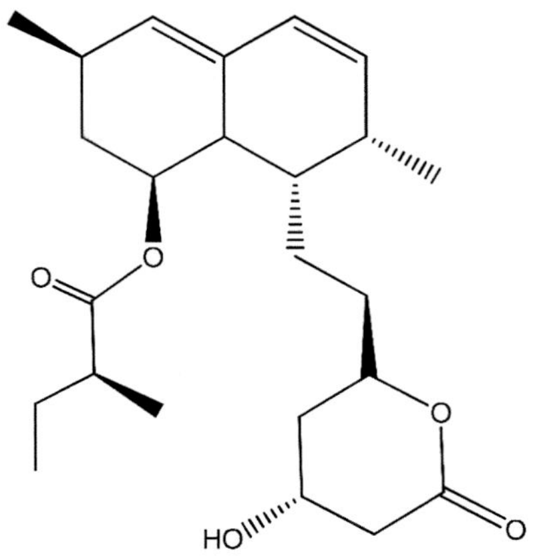

FIGURE 13.25 Structure of mevastatin.

13.2.25 Mevastatin as Prodrug

Mevastatin or compactin isolated from *Penicillium citinium* is a cholesterol-lowering agent and acts as a competitive inhibitor of HMG-CoA reductase with strong interaction affinity (Figure 13.25) (McFarland et al., 2014), Mevastatin, first discovered agent belonging to the class of cholesterol-lowering medications known as statins during a search for antibiotic compounds produced by fungi in 1971, by Akira Endo at Sankyo Co. Japan. *In vivo* conversion of mevastatin into active drug occur through hydrolysis of lactone ring lead to formation of hydrolytic product which mimic the metabolite produced by the enzyme HMG-CoA reductase claimed to be as an one of the lead compounds for the development of the synthetic compounds used today as antiartheroscelorosis. Various prodrug approaches for natural products have been summarized in Table 13.1.

13.3 Conclusion

Billions of phytopharmaceuticals biosynthesized in countless species of thousands of genera are providing natural opportunity to explore the natural prodrugs which can get metabolized by a human body upon catalysis through various bioenzymes, natural gift to mankind by the grace of super healer, universally conquering the pathological battle and empowering the living without disease. The natural secondary metabolites exist for the healthcare of human beings; a universal concept should be acceptable at all levels, with a motive to settle directly the value natural prodrugs in clinical applications. To explore and to accomplish nature-based prodrugs skillful operations are essential. However, mutual cooperation and interaction between natural drugs and synthetic drugs scientists working play an essential role in modifying unexplored countless natural products for commercial use through prodrug research and development.

TABLE 13.1

Data Representing Prodrug Approaches for Natural Product

S.No	Prodrug	Active Moiety	Chemical Nature	Source	Mechanism of Action	Therapeutic Category	References
1.	Mipsagargin (G-202) PSMA targeted	Thapsigargin	Sesquiterpene lactones	*Thapsia garganica* and Mediterranean species (Fruits and roots)	Inhibition of Ca^{2+} ATPase Inhibition of sarcoplasmic and endoplasmic reticulum calcium adenosine triphosphatase (SERCA) pumps protein Induces cell death in both normal and malignant cells	Utilized in treatment of solid tumors, activated by PSMA mediated cleavage of an inert masking peptide.	Shapiro et al. (2001)
2.	Disodium phosphono oxmethyl triptolide	Triptolide	Diterpene triepoxide	*Tripterygium wilfordii* (Leaves)	*In vitro*, triptolide inhibits proliferation and induces apoptosis of various cancer cell lines. *In vivo*, it prevents the tumor growth and metastasis.	To target various cancers like pancreatic, breast and prostate cancers and for the treatment of immune-inflammatory disorders such as psoriasis, asthma and rheumatoid arthritis.	Verma and Singh (2008)
3.	Glycolamide ester of Scutellarin	Scutellarin	Flavone	*Scutellaria lateriflora, Scutellaria barbata* (Aerial parts and roots)	Scutellarin induced apoptosis of cancer cells by down regulating Bcl-2, Bax, and caspase-3. Suppressed migration and invasion of cancer cells by inhibiting the activities of STAT3 and Girdin	Antioxidant anti-inflammation, vascular relaxation, antiplatelet, anti-coagulation, myocardial protection. Scutellarin has been used clinically to treat stroke, myocardial infarction and diabetic complications	Huang et al. (2015)

(*Continued*)

TABLE 13.1 (*Continued*)

Data Representing Prodrug Approaches for Natural Product

S.No	Prodrug	Active Moiety	Chemical Nature	Source	Mechanism of Action	Therapeutic Category	References
4.	Multiarm-polyethylene glycol linker	Betulinic acid	Pentacyclic triterpenoid	*Betula pubescens* (Bark)	Multi-arm polyethylene glycol linkers increase water solubility. High drug loading capacity and *in vitro* anticancer activity	Anti-inflammatory, antiretroviral and antimalarial activities. Human melanoma xenografts and leukemia (*In vitro*)	Farnsworth and Morris (1976)
5.	Heroin	Morphine	Alkaloid	*Papaver somniferum* (Fruit capsule)	Narcotic analgesic Heroin is converted into morphine by deacetylation.	Opoid painkiller which is used as a recreational drug for its euphoric effects, anti-diarrheal and cough suppressant.	Jana et al. (2010)
6.	Quercetin-3-O-acyl esters	Quercitin	Flavonol Quercetin propyl and quercetin butyl esters	Citrus genus Leaves, vegetables, grains and fruits	Quercetin propyl and Quercetin butyl esters	Breast, colon, prostate, ovary, endometrium, and lung tumor cancer	Cirmi et al. (2016)
7.	Aspirin	Salicylic acid	Acetylsalicylic acid compounds	Willow tree (Leaf and Bark)	Reduces pain and decreases the production of prostaglandins and thromboxanes	Antipyretic, analgesic and prevents blood clotting	
8.	Codeine	Morphine	Alkaloid	*Papaver somniferum* and opium poppy (Fruit capsule)	Narcotic analgesic, Codeine is converted into morphine by CYP2D6 2009	Used in mild to moderate pain, in diarrhea, and as a cough suppressant	Lahlou (2013)
9.	Irinotecan	Camptothecin	Quinoline alkaloid	*Camptotheca acuminata* (Stem)	Act on Topoisomerase I, allowing DNA cleavage, inhibiting subsequent ligation, and resulting in DNA strand breaks	Antitumor Anticancer	Lee et al. (2004)
10.	L-Dopa	Dopamine	Catecholamines	*Vicia faba* *Mucuna pruriens* (Green Pods and Seeds)	Levodopa is converted to dopamine catalyzed by DOPA decarboxylase. After crossing the blood brain barrier activates central dopamine receptors	For the treatment of Parkinson's disease.	Li et al. (2015)

(*Continued*)

TABLE 13.1 (*Continued*)

Data Representing Prodrug Approaches for Natural Product

S.No	Prodrug	Active Moiety	Chemical Nature	Source	Mechanism of Action	Therapeutic Category	References
11.	Vidarabine	Spongouridine and Spongothymidine	Analog of adenosine	*Tethya crypta* (Sponge)	Inhibits other host cell enzyme systems, including ribonucleoside reductase, RNA polyadenylation, and S adenosyl homocysteine hydrolase	Active against the varicella zoster virus and herpes simplex virus.	Martin et al. (1990))
12.	Isoliquiritigenin Phosphate	Isoliquiritigenin	Chalcone-type flavonoid	*Glycyrrhiza spp.* (Licorice)	Reduce expression of inflammatory molecules such as prostaglandin E2 (PGE2), tumor necrosis factor-alpha (TNF-α), and inteurlekin-6 (IL-6). Decrease nuclear factor kappa B (NF-κB) activity	Anti-inflammatory, antimicrobial, antioxidant, antitumor, hepatoprotective and cardioprotective effects	Boyapelly et al. (2017)
13.	Palmitoyl Danshensu	Danshensu	3-(3,4-dihydroxy phenyl) lactic acid	*Salvia miltiorrhiza* roots	Inhibit platelet aggregation and thrombus formation. Improve cardiac function, ability to normalize the thromboxane $A_2(TXA_2)$/prostacyclin (PGI_2) balance. Inhibit production of ROS	Cardioprotective anti-inflammatory, antitumor activities	Zhang et al. (2010), Chan et al. (2004), Ding et al. (2005)
14.	Tectoridin	Tectorigenin	Isoflavone, 7-glucoside of tectorigenin	*Pueraria thunbergiana*	Beta-glucuronidase inhibitor	Hepatoprotective	Lee et al. (2003, 2005)
15.	Alpha-hederin	α-hederin	Water-soluble pentacyclic Triterpenoid saponin	*Nigella sativa* seed	*Cytotoxic* and *apoptotic/necrotic effects* inhibits cell proliferation	Anticancer, antioxidant, anti-inflammatory, antiarthritic, bronchiolytic, antifungal, antiparasitic	Rooney and Ryan (2005)
16.	Climacostol	1,3-*bis*(methoxymethoxy)-5-[(2*Z*)-non-2-en-1-yl] benzene)	Cytotoxin	*Climacostomum virens*	DNA damage, activation of Caspase 9-dependent cleavage of Caspase 3 and induced the intrinsic apoptotic pathway	Anticancer Cytotoxic	Catalani et al. (2019)

REFERENCES

Acuña-Castroviejo, D., Escames, G., LeÓn, J., Carazo, A. and Khaldy, H., 2003. Mitochondrial regulation by melatonin and its metabolites. In *Developments in Tryptophan and Serotonin Metabolism* (549–557). Springer, Boston, MA.

Akhtar, N. and Tariq M.H., 2012. Current nutraceuticals in the management of osteoarthritis: a review. *Therapeutic Advances in Musculoskeletal Disease*, *4*:181–207.

Alam, P., Aftab A., Md Khalid A. and Saleh I.A., 2014. Quantitative estimation of hesperidin by HPTLC in different varieties of citrus peels. *Asian Pacific Journal of Tropical Biomedicine*, *4*:262–266.

Albert, A., 1958. Chemical aspects of selective toxicity. *Nature*, *182*:421.

Amalraj, A., Anitha P., Sreerag G. and Sreeraj G., 2017. Biological activities of curcuminoids, other biomolecules from turmeric and their derivatives–A review. *Journal of Traditional and Complementary Medicine*, *7*(2):205–233.

Babizhayev, M.A., 2012. Bioactivation antioxidant and transglycating properties of N-acetylcarnosine autoinduction prodrug of a dipeptide L-carnosine in mucoadhesive drug delivery eye-drop formulation: powerful eye health application technique and therapeutic platform. *Drug Testing and Analysis*, *4*(6): 468–485.

Babizhayev, M.A., Valentina N. Y., Natalya L.S., Rima P.E., Elena A.R. and Galina A.Z., 1996. Nα-Acetylcarnosine is a prodrug of L-carnosine in ophthalmic application as antioxidant. *Clinica Chimica Acta*, 254 (1):1–21.

Balunas, M.J. and Kinghorn, A.D., 2005. Drug discovery from medicinal plants. *Life Sciences*, *78*:431–441.

Bang, S-H., Sang-Jun H. and Dong-Hyun K., 2008. Hydrolysis of arbutin to hydroquinone by human skin bacteria and its effect on antioxidant activity. *Journal of Cosmetic Dermatology*, *7*(3):189–193.

Bhuket, P.R.N., El-Magboub, A., Haworth, I.S. and Rojsitthisak, P., 2017. Enhancement of curcumin bioavailability via the prodrug approach: Challenges and prospects. *European Journal of Drug Metabolism and Pharmacokinetics*, *42*(3):341–353.

Bone, K. and Mils, S., 2013. *Principles and Practice of Phytotherapy*. Churchill Livingstone, New York.

Boyapelly, K., Bonin, M.A., Traboulsi, H., Cloutier, A., Phaneuf, S.C., Fortin, D., et al., 2017. Synthesis and characterization of a phosphate prodrug of isoliquiritigenin. *Journal of Natural Products*, *80*:879–886.

Catalani, E., Buonanno, F., Lupidi, G., Bongiorni, S., Belardi, R., Zecchini, S., Giovarelli, M. et al., 2019. The natural compound climacostol as a prodrug strategy based on pH activation for efficient delivery of cytotoxic small agents. *Frontiers in Chemistry*, *7*.

Chan, K., Chui, S.H., Wong, D.Y.L., Ha, W.Y., Chan, C.L. and Wong, R.N.S., 2004. Protective effects of Danshensu from the aqueous extract of *Salvia miltiorrhiza* (Danshen) against homocysteine-induced endothelial dysfunction. *Life Sciences*, *75*:3157–3171.

Cirmi, S., Ferlazzo, N., Lombardo, G.E., Maugeri, A., Calapai, G., Gangemi, S. and Navarra, M., 2016. Chemopreventive agents and inhibitors of cancer hallmarks: may Citrus offer new perspectives? *Nutrients*, *8*(11):698.

Cragg, G.M. and Newman, D.J., 2013. Natural products: A continuing source of novel drug leads. *Biochimica et Biophysica Acta (BBA)-General Subjects*, *1830*:3670–3695.

Cui, Qingling, Yingni Pan, Xiaotong Xu, Wenjie Zhang, Xiao Wu, Shouhe Qu, and Xiaoqiu Liu, 2016. The metabolic profile of acteoside produced by human or rat intestinal bacteria or intestinal enzyme in vitro employed UPLC-Q-TOF–MS. *Fitoterapia*, *109*:67–74.

Dąbrowska-Maś, Elżbieta, 2017. Insights on Fatty Acids in Lipophilic Prodrug Strategy. *International Research Journal of Pure and Applied Chemistry*, *14*:1–10.

Dahan, A., Zimmermann, E. and Ben-Shabat, S., 2014. Modern prodrug design for targeted oral drug delivery. *Molecules*, *19*:16489–16505.

Daniel, O., Meier, M.S., Schlatter, J. and Frischknecht, P., 1999. Selected phenolic compounds in cultivated plants: Ecologic functions, health implications, and modulation by pesticides. *Environmental Health Perspectives*, *107*:109–114.

Dax, S.L., 2012. *Antibacterial Chemotherapeutic Agents*. Springer Science & Business Media, London, UK.

Dhareshwar, Sundeep S. and Valentino J. Stella, 2007. Prodrugs of alcohols and phenols. In *Prodrugs*, Springer, New York, pp. 731–799.

Ding, M., Ye, T.X., Zhao, G.R., Yuan, Y.J., Guo, Z.X., 2005. Aqueous extract of Salvia miltiorrhiza attenuates increased endothelial permeability induced by tumor necrosis factor-α. *International Immunopharmacology*, *5*:1641–1651.

Downie, D., McFadyen, M.C., Rooney, P.H., Cruickshank, M.E., Parkin, D.E., Miller, I.D., Telfer, C., Melvin, W.T. and Murray, G.I., 2005. Profiling cytochrome P450 expression in ovarian cancer: Identification of prognostic markers. *Clinical Cancer Research*, *11*:7369–7375.

Enne, V.I., Bennett, P.M., Livermore, D.M. and Hall, L.M., 2004. Enhancement of host fitness by the sul2-coding plasmid p9123 in the absence of selective pressure. *Journal of Antimicrobial Chemotherapy*, *53*:958–963.

Farnsworth, N.R. and Morris, R.W., 1976. Higher plants–the sleeping giant of drug development. *American Journal of Pharmacy and the Sciences Supporting Public Health*, *148*:46.

Franke, R. and Schilcher, H., 2006, June. Relevance and use of chamomile (*Matricaria recutita* L.). *I International Symposium on Chamomile Research, Development and Production*, *749*:29–43.

Furumai, R., Matsuyama, A., Kobashi, N., Lee, K.H., Nishiyama, M., Nakajima, H., Tanaka, A., et al., 2002. FK228 (depsipeptide) as a natural prodrug that inhibits class I histone deacetylases. *Cancer Research*, *62*:4916–4921.

Galletti, E., Magnani, M., Renzulli, M.L. and Botta, M., 2007. Paclitaxel and docetaxel resistance: Molecular mechanisms and development of new generation taxanes. *ChemMedChem:Chemistry Enabling Drug Discovery*, *2*:920–942.

Gao, C., Jiang, X. Wang, H. Zhao, Z. and Wang, W., 2013. Drug metabolism and pharmacokinetics of organosulfur compounds from garlic. *Journal of Drug Metabolism and Toxicology*, *4* (5):1–10.

Georgiev, M.I., Nina I., Kalina A., Petya D. and Robert V., 2013. Harpagoside: From Kalahari Desert to pharmacy shelf. *Phytochemistry*, *92*:8–15.

Ghadage, R.V., 2013. Prodrug design for optimized drug delivery systems. *Journal of Biological & Scientific Opinion Review*, *1*:255–262.

Gibson, P., Gill, J.H., Khan, P.A., Seargent, J.M., Martin, S.W., Batman, P.A., Griffith, J. et al., 2003. Cytochrome P450 1B1 (CYP1B1) is overexpressed in human colon adenocarcinomas relative to normal colon: Implications for drug development1. *Molecular Cancer Therapeutics*, *2*:527–534.

Guerrero, A., Romain G., Herranz, N., Anthony U., Withers, D.J., Martinez-Barbera, J.P., et al., 2019. Galactose-modified duocarmycin prodrugs as senolytics. bioRxiv:746669.

Guo, Z., 2017. The modification of natural products for medical use. *Acta Pharmaceutica Sinica B*, *7*:119–136.

Hardeland, R. and Pandi-Perumal, S.R., 2005. Melatonin, a potent agent in antioxidative defense: Actions as a natural food constituent, gastrointestinal factor, drug and prodrug. *Nutrition & Metabolism*, *2*:22.

Huang, S.X., Yun, B.S., Ma, M., Basu, H.S., Church, D.R., Ingenhorst, G., Huang, Y. et al., 2015. Leinamycin E1 acting as an anticancer prodrug activated by reactive oxygen species. *Proceedings of the National Academy of Sciences*, *112*:8278–8283.

Huang, Tom Hsun-Wei, Tran, V.H., Duke, R.K., Sigrun Chrubasik, S.T., Roufogalis, B.D. and Duke, C.C., 2006. Harpagoside suppresses lipopolysaccharide-induced iNOS and COX-2 expression through inhibition of NF-κB activation. *Journal of Ethnopharmacology, 104* (1–2):149–155.

Hung, J.Y., Yang, C.J., Tsai, Y.M., Huang, H.W. and Huang, M.S., 2008. Antiproliferative activity of paeoniflorin is through cell cycle arrest and the Fas/Fas ligand-mediated apoptotic pathway in human non-small cell lung cancer A549 cells. *Clinical and Experimental Pharmacology & Physiology*, *35*:141–147.

Huttunen, K.M., Raunio, H. and Rautio, J., 2011. Prodrugs—from serendipity to rational design. *Pharmacological Reviews*, *63*:750–771.

Jana, S., Mandlekar, S. and Marathe, P., 2010. Prodrug design to improve pharmacokinetic and drug delivery properties: Challenges to the discovery scientists. *Current Medicinal Chemistry*, *17*:3874–3908.

Jia, Y., Zan-Hui J., Jun C., He Z. and Man-Hua C., 2015. Ursolic acid benzaldehyde chalcone leads to inhibition of cell proliferation and arrests cycle in G1/G0 phase in ovarian cancer. *Bangladesh Journal of Pharmacology*, *10*(2):358–365.

Jou, M.J., Peng, T.I., Reiter, R.J., Jou, S.B., Wu, H.Y. and Wen, S.T., 2004. Visualization of the antioxidative effects of melatonin at the mitochondrial level during oxidative stress-induced apoptosis of rat brain astrocytes. *Journal of Pineal Research*, *37*:55–70.

Kalyani, G., Sharma, D., Vaishnav, Y. and Deshmukh, V.S., 2013. A review on drug designing, methods, its applications and prospects. *International Journal of Pharmaceutical Research and Development*, *5*:15–30.

Kumar, S.V., Saravanan, D., Kumar, B. and Jayakumar, A., 2014. An update on prodrugs from natural products. *Asian Pacific Journal of Tropical Medicine*, *7*:S54–S59.

Kumar, V., Sonali S.B. and Ram A.V., 2016. Modulating lipophilicity of rohitukine via prodrug approach: Preparation, characterization, and in vitro enzymatic hydrolysis in biorelevant media. *European Journal of Pharmaceutical Sciences*, *92*:203–211.

Lahlou, M., 2013. The success of natural products in drug discovery. *Pharmacology and Pharmacy*, *4*:17–31.

Lee, H.U., Bae, E.A. and Kim, D.H., 2005. Hepatoprotective effect of tectoridin and tectorigenin on tert-butyl hyperoxide-induced liver injury. *Journal of Pharmacological Sciences*, *97*:541–544.

Lee, H.W., Choo, M.K., Bae, E.A. and Kim, D.H., 2003. β-Glucuronidase inhibitor tectorigenin isolated from the flower of *Pueraria thunbergiana* protects carbon tetrachloride-induced liver injury. *Liver International*, *23*(4): 221–226.

Lee, N.K., Choi, S.H., Park, S.H., Park, E.K. and Kim, D.H., 2004. Antiallergic activity of hesperidin is activated by intestinal microflora. *Pharmacology*, *71*:174–180.

Leng-Peschlow, E., 1988. Effect of sennosides and related compounds on intestinal transit in the rat. *Pharmacology*, *36*:40–48.

Leng-Peschlow, E., 1993. Sennoside-induced secretion and its relevance for the laxative effect. *Pharmacology*, *47*:14–21.

Li, L., Feng, X., Chen, Y., Li, S., Sun, Y. and Zhang, L., 2019. A comprehensive study of eriocitrin metabolism in vivo and in vitro based on an efficient UHPLC-Q-TOF-MS/MS strategy. *RSC Advances* *9*(43):24963–24980.

Li, X.B., Deng, Y.G., Hu, J.P., Wang, Z., Xie, R.Z., Luo, H. and Hu, X.Y., 2015. Resveratrol overcomes TRAIL resistance in human colon cancer cells. *Bangladesh Journal of Pharmacology*, *10*:568–576.

Macfarlane, G., 1984. *Alexander Fleming, The Man and The Myth*. Harvard University Press, Cambridge, MA.

Macielag, M.J., Bush, K. and Weidner-Wells, M.A., 2003. Antibacterial agents, overview. *Kirk-Othmer Encyclopedia of Chemical Technology*.

Manolova, Y., Vera D., Liudmil A., Elena D., Denitsa M. and Nikolay L., 2014. The effect of the water on the curcumin tautomerism: A quantitative approach. Spectrochimica Acta Part A: *Molecular and Biomolecular Spectroscopy* *132*:815–820.

Marinelli, L., Fornasari, E., Eusepi, P., Ciulla, M., Genovese, S., Epifano, F., Fiorito, S., Turkez, H., Örtücü, S., Mingoia, M. and Simoni, S., 2019. Carvacrol prodrugs as novel antimicrobial agents. *European Journal of Medicinal Chemistry*, *178*:515–529.

Marshall, J.L., Rizvi, N., Kauh, J., Dahut, W., Figuera, M., Kang, M.H., Figg, W.D. et al., 2002. A phase I trial of depsipeptide (FR901228) in patients with advanced cancer. *Journal of Experimental Therapeutics and Oncology*, *2*:325–332.

Martin, M.L., San Román, L. and Domínguez, A., 1990. In vitro activity of protoanemonin, an antifungal agent. *Planta Medica*, *56*:66–69.

McFarland, A.J., Anoopkumar-Dukie, S., Arora, D.S., Grant, G.D., McDermott, C.M., Perkins, A.V. and Davey, A.K., 2014. Molecular mechanisms underlying the effects of statins in the central nervous system. *International Journal of Molecular Sciences*, *15*(11):20607–20637.

Mishra, R., Upadhyay, S.K. and Maheshwari, P.N., 2001. Conversion of alliin to allicin in garlic-A kinetic study. *Indian Journal of Chemical Technology*, *8*:107–111.

Miyake, Y., Shimoi, K., Kumazawa, S., Yamamoto, K., Kinae, N. and Osawa, T., 2000. Identification and antioxidant activity of flavonoid metabolites in plasma and urine of eriocitrin-treated rats. *Journal of Agricultural and Food Chemistry*, *48*(8):3217–3224.

Morgan, B.J., 2018. An investigation of enzymes capable of activating cannabinoid prodrugs.

Muangnoi, C., Jithavech, P., Ratnatilaka Na Bhuket, P., Supasena, W., Wichitnithad, W., Towiwat, P., et al., 2018. A curcumin-diglutaric acid conjugated prodrug with improved water solubility and antinociceptive properties compared to curcumin. *Bioscience, Biotechnology, and Biochemistry*, *82*(8):1301–1308.

Muangnoi, C., Ratnatilaka Na Bhuket, P., Jithavech, P., Wichitnithad, W., Srikun, O., Nerungsi, C., et al., 2019. Scale-Up Synthesis and In Vivo Anti-Tumor Activity of Curcumin Diethyl Disuccinate, an Ester Prodrug of Curcumin, in HepG2-Xenograft Mice. *Pharmaceutics*, *11*(8):373.

Murray, G.I., Taylor, M.C., McFadyen, M.C., McKay, J.A., Greenlee, W.F., Burke, M.D. and Melvin, W.T., 1997. Tumor-specific expression of cytochrome P450 CYP1B1. *Cancer Research*, *57*:3026–3031.

Naghdi Badi, H., Abdollahi, M., Mehrafarin, A., Ghorbanpour, M., Tolyat, M., Qaderi, A. and Ghiaci Yekta, M., 2017. An overview on two valuable natural and bioactive compounds, thymol and carvacrol, in medicinal plants. *Journal of Medicinal Plants*, *3*(63):1–32.

Naji, L., Carrillo-Vico, A., Guerrero, J.M. and Calvo, J.R., 2004. Expression of membrane and nuclear melatonin receptors in mouse peripheral organs. *Life Sciences*, *74*:2227–2236.

Nelson, K.M., Dahlin, J.L., Bisson, J., Graham, J., Pauli, G.F. and Walters, M.A., 2017. The essential medicinal chemistry of curcumin: Miniperspective. *Journal of Medicinal Chemistry*, *60*(5):1620–1637.

Oliveira, Ana S., Emília Sousa, Maria Helena Vasconcelos, and Madalena Pinto. 2015. Curcumin: A natural lead for potential new drug candidates. *Current Medicinal Chemistry* *22*(36):4196–4232.

Park, Kyoung Sik. 2016. A systematic review on anti-inflammatory activity of harpagoside. *Journal of Biochemistry and Molecular Biology Research*, *2*(3):166–169.

Patel, D.K., Patel, K. and Tahilyani, V., 2012. Barbaloin: A concise report of its pharmacological and analytical aspects. *Asian Pacific Journal of Tropical Biomedicine*, *2*:835–838.

Peart, M.J., Smyth, G.K., van Laar, R.K., Bowtell, D.D., Richon, V.M., Marks, P.A., Holloway, A.J. and Johnstone, R.W., 2005. Identification and functional significance of genes regulated by structurally different histone deacetylase inhibitors. *Proceedings of the National Academy of Sciences*, *102*:3697–3702.

Peng, L., Qiangsong, T., Fengchao, J., Liduan, Z., Fangmin, C., Fuqing, Z., Jihua, D. and Yuefeng, D., 2005. Preparation of curcumin prodrugs and theirin vitro anti-tumor activities. *Journal of Huazhong University of Science and Technology [Medical Sciences]*, *25*(6):668–670.

Peterson, J., Dwyer, J., Adlercreutz, H., Scalbert, A., Jacques, P. and McCullough, M.L., 2010. Dietary lignans: Physiology and potential for cardiovascular disease risk reduction. *Nutrition Reviews*, *68*:571–603.

Pillaiyar, T., Manickam, M. and Namasivayam, V., 2017. Skin whitening agents: Medicinal chemistry perspective of tyrosinase inhibitors. *Journal of Enzyme Inhibition and Medicinal Chemistry*, *32*(1):403–425.

Potter, G.A. and Burke, M.D., 2006. Salvestrols-natural products with tumour selective activity. *Journal of Orthomolecular Medicine*, *21*:34–36.

Qi, M., Xiong, A., Li, P., Yang, Q., Yang, L. and Wang, Z., 2013. Identification of acteoside and its major metabolites in rat urine by ultra-performance liquid chromatography combined with electrospray ionization quadrupole time-of-flight tandem mass spectrometry. *Journal of Chromatography B*, *940*:77–85.

Ramadan, M., Goeters, S., Watzer, B., Krause, E., Lohmann, K., Bauer, R., Hempel, B. and Imming, P., 2006. Chamazulene carboxylic acid and matricin: A natural profen and its natural prodrug, identified through similarity to synthetic drug substances. *Journal of Natural Products*, *69*:1041–1045.

Rautio, J., Kumpulainen, H., Heimbach, T., Oliyai, R., Oh, D., Järvinen, T. and Savolainen, J., 2008. Prodrugs: Design and clinical applications. *Nature Reviews Drug Discovery*, *7*:255.

Reininger, E. and Bauer, R., 1998. A new PGHS-2 microtiter assay for the screening of herbal drugs. In *Poster on the 46th Annual Congress of the Society for Medicinal Plant Research* (Vol. 31, No. 4.9).

Rooney, S. and Ryan, M.F., 2005. Effects of alpha-hederin and thymoquinone, constituents of *Nigella sativa*, on human cancer cell lines. *Anticancer Research*, *25*(3B): 2199–2204.

Salvador, L A. and Luesch, H., 2012. Discovery and mechanism of natural products as modulators of histone acetylation. *Current Drug Targets*, *13*:1029–1047.

Searcey, Mark. 2002. Duocarmycins-natures prodrugs? *Current Pharmaceutical Design* *8*(15):1375–1389.

Seenivasan, A., Subhagar, S., Aravindan, R. and Viruthagiri, T., 2008. Microbial production and biomedical applications of lovastatin. *Indian Journal of Pharmaceutical Sciences*, *70*:701.

Shapiro, T.A., Fahey, J.W., Wade, K.L., Stephenson, K.K. and Talalay, P., 2001. Chemoprotective glucosinolates and isothiocyanates of broccoli sprouts: Metabolism and excretion in humans. *Cancer Epidemiology and Prevention Biomarkers*, *10*:501–508.

Shiao, Young-Ji, Su, M-H., Lin, H-C. and Wu, C-R., 2017. Acteoside and isoacteoside protect amyloid β peptide induced cytotoxicity, cognitive deficit and neurochemical disturbances in vitro and in vivo. *International Journal of Molecular Sciences* *18*(4):895.

Shirota, O., Hakamata, W. and Goda, Y., 2003. Concise large-scale synthesis of psilocin and psilocybin, principal hallucinogenic constituents of "magic mushroom". *Journal of Natural Products*, *66*:885–887.

Takagi, T., Ramachandran, C., Bermejo, M., Yamashita, S., Yu, L.X. and Amidon, G.L., 2006. A provisional biopharmaceutical classification of the top 200 oral drug products in the United States, Great Britain, Spain, and Japan. *Molecular Pharmaceutics*, *3*:631–643.

Tan, H.L., Butler, P.C., Burke, M.D. and Potter, G.A., 2007. Salvestrols: A new perspective in nutritional research. *Journal of Orthomolecular Medicine*, *22*:39–47.

Tarragó, T., Kichik, N., Claasen, B., Prades, R., Teixidó, M. and Giralt, E., 2008. Baicalin, a prodrug able to reach the CNS, is a prolyl oligopeptidase inhibitor. *Bioorganic & Medicinal Chemistry*, *16*:7516–7524.

Ueda, H., Nakajima, H., Hori, Y., Fujita, T., Nishimura, M., Goto, T. and Okuhara, M., 1994. FR901228, a novel antitumor bicyclic depsipeptide produced by Chromobacterium violaceum no. 968. *The Journal of Antibiotics*, *47*:301–310.

Vander Molen, K.M., McCulloch, W., Pearce, C.J. and Oberlies, N.H., 2011. Romidepsin (Istodax, NSC 630176, FR901228, FK228, depsipeptide): A natural product recently approved for cutaneous T-cell lymphoma. *The Journal of Antibiotics*, *64*:525.

Verma, S. and Singh, S.P., 2008. Current and future status of herbal medicines. *Veterinary World*, *1*:347.

Vijayaraj, S., Omshanthi, B., Anitha, S. and Sampath Kumar, K.P., 2014. Synthesis and characterization of novel sulphoxide prodrug of famotidine. *Indian Journal of Pharmaceutical Education and Research*, *48*:35–44.

Wang, X., Zhang, F., Yang, L., Mei, Y., Long, H., Zhang, X., Zhang, J. and Su, X., 2011. Ursolic acid inhibits proliferation and induces apoptosis of cancer cells in vitro and in vivo. *BioMed Research International*, 2011.

Woods, J.R., Mo, H., Bieberich, A.A., Alavanja, T. and Colby, D.A., 2013. Amino-derivatives of the sesquiterpene lactone class of natural products as prodrugs. *Med Chem Comm*, *4*:27–33.

Xiang, L., Chi, T., Tang, Q., Yang, X., Ou, M., Chen, X., Yu, X., Chen, J., Ho, R.J., Shao, J. and Jia, L., 2015. A pentacyclic triterpene natural product, ursolic acid and its prodrug US597 inhibit targets within cell adhesion pathway and prevent cancer metastasis. *Oncotarget*, *6*(11):9295–9312.

Xiao, J.J., Byrd, J., Marcucci, G., Grever, M. and Chan, K.K., 2003. Identification of thiols and glutathione conjugates of depsipeptide FK228 (FR901228), a novel histone protein deacetylase inhibitor, in the blood. *Rapid Communications in Mass Spectrometry*, *17*:757–766.

Yang, Y.S., Zhang, T., Yu, S.C., Ding, Y., Zhang, L.Y., Qiu, C. and Jin, D., 2011. Transformation of geniposide into genipin by immobilized β-glucosidase in a two-phase aqueous-organic system. *Molecules*, *16*:4295–4304.

Yoon, Y., Lim, J.W., Kim, J., Kim, Y. and Chun, K.H., 2016. Discovery of ursolic acid prodrug (NX-201): Pharmacokinetics and in vivo antitumor effects in PANC-1 pancreatic cancer. *Bioorganic and Medicinal Chemistry Letters*, *26*(22):5524–5527.

Youssefi, M.R., Moghaddas, E., Tabari, M.A., Moghadamnia, A.A., Hosseini, S.M., Farash, B.R.H., Ebrahimi, M.A., Mousavi, N.N., Fata, A., Maggi, F. and Petrelli, R., 2019. In vitro and in vivo effectiveness of carvacrol, thymol and linalool against Leishmania infantum. *Molecules*, *24*(11):2072.

Zawilska, J.B., Wojcieszak, J. and Olejniczak, A.B., 2013. Prodrugs: A challenge for the drug development. *Pharmacological Reports*, *65*:1–14.

Zhang, L.J., Chen, L., Lu, Y., Wu, J.M., Xu, B., Sun, Z.G., et al., 2010. Danshensu has anti-tumor activity in B16F10 melanoma by inhibiting angiogenesis and tumor cell invasion. *European Journal of Pharmacology*, *643*:195–201.

14

Biotechnology-Based Prodrug Approach for Cell-Specific Targeting: Its Theoretical Basis and Application

Sunita Minz, Manju Rawat Singh, Deependra Singh, Arun Singh Parihar, and Madhulika Pradhan

CONTENTS

14.1 Introduction

Many therapeutic entities offer unwanted pharmacological, pharmaceutical, or pharmacokinetic characteristics that limit their clinical application. Among numerous strategies to reduce unwanted characteristics of therapeutic entity with simultaneous retention of desired therapeutic activity, a prodrug approach has been exploited due to its high flexibility and enhanced efficacy.

A prodrug is defined as molecule that does not exhibit inherent biological activity; however, they own capability to yield biologically active drug molecule following diverse stages of its metabolism. In other words, a prodrug could be any compound that endures biotransformation prior to exerting its pharmacological action (Han and Amidon, 2000).

Methenamine was the first purposely fabricated prodrug, which was introduced to pharmacies in 1899 (Rautio et al., 2008). However, acetanilide was the very first drug entity that fulfilled the standards for existence as a prodrug. It has been employed as an anti-inflammatory agent since 1867 (Jornada et al., 2016; Lesniewska-Kowiel and Muszalska, 2017). About 10% of the drugs accessible globally are categorized as prodrugs (Rautio et al., 2017).

A potent prodrug must have unique properties; for example, it has to be hydrophilic enough to meet bioavailability, solubility and transport conditions; furthermore, it must exhibit lipophilic properties to enable transportation of therapeutic moiety across a membrane or metabolic barrier instantaneously. Many active therapeutic agents exhibit inferior bioavailability following oral administration owing to poor drug absorption and proneness to first pass metabolism (Anastasi et al., 2003). This ultimately results in inactivation of active therapeutic entity and the formation of toxic metabolites. A prodrug approach could be used to improve drug delivery at the chosen location of action by alteration of physicochemical characteristics of prodrug or by explicitly directing enzymes or membrane transporters. In recent years, a prodrug approach has gained application as it offers numerous benefits including augmentation of cell permeation, solubility, and stability to endure enzymatic or chemical reaction, bioavailability, blood-brain barrier penetration and reduction of toxicity. In the target tissue, metabolic biotransformation of prodrug into active drug takes place by explicit catalyzing enzymes thus escaping undesirable side effects (Abet et al., 2017). A novel division of prodrugs that can be intended to accomplish site-specific bioconversion of prodrug consist of enzyme-prodrug therapy (EPT). Fabrication of EPT relies in the bioconversion of prodrug, at the desired site of action. In his book chapter type of prodrugs, role of biotechnology in prodrug therapy with special reference to targeting of specific enzymes and targeting of specific membrane transporters have been discussed. In addition, emphasis has been placed on antibody-directed prodrug therapy, gene-directed prodrug therapy, bacterial-directed prodrug therapy and viral-directed prodrug therapy, all of which have been supported by research.

14.2 Types of Prodrugs

Prodrugs have been classified into two distinct classes.

14.2.1 Carrier-Linked Prodrugs

In carrier-linked prodrugs the therapeutic entity is temporarily associated with a carrier molecule by covalent linkage, where the prodrug might be inactive or less active than parent moiety. Hydrolytic cleavage of a carrier-prodrug produces therapeutic moiety of enhanced physicochemical or pharmacokinetic properties along with biologically inactive carrier molecule (Choi-Sledeski and Wermuth, 2015).

An example of carrier-linked prodrugs includes designing of orally active ampicillin derivatives. However, poor oral absorption remains the main constraint in oral delivery of ampicillin and only about 40% of the drug is absorbed. Therefore, to overcome the challenges, pivampicillin and bacampicillin (prodrugs of ampicillin) were produced by esterification of the polar carboxylic group with an ester-possessing enzyme labile feature. The resultant prodrugs have been reported to exhibit better absorption (Wermuth, 2008).

A well-fabricated carrier-prodrug must fulfill the following conditions:

- Covalent bonding must present among therapeutic moiety and the carrier molecule.
- Dissociation between transport carrier and drug moiety should takes place in vivo.
- In vivo liberated carrier and prodrug should not be toxic.
- The formation of drug (active) must occur following rapid kinetics to confirm effective therapeutic level at the preferred site of action.

14.2.2 Bioprecursor Prodrugs

Bioprecursor prodrugs are produced by molecular alteration of the active therapeutic agent. The alteration gives rise to a new entity that is proficient of being used as substrate for metabolic enzymes in order to release expected active compound. Pharmacologically active metabolic entities are usually produced by phase I reactions and phase II conjugation reactions (Figure 14.1) (Testa, 2007; Abet et al., 2017).

A classic illustration of an active bioprecursor prodrug is presented by sulindac. It is a nonsteroidal anti-inflammatory compound exhibiting a wide range of therapeutic activity in human as well as animal models. Sulindac undergoes biotransformations into sulfone by irreversible oxidation followed by reversible reduction to sulphide; the second compound remains an active moiety (Ettmayer et al., 2004; Wermuth, 2008).

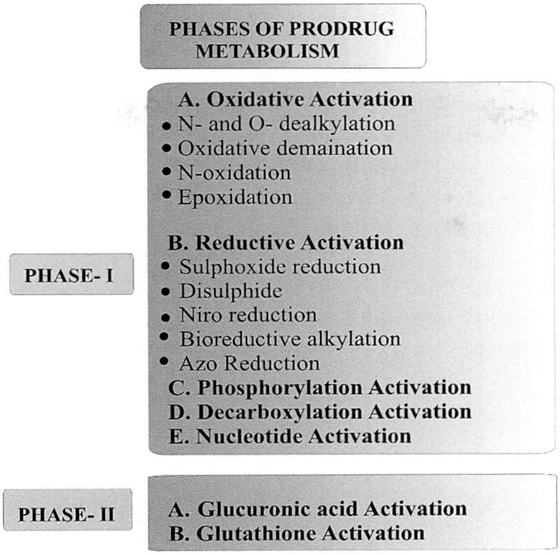

PHASES OF PRODRUG METABOLISM

PHASE- I

A. Oxidative Activation
- N- and O- dealkylation
- Oxidative demaination
- N-oxidation
- Epoxidation

B. Reductive Activation
- Sulphoxide reduction
- Disulphide
- Niro reduction
- Bioreductive alkylation
- Azo Reduction

C. Phosphorylation Activation
D. Decarboxylation Activation
E. Nucleotide Activation

PHASE- II

A. Glucuronic acid Activation
B. Glutathione Activation

FIGURE 14.1 Different phases of prodrug metabolism.

14.3 Biotechnology-Based Prodrug Therapy

Biotechnology has a unique position in prodrug therapy. Progress in gene cloning and meticulous methods of gene expression in mammalian cells have permitted explication of molecular characteristics of carrier proteins and enzymes, thereby providing a more coherent approach of targeted prodrugs.

The role of biotechnology in prodrug therapy could be explained through the following two approaches:

1. Targeting of particular enzymes.
2. Targeting of particular membrane transporters.

14.3.1 Targeting of Particular Enzymes

Earlier the enzyme-targeted prodrug concept was extensively employed to increase oral absorption of therapeutics and to accomplish drug delivery to a specific location considering gastrointestinal enzymes as key targets for fabrication of prodrug.

Following the prodrug concept, location-specific drug delivery could be achieved by stimulation and liberation of prodrug in the target tissue, where the activation occurs due to metabolic reaction catalyzed by the specific enzyme (tissue-specific). For example, aquaphilic glycoside derivatives shows limited absorption from small intestine, but after reaching colon, they could be successfully converted into free drug in presence of bacterial glycosidases, thus it could be easily absorbed by the mucosa of colons (Bohme et al., 1974).

Furthermore, site-specific delivery of prodrugs was reported to be a unique approach for chemotherapy of cancer. These skillfully fabricated prodrugs were effective in animal tumors having increased concentration of an activating enzyme. However, finding of clinical findings suggested that human tumors presenting very high concentration of these activating enzymes were very few. In addition, high level of activating enzymes was not linked upto any specific kind of tumor.

Thus, to overcome such limitations, new biotechnology-based therapies have been suggested.

Newer/modified approach of enzyme-prodrug therapy (EPT) comprises of artificial introduction of enzymes into the body to facilitate conversion of prodrugs into its active form in the desired location within the body (Walther et al., 2017). In contrast to conventional approaches to drug administration,

EPT allows an enhanced local concentration of drug there by favoring increased therapeutic response. In addition, EPT leads to targeted drug delivery with decreased systemic drug distribution thereby reducing side effects. However, the success of EPT specifically relies on the proper selection of the enzyme and extremely specific therapy could be accomplished using proper enzyme (Stadler and Zelikin, 2017).

This concept of EPT was established about 45 years ago by introducing cytosine deaminase enzyme, at the site of tumor tissue, which resulted in localized conversion of 5-fluorocytosine into 5-fluorouracil, a potent anticancer agent. Enzyme targeting at the particular location of action can be accomplished by numerous ways, such as, exploitation of antibody for site-specific targeting of linked enzyme (antibody-directed EPT, ADEPT), for transportation of gene coding for a specific enzyme followed by gene expression within a targeted tissue (gene-directed EPT, GDEPT), use of bacteria-directed EPT where bacteria are genetically fabricated to produce a prodrug-converting enzyme (Lesniewska-Kowiel and Muszalska 2017). All the above stated techniques of EPT have been discussed in detail in the subsequent part of the chapter.

14.3.1.1 Antibody-Directed Enzyme Prodrug Therapy (ADEPT)

Antibodies own exceptional biochemical characteristics which could be altered and utilized as perfect mediators to transport drug contents to the wanted tissue. They have been found expedient for site-specific transport of cytotoxic drugs to tumor cells with decreased systemic toxicity to normal tissues. Progresses in antibody engineering technologies have led to generation of several classes of novel antibody and their derivatives which proposes very specific mechanisms of action (Panowski et al., 2014).

ADEPT is one such smart approach to selectively deliver chemotherapeutic agent for cancer therapy. Previously attempts have been made to explore the existing enzymes within tumors to transform low toxicity prodrug into cytotoxic moiety but unique tumor specific enzymes could not be established. Consequently, the ADEPT approach came into existence.

The elementary principle behind the ADEPT approach has been depicted in Figure 14.2. It involves targeting an antibody conjugated enzyme to a tumor associated antigen which is followed by clearance of unbound enzyme from blood. After removal of unbound enzyme, nontoxic prodrug is administered. The targeted enzyme (in the tumor cells) facilitates conversion of nontoxic prodrug into an active

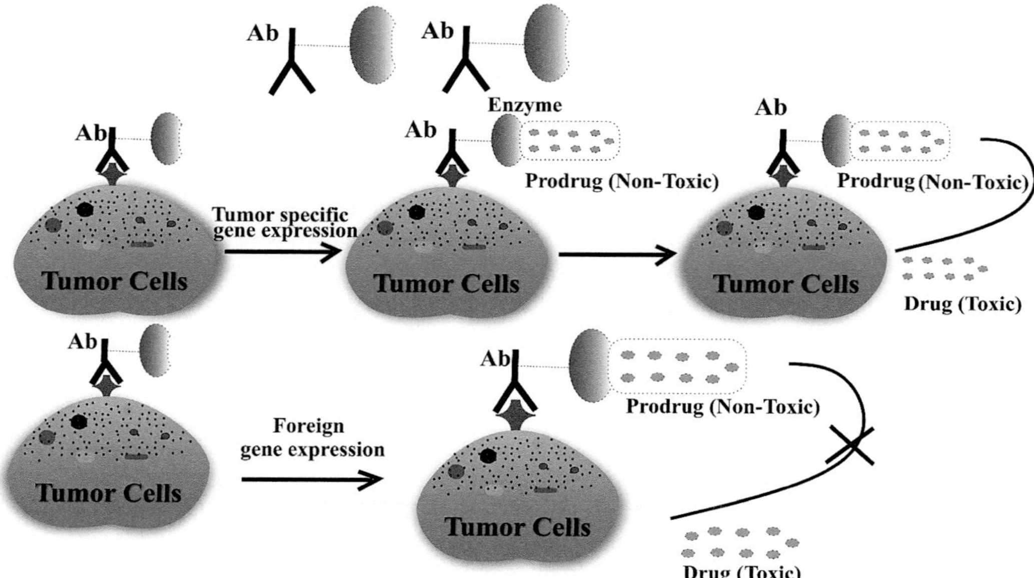

FIGURE 14.2 The elementary principle behind the ADEPT approach.

therapeutic entity, capable of killing agent within tumor cells. Thus, enhanced and effective therapy is accomplished without exhibiting toxicity to the normal cells.

The ADEPT approach utilizing monoclonal antibodies and their fragments have been extensively employed by researchers for safe and effective therapy of cancer and successful results have been presented in the subsequent section of the book chapter.

Recently, Virani and group (2018) have demonstrated the usage of an enzyme-prodrug system and immune stimulation for ovarian cancer therapy. They reported that ADEPT might be employed as a substitute to chemotherapy for initial treatment of patients witnessing recurrent ovarian tumors. ADEPT was suggested to present trifling toxicity to normal tissue as observed in pathological reports. Furthermore, substantial enhancement in patient's survival, decreased tumor burden, and elevated immune response was described using the amalgamation of the enzyme prodrug union with anti-OX40 and anti-CD73.

Kakinuma and group (2002) employed antibody-directed abzyme prodrug therapy (ADAPT) and carried out activation of prodrug by abzymes associated with antitumor antibodies. They produced abzymes by immunization of a vitamin B6 analog (phosphonate transition state analog). The produced antibodies were reported to break down vitamin B6 prodrugs, falling in the category of anti-cancer and anti-inflammatory agents. They also examined prodrug activation catalyzed by abzymes as observed by growth inhibitory effect on human cervical cancer (HeLa) cell. In nutshell, they suggested that the prodrug of vitamin B6 ester possess the ability to exhibit high stability against natural enzymes in serum and could be eliminated by newly developed catalytic antibodies.

Apart from above discussed research finding of ADEPT, numerous other applications justifying the role of biotechnology in ADEPT have been presented in Table 14.1.

14.3.1.2 Gene-Directed Enzyme-Prodrug Therapy (GDEPT)

In last several years extreme attention has been paid for developing novel approaches to achieve effective targeting of prodrugs tumor cells with a view to attain high efficacy of drugs with decreased toxicity. Gene-directed enzyme-prodrug therapy (GDEPT) is an exclusive and successful prodrug delivery approach that has received unique position in cancer therapy. GDEPT is a system that involves gene delivery (encoding for foreign enzyme) to tumor cells where a nonhazardous prodrug administered systemically can be activated following gene expression of enzyme encoded gene (Zhang et al., 2015).

GDEPT usually utilizes engineered viral (i.e., VDEPT) or bacterial (BDEPT) vectors to transport specific enzyme (therapeutic) directly to tumor cells. This enzyme further converts a prodrug into cytotoxic metabolites (Forbes, 2010; Copp et al., 2017). Basic mechanism of GDPET includes the following three steps

- In the first step, gene carrying the codes for a specific enzyme is inserted into an appropriate vector and transferred to targeted location with or without carriers.
- The next step is followed by transcription of transgene into mRNA with subsequent translation of mRNA into coded enzyme inside directed cell.
- The last step is followed by systemic administration of prodrug with subsequent absorption by the same cell; at last the prodrug is converted into active moiety in presence of translated enzyme (Lehouritis et al., 2013).

Gene expression is regulated by tumor cell-specific promoters thus the enzyme which catalyses the biochemical reaction might be directed to target cells so as to keep the healthy cells harmless even in case they guzzle the gene and the prodrug. Thus, favored conversion of prodrugs to drugs/cytotoxic compounds only occurs in tumor cells. Following GDEPT, much higher therapeutic index of prodrugs than conventional chemotherapeutic drugs could be achieved (Both et al., 2009).

However, all tumor cells do not own the ability to take up only one copy of the gene to express the desired enzyme (foreign) thus, an effect called "bystander cytotoxic effect" is favored. In bystander

TABLE 14.1

Application of ADEPT Approach for the Treatment of Cancer

Antibodies	Enzyme	Prodrug	Active Drug	Disease	References
Rabbit polyclonal anti Xen-CPG2 antibody	HSA conjugated glucarpidase/ PEG conjugated glucarpidase	Methotrexate	Glutamate and 4-deoxy-4-amino-N 10-methylpteroic acid (DAMPA)	All type of carcinoma	AlQahtani et al. (2018)
mAb-β-D-Glucosidase conjugate	β-D-galactosidase	Acetal glycoside BE-1	Doxorubicin	Lung carcinoma (A549 Cells)	Tietze and Krewer (2009)
mAb-cytosine deaminase	Cytosine deaminase	**5-fluorocytosine** (5-FC)	5-fluorouracil (5-FU)	Lung adenocarcinoma (H2981)	Bagshawe et al. (2004)
Anti-CEA antibody A5B7	Carboxypeptidase G2 (CPG2)	4-[(2-chloroethyl)(2-mesyloxyethyl) amino] benzoyl-L-glutamic acid (CMDA)	Bifunctional alkylating agent (CJS11)	Colorectal carcinoma	Martin et al. (1997)
Anti-CEA antibody	Human β-glucuronidase	Anthracycline-glucuronide	Epirubicin-glucuronide	All type of carcinoma	Bagshawe et al. (1994)
Anti CEA MAb BW 431	Human β-glucuronidase	4-methyl umbelliferyl P-glucuronide	4-methylumbelliferone and glucuronic acid	Colon carcinoma	Bosslet et al. (1992)
Monoclonal anti-CEA antibody	CPG2	CMDA/4-{[bis(2-iodoethyl)amino]-phenyl}oxycarbonyl-L-glutamic acid	4-(bis(2-iodoethyl)amino) phenol	Colorectal cancer	Mayer et al. (2006)

Abbreviations: CEA: Carcinoembryonic antigen; HSA: Human Serum Albumin; PEG: Poly Ethylene Glycol.

effect, active transport of diffused cytotoxic drug, from one tumor cell to surrounding cells occurs. This effect results into several fold enhancements in the toxic effects of drug. Thus, in order to accomplish substantial regression of tumor along with prolonged clinical response, by stander effect has been extensively employed in GDEPT (Denny, 2003).

However, active metabolite must be easily diffusible or must own the ability to get actively transported to neighboring cells following gap junctions, for accomplishment of bystander effects.

Other conjectured and deliberated mechanisms behind bystander effects includes endocytosis of apoptotic vesicles, liberation of soluble factors, and stimulation of the immune system in vivo (Neschadim and Medin, 2019; Sato et al., 2013).

14.3.1.2.1 Bacterial-Directed Enzyme-Prodrug Therapy (BDEPT)

BDEPT remains a promising approach for enhancing the efficacy of a bacterial vector and minimizing its therapeutic dosage. In BDEPT bacteria is incorporated with a gene that encodes a particular enzyme, capable of converting prodrug into active therapeutic agent and delivering them to the site of tumor. BDEPT is accomplished utilizing two steps: In the first step; bacterial vector incorporated with a specific gene is introduced into the patient where the gene coding for specific enzyme is expressed (Duarte et al., 2012). The expressed enzyme then specifically targets to locality of tumor. Second step includes administration of prodrug, once the optimum concentration of enzyme expression is achieved. The prodrug administered is then converted into cytotoxic therapeutic molecule at the tumor site under the influence of expressed enzyme resulting tumor-selective cytotoxicity (Mowday et al., 2016).

An adequate expression of therapeutic gene is needed for effective conversation prodrug to active therapeutic entity. In BDEPT, targeting by bacterial vector depends on the physical nature of tumor such as nonpathogenic bacteria are usually exploited for gene delivery as they remain intrinsically nontoxic to the host. Furthermore, any sort of probable transgene deadliness in human healthy tissue (that could arise following vector mistargeting) is evaded. In addition, bacterial cell, contains cofactors such as NADH or NADPH available to therapeutic enzymes that acts in reducing milieu; and ultimately, bacteria could be easily scaled up for clinical use (Luo et al., 2001).

Diverse classes of bacteria exploit different mechanisms to achieve tumor specificity. Obligate anaerobes like *Clostridium* species (gram positive) produces spores which grow within anoxic areas of tumors, whereas facultative anaerobes like *Salmonella* and *Escherichia* reside inside tumors to ensure shielding from the immune system and favors chemotactic drive within the tumor micromilieu (Sznol et al., 2000).

Bacteria own numerous benefits as vectors for BDEPT such as: they could be effortlessly modified to produce exogenous products of therapeutic importance; possess the ability to enhance their tumor discernment; ability to express enzymes for prodrug activation and stimulation of reporter proteins for supporting visual approval of tumor site that ultimately leads to accomplishment of desired therapeutic response. In addition, bacteria offer unique benefit as compared to viral vector complements—bacterial infection during cancer therapy can enthusiastically be managed by antibiotics (Bueso et al., 2018).

BDEPT have been successfully exploited by numerous researchers for effective and enhanced therapy of cancer. In this regard, Theys and group performed genetic engineering of *Clostridium* strains and achieved stains having greater tumour colonisation properties with great success rates. Subsequently they followed the same procedure to produce a recombinant *C. sporogenes* strain overexpressing the isolated enzyme-nitroreductase (NTR). Further they reported that intravenous administration of spores of NTR-recombinant *C. sporogenes* exhibited significantly enhanced antitumor efficacy following prodrug administration (Theys et al., 2006).

In recent research Chan-Hyams et al. (2018), examined the capacities of different nitroaromatic prodrug metabolites to exit a model Gram negative vector for bacterial-directed enzyme-prodrug therapy. They used five prodrugs: metronidazole, CB1954, nitro-CBI-DEI, PR-104A and SN27686. They reported that nitroreductase enzymes presented different prejudices for the 2- versus 4-nitro substituents of CB1954. For CB1954, both enzymes exhibited higher effectiveness than O_2-insensitive NfsB_Ec. NfsA_Nm presented comparable levels of activity to the leading NfsA_Ec. Further NfsA_Nm and PnbA_Bh were reported to be more efficient than NfsB_EcIn nutshell they suggested that prodrugs may vary with respect to aptness for BDEPT versus VDEPT.

Rich et al. (2018), illustrated the activities of two different nitroreductases, *Bartonella henselae* (PnbA_Bh) and *Neisseria meningitidis* NfsA (NfsA_Nm), with CB195, metronidazole (the nitro-prodrugs), and 2,4- and 2,6-dinitrotoluene (environmental pollutants). NfsA_Nm and PnbA_Bhunder went over-expression assays in *E. coli* and assessed for His6-tagged proteins in vitro. For CB1954, both enzymes exhibited higher effectiveness than O_2-insensitive nitroreductase *E. coli* NfsB (NfsB_Ec). NfsA_Nm presented similar levels of activity to the leading nitroreductase candidate *E. coli* NfsA (NfsA_Ec). Further, NfsA_Nm and PnbA_Bh were found to be effective than NfsB_Ec following aerobic stimulation of metronidazole into cytotoxic entity. NfsA_Nm were again reported to exhibit enhanced activity as compared to either NfsA_Ec or NfsB_Ec when tested for both the environmental pollutants, however PnbA_Bh remained relatively ineffective with either of the substrate.

Apart from the above discussed research finding of BDEPT, numerous other applications justifying the role of biotechnology in BDEPT have been presented in Table 14.2.

14.3.1.2.2 Virus-Directed Enzyme-Prodrug Therapy (VDEPT)

Virus-directed enzyme-prodrug therapy (VDEPT) approach that exploits viral vectors to transport a gene encoding certain enzyme proficient to convert prodrug (systemically administrated) into active therapeutic entity. Currently, viral vector is the extremely efficient mean of gene transfer with a view to alter particular type of cell or tissue. Numerous types of virus have been investigated till date; including retroviruses (γ-retroviruses and lentiviruses), adenoviruses (Ads), herpes simplex and adeno-associated viruses. Factor affecting choice of virus includes ease of production, competence of transgene expression, stability, toxicity and safety, concern.

VDEPT strategy has gained great potential in cancer therapy and extensive researches have done justifying application of VDEPT in cancer treatment (Szewczuk et al., 2017)

In a recent research, Foloppe and group (2019), prepared vaccinia virus (TG6002) containing a suicide gene FCU1 introduced in the J2R locus that encodes thymidine kinase expressing FCU1. In addition, J2R gene and I4L gene encoding for large subunit of ribonucleotide reductase were deleted. Replication of TG6002 highly reliant on ribonucleotide reductase levels and is less pathogenic compared to single-deleted vaccinia virus. They reported, tumor-specific multiplication of virus, extended 5-fluorouracil level in tumor tissue, and substantial therapeutic effect was reported in multiple human xenograft tumor models following systemic administration 5-fluorocytosine.

In another research Studebaker et al. (2017), examined the therapeutic efficacy oncolytic herpes simplex vector rRp450 in medulloblastoma and teratoid/rhabdoid tumor-derived cell lines. Both the tumor cells were supportive of virus replication and virus-mediated cytotoxicity. Significantly extended survival was reported on orthotropic xenograft models of medulloblastoma and teratoid/rhabdoid tumors following intratumoral injection of rRp450. Besides, increased efficacy of rRp450's reported upon addition of prodrug cyclophosphamide (CPA). Overall, finding suggested that, oncolytic herpes viruses possess ability to biotransform the prodrug (CPA) within the tumor micromilieu for therapy of pediatric brain tumors.

Kasai et al. (2013), perfomed toxicological and biodistribution evaluation on oncolytic virus (MGH2.1) containing genes namely cyclophosphamide (CPA)-activating cytochrome P4502B1 (CYP2B1) and CPT11-activating human intestinal carboxylesterase (shiCE). Expression of viral vector encoded genes was reported to be restricted to brain. Their safety and toxicological findings justified possibility of clinical trial of intratumoral injection of MGH2.1 with concurrent peripheral delivery of prodrugs (CPA and/or CPT11) at the site of malignant gliomas.

In an earlier research Ishida and team (2010) demonstrated efficient and successful introduction and expression of green fluorescent protein (GFP) gene, as a trace maker, employing HSV amplicon in comparison to pHGCX expression vector. The HSV amplicon system was reported to successfully express an active CES enzyme that effectively biotransformed TAX-2′-Et to TAX in Cos7 cells. Conclusively they suggested that introduction of a prodrug-converting enzyme appeared as promising approach for delivery of TAX.

Similarly, Palmer et al. (2004), used virus-directed enzyme-prodrug therapy for effective treatment of liver cancer. They encoded bacterial enzyme nitroreductase in adenovirus to activate short lived, highly toxic CB1954 (5-(aziridin-1-yl)-2,4-dinitrobenzamide). They reported successful clinical trial

TABLE 14.2

Application of BDEPT Approach for the Treatment of Cancer

(Bacterial) Vector	Enzyme	Prodrug	Active Drug	Disease	References
Clostridium sporogenes	Nitroreductase (eNR)	5-aziridinyl-2,4-dinitrobenzamide (CB1954)/ 2-((2-bromoethyl)-2-{[(2-hydroxyethyl)amino]carbonyl}-4,6-dinitroanilino)ethyl methanesulfonate phosphate ester (PR-104)	2-amine and 4-acetoxyamine products	Cervical carcinoma/SiH atumours	Liu et al. (2008)
E. Coli	NR from *Bacillus cereus*	5-aziridinyl-2,4-dinitrobenzamide (CB1954)	2′- and 4′-hydroxylamine metabolites	Human 7 Caucasian ovary adenocarcinoma/SK-OV-3 cell line	Paramasivan et al. (2015)
Salmonella typhimurium (SL7838)	ChrR6 Nitroreductase (eNR)	6-chloro-9-nitro-5-oxo-5H-benzo[a] phenoxazine (CNOB)	9-amino-6-chloro-5H benzo(a)phenoxazine-5-one (MCHB)	Murine mammary cancer/4T1 cells, Human mammary cancer/MCF-7 cells, Human cervical cancer/HeLa, Human colorectal cancer/HCT 116 cells, Human embryonic kidney cancer/293T cells	Thorne et al. (2009)
E. coli (DH5α)	β-Glucuronidase (β-G)	9-aminocamptothecin glucuronide (9ACG)	9-aminocamptothecin (9AC).	Human lung adenocarcinoma/ CL1-0 cells	Cheng et al. (2008)
S. typhimurium (VNP20009)	Purine nucleoside phosphorylase (PNP)	6-methylpurine-2-deoxyriboside (6MePdR)	6-methylpurine (6MeP)	Human skin cancer/ B16F10 melanoma cells	Chen et al. (2013)
C. beijerinckii (NCIMB 805)	Cytosine deaminase (CD)	5-flurocytosine (5-FC)	5-flurouracil (5-FU)	Mammary carcinoma/EMT6 cells	Nuyts et al. (2001)
Bifidobacterium infantis	Cytosine deaminase (CD)	5-flurocytosine (5-FC)	5-flurouracil (5-FU)	Human bladder cancers	Yi et al. (2005)
Bifidobacterium infantis	Herpes simplex virus thymidine kinase (HSV-TK)	Ganciclovir (GCV)	GCV-triphosphate	Bladder cancer	Tang et al. (2009)
Escherichia coli (BL21)	β-glucuronidase (βG)	p-hydroxy aniline mustard β-D-glucuronide, HAMG	p-hydroxy aniline mustard, pHAM	Human colorectal cancer/ HCT116 cells	Cheng et al. (2013)
E. coli NfsA	Nitroreductase (eNR)	PR-104(2-((2-bromoethyl)(2-((2-hydro xyethyl)carbamoyl)-4,6-dinitrophenyl) amino) ethyl methanesulfonate phosphate ester)	DNA alkylating hydroxylamine (PR-104H) or amine (PR-104M)	Colorectal cancer/HCT116	Mowday et al. (2016)

demonstrating effective transgene expression of adenovirus encoding nitroreductase (CTL102) in patients having liver tumors. The vector was reported to be fine tolerated with fewer side effects, and encouraged a strong antibody response. Conclusively, high level of nitroreductase expression observed suggests that advance studies must be carried out in future for the treatment of unworkable liver tumors.

Szewczuk and team (2017) employed virus-directed enzyme-prodrug therapy for enhancing cytotoxicity of bioreductive agents. They examined choosen benzimidazole derivatives as a substrate for nitroreductase, the enzyme capable of producing cytotoxic metabolites. Their finding suggested the pro-apoptotic characteristics of all tested compounds in normoxia and hypoxia, as confirmed by virused A549 cells where the time of exposition decreased from 48 to 4 h. This reduction in exposition period of time, highly enhanced activity was presented by N-oxide compounds with nitro-groups. The apoptosis was further verified by generation of BAX gene and protein and decrease in BCL2 gene and protein.

Apart from the above discussed research finding of VDEPT, numerous other applications justifying the role of biotechnology in VDEPT have been presented in Table 14.3.

14.3.2 Targeting of Specific Membrane Transporters

Prodrugs directed towards membrane transporters present on the surface of epithelial cell on are possibly the most exhilarating existing drug delivery strategies. Numerous nutrient transporters and receptors on expressed over the surface of epithelial cell membrane. Prodrugs directed towards these transporters possess ability to significantly improve the absorption of poorly permeating therapeutic agents thereby enhancing the water solubility of these molecules depending upon promoiety being anchored to the active therapeutic compound (Anand et al., 2002; Yang et al., 2001).

The membrane transporters identify these prodrugs as substrate and are moved across the epithelial membranes followed by either cleavage of prodrug within the intracellular environment, or their arrival to systemic circulation in the intact form. In the second condition, the promoiety follows breakdown by enzymes either available in the systemic circulation or within the targeted tissue, thereby liberating free active entity (Majumdar et al., 2004).

Prodrugs can be fabricated in such a way, so as to get structural resemblance with the intestinal nutrients in order to get easily absorbed by specific carrier proteins. In such condition, prodrugs may have the additional advantage of nontoxic nutrient byproducts in which prodrugs are converted to the parent drug molecules (Balakrishnan et al., 2002; de Koning and Diallinas, 2000).

Numerous types of transporters including peptide transporters, amino acid transporters, nucleoside and nucleobase transporters, bile acid transporters and monocarboxylic acid transporters have been used for targeted delivery of prodrugs (Murakami, 2016; Tao et al., 2018).

In a recent research, Sun et al., developed three 5-FU-fatty acid conjugates (5-FU-octanedioic acid) for targeting monocarboxylate transporter 1 (MCT1) present in intestinal membrane in order to enhance the permeability and oral bioavailability of 5-FU-fatty acid conjugates. They reported that due to activity of MCT1, 5-FU-octanedioic acid (prodrug) presented 13.1-fold enhancement in membrane permeability and 4.1-fold increase in oral bioavailability as compared to 5-FU. Conclusively, they reported superior gastrointestinal stability, enhanced membrane permeability with appropriate biotransformation of prodrug in intestinal cell (Sun et al., 2019).

In this context, Elena Puris and group designed transporter mediated prodrug strategy to efficiently target Ferulic acid (FA) prodrugs into the mouse brain utilizing L-Type amino acid transporter 1 (LAT1) to achieve potential benefits against Alzheimer's disease. FA being a natural antioxidant fined benefits against Alzheimer's disease, but its use is restricted due to its very low permeation across the blood-brain barrier (BBB) thereby providing poor bioavailability. Therefore, in the present work, the research group focused on the method development for designing of transporter-utilizing prodrugs and examined in vitro cellular uptake via LAT1 in ARPE-19 cells. They reported that aromatic ring containing amide-based could efficiently bound to LAT1 followed by utilization of membrane transporter resulting into high uptake by the cells (in vitro). Furthermore, they suggested that amide prodrug with the promoiety linked directly in the meta-position to FA successfully underwent to the parent drug in mouse brain (Puris et al., 2019).

TABLE 14.3

Application of VDEPT Approach for the Treatment of Cancer

(Virus) Vector	Enzyme	Prodrug	Active Drug	Disease	References
Recombinant baculoviruses	Herpes simplex virus thymidine kinase (HSV-TK)	Gancyclovir (GCV)	GCV-triphosphate	Glioblastoma	Chen et al. (2014)
Recombinant adenovirus	NAD(P)H dehydrogenase (quinone)	5-(aziridin-1-yl)-2,4-dinitrobenzamide (CB 1954)	5-(aziridin-1-yl)-4-hydroxylamino-2-nitrobenzamide	Breast carcinoma/Walker carcinoma cells	Knox et al. (1998)
OAdV220, an ovine atadenovirus	*E. coli* enzyme purine nucleoside phosphorylase (PNP)	Fludarabine phosphate	2-Fluoroadenine (2FA)	Transgenic adenocarcinoma of the prostate (TRAMP)	Martiniello-Wilks et al. (2004)
Adenovirus	*E. coli* Nitroimidazole reductase (NTR)	CB1954 [5-(aziridin-1-yl)-2,4-dinitrobenzamide]	5-(aziridin-1-yl)-4-N-acetoxy-2-nitrobenzamide; 5-(aziridin-1-yl) -2-hydroxylamino-4-nitrobenzamide	Gastrointestinal malignancies	Chung-Faye et al. (2001)
Recombinant adenovirus	Cytochrome P450 2B6/NADPH cytochrome P450 reductase fusion protein (CYP2B6/RED)	Cyclophosphamide (CPA)	Acrolein and phosphoramide mustard	Pulmonary tumor cell lines; Calu-6	Tychopoulos et al. (2005)
Recombinant adenovirus	Nitroreductase	Benzimidazole derivatives	N-oxide compounds	Human lung adenocarcinoma A549 cell line	Szewczuk et al. (2017)
Recombinant adenovirus	Herpes simplex virus thymidine kinase (HSV-TK)	Ganciclovir (GCV)	GCV-triphosphate	Hepatocellular carcinoma	Kakinoki et al. (2010)
Recombinant adenovirus	*E. coli* cytosine deaminase (AdCD)	5-Fluorocytosine (5-FC)	5-Fluorouracil(**5-FU**)	Human breast cancer cell line, MDA-MB-231	Li et al. (1997)
Recombinant adenovirus	Nitroreductase	Benzimidazoles	Benzimidazole derivatives, the Annexin V+ propidium iodide (PI)	Human adenocarcinoma A549 cell line	Szewczuk et al. (2017)
Recombinant adenovirus	Cytosine deaminase (CD)/herpes simplex virus thymidine kinase (HSV-1 TK)	5-Fluorocytosine (5-FC)	5-fluorouracil (5-FU)	Prostate cancer	Freytag et al. (2003)

In a research by Katragadda and team reported enhanced affinity of amino acid ester prodrugs towards various amino acid and peptide transporter on the Caco-2 cells. They studied systemic absorption (In vivo) of amino acid prodrugs of acyclovir (ACV) following oral absorption in rats. They also determined that stability of the prodrugs, l-serine-ACV (SACV), l-isoleucine-ACV (IACV), γ-glutamate-ACV (EACV) l-alanine-ACV (AACV), and l-valine-ACV (VACV) in different tissues. EACV was reported to be most enzymatically stable compound in comparison with other prodrugs. SACV and VACV were reported to present about five-time enhancement in area under the curve (AUC), 2 times and 15 times enhanced Cmax(T) than VACV and ACV respectively. Conclusively the suggested SACV to be a promising amino acid ester prodrug for the oral treatment of herpes infections due to its enhanced stability, higher AUC and better Cmax(T) (Katragadda et al., 2008).

Similarly in a recent work, Foley et al. (2018), offered proof-of-concept for the knowledge of targeting poor bioavailable drugs towards PepT1 transporter in order to achieve enhanced oral permeability.

14.4 Conclusion

Prodrug strategy has attained several useful applications in the field of drug research especially for targeted delivery of therapeutics. Earlier, poor rate of drug absorption, low aqueous solubility, quick metabolic breakdown of active drug remained major confront for targeted delivery. Prodrug strategy has been therefore extensively exploited in few decades to overcome such issues. Prodrugs have given the possibility of developing selective drugs towards the therapeutic target, e.g., a cancer cell or a micro-organism.

This chapter is not meant to deliver most impressive trends in the development of biotechnology-based prodrug therapeutic strategy to achieve effective targeted/site-specific delivery. Numerous new prodrug delivery systems including ADEPT, gene prodrug therapy (bacterial vector-based gene prodrug therapy and viral vector-based enzyme-prodrug therapy) and membrane transporters assisted prodrug therapy have proven effectiveness in preclinical and clinical experiments. Still, extensive work is required to be devoted to this field with a view to utilize this promising approach for safe and efficacious for treatment of cancer, microbial infections and other diseases.

ACKNOWLEDGMENTS

The authors want to acknowledge Rungta College of Pharmaceutical Sciences and Research, Bhilai, C.G., University Institute of Pharmacy, Pt Ravishankar Shukla University, Raipur, C.G., and Department of Pharmacy, The Indira Gandhi National Tribal University, Amarkantak, M.P. for providing necessary literature and infrastructural facilities required for the compilation of work. One of the author Dr. MP wants to acknowledge Science and Engineering Research Board (SERB), New Delhi, India for providing financial assistance under National Post-Doctoral Fellowship scheme (File no. PDF/2015/000380). Dr. MRS is thankful to ICMR for DHR-HRD fellowship. Dr. DS is thankful to UGC Raman fellowship.

REFERENCES

Abet, V., Filace, F., Recio, J., Alvarez-Builla, J. and Burgos, C., 2017. Prodrug approach: An overview of recent cases. *European Journal of Medicinal Chemistry*, *127*, pp. 810–827.

AlQahtani, A.D., Al-mansoori, L., Bashraheel, S.S., Rashidi, F.B., Al-Yafei, A., Elsinga, P., Domling, A. and Goda, S.K., 2019. Production of "biobetter" glucarpidase variants to improve drug detoxification and antibody directed enzyme prodrug therapy for cancer treatment. *European Journal of Pharmaceutical Sciences*, *127*, pp. 79–91.

Anand, B.S., Dey, S. and Mitra, A.K., 2002. Current prodrug strategies via membrane transporters/receptors. *Expert Opinion on Biological Therapy*, *2*(6), pp. 607–620.

Anastasi, C., Quelever, G., Burlet, S., Garino, C., Souard, F. and Kraus, J.L., 2003. New antiviral nucleoside prodrugs await application. *Current Medicinal Chemistry*, *10*(18), pp. 1825–1843.

Bagshawe, K.D., Sharma, S.K. and Begent, R.H., 2004. Antibody-directed enzyme prodrug therapy (ADEPT) for cancer. *Expert Opinion on Biological Therapy*, *4*(11), pp. 1777–1789.

Bagshawe, K.D., Sharma, S.K., Springer, C.J. and Rogers, G.T., 1994. Antibody directed enzyme prodrug therapy (ADEPT). A review of some theoretical, experimental and clinical aspects. *Annals of Oncology*, *5*(10), pp. 879–891.

Balakrishnan, A., Jain, V.B., Yang, C., Pal, D., Mitra, A.K., 2002. Carrier mediated uptake of L-tyrosine and its competitive inhibition by model linked compounds in a rabbit corneal cell line (SIRC)-strategy for the design of transporter/receptor targeted prodrugs. *International Journal of Pharmaceutics*, *247*, pp. 115–125.

Bohme, H., Ahrens, K.H., Hotzel, H.H., 1974. Properties and reactions of N-(alphahydroxyalkyl)-thionamides. *Archiv DerPharmazie (Weinheim)*, *307*, pp. 748–755.

Bosslet, K., Czech, J., Lorenz, P., Sedlacek, H.H., Schuermann, M. and Seemann, G., 1992. Molecular and functional characterisation of a fusion protein suited for tumour specific prodrug activation. *British Journal of Cancer*, *65*(2), p. 234.

Both, G.W., 2009. Recent progress in gene-directed enzyme prodrug therapy: An emerging cancer treatment. *Current Opinion in Molecular Therapeutics*, *11*(4), pp. 421–432.

Bueso, Y.F., Lehouritis, P. and Tangney, M., 2018. In situ biomolecule production by bacteria; A synthetic biology approach to medicine. *Journal of Controlled Release*, *275*, pp. 217–228.

Chan-Hyams, J.V., Copp, J.N., Smaill, J.B., Patterson, A.V. and Ackerley, D.F., 2018. Evaluating the abilities of diverse nitroaromatic prodrug metabolites to exit a model Gram negative vector for bacterial-directed enzyme-prodrug therapy. *Biochemical Pharmacology*, *158*, pp. 192–200.

Chen H, Beardsley GP, CoenDM., 2014. Mechanism of ganciclovir induced chain termination revealed by resistant viral polymerase mutants with reduced exonuclease activity. *Proceedings of the National Academy of Sciences USA*, *111*(49), pp. 17462–17467.

Chen, G., Tang, B., Yang, B.Y., Chen, J.X., Zhou, J.H., Li, J.H. and Hua, Z.C., 2013. Tumor-targeting *Salmonella typhimurium*, a natural tool for activation of prodrug 6MePdR and their combination therapy in murine melanoma model. *Applied Microbiology and Biotechnology*, *97*(10), pp. 4393–4401.

Cheng, C.M., Chen, F.M., Lu, Y.L., Tzou, S.C., Wang, J.Y., Kao, C.H., Liao, K.W. et al., 2013. Expression of β-glucuronidase on the surface of bacteria enhances activation of glucuronide prodrugs. *Cancer Gene Therapy*, *20*(5), p. 276.

Cheng, C.M., Lu, Y.L., Chuang, K.H., Hung, W.C., Shiea, J., Su, Y.C., Kao, C.H., Chen, B.M., Roffler, S. and Cheng, T.L., 2008. Tumor-targeting prodrug-activating bacteria for cancer therapy. *Cancer Gene Therapy*, *15*(6), p. 393.

Choi-Sledeski, Y.M. and Wermuth, C.G., 2015. Designing prodrugs and bioprecursors. In *The Practice of Medicinal Chemistry* (pp. 657–696). Academic Press, London, UK.

Chung-Faye, G., Palmer, D., Anderson, D., Clark, J., Downes, M., Baddeley, J., Hussain, S. et al., 2001. Virus-directed, enzyme prodrug therapy with nitroimidazole reductase: A phase I and pharmacokinetic study of its prodrug, CB1954. *Clinical Cancer Research*, *7*, pp. 2662–2668.

Copp, J.N., Mowday, A.M., Williams, E.M., Guise, C.P., Ashoorzadeh, A., Sharrock, A.V., Flanagan, J.U., Smaill, J.B., Patterson, A.V. and Ackerley, D.F., 2017. Engineering a multifunctional nitroreductase for improved activation of prodrugs and PET probes for cancer gene therapy. *Cell Chemical Biology*, *24*(3), pp. 391–403.

De Koning, H. and Diallinas, G., 2000. Nucleobase transporters. *Molecular Membrane Biology*, *17*(2), pp. 75–94.

Denny, W.A., 2003. Prodrugs for gene-directed enzyme-prodrug therapy (suicide gene therapy). *BioMed Research International*, *2003*(1), pp. 48–70.

Duarte, S., Carle, G., Faneca, H., De Lima, M.C.P. and Pierrefite-Carle, V., 2012. Suicide gene therapy in cancer: Where do we stand now? *Cancer Letters*, *324*(2), pp. 160–170.

Ettmayer, P., Amidon, G.L., Clement, B. and Testa, B., 2004. Lessons learned from marketed and investigational prodrugs. *Journal of Medicinal Chemistry*, *47*(10), pp. 2393–2404.

Foley, D.W., Pathak, R.B., Phillips, T.R., Wilson, G.L., Bailey, P.D., Pieri, M., Senan, A. and Meredith, D., 2018. Thiodipeptides targeting the intestinal oligopeptide transporter as a general approach to improving oral drug delivery. *European Journal of Medicinal Chemistry*, *156*, pp. 180–189.

Foloppe, J., Kempf, J., Futin, N., Kintz, J., Cordier, P., Pichon, C., Findeli, A., Vorburger, F., Quemeneur, E. and Erbs, P., 2019. The enhanced tumor specificity of TG6002, an armed oncolytic vaccinia virus deleted in two genes involved in nucleotide metabolism. *Molecular Therapy-Oncolytics*, *14*, pp. 1–14.

Forbes, N.S., 2010. Engineering the perfect (bacterial) cancer therapy. *Nature Reviews Cancer*, *10*(11), p. 785.

Freytag, S.O., Stricker, H., Pegg, J., Paielli, D., Pradhan, D.G., Peabody,J., DePeralta-Venturina, M., Xia, X., Brown, S., Lu, M. and Kim, J.H., 2003. Phase I study of replication-competent adenovirus-mediated double-suicide gene therapy in combination with conventional-dose three-dimensional conformal radiation therapy for the treatment of newly diagnosed, intermediate- to high-risk prostate cancer. *Cancer Research*, *63*, pp. 7497–7506.

Han, H.K. and Amidon, G.L., 2000. Targeted prodrug design to optimize drug delivery. *AAPS PharmSciTech.*, *2*(1), pp. 48–58.

Ishida, D., Nawa, A., Tanino, T., Goshima, F., Luo, C.H., Iwaki, M., Kajiyama, H., Shibata, K., Yamamoto, E., Ino, K. and Tsurumi, T., 2010. Enhanced cytotoxicity with a novel system combining the paclitaxel-2′-ethylcarbonate prodrug and an HSV amplicon with an attenuated replication-competent virus, HF10 as a helper virus. *Cancer Letters*, *288*(1), pp. 17–27.

Jornada, D.H., dos Santos Fernandes, G.F., Chiba, D.E., de Melo, T.R., dos Santos, J.L. and Chung, M.C., 2015. The prodrug approach: A successful tool for improving drug solubility. *Molecules*, *21*(1), p. 42.

Kakinoki, K., Nakamoto, Y., Kagaya, T., Tsuchiyama, T., Sakai, Y., Nakahama, T., Mukaida, N. and Kaneko, S. 2010. Prevention of intrahepatic metastasis of liver cancer by suicide gene therapy and chemokine ligand 2/ monocyte chemoattractant protein-1 delivery in mice. *The Journal of Gene Medicine*, *12*(12), pp. 1002–1013.

Kakinuma, H., Fujii, I. and Nishi, Y., 2002. Selective chemotherapeutic strategies using catalytic antibodies: A common pro-moiety for antibody-directed abzyme prodrug therapy. *Journal of Immunological Methods*, *269*(1–2), pp. 269–281.

Kasai, K., Nakashima, H., Liu, F., Kerr, S., Wang, J., Phelps, M., Potter, P.M., Goins, W.B., Fernandez, S.A. and Chiocca, E.A., 2013. Toxicology and biodistribution studies for MGH2. 1, an oncolytic virus that expresses two prodrug-activating genes, in combination with prodrugs. *Molecular Therapy-Nucleic Acids*, *2*, p. e113.

Katragadda, S., Jain, R., Kwatra, D., Hariharan, S. and Mitra, A.K., 2008. Pharmacokinetics of amino acid ester prodrugs of acyclovir after oral administration: Interaction with the transporters on Caco-2 cells. *International Journal of Pharmaceutics*, *362*(1–2), pp. 93–101.

Knox, R.J., Friedlos, F., Jarman, M. and Roberts, J.J., 1988. A new cytotoxic, DNA interstrand crosslinking agent, 5-(aziridin-1-yl)-4-hydroxylamino-2-nitrobenzamide, is formed from 5-(aziridin-1-yl)-2,4-dinitrobenzamide (CB 1954) by a nitroreductase enzyme in Walker carcinoma cells. *Biochemical Pharmacology*, *37*, pp. 4661–4669.

Lehouritis, P., Springer, C. and Tangney, M., 2013. Bacterial-directed enzyme prodrug therapy. *Journal of Controlled Release*, *170*(1), pp. 120–131.

Lesniewska-Kowiel, M.A. and Muszalska, I., 2017. Strategies in the designing of prodrugs, taking into account the antiviral and anticancer compounds. *European Journal of Medicinal Chemistry*, *129*, pp. 53–71.

Li Z., Shanmugam, N., Katayose, D., Huber, B., Srivastava, S., Cowan, K. and Seth, P., 1997. Enzyme/prodrug gene therapy approach for breast cancer using a recombinant adenovirus expressing Escherichia coli cytosine deaminase. *Cancer Gene Therapy*, *4*(2), pp. 113–117.

Liu, S.C., Ahn, G.O., Kioi, M., Dorie, M.J., Patterson, A.V. and Brown, J.M., 2008. Optimized clostridium-directed enzyme prodrug therapy improves the antitumor activity of the novel DNA cross-linking agent PR-104. *Cancer Research*, *68*(19), pp. 7995–8003.

Luo, X., Li, Z., Lin, S., Le, T., Ittensohn, M., Bermudes, D., Runyab, J.D., Shen, S.Y., Chen, J., King, I.C. and Zheng, L.M., 2001. Antitumor effect of VNP20009, an attenuated Salmonella, in murine tumor models. *Oncology Research Featuring Preclinical and Clinical Cancer Therapeutics*, *12*(11–12), pp. 501–508.

Majumdar, S., Duvvuri, S. and Mitra, A.K., 2004. Membrane transporter/receptor-targeted prodrug design: Strategies for human and veterinary drug development. *Advanced Drug Delivery Reviews*, *56*(10), pp. 1437–1452.

Martin, J., Stribbling, S.M., Poon, G.K., Begent, R.H., Napier, M., Sharma, S.K. and Springer, C.J., 1997. Antibody-directed enzyme prodrug therapy: Pharmacokinetics and plasma levels of prodrug and drug in a phase I clinical trial. *Cancer Chemotherapy and Pharmacology*, *40*(3), pp. 189–201.

Martiniello-Wilks, R., Dane, A., Voeks, D.J., Jeyakumar, G., Mortensen, E., Shaw, J.M., Wang, X.Y., Both, G.W. and Russell, P.J., 2004. Gene-directed enzyme prodrug therapy for prostate cancer in a mouse model that imitates the development of human disease. *The Journal of Gene Medicine: A Cross Disciplinary Journal for Research on the Science of Gene Transfer and its Clinical Applications*, *6*(1), pp. 43–54.

Mayer, A., Francis, R.J., Sharma, S.K., Tolner, B., Springer, C.J., Martin, J., Boxer, G.M. et al., 2006. A phase I study of single administration of antibody-directed enzyme prodrug therapy with the recombinant anti–carcinoembryonic antigen antibody-enzyme fusion protein MFECP1 and a bis-iodo phenol mustard prodrug. *Clinical Cancer Research*, *12*(21), pp. 6509–6516.

Mowday, A.M., Ashoorzadeh, A., Williams, E.M., Copp, J.N., Silva, S., Bull, M.R., Abbattista, M.R., Anderson, R.F., Flanagan, J.U., Guise, C.P. and Ackerley, D.F., 2016. Rational design of an AKR1C3-resistant analog of PR-104 for enzyme-prodrug therapy. *Biochemical Pharmacology*, *116*, pp. 176–187.

Murakami, T., 2016. A minireview: Usefulness of transporter-targeted prodrugs in enhancing membrane permeability. *Journal of Pharmaceutical Sciences*, *105*(9), pp. 2515–2526.

Neschadim, A. and Medin, J.A., 2019. Engineered thymidine-active deoxycytidine kinase for bystander killing of malignant cells. In *Suicide Gene Therapy* (pp. 149–163). Humana Press, New York.

Nuyts, S., Theys, J., Landuyt, W., Lambin, P. and Anné, J., 2001. Increasing specificity of anti-tumor therapy: Cytotoxic protein delivery by non-pathogenic clostridia under regulation of radio-induced promoters. *Anticancer Research*, *21*(2A), pp. 857–861.

Palmer, D.H., Mautner, V., Mirza, D., Oliff, S., Gerritsen, W., Hubscher, S., Reynolds, G. et al., 2004. Virus-directed enzyme prodrug therapy: Intratumoral administration of a replication-deficient adenovirus encoding nitroreductase to patients with resectable liver cancer. *Journal of Clinical Oncology*, *22*(9), pp. 1546–1552.

Panowski, S., Bhakta, S., Raab, H., Polakis, P. and Junutula, J.R., 2014. Site-specific antibody drug conjugates for cancer therapy. In *mAbs* (Vol. 6, No. 1, pp. 34–45). Taylor & Francis Group.

Paramasivan, P., Halliwell, J.H., Gwenin, V.V., Poornima, P., Halliwell, J., Ball, P., Robinson, G. and Gwenin, C.D., 2015. Identification of novel nitroreductases from Bacillus cereus and their interaction with the CB1954 prodrug. *Biochemical Pharmacology*, *98*(3), pp. 392–402.

Puris, E., Gynther, M., Huttunen, J., Auriola, S. and Huttunen, K.M., 2019. L-type amino acid transporter 1 utilizing prodrugs of ferulic acid revealed structural features supporting the design of prodrugs for brain delivery. *European Journal of Pharmaceutical Sciences*, *129*, pp. 99–109.

Rautio, J., Kärkkäinen, J. and Sloan, K.B., 2017. Prodrugs–Recent approvals and a glimpse of the pipeline. *European Journal of Pharmaceutical Sciences*, *109*, pp. 146–161.

Rautio, J., Kumpulainen, H., Heimbach, T., Oliyai, R., Oh, D., Järvinen, T. and Savolainen, J., 2008. Prodrugs: Design and clinical applications. *Nature Reviews Drug Discovery*, *7*(3), pp. 255–270.

Rich, M.H., Sharrock, A.V., Hall, K.R., Ackerley, D.F. and MacKichan, J.K., 2018. Evaluation of NfsA-like nitroreductases from *Neisseria meningitidis* and *Bartonella henselae* for enzyme-prodrug therapy, targeted cellular ablation, and dinitrotoluene bioremediation. *Biotechnology Letters*, *40*(2), pp. 359–367.

Sato, T., Neschadim, A., Lavie, A., Yanagisawa, T. and Medin, J.A., 2013. The engineered thymidylate kinase (TMPK)/AZT enzyme-prodrug axis offers efficient bystander cell killing for suicide gene therapy of cancer. *PLoS One*, *8*(10), p. e78711.

Stadler, B., Zelikin, A.N., 2017. Enzyme prodrug therapies and therapeutic enzymes. *Advanced Drug Delivery Reviews*, 118(1), p. 1.

Studebaker, A.W., Hutzen, B.J., Pierson, C.R., Haworth, K.B., Cripe, T.P., Jackson, E.M. and Leonard, J.R., 2017. Oncolytic herpes virus rRp450 shows efficacy in orthotopic xenograft group 3/4 medulloblastomas and atypical teratoid/rhabdoid tumors. *Molecular Therapy-Oncolytics*, *6*, pp. 22–30.

Sun, Y., Zhao, D., Wang, G., Jiang, Q., Guo, M., Kan, Q., He, Z. and Sun, J., 2019. A novel oral prodrug-targeting transporter MCT 1: 5-fluorouracil-dicarboxylate monoester conjugates. *Asian Journal of Pharmaceutical Sciences*, *14*(6), pp. 631–639.

Szewczuk, M., Boguszewska, K., Żebrowska, M., Balcerczak, E., Stasiak, M., Świątkowska, M. and Błaszczak-Świątkiewicz, K., 2017. Virus-directed enzyme prodrug therapy and the assessment of the cytotoxic impact of some benzimidazole derivatives. *Tumor Biology*, *39*(7), p. 1010428317713675.

Sznol, M., Lin, S.L., Bermudes, D., Zheng, L.M. and King, I., 2000. Use of preferentially replicating bacteria for the treatment of cancer. *Journal of Clinical Investigation*, *105*(8), pp. 1027–1030.

Tang, W., He, Y., Zhou, S., Ma, Y. and Liu, G., 2009. A novel Bifidobacterium infantis-mediated TK/GCV suicide gene therapy system exhibits antitumor activity in a rat model of bladder cancer. *Journal of Experimental & Clinical Cancer Research*, *28*(1), p. 155.

Tao, W., Zhao, D., Sun, M., Wang, Z., Lin, B., Bao, Y., Li, Y., He, Z., Sun, Y. and Sun, J., 2018. Intestinal absorption and activation of decitabine amino acid ester prodrugs mediated by peptide transporter PEPT1 and enterocyte enzymes. *International Journal of Pharmaceutics*, *541*(1–2), pp. 64–71.

Testa, B., 2007. Prodrug objectives and design. In *Comprehensive Medicinal Chemistry II*, Elsevier, the Netherlands, *5*, pp. 1009–1041.

Theys, J., Pennington, O., Dubois, L., Anlezark, G., Vaughan, T., Mengesha, A., Landuyt, W. et al., 2006. Repeated cycles of clostridium-directed enzyme prodrug therapy result in sustained antitumour effects in vivo. *British Journal of Cancer*, *95*(9), p. 1212.

Thorne, S.H., Barak, Y., Liang, W., Bachmann, M.H., Rao, J., Contag, C.H. and Matin, A., 2009. CNOB/ChrR6, a new prodrug enzyme cancer chemotherapy. *Molecular Cancer Therapeutics*, *8*(2), pp. 333–341.

Tietze, L.F. and Krewer, B., 2009. Antibody-directed enzyme prodrug therapy: A promising approach for a selective treatment of cancer based on prodrugs and monoclonal antibodies. *Chemical Biology & Drug Design*, *74*(3), pp. 205–211.

Tychopoulos, M., Corcos, L., Genne, P., Beaune, P. and de Waziers, I., 2005. A virus-directed enzyme prodrug therapy (VDEPT) strategy for lung cancer using a CYP2B6/NADPH cytochrome P450 reductase fusion protein. *Cancer Gene Therapy*, *12*(5), p. 497.

Virani, N.A., Thavathiru, E., McKernan, P., Moore, K., Benbrook, D. M., Harrison, R.G., 2018. Anti-CD73 and anti-OX40 immunotherapy coupled with a novel biocompatible enzyme prodrug system for the treatment of recurrent, metastatic ovarian cancer. *Cancer Letters*, *425*, pp. 174–182.

Walther, R., Rautio, J. and Zelikin, A.N., 2017. Prodrugs in medicinal chemistry and enzyme prodrug therapies. *Advanced Drug Delivery Reviews*, *118*, pp. 65–77.

Wermuth, C.G., 2008. Designing prodrugs and bioprecursors. In *The Practice of Medicinal Chemistry* (pp. 721–746). Academic Press, Amsterdam, the Netherlands.

Yang, C., Tirucherai, G.S. and Mitra, A.K., 2001. Prodrug based optimal drug delivery via membrane transporter/receptor. *Expert Opinion on Biological Therapy*, *1*(2), pp. 159–175.

Yi, C., Huang, Y., GUO, Z.Y. and WANG, S.R., 2005. Antitumor effect of cytosine deaminase/5-fluorocytosine suicide gene therapy system mediated by *Bifidobacterium infantis* on melanoma 1. *Acta Pharmacologica Sinica*, *26*(5), pp. 629–634.

Zhang, J., Kale, V., Chen M., 2015. Gene-directed enzyme prodrug therapy. *AAPS Journal*, *17*(1), pp. 102–110.

Index

Note: Page numbers in italic refer to figures, and those in bold to tables.

G

H

I

K

L

M

N

O

P